MicroRNAs in Diseases and Disorders

Emerging Therapeutic Targets

Drug Discovery Series

Titles in the Series:
1: Metabolism, Pharmacokinetics and Toxicity of Functional Groups
2: Emerging Drugs and Targets for Alzheimer's Disease; Volume 1
3: Emerging Drugs and Targets for Alzheimer's Disease; Volume 2
4: Accounts in Drug Discovery
5: New Frontiers in Chemical Biology
6: Animal Models for Neurodegenerative Disease
7: Neurodegeneration
8: G Protein-coupled Receptors
9: Pharmaceutical Process Development
10: Extracellular and Intracellular Signaling
11: New Synthetic Technologies in Medicinal Chemistry
12: New Horizons in Predictive Toxicology
13: Drug Design Strategies: Quantitative Approaches
14: Neglected Diseases and Drug Discovery
15: Biomedical Imaging
16: Pharmaceutical Salts and Cocrystals
17: Polyamine Drug Discovery
18: Proteinases as Drug Targets
19: Kinase Drug Discovery
20: Drug Design Strategies: Computational Techniques and Applications
21: Designing Multi-target Drugs
22: Nanostructured Biomaterials for Overcoming Biological Barriers
23: Physico-chemical and Computational Approaches to Drug Discovery
24: Biomarkers for Traumatic Brain Injury
25: Drug Discovery from Natural Products
26: Anti-inflammatory Drug Discovery
27: New Therapeutic Strategies for Type 2 Diabetes: Small Molecules
28: Drug Discovery for Psychiatric Disorders

29: Organic Chemistry of Drug Degradation
30: Computational Approaches to Nuclear Receptors
31: Traditional Chinese Medicine
32: Successful Strategies for the Discovery of Antiviral Drugs
33: Comprehensive Biomarker Discovery and Validation for Clinical Application
34: Emerging Drugs and Targets for Parkinson's Disease
35: Pain Therapeutics; Current and Future Treatment Paradigms
36: Biotherapeutics: Recent Developments using Chemical and Molecular Biology
37: Inhibitors of Molecular Chaperones as Therapeutic Agents
38: Orphan Drugs and Rare Diseases
39: Ion Channel Drug Discovery
40: Macrocycles in Drug Discovery
41: Human-based Systems for Translational Research
42: Venoms to Drugs: Venom as a Source for the Development of Human Therapeutics
43: Carbohydrates in Drug Design and Discovery
44: Drug Discovery for Schizophrenia
45: Cardiovascular and Metabolic Disease: Scientific Discoveries and New Therapies
46: Green Chemistry Strategies for Drug Discovery
47: Fragment-based Drug Discovery
48: Epigenetics for Drug Discovery
49: New Horizons in Predictive Drug Metabolism and Pharmacokinetics
50: Privileged Scaffolds in Medicinal Chemistry: Design, Synthesis, Evaluation
51: Nanomedicines: Design, Delivery and Detection
52: Synthetic Methods in Drug Discovery: Volume 1
53: Synthetic Methods in Drug Discovery: Volume 2
54: Drug Transporters: Role and Importance in ADME and Drug Development
55: Drug Transporters: Recent Advances and Emerging Technologies
56: Allosterism in Drug Discovery
57: Anti-aging Drugs: From Basic Research to Clinical Practice
58: Antibiotic Drug Discovery: New Targets and Molecular Entities
59: Peptide-based Drug Discovery: Challenges and New Therapeutics
60: Drug Discovery for Leishmaniasis
61: Biophysical Techniques in Drug Discovery
62: Acute Brain Impairment Through Stroke: Drug Discovery and Translational Research
63: Theranostics and Image Guided Drug Delivery

64: Pharmaceutical Formulation: The Science and Technology of Dosage Forms
65: Small-molecule Transcription Factor Inhibitors in Oncology
66: Therapies for Retinal Degeneration: Targeting Common Processes
67: Kinase Drug Discovery: Modern Approaches
68: Advances in Nucleic Acid Therapeutics
69: MicroRNAs in Diseases and Disorders: Emerging Therapeutic Targets

How to obtain future titles on publication:
A standing order plan is available for this series. A standing order will bring delivery of each new volume immediately on publication.

For further information please contact:
Book Sales Department, Royal Society of Chemistry, Thomas Graham House, Science Park, Milton Road, Cambridge, CB4 0WF, UK
Telephone: +44 (0)1223 420066, Fax: +44 (0)1223 420247,
Email: booksales@rsc.org
Visit our website at www.rsc.org/books

MicroRNAs in Diseases and Disorders

Emerging Therapeutic Targets

Edited by

Philip V. Peplow
University of Otago, New Zealand
Email: phil.peplow@otago.ac.nz

Bridget Martinez
Los Alamos National Laboratory, USA
Email: bmartinez26@ucmerced.edu

George A. Calin
MD Anderson Cancer Center, USA
Email: gcalin@mdanderson.org

and

Aurora Esquela-Kerscher
Eastern Virginia Medical School, USA
Email: kerschae@evms.edu

ROYAL SOCIETY OF CHEMISTRY

Drug Discovery Series No. 69

Print ISBN: 978-1-78801-394-9
PDF ISBN: 978-1-78801-642-1
EPUB ISBN: 978-1-78801-781-7
Print ISSN: 2041-3203
Electronic ISSN: 2041-3211

A catalogue record for this book is available from the British Library

The Royal Society of Chemistry is a charity, registered in England and Wales, Number 207890, and a company incorporated in England by Royal Charter (Registered No. RC000524), registered office: Burlington House, Piccadilly, London W1J 0BA, UK, Telephone: +44 (0) 20 7437 8656.

For further information see our web site at www.rsc.org

Printed in the United Kingdom by CPI Group (UK) Ltd, Croydon, CR0 4YY, UK

Preface

MicroRNAs have been found to be associated with the development and progression of many different diseases and disorders. The increasing effort devoted to understanding the roles of these molecules in the pathophysiology of these conditions, together with identifying specific microRNAs as markers that can improve diagnostic and prognostic accuracy, and serve as indicators of response to therapy, has provided the impetus for the construction of this book.

It is the goal of this book to provide a forum for clinical experts to present recent data on the design and testing of drugs in relation to microRNAs, including drug targets and response, synthesis of microRNA mimics and inhibitors, preclinical studies, and clinical trials. It also brings together scientific experts to present the state of the art in their fields of health and disease. It is my hope that the topics covered herein will provide new information on the role of microRNAs in diseases and disorders and that this knowledge will lead to a more effective approach and clinical management of these conditions.

I wish to express my deep appreciation to each of the chapter authors for the time and effort spent on writing informative reviews on their respective areas of clinical and research interest. Also I wish to express my gratitude to my co-editors Bridget Martinez, George Calin and Aurora Esquela-Kerscher. Finally, I wish to thank the staff of the Royal Society of Chemistry for their help, support, and patience in the course of putting together this book.

Philip V. Peplow

Drug Discovery Series No. 69
MicroRNAs in Diseases and Disorders: Emerging Therapeutic Targets
Edited by Philip V. Peplow, Bridget Martinez, George A. Calin and Aurora Esquela-Kerscher

Published by the Royal Society of Chemistry, www.rsc.org

Dedication

To the editors and chapter authors for their hard work to put together this exciting book summarizing the bright future ahead for non-coding RNAs in translational medicine. Many thanks to members of the Esquela-Kerscher laboratory and faculty, students and staff of the Leroy T. Canoles Jr. Cancer Research Center for their support and spirited discussions about how our research can better impact the cancer biology field and the community.

Aurora Esquela-Kerscher

This book is dedicated to our universe, which inspires, challenges, creates and molds beauty, even from the most unexpected corners and circumstances. The pursuit of knowledge is a great adventure. In loving memory of Elmer Martinez, my late father. Endless gratitude to my mother Eda Noguera, my sister Karla "Zeo" Sanchez and my brother Elmer Martinez.

Bridget Martinez

To the many students, fellows, scientists and clinicians who contributed to the extraordinary expansion of the field of non-coding RNA biology and medicine. A special thanks to the past and present members of Calin's laboratory. Without their dedication and hard work no discoveries or papers would come out of my laboratory.

George A. Calin

To my family, friends, staff and students at the Otago Medical School, thank you for your support and encouragement as I have travelled on my journey in teaching and biomedical research. And special thanks to you Bridget for shining your light on me in these past few years and helping me achieve so much more.

Philip Peplow

Drug Discovery Series No. 69
MicroRNAs in Diseases and Disorders: Emerging Therapeutic Targets
Edited by Philip V. Peplow, Bridget Martinez, George A. Calin and Aurora Esquela-Kerscher

Published by the Royal Society of Chemistry, www.rsc.org

Contents

Drug Discovery Series No. 69
MicroRNAs in Diseases and Disorders: Emerging Therapeutic Targets
Edited by Philip V. Peplow, Bridget Martinez, George A. Calin and Aurora Esquela-Kerscher

Published by the Royal Society of Chemistry, www.rsc.org

CHAPTER 1

Role of Nutrition, the Epigenome, and MicroRNAs in Cancer Pathogenesis

ZACHARY CADIEUX, HOLLY LEWIS AND AURORA ESQUELA-KERSCHER*

Department of Microbiology & Molecular Cell Biology, Leroy T. Canoles Jr Cancer Research Center Eastern Virginia Medical School, Norfolk, Virginia, USA
*E-mail: kerschae@evms.edu

1.1 Introduction

Good nutrition and the daily consumption of fresh fruit and vegetables have long been encouraged to maintain human health and extend lifespan. In recent years, accumulating experimental evidence supports the notion that dietary factors confer protective effects against diseases such as cancer. Scientists have identified specific bioactive factors present in dietary foods such as green tea, soy, whole grains, fruits, and cruciferous vegetables (broccoli, kale, cauliflower) that impact epigenetic mechanisms regulating chromatin remodeling and gene expression, and which are closely associated with human cancers. An important mechanism employed by these nutritional agents to mediate their cellular effects is the regulation of microRNAs (miRNAs). miRNAs are a major class of ~22-nucleotide long non-coding RNAs that generally function to block protein translation and/

Drug Discovery Series No. 69
MicroRNAs in Diseases and Disorders: Emerging Therapeutic Targets
Edited by Philip V. Peplow, Bridget Martinez, George A. Calin and Aurora Esquela-Kerscher

Published by the Royal Society of Chemistry, www.rsc.org

or degrade their messenger RNA (mRNA) targets. These small RNAs direct many essential processes related to cellular growth, apoptosis, differentiation, metabolism, and the immune response. miRNAs are often aberrantly expressed in human tumors. Many of these cancer-associated miRNAs act as tumor suppressors or pro-oncogenic factors that directly impact cancer progression and metastasis. In this chapter, we review how nutritional factors influence the epigenome and miRNA expression to confer their cancer-protective effects based on emerging *in vitro* data and animal studies. The diet–epigenome interactions and their influence on cancer-associated miRNA-mediated gene regulation are also explored in the context of the gut microbiota. Finally, the controversial field of ingested xenomiRs, particularly plant-based miRNAs, is discussed as a novel method of miRNA delivery to control tumorigenesis. A complete understanding of how miRNAs respond to nutritional cues and cancer-related signaling pathways will lead to promising diagnostic biomarkers as well as affordable and easily obtained therapeutic dietary supplements and/or edible vaccines to improve human health.

1.2 Epigenetic Link to Cancer

A hallmark of cancer is the uncontrolled growth and survival of damaged cells. This is caused by inappropriate activation or inhibition of RNA and protein factors residing within signaling pathways that control proliferation, differentiation, and apoptosis. These pathways can be altered by exposure to environmental factors, such as stress, drugs, and nutrition, leading to genomic mutations or alterations of the epigenome. Epigenetic modifications are heritable and often reversible changes in gene expression that do not alter the DNA sequence. These epigenetic alterations control gene expression in both positive and negative ways, commonly involving DNA hypo- and hypermethylation (*i.e.*, CpG islands within promoters), chromatin remodeling, histone protein modifications (*e.g.*, acetylation, methylation, phosphorylation), and non-coding RNAs (*i.e.*, miRNAs). Epigenetic modifications can influence DNA stability as well as the ability of transcription factors to interact with genomic elements, and thus ultimately determine if genes will be active or silenced. Therefore, the epigenome can dictate the overall protein profile in a cell that can have important and lasting biological and/or pathological ramifications. Epigenetic control of developmental events and differentiation processes include X-chromosome inactivation, genomic imprinting, genomic reprogramming, and stem cell maintenance.

An epigenetic link to cancer was first proposed in 1983 by Feinberg and Volgelstein, who observed that certain genes in tumor cells of cancer patients were hypomethylated compared to cells from normal adjacent tissues.[1] The theory that changes in gene expression due to epigenetic alterations (*e.g.*, DNA methylation status) predisposed individuals to cancer was verified with the discovery in 1989 that hypermethylation of the tumor suppressor gene Retinoblastoma (RB) was the driver of disease initiation and spontaneous regression.[2] It is now well established that DNA methylation and histone modifications that result in gene silencing occur in a variety of human cancers.[3,4] Recently, epigenetic mechanisms associated with miRNAs have gained considerable attention.[5] These small ~22-nucleotide non-coding RNA transcripts are often aberrantly expressed in a wide array of human cancers.[6] A growing subset of this non-coding RNA class, designated as "oncomirs" (miRNAs associated with cancer), mediate tumor formation and disease progression. A greater understanding of how cancer-associated miRNAs intersect with the epigenome and the nutritional axis, discussed in this chapter, will aid in their development as novel diagnostic and therapeutic tools for cancer.

1.3 miRNAs are Closely Associated with Human Cancer

1.3.1 miRNA Biogenesis is Complex

There are more than 2600 miRNAs in the human genome to date, largely identified *via* cloning and RNAseq methods (miRBase, release 22). miRNAs do not encode for proteins, but exert their biological effects as non-coding RNAs, and generally act to block target gene expression post-transcriptionally. The biogenesis of miRNAs is complex.[7] miRNAs are transcribed in the nucleus by RNA polymerase II (although Pol III transcription has been observed) to generate a precursor, pri-miRNA, which are often 5′ capped and 3′ polyadenylated. A pri-miRNA transcript can encode for multiple miRNA genes and each gene is processed into a ~70-nucleotide pre-miRNA hairpin precursor by the RNase III enzyme Drosha and its cofactor, DiGeorge syndrome critical region 8 (DGCR8). An independent subclass of pre-miRNAs, termed "mirtrons", do not rely on Drosha processing and rather are generated from mRNA transcripts as by-products of exon splicing and intron disbranching events.[8] Pre-miRNAs are subsequently exported out of the nucleus by RAN-GTP and Exportin 5. Once in the cytoplasm, pre-miRNAs are processed further by the RNase III enzyme, Dicer, and its co-factor, TAR RNA binding protein (TRBP) to generate a ~22-nucleotide double-stranded RNA duplex. One strand of this duplex is preferentially loaded into a large multiprotein miRNA-associated RNA-induced silencing complex (miRISC). Argonaute (AGO) is the key catalytic component of this complex responsible for miRNA strand selection and mediating miRNA-based alterations of gene expression. The loaded miRNA serves as a guide to escort the AGO/miRISC complex to the targeted site *via* miRNA binding to complementary sequences within the mRNA transcript, and ultimately results in target mRNA degradation and/or blocked protein translation.

Human miRNAs have been well characterized to bind with imperfect complementarity to the 3′ untranslated region (3′ UTR) of the target mRNA, although miRNA binding within the 5′ UTR and coding regions of the mRNA target can also modulate target gene expression.[7] Perfect base pairing of a highly conserved "seed sequence" (nucleotides 2–8) in the miRNA is important for proper miRNA targeting.[9] A few notable exceptions exist where miRNAs activate target gene expression *via* epigenetic regulation of enhancer regions in the nucleus or post-transcriptionally to induce protein translation.[10,11] Bioinformatic estimates indicate that a single miRNA can recognize upwards of 100 distinct targets, and therefore can regulate multiple non-overlapping biological pathways simultaneously. From a therapeutic standpoint, single miRNA therapy could be a powerful tool to treat cancer in patients that harbor an accumulation of genetic alterations and be effective without identifying the key mutations that led to tumor formation. This strategy could be particularly useful in diseases that exhibit extremely heterogeneous tumor populations carrying multiple genetic mutations within the same cancer patient.

1.3.2 miRNAs as Tumor Suppressors and Oncogenic Factors

miRNAs often exhibit abnormal expression profiles in fluids and tissues obtained from cancer patients compared to non-cancer patients.[6] A large proportion of these cancer-associated miRNAs are located in genomic loci referred to as "fragile sites", which are unstable regions of human chromosomes subject to gaps, breaks, or DNA rearrangements during replicative stress, and are closely associated with cancer.[12] For example, miR-125 and *let-7* family members reside within fragile sites often deleted in lung, breast, ovary, and cervical cancers.[12] miRNA dysregulation in cancerous tissues can also be caused by genetic defects outside of fragile sites, epigenetic alterations, or disruptions in miRNA biogenesis or miRNA–target interactions. Research using *in vitro* cell culture systems and animal disease models indicates that aberrant miRNA expression in turn can directly impact chromatin remodeling, tumor initiation and progression, epithelia to mesenchymal (EMT) transitions related to cell migration and invasion, metastasis and drug resistance *via* dysregulation of miRNA cancer-related targets, such as RAS and TP53.[6,13,14] In this way, patterns of miRNA expression detected in diseased patients compared to normal control populations likely reflect abnormalities in mechanistic pathways related to cancer progression, patient survival, and response to clinical therapies. These miRNA expression signatures are promising biomarkers for detection of early-stage cancer as well as discriminatory markers to identify patients at highest risk for aggressive and lethal forms of cancer and who may benefit most from therapeutic and/or surgical interventions.

Calin *et al.* first proposed that miRNAs play a direct role in cancer progression in 2002, and found that the clustered miRNAs miR-15a and miR-16-1 were often deleted in patients with B-cell chronic lymphocytic leukemia.[15] Indeed, these miRNAs have been well characterized and belong to a subclass of tumor suppressor miRNAs that act *in vitro* and *in vivo* to block proliferation, invasion, and tumorigenesis of leukemias (as well as epithelial cancers) by repressing downstream miRNA targets, which include anti-apoptotic gene B-cell lymphoma 2 (BCL2), G1/S cell cycle progression factor Cyclin D2 (CCND2), and cell growth and survival factor insulin-like growth factor 1 receptor (IGF1R).[16–18] Conversely, pro-oncogenic miRNAs are overexpressed in cancer patients and experimentally verified to promote cancer progression pathways and accelerate tumor formation and metastasis. miR-21 is one of the most commonly overexpressed miRNAs in human cancers and promotes tumor growth by blocking the expression of tumor suppressor genes, including phosphatase and tensin homologue (PTEN) and programmed cell death protein 4 (PDCD4).[19] Furthermore, miRNA expression profiling more accurately classifies human cancers based on tissue origin than the expression of protein-coding genes, reflecting the utility of miRNAs as clinical diagnostic tools.[20] In addition, miRNAs show promise as prognostic biomarkers.[21] For example, lung cancer patients exhibiting reduced *let-7* expression and elevated miR-155 expression have worse post-operative survival than patients exhibiting reciprocal expression patterns.[22,23] Thus, miRNAs have emerged as major regulators of cancer progression with immense clinical potential. The ability of dietary factors to modulate miRNA expression will have a huge impact on disease prevention and therapeutic development.

1.3.3 Extracellular Delivery of miRNAs

miRNA activity is not confined to the cells in which they are produced. Rather, miRNAs are found to circulate in the blood and can be taken up by other cells and tissues at distant sites.[24] These small non-coding RNAs are resistant to degradation in body fluids, presumably when complexed with proteins such as AGO2.[25] Mitchell *et al.* were the first to isolate cell-free miRNAs from human serum and found that miR-141 was elevated in the blood of metastatic prostate cancer patients compared to controls.[26] miRNAs have also been reported in human urine and Bryant *et al.* reported significantly higher miR-107 and miR-573-3p levels in urine of patients with prostate cancer, compared with controls.[27]

The majority of miRNAs found in body fluids are concentrated in small membrane vesicles referred to as "exosomes" or "microvesicles".[28] Exosomes are 40–100 nm vesicles generated by a large range of cell types, including tumor cells and are released extracellularly *via* fusion of multivesicular bodies with the plasma membrane.[29] These secreted microvesicles have been detected in a variety of human body fluids, *e.g.*, blood (serum and plasma), bronchoalveolar lavage, urine, bile, ascites, breast milk, and cerebrospinal fluid.[30] Delivery of exosomes to other cells is not well understood and may occur by receptor-mediated internalization, docking and fusion to target cells, or endocytotic mechanisms. Exosomes probably represent a novel mechanism of miRNA transfer from one cell to another and evidence suggests that these miRNAs can mediate downregulation of gene expression and influence cellular activities within the target cell.[31,32] Therefore, miRNAs selectively packaged in exosomes and secreted by tumor cells into the circulation may be useful fluid-based biomarkers reflecting a patient's cancerous state and disease prognosis.

From the perspective of this chapter, exosomal miRNAs delivered through the diet may be used to mediate cellular effects in the human body (Figure 1.2). Exosomes containing miRNAs are in fact present in human breast milk and bovine milk.[33,34] Novel (and somewhat controversial) evidence indicates that exosomes from bovine milk are taken up by human intestinal cells and endothelial vascular cells *via* endocytosis.[35,36] Wolf *et al.* showed that bovine exosomes are transported into human colon carcinoma Caco-2 cells in culture and result in elevated miR-29b and miR-200c levels.[35] Interestingly, both of these miRNAs have been implicated in human cancers. Exosomal miRNAs are unusually stable and are resistant to acidic conditions and RNase degradation in the gut.[37] Bovine exosomes can enter into the body's circulation and are transported to peripheral tissues—likely due to exosomal uptake by circulating immune cells (*i.e.* human macrophages) in the blood.[36–38] Taken together, dietary uptake of exosomes could have major ramifications in our discussion of how miRNAs are ingested through the diet (milk, plants, microbiota) and regulate gene expression in recipient cells. Continued work to understand the mechanisms of exosome-based delivery of cancer-protective miRNAs is required to translate this research into the clinic.

1.4 miRNAs and the Epigenome

Dietary factors can impact both the epigenome and miRNA expression to mediate cancer progression pathways in human cells. It is therefore important to briefly review the overall role miRNAs play in epigenetics and chromatin remodeling (Figure 1.1).

Epigenetic mechanisms involving DNA modifications such as methylation and changes to chromatin structure can have a significant impact on gene expression in both positive and negative ways by regulating access of the transcriptional machinery to the DNA (for a comprehensive review, see[39]). For example, DNA methylation is a key means of silencing gene expression in human cells (Figure 1.1A). This involves the addition of a methyl group exclusively on cytosine residues adjacent to guanines (clustered in CpG islands) and generally located within gene promoter regions. The DNA methyltransferase (DNMT) enzymes, DNMT1, DNMT3A and DNMT3B, catalyze this reaction *de novo* by transferring a methyl ($-CH_3$) group from *S*-adenosyl methionine (a methyl donor) to the 5′ position on the cytosine ring. The promoters of active genes are sparsely methylated (hypomethylated), whereas silenced genes are commonly associated with hypermethylation.

Another method to regulate gene expression is the control of chromatin structure (Figure 1.1B). DNA is compactly packaged into the nucleus. Approximately two turns of DNA are wrapped around a nucleosome, which is composed of an octamer of histone core proteins. Modifications to these histone proteins (*e.g.*, methylation, acetylation, phosphorylation, ubiquitination,

SUMOylation, and ADPribosylation) will alter the compactness of the DNA and thus the ability of DNA-binding factors and cellular machinery to access the chromatin and initiate/maintain gene transcription. For instance, acetylation of histones is one of the primary mechanisms used to open up the chromatin and activate gene transcription. These modifications are dynamic. Specific enzymes called histone acetyltransferases (HATs), which add acetyl ($-CH_3CO$) groups onto histone tails (*i.e.*, H3K9ac), and histone deacetylases (HDACs), which catalyze the removal of these acetyl groups, and are used to fine-tune gene expression. The effects of histone methylation, mediated by the enzymes histone methyltransferases (HMTs) and histone demethylases (HDMs) are more complex then histone acetylation. In some cases, histone methylation can enhance gene transcription (*i.e.*, H3K4me3); however, in other cases, histone methylation represses transcription (*i.e.*, H3K9me3, H3K27me3). In this section, we focus on how miRNAs associated with human cancers can directly alter the enzymes responsible for such epigenetic modifications, as well as how miRNAs are themselves regulated by the epigenetic machinery, and in turn impact cancer progression.

1.4.1 miRNAs that Regulate the Epigenetic Machinery

miRNAs can contribute to cell transformations and cancer progression by targeting DNA methyltransferase enzymes. Hypermethylation of DNA promoters is a common feature in human cancers.[4] The DNMT enzymes DNMT1, DMNT3A, and DNMT3B, are often elevated in liver, prostate, breast, colorectal, and lung cancers.[40–43] Expression of the miR-29 family (miR-29a, miR-29b, and miR-29c) is downregulated in tumors from non-small cell lung cancer patients and inversely correlates with DNMT3A and DNMT3B expression. Work by Fabbri *et al.* validated that these DNMT enzymes are direct miR-29 family targets.[44] Enforced miR-29 expression in human A549 lung cancer cell lines resulted in global reduction of DNA methylation patterns and blocked tumorigenesis *in vitro* and *in vivo* due to the re-expression of the tumor suppressor genes fragile histidine triad protein (FHIT) and WW domain-containing oxidoreductase (WWOX).[44] Overexpression of miR-29b also causes global DNA hypomethylation in acute myleloid leukemia by repressing DNMT3A and DNMT3B.[13] Interestingly, miR-143 (but not miR-29) targets DNMT3A in colorectal tissues, indicating that these miRNAs function in a tissue-specific manner.[45]

In addition, miRNAs can repress enzymes involved in histone modifications. For example, miR-101 directly targets the histone methyltransferase enhancer of Zeste homolog 2 (EZH2), a subunit of the polycomb-repressive complex 2 (PRC2) which catalyzes methylation of lysine 27 of histone H3 (H3K27me3) and lysine 9 of histone H3 (H3K9me3) to promote transcriptional silencing.[46] EZH2 is often overexpressed in cancers (*i.e.* prostate, breast, bladder, gastric) and functions to promote cell proliferation, invasive disease, and stem cell maintenance.[46] In prostate cancer, reduction of miR-101 expression inversely correlates with increased EZH2 levels, and in many

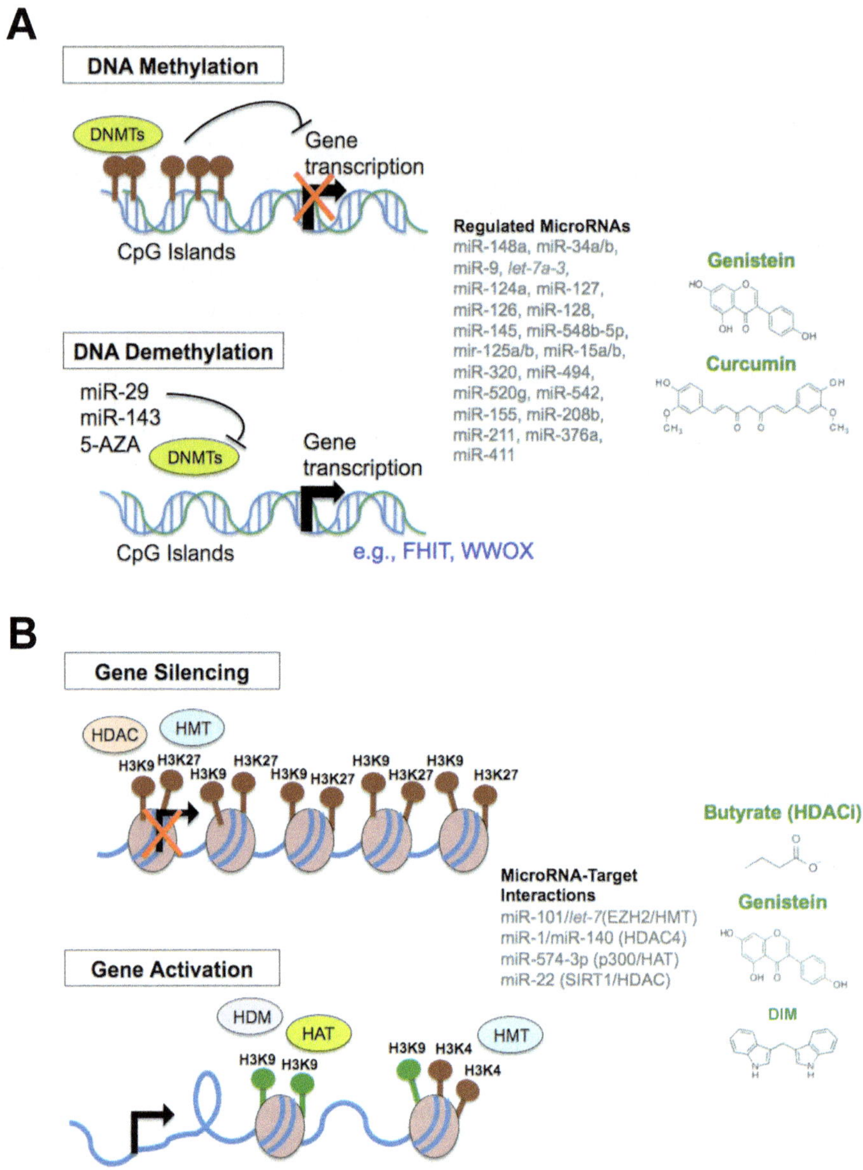

Figure 1.1 MicroRNAs and epigenetic regulation of gene expression. (A). DNA methylation results in gene silencing. DNA methyltransferases (DNMTs) catalyze the addition of methyl groups (red) onto cytosine residues adjacent to guanines, which are clustered in CpG islands located in promoter regions. These epigenetic alterations can impact miRNA expression (examples listed). miR-29 and miR-143 directly target DNMTs to activate tumor suppressor gene expression. Edible polyphenols, *i.e.*, genistein and curcumin act similarly to DNA methyltransferase inhibitory drugs (*e.g.*, 5-AZA) to regulate expression of cancer-associated miRNAs. (B). Histone modifications alter

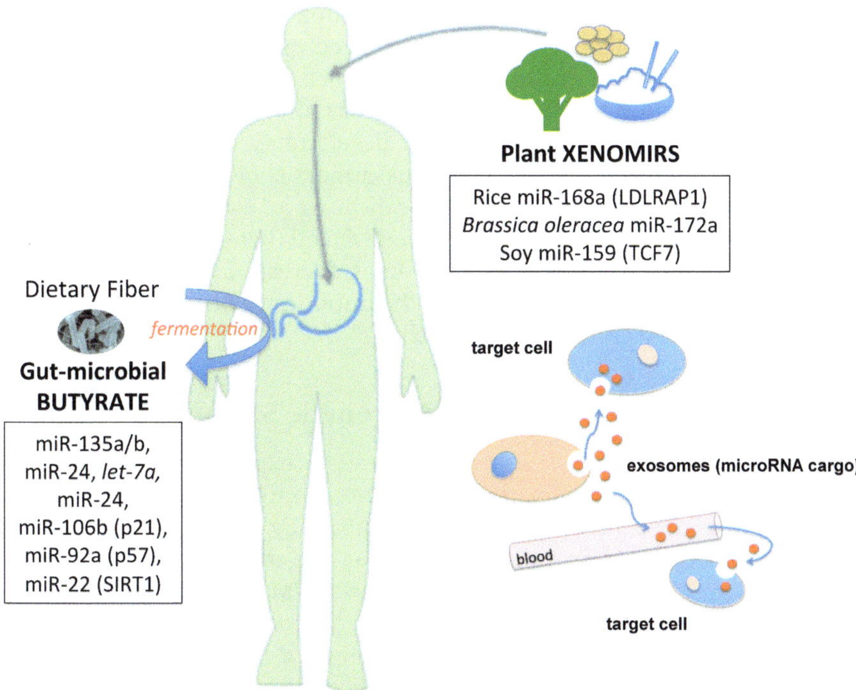

Figure 1.2 Dietary factors associated with the gut microbiota and ingested plant xenomiRs impact cancer progression in humans. Dietary fiber is fermented in the gut by bacterial anaerobic fermentation into short chain fatty acids (SCFAs), including butyrate. Colon cells use this for energy. Cancer cells do not metabolize butyrate and rather this microbial metabolite acts as a Histone deacetylase inhibitor to regulate the expression of cancer-associated miRNAs (examples listed). Exogenous miRNAs (termed xenomiRs) ingested from the diet (*e.g.*, rice, cultivars like broccoli, soy beans) can survive digestion and impact gene expression in recipient cells. Circulating plant xomiRs have been isolated in the blood of animals and humans from exosomes and are hypothesized to play a role in cancer progression.

chromatin structure and gene expression. Methylation (red) of lysine 9 on histone 3 (H3K9) and lysine 27 on histone 3 (H3K27) causes condensed chromatin and gene silencing. Acetylation (green) of lysine 9 on histone 3 (H3K9) and methylation of lysine 4 on histone 3 (H3K4) allow cellular machinery to access the chromatin and activate/maintain gene expression. Histone acetyltransferases (HATs), histone deacetylases (HDACs), histone methyltransferases (HMTs) and histone demethylases (HDMs) catalyze these modifications. miRNAs can target epigenetic enzymes (examples shown). Dietary factors, *e.g.*, genistein, DIM, and butyrate (a histone deacetylase inhibitor, HDACi) employ histone modifications to regulate miRNA expression and cancer progression.

cases this is due to miR-101 genomic loci deletions.[46,47] Furthermore, miR-1 and miR-140 have both been validated as direct targets for HDAC4. miR-1 targets HDAC4 in muscle and chondrocyte tissues and is responsible for promoting differentiation of these tissues, whereas miR-140 is cartilage-specific and targets HDAC4 to regulate long bone development.[48,49] Interestingly, an estimated 93% of human chordoma bone tumors show low levels of miR-1 expression compared to normal tissue, indicating its use as a biomarker for this cancer type.[50] In addition, Duan *et al.* revealed that forced expression of miR-1 in human chordoma cancer cell lines decreased HDAC4 expression and inhibited cell growth—implicating the importance of tumor-suppressor miRNAs to fine-tune chromatin remodeling.[50]

1.4.2 miRNAs Targeted by the Epigenetic Machinery

The epigenetic machinery can regulate miRNA expression and, in turn, impact cancer progression pathways. 5–10% of miRNA loci reside within CpG islands and are proposed to be epigenetically regulated. The expression of miRNAs such as miR-148a, miR-34b/c, miR-9, and *let-7a-3* are dependent on their methylation status and DNMT1 and DNMT3b DNA methyltransferases.[51–55]

Promoter CpG hypermethylation is associated with a global loss of miRNA expression in colon cancer cell lines.[3] Many of these miRNAs play tumor-suppressive roles. For example, DNA hypermethylation of miR-124a leads to increased activation of cyclin dependent kinase 6 (CDK6) and the phosphorylation of RB, both of which promote cell cycle progression and tumor growth.[56] Furthermore, in a screen using T24 bladder cancer cells, Saito *et al.* compared miRNA expression profiles in the presence and absence of DNMT inhibitor 5-aza-2′-deoxycytidine (5-AZA) and HDAC inhibitor 4-phenylbutyric acid (4-PBA).[53] 17 out of 313 miRNA genes studied were upregulated following this treatment. This group noted that the *miR-127* gene resides within a CpG island and miR-127 expression was upregulated due to DNA demethylation as well as histone H3 acetylation and histone H3 lysine 4 trimethylation (H3K4me3). Therefore, epigenetic mechanisms control miRNA expression and ultimately regulate the miR-127 target, pro-oncogene *B-cell* lymphoma 6 (*BCL6*).[53] In addition, miR-148a, miR-34b/c, and miR-9 hypermethylation have all been linked to metastasis in patients with melanoma, colon, lung, breast, head, and neck carcinomas, and thus could be employed as a prognostic tool. In animal models, these miRNAs function as anti-metastatic agents to reduce migration, tumor formation, and metastasis.[55]

miRNA promoter hypomethylation has also been linked to cancer progression and can result in overexpression of pro-oncogenic miRNAs to promote tumor formation. miRNA promoter demethylation is attributed to increased expression of miR-126 and miR-128 in acute myeloid leukemia and acute lymphocytic leukemia, respectively.[57] Therefore, chromatin-modifying drugs may be a powerful approach to regulate oncomiR expression and treat a large range of human cancers.

1.5 miRNAs, Dietary Factors, and the Epigenome

We have discussed the impact miRNAs and epigenetic modifications have on cancer progression. Diet is also a key player in the development of cancer. In this section, we discuss how the dietary intake of fresh fruits, vegetables, nuts, and beverages (*i.e.*, tea, wine), which are rich in polyphenols modulate miRNA expression, and in some cases the epigenome, in order to mediate their cancer-protective effects.

Polyphenols are a major class of metabolites that possess antioxidant, free-radical scavenging properties, and notable cancer-protective effects. More than 500 polyphenols have been identified, including classes of phenolic acids, flavonoids, stilbenes, tannins, lignans, quinones, and curcuminoids. Many of these compounds directly regulate miRNAs that are closely associated with human cancers.[58,59] Specifically, the dietary polyphenols resveratrol, ellagitannins and ellagic acid, epigallocatechin-3-gallate, genistein, curcumin, and diindolylmethane have been studied intensely and shown mechanistically to exert their tumor-suppressive and cancer-protective effects by regulating miRNA expression (Table 1.1). These edible compounds are promising therapeutic drugs and disease-prevention agents for administration to the general population, since foods rich in these anti-cancer polyphenols are easily accessible, non-toxic, and affordable.

1.5.1 Resveratrol

Resveratrol belongs to the stilbenoid class of polyphenol compounds commonly found in grapes, wine, peanuts, cocoa, blueberries, and cranberries. This polyphenol has antioxidant and protective effects against a variety of malignancies, including pancreas, colon, and breast cancers.[60] Resveratrol has been shown to inhibit pancreatic and bladder cancer cell growth and induce apoptosis by suppressing pro-oncogenic miR-21 expression and anti-apoptotic factor BCL2.[61,62] In the context of colon cancer, treatment of human SW480 colon cancer cells with resveratrol decreased several pro-oncogenic miRNAs, including miR-17, miR-21, miR-25, and miR-92a-2, which are often overexpressed in colorectal cancer patients.[63] Resveratrol delivery also elevated the expression of tumor-suppressive miRNAs, including miR-34a and miR-663, in colon cancer cells.[63,64] Reciprocal repression of miR-663 target transforming growth factor-β receptor type I (TGFβR1) in treated cells probably mediates the protective effects of resveratrol against cancer progression phenotypes associated with angiogenesis, cell motility, invasion, and metastasis.[63] Furthermore, Hagiwara *et al.* found that resveratrol exposure increased tumor-suppressive miRNAs miR-141 and miR-200c in MDA-MB-231-luc-D3H2LN breast cancer cells *in vitro*, resulting in the inhibition of cancer cell invasiveness.[65] Consequently, miR-200c targeted and repressed pro-oncogenic transcription factor zinc finger E-box binding homeobox 1 (ZEB1), an E-cadherin regulator and promoter of tumor invasion and metastasis. Interestingly, resveratrol is likely to have a more

Table 1.1 MicroRNAs regulated by dietary polyphenols associated with human cancer.[a]

Polyphenol	MicroRNAs regulated	Targets of regulated microRNAs	Cancer type(s)
Resveratrol 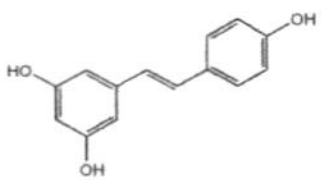	**miR-21**, **miR-17** **miR-25**, **miR-92a-2** *miR-34a*, *miR-663* *miR-141*, *miR-200c*	**TGFβR1**, **ZEB1**	Pancreas, bladder, colon, breast
Ellagitannins and ellagic acid 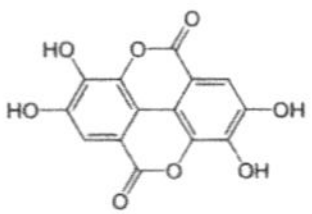	miR-646, miR-1249 miR-135b-5p miR-135b-3p miR-92b-5p, miR-765, miR-496, miR-181c-3p, miR-18a-3p	Unknown	Colon
Epigallocatechin-3-gallate 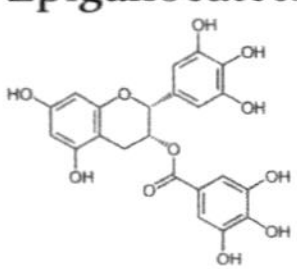	**miR-21**, **miR-98-5p** *miR-16*, *miR-330* *miR-210*	**BCL2**	Liver, prostate, lung
Genistein 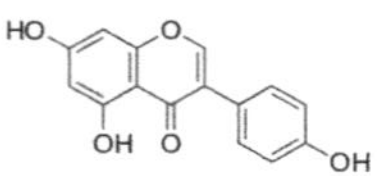	**miR-155**, **miR-221** **miR-222**, **miR-125a** **miR-125b**, **miR-15b**, **miR-320**, **miR-494** **miR-520g**, **miR-542**, **miR-208b**, **miR-376a**, **miR-411**, **miR-95**, **miR-21**, *miR-23b*, *miR-574-3p*, *miR-1296*, *miR-145* *miR-548b-5p*, *miR-15a* miR-27a	*FOXO3*, *PTEN*, *CK1∞*, *p27*, *SPRY2*, *ARH1*, **MCM2**, **RAC1**, **EGFR**, **p300**, **MET**	Breast, ovary, prostate, lung, colon, kidney
Curcumin 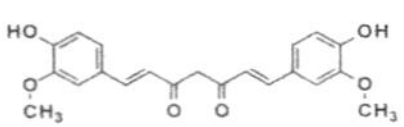	**miR-19a**, **miR-19b** **miR-21**, **miR-770-5p**, **miR-1247** *miR-181b*, *miR-34a* *miR-16*, *miR-15a* *miR-146-5p*, *miR-98*	**CXCL1**, **CXCL2**, **BCL2**, **WT1** *PDCD4*, *PTEN*, **NOTCH1**, **LIN28**	Larynx, colon, blood (leukemia), brain (glioblastoma)
Diindolylmethane 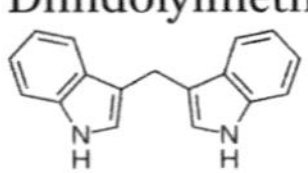	**miR-221** *miR-200a*, *miR-200b*, *miR-200c*, *let-7a*, *let-7b*, *let-7c*, *let-7d*, *miR-146a* miR-21	**EGFR**, **NF-κB**, **IRAK1**, **MTA2**, **EZH2**, **CDC25A**, *PTEN*, *p27*, *p57*, *PUMA*	Pancreas, breast, prostate

global impact on miRNA activity in cancer cells. Resveratrol treatment can increase transcription of AGO2, the catalytic component of miRISC, in breast cancer cells.[65] Since cancer cells often exhibit widespread miRNA underexpression, this work indicates that resveratrol treatment could normalize and restore miRNA activity indirectly *via* AGO2 to confer its protective cancer effects.

1.5.2 Ellagitannins and Ellagic Acid

Ellagitannins and ellagic acid are the main bioreactive polyphenol compounds found in fruits and nuts, such as pomegranates, black raspberries, raspberries, strawberries, walnuts, and almonds. Colon cancer cells treated with pomegranate extract containing ellagitannins and ellagic acid indicated that these polyphenol compounds have anti-proliferative and apoptotic effects *in vitro* and *in vivo*.[66] Nunez-Sanchez *et al.* investigated miRNA gene modulation in 35 colorectal cancer patients who consumed 900 mg of pomegranate extract daily before surgery compared to patients who did not consume the extract.[67] Specifically, miR-646, miR-1249, miR-135b-5p, miR-135b-3p, miR-92b-5p, miR-765, miR-496, miR-181c-3p, and miR-18a-3p were differentially expressed in colon tumors *versus* normal adjacent colon tissues in the treated patients. Unfortunately, the functional significance of these miRNAs was not further explored, and thus the contribution of non-coding RNAs in mediating the protective effects of pomegranate derived-ellagitannins and ellagic acids are unknown.

1.5.3 Epigallocatechin-3-gallate

Epigallocatechin-3-gallate (EGCG) is a polyphenol catechin found in green tea (*Camellia sinensis*), which is a popular beverage in China and Japan. Epidemiologically, EGCG ingestion is attributed to the prevention of human oral and lung cancers.[68,69] EGCG effectively blocks tumor growth and promotes apoptotic cell death for a range of cancers, including oral, lung, digestive tract, skin, liver, prostate, and breast cancers.[70] EGCG has emerged as an important regulator of cancer-associated miRNAs. In the context of liver cancer, EGCG treatment of human hepatocellular carcinoma HepG2 cells *in vitro* resulted in the upregulation of tumor-suppressive miRNA miR-16 and reciprocal downregulation of the miR-16 target anti-apoptotic protein BCL2. Indeed, treatment of HepG2 cells with miR-16 antimiRs blocked EGCG-mediated apoptotic effects and increased BCL2 expression.[71] In addition, EGCG impacts prostate cancer progression. EGCG can block the nuclear translocation of the androgen receptor, a major signaling pathway for prostate growth. EGCG suppressed prostate tumor growth in mouse xenograft models, resulting in the down-regulation of oncogenic miRNA miR-21 as well as elevated expression of tumor suppressive and pro-apoptotic miR-330.[72] Therefore, EGCG controls the expression of androgen receptor-regulated miRNAs and holds promise as a clinical tool for castration-resistant prostate cancer, which remains an incurable and lethal disease. In lung cancer, Wang *et al.* showed that EGCG exposure in human H1299 and H460 lung cancer cell lines caused hypoxia-induced factor 1α (HIF-1α) accumulation, as well as upregulation of the hypoxia-inducible miRNA miR-210.[73] As predicted, miR-210 treatment mirrored the ability of EGCG to reduce cell proliferation and block anchorage-independent growth in lung cancer cells. Furthermore, EGCG treatment of human lung cancer A549 cells resulted in selective repression of miR-98-5p.[74] Interestingly, Zhou *et al.* found that EGCG enhanced the efficacy of cisplatin-induced apoptosis in these cells, which was mediated *via* miR-98-5p downregulation.[74] Therefore, the combined treatment of chemotherapeutic agents, such as cisplatin, with green tea extracts could be a novel clinical strategy for lung cancer.

1.5.4 Genistein

Genistein is an isoflavinoid present in soybeans with antioxidant, antimicrobial, and protein tyrosine activation enzyme inhibitor activities. This plant polyphenol shares a similar chemical structure to estrogen and is shown to modulate cancer progression of hormone-responsive tissues such as the breast, ovary, and prostate. Although genistein is reported to have oncogenic estrogenic properties in estrogen receptor positive (ER+) breast cancers, dietary levels of genistein inhibited cell growth and induced apoptosis in estrogen receptor negative (ER−) human breast cancer cell lines MDA-MB-435 and Hs578t (but not in ER + MCF-7 cells).[75,76] de la Parra *et al.* found that anti-cancer activity correlated with pro-oncogenic miR-155 expression in the ER− cell lines and genistein treatment of MDA-MB-435 and Hs578t cells decreased endogenous miR-155 levels.[76] Reciprocally, treated ER− cells showed increased expression of miR-155 targets, *i.e.*, tumor suppressors forkhead box O3 (FOXO3) transcription factor, PTEN, serine-threonine kinase β-catenin inhibitor casein kinase 1alpha (CK1α), and cyclin dependent cell cycle inhibitor p27.[76] Additional miRNAs expressed in the mammary gland, notably miR-23b, were induced upon genistein treatment and acted to inhibit breast cancer cell growth.[77] In the context of ovarian cancer, Xu *et al.* found that genistein downregulated miR-27a levels in ovarian carcinoma SKOV3 cells and subsequently blocked cell growth and migration.[78] This was accompanied by reciprocal elevation of putative miR-27a target sprouty homologue 2 (SPRY2), an extracellular receptor kinase 1/2 (ERK1/2) signaling inhibitor.[78] Furthermore, genistein reduced miR-221 and miR-222 expression in human prostate cancer PC3 cells and conversely elevated levels of the validated miR-221/miR-222 target aplysia ras homology member I (ARH1), a tumor suppressor that inhibits prostate proliferation, invasion, and promotes apoptosis.[79] Genistein delivery to human PC3 and LNCaP prostate cancer cell lines also induced miR-1296 expression, which in turn suppressed the DNA replication factor minichromosome maintenance 2 (MCM2).[80]

Interestingly, a link between genistein's tumor suppressive effects and its ability to regulate both miRNA expression and the epigenome has been well established in the prostate. For example, Chiyomaru *et al.* showed that genistein treatment elevated miR-574-3p levels in prostate cancer cells, and reciprocally downregulated the pro-oncogenic miR-574-3p targets that control prostate cell proliferation, migration, invasion, and metastasis. These included Rho family GTPase RAS-related C3 botulinum toxin substrate 1 (RAC1), epidermal growth factor receptor (EGFR), and histone acetyltransferase p300.[81] In another study, Zaman *et al.* noted that treatment with genistein or demethylation drug 5-azacytidine (5-AZA) of human PC3 prostate cancer cells induced expression of miR-145 *via* demethylation of the miR-145 promoter, and subsequently resulted in cell cycle arrest and apoptosis.[82] Similarly, Rabiau *et al.* compared genistein treatment in a panel of human prostate cancer cell lines with 5-AZA exposure.[83] They found that either condition resulted in similar miRNA expression profiles. Specifically, either genistein or 5-AZA treatment significantly increased miR-548b-5p levels (and decreased miR-125a, miR-125b, miR-15b, and miR-320 expression) in PC3 cells, increased the tumor suppressor miR-15a (and decreased miR-494, miR-520g, and miR-542) in LNCaP cells; and reduced miR-155, miR-208b, miR-211, miR-376a, and miR-411 levels in DU145 cells. Therefore, the soy-based isoflavone genistein relies on DNA methylation and histone protein alterations to regulate miRNA expression and cancer progression.

In addition, genistein is an important cancer-associated miRNA regulator in tissues outside of the reproductive system. For example, application of genistein to human A549 lung cancer cells resulted in increased caspase-3/9 driven apoptosis and suppressed cell proliferation by activating miR-27a expression and reciprocally shutting down the miR-27a target MET, a pro-oncogenic receptor tyrosine kinase.[84] (Potentially, miR-27a has oncogenic activities in the ovary, and genistein had opposite effects on miRNA expression in this cell type, as discussed earlier.) In colorectal cancers, genistein was found to reduce cell growth and induce apoptosis in HCT-116 cells both *in vitro* and *in vivo*.[85] These effects were mediated by downregulation of miR-95 as well as the oncogenic serine/threonine-protein kinases Ak strain transforming (AKT) and serum glucocorticoid regulated kinase 1 (SGK1).[85] Furthermore, genistein reduced oncogenic miR-21 in renal cancer cells in order to inactive NOTCH signaling.[86] Taken together, genistein has potential as a natural anti-cancer agent and may prove especially useful in drug-resistant ER− breast cancers.

1.5.5 Curcumin

Curcumin is a natural curcuminoid phenol derived from turmeric root (*Curcuma longa*), a plant native to South Asia that has been used as a cooking spice and an herbal medicine for centuries. Scientific evidence demonstrates that curcumin has anti-oxidant, anti-inflammatory, and anti-cancer activities. This dietary compound is particularly toxic to tumor cells, but ineffective in normal cells and cellular uptake is notably higher in cancer cells for reasons not well understood.[87] Curcumin regulates miRNA expression in a large range of tumors. Curcumin treatment of breast cancer cell lines, for instance, induces miR-181b, miR-34a, miR-16, miR-15a, and miR-146b-5p expression, and downregulates miR-19a and miR-19b.[88] This treatment resulted in apoptosis, suppressed tumor formation, and metastasis. Many of these modulated miRNAs have been functionally characterized. For example, Kronski *et al.* observed that curcumin treatment of human breast cancer MDA-MB-231 cells elevated miR-181b and consequently suppressed the miR-181b pro-metastatic targets cytokines CXCL1 and CXCL2.[89] Indeed, forced miR-181b expression in breast cancer cells similarly blocked proliferation and invasion, promoted apoptosis, and inhibited metastasis in mouse models.[89] Conversely, miR-181b anti-miRs abolished curcumin's anti-cancer effects.[89] Curcumin appears to use miR-15-based regulation as a common mechanism to suppress cancer cell growth and induce apoptosis. Curcumin-induced miR-15a expression and reciprocal downregulation of miR-15a target anti-apoptotic factor BCL2 was employed in human breast cancer and laryngeal cancer cells to curb their growth.[90,91] In leukemia cells, curcumin elevated both miR-15 and miR-16 levels to suppress growth *via* the miR15/miR-16 target, oncogene Wilms' tumor 1 protein (WT1).[92]

Pro-oncogenic miRNA miR-21 is another important mechanism used by curcumin to mediate cancer protective effects in colorectal cancers. Evidence indicates that curcumin blocks miR-21 expression *via* transcriptional regulation. Curcumin treatment of colon cancer Rko and HCT116 cells reduced miR-21 promoter activity by inhibiting activator protein 1 transcription factor (AP1) binding to the miRNA promoter.[93] This resulted in elevated expression of miR-21 targets, programmed cell death protein 4 (PDCD4) and PTEN, tumor suppressors known to block colon cancer cell growth, invasion and metastasis.[93,94] In chemotherapy (5-flurouracil + oxaliplatin)-resistant HCT116 and HT-29 colon cancer cell lines, curcumin-mediated reduction of miR-21 restored PTEN levels, reduced oncogenic AKT-phosphorylation, and inhibited colon cell growth.[94] Therefore, curcumin-based control of pro-oncogenic miR-21 may be a promising approach to treat drug-resistant colorectal cancers.

Curcumin may also use epigenetic mechanisms to elevate tumor suppressor miRNA expression. Roy *et al.* specifically investigated how curcumin restored tumor suppressor miR-34a expression in colon cancer.[95] miR-34a is silenced in colon cancer cells due to CpG island hypermethylation within the miR-34 promoter. They noted that treatment of human colon cancer SW620 cells with curcumin or demethylating drug azacitidine (Aza-dC) reverted the miR-34a promoter to a CpG hypomethylated state. Both treatments increased miR-34a levels and reciprocally decreased miR-34a target NOTCH1 expression in order to inhibit colon cancer growth.[95] Interestingly, curcumin can suppress human prostate cancer DU145 and 22Rv1 proliferation and invasion *in vitro* by inhibiting miR-770-5p and miR-1247 expression. These miRNAs are located within the DLK1-DIO3 imprinted gene cluster and further indicate epigenetic control by curcumin in this cell type.[96]

Curcumin also regulates non-coding RNA expression at the level of miRNA biogenesis. Work by Liu *et al.* in A549 lung cancer cells showed that curcumin upregulated expression of the *let-7* family member, miR-98, and reciprocally downregulated miR-98 target LIN28.[97] Suppressed lung cancer cell migration and invasion following curcumin treatment were mediated by blocking LIN28A-induced matrix metalloproteinase proteins MMP2 and MMP9 expression. Intriguingly, LIN28 inhibits *let-7* family pri-miRNA and pre-miRNA processing.[98,99] It is conceivable that the *let-7*/miR-98/LIN28 negative feedback loop is broken in cancerous colon cells, allowing for elevated LIN28 protein and thereby blocking *let-7*/miR-98 activity *via* interference with miRNA biogenesis. The re-establishment of *let-7*/miR-98 expression by curcumin treatment is probably cancer-protective by keeping LIN28 levels in check.

Therapeutic development of curcumin to regulate miRNA expression in order to suppress tumor progression and metastasis is currently being explored in clinical trials. Use of chemically modified curcumin analogs, such as difluorinated curcumin (CDF) designed to increase stability and bioavailability, will facilitate these studies.[94] Combined treatment approaches involving curcumin and miRNA delivery may also be on the horizon. Indeed, studies by Li *et al.* to treat glioblastoma verified that co-delivery of miR-138a (an established tumor suppressive cell cycle inhibitor in glioblastoma) with curcumin enhanced the sensitivity of curcumin treatment in human U87 glioblastoma cells by targeting the p38 mitogen-activated protein kinase (MAPK) signaling pathway.[100]

1.5.6 Diindolylmethane

Indole-3-carbinol (I3C) is a glucosinolate found in cruciferous vegetables (*i.e.*, broccoli, cauliflower, and cabbage). Upon ingestion and contact with gastric acids in the stomach, I3C undergoes *in vivo* self-condensation to form the bioactive metabolite 3,3′-diindolylmethane (DIM). DIM is well characterized for its antioxidant and anti-inflammatory activities *via* inhibiting the nuclear factor-κB (NF-κB) and AKT pathways to reduce oxidative stress.[101] In addition, DIM is shown to have anti-tumor properties in various tissues, in part by regulating the expression of cancer-associated miRNAs. For example, Li *et al.* performed miRNA profiling on gemcitabine-sensitive and gemcitabine-resistant pancreatic cancer cell lines. They noted significant downregulation of tumor suppressor miRNAs miR-200b, miR-200c, *let-7b*, *let-7c*, *let-7d*, and *let-7e* in gemcitabine-resistant cells that possessed pronounced EMT phenotypes.[102] Treatment of DIM using a modified bioresponse-formulated DIM (BR-DIM) (conferring greater bioavailability) to gemcitabine-resistant MiaPaCa-2, Panc-1, and Aspc-1 human pancreatic cancer cell lines significantly increased the expression of miR-200 (miR-200a, miR-200b, miR-200c), the *let-7* family, and miR-146a.[102,103] Forced expression of miR-200 using miRNA mimics in gemcitabine-resistant MiaPaCa-2 cells reversed EMT phenotypes, marked by increased epithelial E-cadherin and suppressed mesenchymal vimentin proteins, in a similar manner as cells exposed to DIM alone.[102]

Similarly, Li *et al.* verified that upregulation of miR-146a by DIM plays an important role in pancreatic cancer progression. miR-146a is commonly repressed in human pancreatic cancer cells compared to normal human pancreatic duct epithelial cells.[103] When pancreatic cancer cells were either treated with DIM or with miR-146a mimics, cancer cell invasion was suppressed through inhibition of the NF-κB and EGFR pathways. Anti-cancer effects were mediated *via* downregulation of the miR-146a targets EGFR, NF-κB, NF-κB regulatory kinase interleukin 1 receptor-associated kinase 1 (IRAK-1), and metastasis-associated protein 2 (MTA-2).[103] DIM was previously reported to sensitize pancreatic cancer cells to the EGFR inhibitor erlotinib and induce apoptotic cell death.[104]

These miRNA studies may provide insight into how DIM mechanistically sensitizes cancer cells to chemotherapeutic agents (*i.e.*, erlotinib) through miRNA-based regulation (*e.g.*, miR-146a–EGFR target interactions). Indeed, Li *et al.* found that treatment of miR-200b or DIM could increase the sensitivity of human MiaPaCa-2 cells to gemcitabine and significantly decrease pancreatic cancer cell growth.[102] The combined efficacy of DIM and miR-200 with the chemotherapeutic drug Herceptin (trastuzumab) was also noted. DIM treatment of breast cancer cells induces miR-200 family (miR-200a, miR-200b, and miR-200c) levels.[102] These tumor-suppressive miRNAs are often decreased in breast cancer tumors and shown to block EMT phenotypes associated with migration and invasion.[105] Interestingly, although Herceptin treatment alone could not modulate miR-200 levels in human MBA-MB-468 breast cancer cells, DIM and Herceptin treatment combined had synergistic effects to significantly increase miR-200a and miR-200b expression.[105]

DIM is also shown to have cancer-protective effects in the prostate using both miRNA and epigenetic mechanisms. Kong *et al.* found that treatment of formulated BR-DIM to human LNCaP, C4-2B, and PC3 prostate cancer cell lines increased *let-7a*, *let-7b*, *let-7c*, and *let-7d* levels and reciprocally reduced *let-7* target EZH2, a histone methyltransferase that induces cancer stem cell maintenance, EMT transitions, and cancer progression.[106] The *let-7*-EZH2 target interaction in BR-DIM-treated cells resulted in reduced prostate cancer cell growth in clonogenic assays and suppressed sphere-forming capacity. Subsequently, this group tested prostate tissue from patients enrolled in phase II clinical trials for prostate cancer prior to radical prostatectomy to test BR-DIM intervention, and confirmed *let-7* upregulation and reciprocal EZHA expression following DIM treatment occurred in a clinical setting.[106]

In addition to the ability of DIM to downregulate miRNA expression in cancer cells, this polyphenol agent is reported to repress the expression of pro-oncogenic miRNAs. DIM treatment decreased miR-21 expression in liver cells.[107] Since miR-21 is commonly overexpressed in a broad range of human cancers, this work could impact therapeutic treatment in other tissues.[19] In fact, DIM may be a natural and potent agent to specifically antagonize miR-21 activity. Molecular docking and dynamics simulations revealed that DIM binds covalently to the precursor pre-miR-21 transcript at a guanine at the sixth position, which is postulated to interfere with DICER processing and miR-21 maturation.[108] Intriguingly, Jin reported that DIM increased miR-21 in MCF-7 breast cancer cells and functioned to inhibit cell growth *via* suppression of miR-21 target Cell division cycle 25A (CDC25A), a G1/S cell cycle factor.[109] This result underscores the complexity of miR-21 regulation in breast cancer. Contrary reports exist that indicate forced miR-21 expression in MCF-7 cells is oncogenic and promotes breast cancer cell proliferation, migration, and tumor formation.[110–112] DIM is also shown to regulate the oncogenic miR-221, which is elevated in most epithelial cancers (*e.g.*, liver, lung, breast, prostate, brain, stomach, and pancreas).[113] Sarkar *et al.* found that BR-DIM treatment of MiaPaCa-2 and Panc-1 pancreatic cancer cells resulted in decreased miR-221 expression.[114] The subsequent block in pancreatic cell growth and migration following treatment was attributed to an increase in the miR-221 target proteins PTEN, cyclin dependent kinases (CDKs) $p27^{kip1}$ and $p57^{kip2}$, as well as p53 upregulated modulator of apoptosis (PUMA). Interestingly, this study showed that a synthetic curcumin analogue produced similar results *via* miR-221-target regulation. Therefore, distinct subclasses of natural polyphenol agents could employ overlapping miRNA mechanisms to achieve cancer-protective effects. These findings further support use of dietary compounds and miRNAs as effective treatment regimens to sensitize cancer cells and inhibit drug-resistant cancers.

1.6 miRNAs, the Microbiota, and Colon Cancer

We have discussed how dietary factors regulate miRNAs and impact cancer progression in a large range of tissues. Research also indicates that dietary factors have a pronounced influence on the gut microbiota. In turn, gut microbial metabolites, such as butyrate, exert physiological effects on the host *via* miRNA expression and the epigenome (Figure 1.2). Whole-genome sequencing studies of clinical fecal samples revealed that human individuals across nationalities harbor 10–100 trillion microorganisms, which make up a surprisingly diverse colonization of microbes that include fungi, archaea, protists, and viruses. The vast majority of the microbiome, however, is composed of bacteria, predominantly Firmicutes and Bacteroidetes.[115,116] Selective enrichment of particular subgroups of microbes appears to vary across individuals and likely correlates with diet, drug intake, and health status. Bacterial enterotypes analyzed by Qin *et al.* indicate that gut-harbored bacterial strains are those that predominantly utilize fermentable substrates in the colon for energy.[117] This relationship is symbiotic and the human host benefits from microbial colonization in the gut due to the need for microbial conversion of complex carbohydrates into absorbable substrates, vitamin production, and maintenance of physiological homeostasis. Humans exhibit a wide range of metabolic, immune, neurological, and cognitive disorders when these synergistic interrelationships are disrupted. There is a strong association between gut microbiota dysregulation and human disease, *i.e.*, obesity, diabetes, autoimmune disorders, and, of course, cancer.[118]

The interplay of miRNA regulation and the microbiota has only recently been appreciated. Evidence indicates that human fecal miRNAs, such miR-515-5p and miR-1226-5p, specifically target bacterial genes that control bacterial growth, and thus directly shape the overall composition of the gut microbiota.[119] Conversely, studies reveal that gut microbes can control host miRNA expression and influence cancer progression. For instance, gut microbiota generated butyrate induces the expression of miRNAs, such as miR-22, and decreases others, including miR-106b and miR-92a.[120–122] Therefore, non-coding RNA mechanisms involving inter-species gene regulation are extremely important to maintain homeostasis and human health. In this section, we explore how nutrition, the gut microbiota, and the epigenome are closely associated with the role miRNAs play during disease progression, primarily in colorectal cancer.

1.6.1 Microbial-produced Butyrate Influences Host miRNA Expression

Colorectal cancer (CRC) is the third most common cancer and possesses the fourth highest cancer mortality rate worldwide.[123] Epidemiological studies show that Western-style diet (consumption of energy-dense foods, red and processed meats and low consumption of fresh fruits, vegetables, or whole grains) is one of the highest risk factors for this disease.[124,125] Conversely, adherence to the Mediterranean diet correlates with a 14% reduced CRC risk and diets high in plant-based fiber contribute to decreased rates of CRC.[126,127] Dietary fibers are broken down in the distal intestine by microbial anaerobic fermentation into short-chain fatty acids (SCFAs), namely butyrate, acetate, and propionate. The colon and liver absorb these metabolites as a nutrient energy source. Specifically, high levels of butyrate strongly correlate with decreased CRC risk, and this dietary factor protects against cancer progression.[128] Indeed, cancer patients harbor a lower number of butyrate-producing bacteria than healthy volunteers.[129,130]

1.6.2 Butyrate Acts as a Histone Deacetylase Inhibitor in Cancer Cells

Butyrate is the principle source of energy in normal colon epithelial cells, rather than glucose, and promotes cell growth.[131] Due to the Warburg effect in colorectal cancer cells however, butyrate is not metabolized. Instead, malignant cells accumulate butyrate in the nucleus and this microbial metabolite functions as a HDAC inhibitor to impact gene expression.[132]

Butyrate influences the epigenome of tumor cells to suppress colon cancer cell proliferation and induce apoptosis *via* controlling gene expression of the histone deacetylase sirtuin 1 (SIRT1), caspase 3, and NF-κB, as well as the wingless/integrated (WNT) and ERK/MAPK cancer signaling pathways.[118,133] Histone deacetylases (HDAC 1,2,3) are elevated in gastric and colorectal cancers, and therefore inhibition of HDACs by butyrate likely confers protective anti-cancer effects.[134,135]

1.6.3 Anti-tumor Effects of Butyrate and miRNA Expression in the Colon

The anti-tumor effects of the gut microbiota generated butyrate are also associated with the regulation of miRNA expression in CRC cells. Research by Schlormann *et al.* showed that treatment of LT97 colon adenoma cells with physiological doses of butyrate reduced proliferation and resulted in downregulated miRNA expression of miR-135a/b, *let-7a*, miR-24, and miR-106b.[136] Hu *et al.* found that HCT-116 human colon cancer cells treated with butyrate significantly repressed miR-106b family levels and this correlated with elevated expression of the miR-106b target p21, an established cell cycle progression inhibitor.[122]

Hu *et al.* also investigated the correlation between patients with sporadic colon cancer and increased expression of the oncogenic miR-17-92a cluster.[121] Butyrate treatment of HCT116 and HT29 human colon cancer cells reduced precursor miR-17-92a pri-miRNA transcript levels and mature miR-17-92a cluster members, miR-17, miR-18a, miR-19a/b, miR-20a, and most predominantly miR-92a. (Similar results were obtained using the HDAC inhibitors suberoylanilide hydroxamic acid and valproic acid.) This study indicated that butyrate regulates the miR-17-92a cluster indirectly by epigenetically suppressing MYC expression. MYC is an established transcriptional activator that binds directly to the miR-17-92a promoter C13*orf25* transcription.[137] *In vitro* experiments to overexpress miR-92a in colon cancer cell lines revealed that this miRNA effectively blocked butyrate's ability to slow the growth and induce apoptosis by targeting the tumor suppressor p57.[121] p57 regulation by butyrate was noted to be extraordinarily complex. p57 transcription is epigenetically modulated due to butyrate's HDAC inhibitor activity but butyrate also promotes p57 translation by reducing miR-92a activity (*via* c-MYC suppression).[121]

1.6.4 Butyrate's Role in Liver Cancer, the Epigenome, and miRNAs

Butyrate is also shown to play anti-cancer activities in the liver by regulating miRNA expression. Pant *et al.* found that butyrate incubation significantly increased miR-22 expression in human hepatic cancer Huh 7 cells and resulted in suppression of the miR-22 target SIRT1.[120] Indeed, miR-22 acts as a tumor suppressive miRNA in a variety of cell types to block proliferation, tumor formation, and metastasis.[138] Interestingly, the miR-22 target SIRT1 is a histone deacetylase directly suppressed by butyrate.[139] SIRT1 is highly expressed in hepatoma and colorectal adenocarcinomas and promotes tumor growth in mouse models.[140] The pro-oncogenic activity of SIRT1 is mediated by superoxide dismutase (SOD), which serves to quench reactive oxygen species (ROS) production and suppress apoptosis. Researchers found that miR-22 suppresses SIRT1 expression in hepatic cancer cells resulting in increased ROS levels and apoptosis (*via* mitochondrial cytochrome c/caspase pathway) and decreased proliferation (*via* PTEN/p-AKT pathway). Therefore, the HDAC inhibitor activity of butyrate is likely mediated by miR-22 in the liver. Taken together, gut microbial metabolites are potent cancer protective agents against colon and liver cancers by employing epigenetic and non-coding RNA regulation, and which could be translated into effective therapeutics. This work highlights the complex interplay between nutrition (high-fiber diet), the gut microbiota, and their impact on human health.

1.7 Plant-derived XenomiRs Impact Cancer Pathways in Human Cells

1.7.1 Ingested Plant miRNAs Regulate Recipient Cell Targets

As discussed, extrinsic factors, such as diet and the gut microbiota, have a huge impact on endogenous miRNAs expression in human cells and cancer progression pathways. Recent evidence indicates that the intake of exogenous miRNAs—those ingested from the diet as food (termed xenomiRs)—can also impact human health and disease pathology (Figure 1.2).[141] Foods, including rice, potatoes, cabbage, and breast milk, contain miRNAs that are hypothesized to survive digestion and be available for uptake into recipient cells.[142] It is proposed that exogenous plant miRNAs enter the body's circulation in their free form or are incorporated into exosomes, and once delivered into recipient cells, plant xenomiRs regulate endogenous mRNA targets to modulate cancer progression. Plant miRNAs are particularly stable because of their unique 2′*O*-methylated 3′terminal nucleotide modifications, and therefore are well suited to withstand gastric low-pH environments and remain stable in fluids such as the blood.[143] Studies by Philip *et al.* verified that prolonged food storage, processing, cooking, and simulated digestion did not destroy plant miRNA activity.[143] However, the concept of exogenous miRNA delivery from dietary foods remains controversial. O'Neill *et al.* found that oral administration of plant miRNA resulted in rapid degradation upon digestion.[144] Other groups have reported that plant miRNAs are not expressed in recipient cells at high enough levels to be reliably detected in the blood or tissues and failed to modulate target gene expression in their mouse models.[145–148]

There is growing experimental evidence to support the notion that exogenous miRNAs can be delivered through the diet, which we review in this section. An early study in Chinese subjects reported high levels of plant miRNAs in human sera, including miR-168a, which is enriched in rice (*Oryza sativa*).[149] Zhang *et al.* tested *in vivo* if exogenous rice miRNAs could be delivered to mice through their diet.[149] Animals were fed a chow diet of fresh rice or were gavage-fed with total RNA extracted from fresh rice. As predicted, animals exhibited elevated levels of miR-168a in the stomach, small intestine, and liver, but not in the kidney. Furthermore, *in vitro* and *in vivo* evidence indicated that dietary plant miR-168a was selectively loaded into microvesicles (exosomes), entered the blood *via* the animal gastrointestinal tract, and targeted the low-density lipoprotein receptor adapter protein 1 (LDLRAP1) mRNA in liver cells, which ultimately increased plasma LDL levels.[149]

In a similar study, Liang *et al.* administered cruciferous miRNAs extracted from *Brassica oleracea* (a member of the cabbage family) in the food or as oral gavage to mice.[150] *Brassica oleracea* miR-172 was resistant to gastrointestinal tract degradation and this miRNA was recovered in the blood, stomach, intestines, spleen, and feces of treated animals up to 72 h after feeding. This study is particularly intriguing from the perspective of nutritional therapeutics, because epidemiological data support a cancer-protective role for cruciferous vegetables such as broccoli.[151] Additional studies indicate that miRNAs from *Zea mays* (corn), *Solanum lycopersicum* (tomato), *Vitis vinifera* (grapes), and *Glycine max* (soybean) are present in the blood of animals exposed to these plants in their diet.[152–154] Therefore, dietary intake of exogenous plant miRNAs may be more widespread than initially appreciated and contribute to recipient gene expression and human health.

1.7.2 Plant miRNAs Impact Cancer Gene Expression

Introduction of exogenous miRNAs *via* the diet is very intriguing both in regards to maintaining human health and as a novel therapeutic approach, particularly plant xenomiRs carrying 2′-*O*-methylated modifications, which are unusually stable. A study by Chin *et al.* revealed that plant miRNA miR-159 (present in *Arabidopsis thaliana* and *Glycine max*) was detected in microvesicles of human sera.[155] miR-159 levels were inversely correlated with breast cancer incidence and progression in human subjects, highlighting a use for miR-159 as a prognostic marker for breast cancer risk. *In vitro* and *in vivo* studies revealed that miR-159 played a tumor suppressive role to block breast cancer cell growth by targeting the WNT signaling transcription factor 7 (TCF7) and lowering MYC oncogenic protein levels.[155] Furthermore, Chin *et al.* employed an MDA-MB-231 breast cancer xenograft mouse model, in which animals were gavage-fed synthetic 2′-*O*-methylated miR-159 daily. The treated animals exhibited significantly decreased tumor growth compared to controls. These findings could have major implications for the development of oral vaccines against cancer.

Mlotshwa *et al.* tested the feasibility of xenomiRs as therapeutics.[156] Delivery of a cocktail of total plant RNA spiked with three established tumor suppressive mammalian miRNAs, miR-34a, miR-143, and miR-145, that were chemically modified to contain 2′-*O*-methylated 3′-terminal nucleotide plant modifications. This cocktail was delivered by oral gavage to $Apc^{Min/+}$ mice, an established animal model for colon cancer. Treated animals showed elevated levels of these miRNAs in the intestine and exhibited significant reduction in tumor load. In short, these results indicate that edible plants could be specifically engineered to produce tumor-suppressive miRNAs for treatment of a wide range of cancers.

1.8 Conclusion

The growing list of promising "nutraceuticals" (pharmaceutical-grade and standardized nutrients) discussed in this chapter all have the remarkable ability to protect against a large range of human cancers. These natural dietary agents, which include edible polyphenols, gut microbial metabolites, and plant xenomiRs, mediate their anti-cancer effects *via* epigenetic regulation and miRNA expression. A greater molecular understanding of how nutritional compounds influence cancer progression pathways is needed as these nutraceuticals move into clinical trials.[157] Development of nutritional compounds and metabolites as disease biomarkers and therapeutic drugs will be transformative in the clinic due to their overall affordability, proven non-toxicity, and ease of administration.

Acknowledgements

Support was provided to AE-K (PI) by the National Institutes of Health (R21CA175894), Department of Defense (PC131691), EVMS Grant Enhancement Fund, Breedan Adams Foundation, Edmondson Fund, Coach Ray Barlow Prostate Cancer Research Funds, and EVMS Prostate Cancer Research Funds.

References

1. A. P. Feinberg, B. Vogelstein, M. J. Droller, S. B. Baylin and B. D. Nelkin, Mutation affecting the 12th amino acid of the c-Ha-ras oncogene product occurs infrequently in human cancer, *Science*, 1983, **220**(4602), 1175–1177.
2. V. Greger, E. Passarge, W. Hopping, E. Messmer and B. Horsthemke, Epigenetic changes may contribute to the formation and spontaneous regression of retinoblastoma, *Hum. Genet.*, 1989, **83**(2), 155–158.
3. J. Veeck and M. Esteller, Breast cancer epigenetics: from DNA methylation to microRNAs, *J. Mammary Gland Biol. Neoplasia*, 2010, **15**(1), 5–17.
4. P. A. Jones, DNA methylation and cancer, *Oncogene*, 2002, **21**(35), 5358–5360.

5. H. Lewis and A. Esquela-Kerscher, MicroRNA Epigenetic Systems and cancer, in *Systems Biology of Cancer*, ed. S. Thiagalingam, Cambridge University Press, Cambridge, U.K., 2015, pp. 134–153.
6. A. Esquela-Kerscher and F. J. Slack, Oncomirs - microRNAs with a role in cancer, *Nat. Rev. Cancer*, 2006, **6**(4), 259–269.
7. K. Breving and A. Esquela-Kerscher, The complexities of microRNA regulation: mirandering around the rules, *Int. J. Biochem. Cell Biol.*, 2010, **42**(8), 1316–1329.
8. S. P. Chan and F. J. Slack, And now introducing mammalian mirtrons, *Dev. Cell*, 2007, **13**(5), 605–607.
9. J. G. Doench and P. A. Sharp, Specificity of microRNA target selection in translational repression, *Genes Dev.*, 2004, **18**(5), 504–511.
10. A. Valinezhad Orang, R. Safaralizadeh and M. Kazemzadeh-Bavili, Mechanisms of miRNA-Mediated Gene Regulation from Common Downregulation to mRNA-Specific Upregulation, *Int. J. Genomics*, 2014, **2014**, 970607.
11. M. Xiao, J. Li, W. Li, Y. Wang, F. Wu and Y. Xi, *et al.*, MicroRNAs activate gene transcription epigenetically as an enhancer trigger, *RNA Biol.*, 2016, 1–9.
12. G. A. Calin, C. Sevignani, C. D. Dumitru, T. Hyslop, E. Noch and S. Yendamuri, *et al.*, Human microRNA genes are frequently located at fragile sites and genomic regions involved in cancers, *Proc. Natl. Acad. Sci. U. S. A.*, 2004, **101**(9), 2999–3004.
13. R. Garzon, G. A. Calin and C. M. Croce, MicroRNAs in Cancer, *Annu. Rev. Med.*, 2009, **60**, 167–179.
14. J. Ma, C. Dong and C. Ji, MicroRNA and drug resistance, *Cancer Gene Ther.*, 2010, **17**(8), 523–531.
15. G. A. Calin, C. D. Dumitru, M. Shimizu, R. Bichi, S. Zupo and E. Noch, *et al.*, Frequent deletions and down-regulation of micro- RNA genes *miR15* and *miR16* at 13q14 in chronic lymphocytic leukemia, *Proc. Natl. Acad. Sci. U. S. A.*, 2002, **99**(24), 15524–15529.
16. A. Cimmino, G. A. Calin, M. Fabbri, M. V. Iorio, M. Ferracin and M. Shimizu, *et al.*, miR-15 and miR-16 induce apoptosis by targeting BCL2, *Proc. Natl. Acad. Sci. U. S. A.*, 2005, **102**(39), 13944–13949.
17. G. A. Calin, A. Cimmino, M. Fabbri, M. Ferracin, S. E. Wojcik and M. Shimizu, *et al.*, MiR-15a and miR-16-1 cluster functions in human leukemia, *Proc. Natl. Acad. Sci. U. S. A.*, 2008, **105**(13), 5166–5171.
18. F. Lovat, M. Fassan, P. Gasparini, L. Rizzotto, L. Cascione and M. Pizzi, *et al.*, miR-15b/16-2 deletion promotes B-cell malignancies, *Proc. Natl. Acad. Sci. U. S. A.*, 2015, **112**(37), 11636–11641.
19. Y. H. Feng and C. J. Tsao, Emerging role of microRNA-21 in cancer, *Biomed. Rep.*, 2016, **5**(4), 395–402.
20. J. Lu, G. Getz, E. A. Miska, E. Alvarez-Saavedra, J. Lamb and D. Peck, *et al.*, MicroRNA expression profiles classify human cancers, *Nature*, 2005, **435**(7043), 834–838.
21. G. A. Calin and C. M. Croce, MicroRNA signatures in human cancers, *Nat. Rev. Cancer*, 2006, **6**(11), 857–866.

22. J. Takamizawa, H. Konishi, K. Yanagisawa, S. Tomida, H. Osada and H. Endoh, *et al.*, Reduced expression of the *let-7* microRNAs in human lung cancers in association with shortened postoperative survival, *Cancer Res.*, 2004, **64**(11), 3753–3756.
23. N. Yanaihara, N. Caplen, E. Bowman, M. Seike, K. Kumamoto and M. Yi, *et al.*, Unique microRNA molecular profiles in lung cancer diagnosis and prognosis, *Cancer Cell*, 2006, **9**(3), 189–198.
24. T. Hasegawa, H. Lewis and A. Esquela-Kerscher, Chapter 12 – The Role of Noncoding RNAs in Prostate Cancer, in *Translating MicroRNAs to the Clinic*, ed. J. Laurence, Academic Press, Boston, Massachusetts, 1st edn, 2017, pp. 329–369.
25. J. D. Arroyo, J. R. Chevillet, E. M. Kroh, I. K. Ruf, C. C. Pritchard and D. F. Gibson, *et al.*, Argonaute2 complexes carry a population of circulating microRNAs independent of vesicles in human plasma, *Proc. Natl. Acad. Sci. U. S. A.*, 2011, **108**(12), 5003–5008.
26. P. S. Mitchell, R. K. Parkin, E. M. Kroh, B. R. Fritz, S. K. Wyman and E. L. Pogosova-Agadjanyan, *et al.*, Circulating microRNAs as stable blood-based markers for cancer detection, *Proc. Natl. Acad. Sci. U. S. A.*, 2008, **105**(30), 10513–10518.
27. R. J. Bryant, T. Pawlowski, J. W. Catto, G. Marsden, R. L. Vessella and B. Rhees, *et al.*, Changes in circulating microRNA levels associated with prostate cancer, *Br. J. Cancer*, 2012, **106**(4), 768–774.
28. A. Gallo, M. Tandon, I. Alevizos and G. G. Illei, The majority of microRNAs detectable in serum and saliva is concentrated in exosomes, *PLoS One*, 2012, **7**(3), e30679.
29. N. P. Hessvik, K. Sandvig and A. Llorente, Exosomal miRNAs as Biomarkers for Prostate Cancer, *Front. Genet.*, 2013, **4**, 36.
30. A. K. Rupp, C. Rupp, S. Keller, J. C. Brase, R. Ehehalt and M. Fogel, *et al.*, Loss of EpCAM expression in breast cancer derived serum exosomes: role of proteolytic cleavage, *Gynecol. Oncol.*, 2011, **122**(2), 437–446.
31. H. Valadi, K. Ekstrom, A. Bossios, M. Sjostrand, J. J. Lee and J. O. Lotvall, Exosome-mediated transfer of mRNAs and microRNAs is a novel mechanism of genetic exchange between cells, *Nat. Cell Biol.*, 2007, **9**(6), 654–659.
32. M. Mittelbrunn, C. Gutierrez-Vazquez, C. Villarroya-Beltri, S. Gonzalez, F. Sanchez-Cabo and M. A. Gonzalez, *et al.*, Unidirectional transfer of microRNA-loaded exosomes from T cells to antigen-presenting cells, *Nat. Commun.*, 2011, **2**, 282.
33. J. Zempleni, Milk exosomes: beyond dietary microRNAs, *Genes Nutr.*, 2017, **12**, 12.
34. K. M. Howard, R. Jati Kusuma, S. R. Baier, T. Friemel, L. Markham and J. Vanamala, *et al.*, Loss of miRNAs during processing and storage of cow's (Bos taurus) milk, *J. Agric. Food Chem.*, 2015, **63**(2), 588–592.
35. T. Wolf, S. R. Baier and J. Zempleni, The Intestinal Transport of Bovine Milk Exosomes Is Mediated by Endocytosis in Human Colon Carcinoma Caco-2 Cells and Rat Small Intestinal IEC-6 Cells, *J. Nutr.*, 2015, **145**(10), 2201–2206.

36. R. J. Kusuma, S. Manca, T. Friemel, S. Sukreet, C. Nguyen and J. Zempleni, Human vascular endothelial cells transport foreign exosomes from cow's milk by endocytosis, *Am. J. Physiol., Cell Physiol.*, 2016, **310**(10), C800–C807.
37. H. Izumi, N. Kosaka, T. Shimizu, K. Sekine, T. Ochiya and M. Takase, Bovine milk contains microRNA and messenger RNA that are stable under degradative conditions, *J. Dairy Sci.*, 2012, **95**(9), 4831–4841.
38. R. Munagala, F. Aqil, J. Jeyabalan and R. C. Gupta, Bovine milk-derived exosomes for drug delivery, *Cancer Lett.*, 2016, **371**(1), 48–61.
39. A. Portela and M. Esteller, Epigenetic modifications and human disease, *Nat. Biotechnol.*, 2010, **28**(10), 1057–1068.
40. I. Girault, S. Tozlu, R. Lidereau and I. Bieche, Expression analysis of DNA methyltransferases 1, 3A, and 3B in sporadic breast carcinomas, *Clin. Cancer Res.*, 2003, **9**(12), 4415–4422.
41. Y. Saito, Y. Kanai, T. Nakagawa, M. Sakamoto, H. Saito and H. Ishii, *et al.*, Increased protein expression of DNA methyltransferase (DNMT) 1 is significantly correlated with the malignant potential and poor prognosis of human hepatocellular carcinomas, *Int. J. Cancer*, 2003, **105**(4), 527–532.
42. S. K. Patra, A. Patra, H. Zhao and R. Dahiya, DNA methyltransferase and demethylase in human prostate cancer, *Mol. Carcinog.*, 2002, **33**(3), 163–171.
43. C. A. Eads, K. D. Danenberg, K. Kawakami, L. B. Saltz, P. V. Danenberg and P. W. Laird, CpG island hypermethylation in human colorectal tumors is not associated with DNA methyltransferase overexpression, *Cancer Res.*, 1999, **59**(10), 2302–2306.
44. M. Fabbri, R. Garzon, A. Cimmino, Z. Liu, N. Zanesi and E. Callegari, *et al.*, MicroRNA-29 family reverts aberrant methylation in lung cancer by targeting DNA methyltransferases 3A and 3B, *Proc. Natl. Acad. Sci. U. S. A.*, 2007, **104**(40), 15805–15810.
45. E. K. Ng, W. P. Tsang, S. S. Ng, H. C. Jin, J. Yu and J. J. Li, *et al.*, MicroRNA-143 targets DNA methyltransferases 3A in colorectal cancer, *Br. J. Cancer*, 2009, **101**(4), 699–706.
46. S. Varambally, Q. Cao, R. S. Mani, S. Shankar, X. Wang and B. Ateeq, *et al.*, Genomic loss of microRNA-101 leads to overexpression of histone methyltransferase EZH2 in cancer, *Science*, 2008, **322**(5908), 1695–1699.
47. P. Cao, Z. Deng, M. Wan, W. Huang, S. D. Cramer and J. Xu, *et al.*, MicroRNA-101 negatively regulates Ezh2 and its expression is modulated by androgen receptor and HIF-1alpha/HIF-1beta, *Mol. Cancer*, 2010, **9**, 108.
48. L. Tuddenham, G. Wheeler, S. Ntounia-Fousara, J. Waters, M. K. Hajihosseini and I. Clark, *et al.*, The cartilage specific microRNA-140 targets histone deacetylase 4 in mouse cells, *FEBS Lett.*, 2006, **580**(17), 4214–4217.
49. J. F. Chen, E. M. Mandel, J. M. Thomson, Q. Wu, T. E. Callis and S. M. Hammond, *et al.*, The role of microRNA-1 and microRNA-133 in skeletal muscle proliferation and differentiation, *Nat. Genet.*, 2006, **38**(2), 228–233.

50. Z. Duan, E. Choy, G. P. Nielsen, A. Rosenberg, J. Iafrate and C. Yang, *et al.*, Differential expression of microRNA (miRNA) in chordoma reveals a role for miRNA-1 in Met expression, *J. Orthop. Res.*, 2010, **28**(6), 746–752.
51. M. Toyota, H. Suzuki, Y. Sasaki, R. Maruyama, K. Imai and Y. Shinomura, *et al.*, Epigenetic silencing of microRNA-34b/c and B-cell translocation gene 4 is associated with CpG island methylation in colorectal cancer, *Cancer Res.*, 2008, **68**(11), 4123–4132.
52. B. Brueckner, C. Stresemann, R. Kuner, C. Mund, T. Musch and M. Meister, *et al.*, The human *let-7a-3* locus contains an epigenetically regulated microRNA gene with oncogenic function, *Cancer Res.*, 2007, **67**(4), 1419–1423.
53. Y. Saito, G. Liang, G. Egger, J. M. Friedman, J. C. Chuang and G. A. Coetzee, *et al.*, Specific activation of microRNA-127 with downregulation of the proto-oncogene BCL6 by chromatin-modifying drugs in human cancer cells, *Cancer Cell*, 2006, **9**(6), 435–443.
54. L. Han, P. D. Witmer, E. Casey, D. Valle and S. Sukumar, DNA methylation regulates MicroRNA expression, *Cancer Biol. Ther.*, 2007, **6**(8), 1284–1288.
55. A. Lujambio, G. A. Calin, A. Villanueva, S. Ropero, M. Sanchez-Cespedes and D. Blanco, *et al.*, A microRNA DNA methylation signature for human cancer metastasis, *Proc. Natl. Acad. Sci. U. S. A.*, 2008, **105**(36), 13556–13561.
56. A. Lujambio, S. Ropero, E. Ballestar, M. F. Fraga, C. Cerrato and F. Setien, *et al.*, Genetic unmasking of an epigenetically silenced microRNA in human cancer cells, *Cancer Res.*, 2007, **67**(4), 1424–1429.
57. Z. Li, P. Chen, R. Su, Y. Li, C. Hu and Y. Wang, *et al.*, Overexpression and knockout of miR-126 both promote leukemogenesis, *Blood*, 2015, **126**(17), 2005–2015.
58. J. Perez-Jimenez, V. Neveu, F. Vos and A. Scalbert, Systematic analysis of the content of 502 polyphenols in 452 foods and beverages: an application of the phenol-explorer database, *J. Agric. Food Chem.*, 2010, **58**(8), 4959–4969.
59. S. K. Srivastava, S. Arora, C. Averett, S. Singh and A. P. Singh, Modulation of microRNAs by phytochemicals in cancer: underlying mechanisms and translational significance, *BioMed Res. Int.*, 2015, **2015**, 848710.
60. M. Athar, J. H. Back, X. Tang, K. H. Kim, L. Kopelovich and D. R. Bickers, *et al.*, Resveratrol: a review of preclinical studies for human cancer prevention, *Toxicol. Appl. Pharmacol.*, 2007, **224**(3), 274–283.
61. P. Liu, H. Liang, Q. Xia, P. Li, H. Kong and P. Lei, *et al.*, Resveratrol induces apoptosis of pancreatic cancers cells by inhibiting miR-21 regulation of BCL-2 expression, *Clin. Transl. Oncol.*, 2013, **15**(9), 741–746.
62. C. Zhou, J. Ding and Y. Wu, Resveratrol induces apoptosis of bladder cancer cells *via* miR21 regulation of the Akt/Bcl2 signaling pathway, *Mol. Med. Rep.*, 2014, **9**(4), 1467–1473.
63. E. Tili, J. J. Michaille, H. Alder, S. Volinia, D. Delmas and N. Latruffe, *et al.*, Resveratrol modulates the levels of microRNAs targeting genes

encoding tumor-suppressors and effectors of TGFbeta signaling pathway in SW480 cells, *Biochem. Pharmacol.*, 2010, **80**(12), 2057–2065.
64. M. Kumazaki, S. Noguchi, Y. Yasui, J. Iwasaki, H. Shinohara and N. Yamada, *et al.*, Anti-cancer effects of naturally occurring compounds through modulation of signal transduction and miRNA expression in human colon cancer cells, *J. Nutr. Biochem.*, 2013, **24**(11), 1849–1858.
65. K. Hagiwara, N. Kosaka, Y. Yoshioka, R. U. Takahashi, F. Takeshita and T. Ochiya, Stilbene derivatives promote Ago2-dependent tumour-suppressive microRNA activity, *Sci. Rep.*, 2012, **2**, 314.
66. N. P. Seeram, L. S. Adams, S. M. Henning, Y. Niu, Y. Zhang and M. G. Nair, *et al.*, *In vitro* antiproliferative, apoptotic and antioxidant activities of punicalagin, ellagic acid and a total pomegranate tannin extract are enhanced in combination with other polyphenols as found in pomegranate juice, *J. Nutr. Biochem.*, 2005, **16**(6), 360–367.
67. M. A. Nunez-Sanchez, A. Davalos, A. Gonzalez-Sarrias, P. Casas-Agustench, F. Visioli and T. Monedero-Saiz, *et al.*, MicroRNAs expression in normal and malignant colon tissues as biomarkers of colorectal cancer and in response to pomegranate extracts consumption: Critical issues to discern between modulatory effects and potential artefacts, *Mol. Nutr. Food Res.*, 2015, **59**(10), 1973–1986.
68. H. Fritz, D. Seely, D. A. Kennedy, R. Fernandes, K. Cooley and D. Fergusson, Green tea and lung cancer: a systematic review, *Integr. Cancer Ther.*, 2013, **12**(1), 7–24.
69. W. Zhang, T. Geng, W. Han and H. Dou, Tea intake and risk of oral, pharyngeal, and laryngeal carcinoma: a meta-analysis, *Med. Sci. Monit.*, 2014, **20**, 2142–2150.
70. C. S. Yang, X. Wang, G. Lu and S. C. Picinich, Cancer prevention by tea: animal studies, molecular mechanisms and human relevance, *Nat. Rev. Cancer*, 2009, **9**(6), 429–439.
71. W. P. Tsang and T. T. Kwok, Epigallocatechin gallate up-regulation of miR-16 and induction of apoptosis in human cancer cells, *J. Nutr. Biochem.*, 2010, **21**(2), 140–146.
72. I. A. Siddiqui, M. Asim, B. B. Hafeez, V. M. Adhami, R. S. Tarapore and H. Mukhtar, Green tea polyphenol EGCG blunts androgen receptor function in prostate cancer, *FASEB J.*, 2011, **25**(4), 1198–1207.
73. H. Wang, S. Bian and C. S. Yang, Green tea polyphenol EGCG suppresses lung cancer cell growth through upregulating miR-210 expression caused by stabilizing HIF-1alpha, *Carcinogenesis*, 2011, **32**(12), 1881–1889.
74. D. H. Zhou, X. Wang and Q. Feng, EGCG enhances the efficacy of cisplatin by downregulating hsa-miR-98-5p in NSCLC A549 cells, *Nutr. Cancer*, 2014, **66**(4), 636–644.
75. M. B. van Duursen, S. M. Nijmeijer, E. S. de Morree, P. C. de Jong and M. van den Berg, Genistein induces breast cancer-associated aromatase and stimulates estrogen-dependent tumor cell growth in *in vitro* breast cancer model, *Toxicology*, 2011, **289**(2–3), 67–73.

76. C. de la Parra, L. Castillo-Pichardo, A. Cruz-Collazo, L. Cubano, R. Redis and G. A. Calin, *et al.*, Soy Isoflavone Genistein-Mediated Downregulation of miR-155 Contributes to the Anticancer Effects of Genistein, *Nutr. Cancer*, 2016, **68**(1), 154–164.
77. C. B. Avci, S. Y. Susluer, H. O. Caglar, T. Balci, D. Aygunes and Y. Dodurga, *et al.*, Genistein-induced mir-23b expression inhibits the growth of breast cancer cells, *Contemp. Oncol.*, 2015, **19**(1), 32–35.
78. L. Xu, J. Xiang, J. Shen, X. Zou, S. Zhai and Y. Yin, *et al.*, Oncogenic MicroRNA-27a is a target for genistein in ovarian cancer cells, *Anticancer Agents Med. Chem.*, 2013, **13**(7), 1126–1132.
79. Y. Chen, M. S. Zaman, G. Deng, S. Majid, S. Saini and J. Liu, *et al.*, MicroRNAs 221/222 and genistein-mediated regulation of ARHI tumor suppressor gene in prostate cancer, *Cancer Prev. Res.*, 2011, **4**(1), 76–86.
80. S. Majid, A. A. Dar, S. Saini, Y. Chen, V. Shahryari and J. Liu, *et al.*, Regulation of minichromosome maintenance gene family by microRNA-1296 and genistein in prostate cancer, *Cancer Res.*, 2010, **70**(7), 2809–2818.
81. T. Chiyomaru, S. Yamamura, S. Fukuhara, H. Hidaka, S. Majid and S. Saini, *et al.*, Genistein up-regulates tumor suppressor microRNA-574-3p in prostate cancer, *PLoS One*, 2013, **8**(3), e58929.
82. M. S. Zaman, Y. Chen, G. Deng, V. Shahryari, S. O. Suh and S. Saini, *et al.*, The functional significance of microRNA-145 in prostate cancer, *Br. J. Cancer*, 2010, **103**(2), 256–264.
83. N. Rabiau, H. K. Trraf, M. Adjakly, R. Bosviel, L. Guy and L. Fontana, *et al.*, miRNAs differentially expressed in prostate cancer cell lines after soy treatment, *In Vivo*, 2011, **25**(6), 917–921.
84. Y. Yang, A. Zang, Y. Jia, Y. Shang, Z. Zhang and K. Ge, *et al.*, Genistein inhibits A549 human lung cancer cell proliferation *via* miR-27a and MET signaling, *Oncol. Lett.*, 2016, **12**(3), 2189–2193.
85. J. Qin, J. X. Chen, Z. Zhu and J. A. Teng, Genistein inhibits human colorectal cancer growth and suppresses miR-95, Akt and SGK1, *Cell. Physiol. Biochem.*, 2015, **35**(5), 2069–2077.
86. M. S. Zaman, V. Shahryari, G. Deng, S. Thamminana, S. Saini and S. Majid, *et al.*, Up-regulation of microRNA-21 correlates with lower kidney cancer survival, *PLoS One*, 2012, **7**(2), e31060.
87. A. Kunwar, A. Barik, B. Mishra, K. Rathinasamy, R. Pandey and K. I. Priyadarsini, Quantitative cellular uptake, localization and cytotoxicity of curcumin in normal and tumor cells, *Biochim. Biophys. Acta*, 2008, **1780**(4), 673–679.
88. S. Norouzi, M. Majeed, M. Pirro, D. Generali and A. Sahebkar, Curcumin as an Adjunct Therapy and microRNA Modulator in Breast Cancer, *Curr. Pharm. Des.*, 2018, **24**(2), 171–177.
89. E. Kronski, M. E. Fiori, O. Barbieri, S. Astigiano, V. Mirisola and P. H. Killian, *et al.*, miR181b is induced by the chemopreventive polyphenol curcumin and inhibits breast cancer metastasis *via* down-regulation of the inflammatory cytokines CXCL1 and -2, *Mol. Oncol.*, 2014, **8**(3), 581–595.

90. J. Yang, Y. Cao, J. Sun and Y. Zhang, Curcumin reduces the expression of Bcl-2 by upregulating miR-15a and miR-16 in MCF-7 cells, *Med. Oncol.*, 2010, **27**(4), 1114–1118.
91. S. Mou, Z. Zhou, Y. He, F. Liu and L. Gong, Curcumin inhibits cell proliferation and promotes apoptosis of laryngeal cancer cells through Bcl-2 and PI3K/Akt, and by upregulating miR-15a, *Oncol. Lett.*, 2017, **14**(4), 4937–4942.
92. S. M. Gao, J. J. Yang, C. Q. Chen, J. J. Chen, L. P. Ye and L. Y. Wang, *et al.*, Pure curcumin decreases the expression of WT1 by upregulation of miR-15a and miR-16-1 in leukemic cells, *J. Exp. Clin. Cancer Res.*, 2012, **31**, 27.
93. G. Mudduluru, J. N. George-William, S. Muppala, I. A. Asangani, R. Kumarswamy and L. D. Nelson, *et al.*, Curcumin regulates miR-21 expression and inhibits invasion and metastasis in colorectal cancer, *Biosci. Rep.*, 2011, **31**(3), 185–197.
94. S. Roy, Y. Yu, S. B. Padhye, F. H. Sarkar and A. P. Majumdar, Difluorinated-curcumin (CDF) restores PTEN expression in colon cancer cells by down-regulating miR-21, *PLoS One*, 2013, **8**(7), e68543.
95. S. Roy, E. Levi, A. P. Majumdar and F. H. Sarkar, Expression of miR-34 is lost in colon cancer which can be re-expressed by a novel agent CDF, *J. Hematol. Oncol.*, 2012, **5**, 58.
96. H. Zhang, J. Zheng, H. Shen, Y. Huang, T. Liu and H. Xi, *et al.*, Curcumin Suppresses *In Vitro* Proliferation and Invasion of Human Prostate Cancer Stem Cells by Modulating DLK1-DIO3 Imprinted Gene Cluster MicroRNAs, *Genet. Test. Mol. Biomarkers*, 2018, **22**(1), 43–50.
97. W. L. Liu, J. M. Chang, I. W. Chong, Y. L. Hung, Y. H. Chen and W. T. Huang, *et al.*, Curcumin Inhibits LIN-28A through the Activation of miRNA-98 in the Lung Cancer Cell Line A549, *Molecules*, 2017, **22**(6), 929.
98. I. Heo, C. Joo, J. Cho, M. Ha, J. Han and V. N. Kim, Lin28 mediates the terminal uridylation of *let-7* precursor MicroRNA, *Mol. Cell*, 2008, **32**(2), 276–284.
99. S. R. Viswanathan, G. Q. Daley and R. I. Gregory, Selective blockade of microRNA processing by Lin28, *Science*, 2008, **320**(5872), 97–100.
100. W. Li, W. Yang, Y. Liu, S. Chen, S. Chin and X. Qi, *et al.*, MicroRNA-378 enhances inhibitory effect of curcumin on glioblastoma, *Oncotarget*, 2017, **8**(43), 73938–73946.
101. Y. Li, D. Kong, A. Ahmad, B. Bao and F. H. Sarkar, Antioxidant function of isoflavone and 3,3′-diindolylmethane: are they important for cancer prevention and therapy?, *Antioxid. Redox Signaling*, 2013, **19**(2), 139–150.
102. Y. Li, T. G. VandenBoom 2nd, D. Kong, Z. Wang, S. Ali and P. A. Philip, *et al.*, Up-regulation of miR-200 and *let-7* by natural agents leads to the reversal of epithelial-to-mesenchymal transition in gemcitabine-resistant pancreatic cancer cells, *Cancer Res.*, 2009, **69**(16), 6704–6712.

103. Y. Li, T. G. Vandenboom 2nd, Z. Wang, D. Kong, S. Ali and P. A. Philip, *et al.*, miR-146a suppresses invasion of pancreatic cancer cells, *Cancer Res.*, 2010, **70**(4), 1486–1495.
104. S. Ali, S. Banerjee, A. Ahmad, B. F. El-Rayes, P. A. Philip and F. H. Sarkar, Apoptosis-inducing effect of erlotinib is potentiated by 3,3′-diindolylmethane *in vitro* and *in vivo* using an orthotopic model of pancreatic cancer, *Mol. Cancer Ther.*, 2008, **7**(6), 1708–1719.
105. A. Ahmad, S. Ali, A. Ahmed, A. S. Ali, A. Raz and W. A. Sakr, *et al.*, 3,3′-Diindolylmethane enhances the effectiveness of herceptin against HER-2/neu-expressing breast cancer cells, *PLoS One*, 2013, **8**(1), e54657.
106. D. Kong, E. Heath, W. Chen, M. L. Cher, I. Powell and L. Heilbrun, *et al.*, Loss of *let-7* up-regulates EZH2 in prostate cancer consistent with the acquisition of cancer stem cell signatures that are attenuated by BR-DIM, *PLoS One*, 2012, **7**(3), e33729.
107. Z. Zhang, Z. Gao, W. Hu, S. Yin, C. Wang and Y. Zang, *et al.*, 3,3′-Diindolylmethane ameliorates experimental hepatic fibrosis *via* inhibiting miR-21 expression, *Br. J. Pharmacol.*, 2013, **170**(3), 649–660.
108. M. Junaid, R. Dash, N. Islam, J. Chowdhury, M. J. Alam and S. D. Nath, *et al.*, Molecular Simulation Studies of 3,3′-Diindolylmethane as a Potent MicroRNA-21 Antagonist, *J. Pharm. BioAllied Sci.*, 2017, **9**(4), 259–265.
109. Y. Jin, 3,3′-Diindolylmethane inhibits breast cancer cell growth *via* miR-21-mediated Cdc25A degradation, *Mol. Cell. Biochem.*, 2011, **358**(1–2), 345–354.
110. L. X. Yan, Q. N. Wu, Y. Zhang, Y. Y. Li, D. Z. Liao and J. H. Hou, *et al.*, Knockdown of miR-21 in human breast cancer cell lines inhibits proliferation, *in vitro* migration and *in vivo* tumor growth, *Breast Cancer Res.*, 2011, **13**(1), R2.
111. M. Han, Y. Wang, M. Liu, X. Bi, J. Bao and N. Zeng, *et al.*, MiR-21 regulates epithelial-mesenchymal transition phenotype and hypoxia-inducible factor-1alpha expression in third-sphere forming breast cancer stem cell-like cells, *Cancer Sci.*, 2012, **103**(6), 1058–1064.
112. M. Han, M. Liu, Y. Wang, Z. Mo, X. Bi and Z. Liu, *et al.*, Re-expression of miR-21 contributes to migration and invasion by inducing epithelial-mesenchymal transition consistent with cancer stem cell characteristics in MCF-7 cells, *Mol. Cell. Biochem.*, 2012, **363**(1–2), 427–436.
113. M. Garofalo, C. Quintavalle, G. Romano, C. M. Croce and G. Condorelli, miR221/222 in cancer: their role in tumor progression and response to therapy, *Curr. Mol. Med.*, 2012, **12**(1), 27–33.
114. S. Sarkar, H. Dubaybo, S. Ali, P. Goncalves, S. L. Kollepara and S. Sethi, *et al.*, Down-regulation of miR-221 inhibits proliferation of pancreatic cancer cells through up-regulation of PTEN, p27(kip1), p57(kip2), and PUMA, *Am. J. Cancer Res.*, 2013, **3**(5), 465–477.
115. M. Arumugam, J. Raes, E. Pelletier, D. Le Paslier, T. Yamada and D. R. Mende, *et al.*, Enterotypes of the human gut microbiome, *Nature*, 2011, **473**(7346), 174–180.

116. V. Tremaroli and F. Backhed, Functional interactions between the gut microbiota and host metabolism, *Nature*, 2012, **489**(7415), 242–249.
117. J. Qin, R. Li, J. Raes, M. Arumugam, K. S. Burgdorf and C. Manichanh, *et al.*, A human gut microbial gene catalogue established by metagenomic sequencing, *Nature*, 2010, **464**(7285), 59–65.
118. Q. Feng, W. D. Chen and Y. D. Wang, Gut Microbiota: An Integral Moderator in Health and Disease, *Front. Microbiol.*, 2018, **9**, 151.
119. S. Liu, A. P. da Cunha, R. M. Rezende, R. Cialic, Z. Wei and L. Bry, *et al.*, The Host Shapes the Gut Microbiota *via* Fecal MicroRNA, *Cell Host Microbe*, 2016, **19**(1), 32–43.
120. K. Pant, A. K. Yadav, P. Gupta, R. Islam, A. Saraya and S. K. Venugopal, Butyrate induces ROS-mediated apoptosis by modulating miR-22/SIRT-1 pathway in hepatic cancer cells, *Redox Biol.*, 2017, **12**, 340–349.
121. S. Hu, L. Liu, E. B. Chang, J. Y. Wang and J. P. Raufman, Butyrate inhibits pro-proliferative miR-92a by diminishing c-Myc-induced miR-17-92a cluster transcription in human colon cancer cells, *Mol. Cancer*, 2015, **14**, 180.
122. S. Hu, T. S. Dong, S. R. Dalal, F. Wu, M. Bissonnette and J. H. Kwon, *et al.*, The microbe-derived short chain fatty acid butyrate targets miRNA-dependent p21 gene expression in human colon cancer, *PLoS One*, 2011, **6**(1), e16221.
123. R. L. Siegel, K. D. Miller, S. A. Fedewa, D. J. Ahnen, R. G. S. Meester and A. Barzi, *et al.*, Colorectal cancer statistics, 2017, *Ca-Cancer J. Clin.*, 2017, **67**(3), 177–193.
124. P. Terry, E. Giovannucci, K. B. Michels, L. Bergkvist, H. Hansen and L. Holmberg, *et al.*, Fruit, vegetables, dietary fiber, and risk of colorectal cancer, *J. Natl. Cancer Inst.*, 2001, **93**(7), 525–533.
125. F. A. Haggar and R. P. Boushey, Colorectal cancer epidemiology: incidence, mortality, survival, and risk factors, *Clin. Colon Rectal Surg.*, 2009, **22**(4), 191–197.
126. L. Schwingshackl and G. Hoffmann, Adherence to Mediterranean diet and risk of cancer: a systematic review and meta-analysis of observational studies, *Int. J. Cancer*, 2014, **135**(8), 1884–1897.
127. Q. Ben, Y. Sun, R. Chai, A. Qian, B. Xu and Y. Yuan, Dietary fiber intake reduces risk for colorectal adenoma: a meta-analysis, *Gastroenterology*, 2014, **146**(3), 689–699 e6.
128. H. M. Chen, Y. N. Yu, J. L. Wang, Y. W. Lin, X. Kong and C. Q. Yang, *et al.*, Decreased dietary fiber intake and structural alteration of gut microbiota in patients with advanced colorectal adenoma, *Am. J. Clin. Nutr.*, 2013, **97**(5), 1044–1052.
129. T. Wang, G. Cai, Y. Qiu, N. Fei, M. Zhang and X. Pang, *et al.*, Structural segregation of gut microbiota between colorectal cancer patients and healthy volunteers, *ISME J.*, 2012, **6**(2), 320–329.
130. N. Wu, X. Yang, R. Zhang, J. Li, X. Xiao and Y. Hu, *et al.*, Dysbiosis signature of fecal microbiota in colorectal cancer patients, *Microb. Ecol.*, 2013, **66**(2), 462–470.

131. A. Belcheva, T. Irrazabal, S. J. Robertson, C. Streutker, H. Maughan and S. Rubino, *et al.*, Gut microbial metabolism drives transformation of MSH2-deficient colon epithelial cells, *Cell*, 2014, **158**(2), 288–299.
132. D. R. Donohoe, L. B. Collins, A. Wali, R. Bigler, W. Sun and S. J. Bultman, The Warburg effect dictates the mechanism of butyrate-mediated histone acetylation and cell proliferation, *Mol. Cell*, 2012, **48**(4), 612–626.
133. A. A. Lane and B. A. Chabner, Histone deacetylase inhibitors in cancer therapy, *J. Clin. Oncol.*, 2009, **27**(32), 5459–5468.
134. C. A. Godman, R. Joshi, B. R. Tierney, E. Greenspan, T. P. Rasmussen and H. W. Wang, *et al.*, HDAC3 impacts multiple oncogenic pathways in colon cancer cells with effects on Wnt and vitamin D signaling, *Cancer Biol. Ther.*, 2008, **7**(10), 1570–1580.
135. M. Mottamal, S. Zheng, T. L. Huang and G. Wang, Histone deacetylase inhibitors in clinical studies as templates for new anticancer agents, *Molecules*, 2015, **20**(3), 3898–3941.
136. W. Schlormann, S. Naumann, C. Renner and M. Glei, Influence of miRNA-106b and miRNA-135a on butyrate-regulated expression of p21 and Cyclin D2 in human colon adenoma cells, *Genes Nutr.*, 2015, **10**(6), 50.
137. K. A. O'Donnell, E. A. Wentzel, K. I. Zeller, C. V. Dang and J. T. Mendell, c-Myc-regulated microRNAs modulate E2F1 expression, *Nature*, 2005, **435**(7043), 839–843.
138. J. Wang, Y. Li, M. Ding, H. Zhang, X. Xu and J. Tang, Molecular mechanisms and clinical applications of miR-22 in regulating malignant progression in human cancer (Review), *Int. J. Oncol.*, 2017, **50**(2), 345–355.
139. A. Rada-Iglesias, S. Enroth, A. Ameur, C. M. Koch, G. K. Clelland and P. Respuela-Alonso, *et al.*, Butyrate mediates decrease of histone acetylation centered on transcription start sites and down-regulation of associated genes, *Genome Res.*, 2007, **17**(6), 708–719.
140. T. Zhang, N. Rong, J. Chen, C. Zou, H. Jing and X. Zhu, *et al.*, SIRT1 expression is associated with the chemotherapy response and prognosis of patients with advanced NSCLC, *PLoS One*, 2013, **8**(11), e79162.
141. M. L. Motti and D. A. S. R. Meccariello, MicroRNAs, Cancer and Diet: Facts and New Exciting Perspectives, *Curr. Mol. Pharmacol.*, 2018, **11**(2), 90–96.
142. K. W. Witwer, XenomiRs and miRNA homeostasis in health and disease: evidence that diet and dietary miRNAs directly and indirectly influence circulating miRNA profiles, *RNA Biol.*, 2012, **9**(9), 1147–1154.
143. A. Philip, V. A. Ferro and R. J. Tate, Determination of the potential bioavailability of plant microRNAs using a simulated human digestion process, *Mol. Nutr. Food Res.*, 2015, **59**(10), 1962–1972.
144. M. J. O'Neill, L. Bourre, S. Melgar and C. M. O'Driscoll, Intestinal delivery of non-viral gene therapeutics: physiological barriers and preclinical models, *Drug Discovery Today*, 2011, **16**(5–6), 203–218.
145. J. W. Snow, A. E. Hale, S. K. Isaacs, A. L. Baggish and S. Y. Chan, Ineffective delivery of diet-derived microRNAs to recipient animal organisms, *RNA Biol.*, 2013, **10**(7), 1107–1116.

146. J. S. Petrick, B. Brower-Toland, A. L. Jackson and L. D. Kier, Safety assessment of food and feed from biotechnology-derived crops employing RNA-mediated gene regulation to achieve desired traits: a scientific review, *Regul. Toxicol. Pharmacol.*, 2013, **66**(2), 167–176.
147. K. W. Witwer, M. A. McAlexander, S. E. Queen and R. J. Adams, Real-time quantitative PCR and droplet digital PCR for plant miRNAs in mammalian blood provide little evidence for general uptake of dietary miRNAs: limited evidence for general uptake of dietary plant xenomiRs, *RNA Biol.*, 2013, **10**(7), 1080–1086.
148. C. Pastrello, M. Tsay, R. McQuaid, M. Abovsky, E. Pasini and E. Shirdel, *et al.*, Circulating plant miRNAs can regulate human gene expression *in vitro*, *Sci. Rep.*, 2016, **6**, 32773.
149. L. Zhang, D. Hou, X. Chen, D. Li, L. Zhu and Y. Zhang, *et al.*, Exogenous plant MIR168a specifically targets mammalian LDLRAP1: evidence of cross-kingdom regulation by microRNA, *Cell Res.*, 2012, **22**(1), 107–126.
150. G. Liang, Y. Zhu, B. Sun, Y. Shao, A. Jing and J. Wang, *et al.*, Assessing the survival of exogenous plant microRNA in mice, *Food Sci. Nutr.*, 2014, **2**(4), 380–388.
151. J. V. Higdon, B. Delage, D. E. Williams and R. H. Dashwood, Cruciferous vegetables and human cancer risk: epidemiologic evidence and mechanistic basis, *Pharmacol. Res.*, 2007, **55**(3), 224–236.
152. K. Wang, H. Li, Y. Yuan, A. Etheridge, Y. Zhou and D. Huang, *et al.*, The complex exogenous RNA spectra in human plasma: an interface with human gut biota?, *PLoS One*, 2012, **7**(12), e51009.
153. J. Yang, L. M. Farmer, A. A. Agyekum and K. D. Hirschi, Detection of dietary plant-based small RNAs in animals, *Cell Res.*, 2015, **25**(4), 517–520.
154. J. Yang, L. M. Farmer, A. A. Agyekum, I. Elbaz-Younes and K. D. Hirschi, Detection of an Abundant Plant-Based Small RNA in Healthy Consumers, *PLoS One*, 2015, **10**(9), e0137516.
155. A. R. Chin, M. Y. Fong, G. Somlo, J. Wu, P. Swiderski and X. Wu, *et al.*, Cross-kingdom inhibition of breast cancer growth by plant miR159, *Cell Res.*, 2016, **26**(2), 217–228.
156. S. Mlotshwa, G. J. Pruss, J. L. MacArthur, M. W. Endres, C. Davis and L. J. Hofseth, *et al.*, A novel chemopreventive strategy based on therapeutic microRNAs produced in plants, *Cell Res.*, 2015, **25**(4), 521–524.
157. N. H. Phuah and N. H. Nagoor, Regulation of microRNAs by natural agents: new strategies in cancer therapies, *BioMed Res. Int.*, 2014, **2014**, 804510.

CHAPTER 2

Role of MicroRNAs in Metabolic Diseases

RAFAY M. SYED[a] AND JEFFREY A. DEIULIIS*[b]

[a]Rush Medical College, 600 S. Paulina St, Chicago, IL 60612, USA; [b]Case Western Reserve University, 2103 Cornell Rd, WRB Rm 4523, Cleveland, OH 44106, USA
*E-mail: jeffrey.deiuliis@gmail.com; jeffrey.deiuliis@case.edu

2.1 Introduction

Type 2 diabetes mellitus (T2DM) cases are estimated to account for 90–95% of diabetes (T1DM + T2DM) in the USA, as reported by the National Institutes of Health. Native American, Black and Hispanic groups suffer from the highest incidence.[1,2] The rate of T2DM between sexes is not significantly different. About 7% of adults aged >18 years in the United States are diagnosed with T2DM [fasting glucose ≥126 mg dL^{-1}, or a hemoglobin A1c (HbA1c) level ≥6.5%].[2] Pre-diabetes occurs in ~34% of US adults, as defined by HbA1c 5.7–6.4% or fasting glucose of 100–125 mg dL^{-1}. A significant body of work attempts to understand the role of microRNAs (miRNAs) in the pathogenesis of metabolic disease associated with obesity in humans; we summarize herein the most significant findings.

Drug Discovery Series No. 69
MicroRNAs in Diseases and Disorders: Emerging Therapeutic Targets
Edited by Philip V. Peplow, Bridget Martinez, George A. Calin and Aurora Esquela-Kerscher

Published by the Royal Society of Chemistry, www.rsc.org

2.1.1 The Innate Immune System in Obesity-mediated Insulin Resistance and T2DM

The development of metabolic disease is mainly the result of insulin insensitivity in peripheral tissues such as liver, skeletal muscle, and adipose tissue due to overnutrition and the effects of immune system-generated inflammation. Obesity is characterized by an increase in circulating neutrophils and monocytes,[3,4] which has been reported in mouse models of metabolic disease[5] as well as in obese humans.[6,7] More specifically, macrophage-mediated visceral adipose tissue (VAT) inflammation dramatically increases in obesity and metabolic disease.[8,9] Interventions that treat inflammation, or more specifically VAT inflammation, greatly reduce or completely abrogate the progression of metabolic disease and peripheral tissue insulin resistance in the presence of obesity and/or high fat feeding in mice.[10] It is theorized that during increased body fat accumulation, adipocyte cell death is induced, which "pushes" macrophages toward a classically activated M1 phenotype.[11,12] However, the concept of M1 activation is debated and has been highly criticized for being an oversimplification. In addition, it is likely that elevated fatty acids [from the hydrolysis of stored triglyceride (adipose) as well as from lipolysis of dietary neutral fats] contribute to macrophage M1 activation,[13–15] and this may potentiate adipocyte death and insulin resistance.[16] Saturated fatty acids such as palmitic acid can ligate the toll-like receptor 4 (TLR4), initiating c-Jun N-terminal kinase (JNK)[17] and nuclear factor κB (NFκB) signaling in liver and adipose tissue, resulting in impaired insulin signaling in these metabolically important tissues.

The relative contribution of "resident" VAT innate immune cells *vs.* infiltrating macrophages to the inflammatory state of VAT in obesity is also unclear, with evidence of contributions by both mechanisms.[4,18–22] Inflamed liver tissue produces significant systemic increases in inflammatory chemokines,[23,24] and increased immature myeloid cell populations have been reported in the livers of obese mice, similar to the phenotype observed in adipose tissue.[25,26] In addition, it is clear that inflamed VAT in obesity is enriched in immature macrophage populations ($CD11b^{+}F4/80^{lo}Ly6C^{+}$),[3,26,27] suggesting that obesity alters hematopoiesis or monocyte differentiation within the VAT. The complex regulation of adipose tissue inflammation, including hematopoietic recruitment and monocyte-to-macrophage differentiation within the adipose tissue, is modulated heavily by miRNAs such as miR-223.

2.1.2 miRNAs and the Etiology of T2DM

miRNAs are small non-coding RNAs that repress messenger RNAs (mRNAs) by increasing mRNA decay rates and impairing translational efficiency in a targeted manner *via* the RNA-induced silencing complex. miRNA-mediated repression of genes is present in most eukaryotes. Due to the near-global regulation of gene expression by miRNAs, all metabolic pathways are affected to some degree by miRNAs that are differentially expressed in the pathophysiology of obesity; the full magnitude is unknown even for well-studied pathways due to the combinatorial effects of multiple miRNA species on multiple targets. The role of miRNAs in insulin-sensitive tissues—adipose, liver, and skeletal muscle—are covered in this chapter. Obesity-associated metabolic disease is associated with reduced insulin sensitivity, increased hepatic lipid accumulation, and diabetic kidney disease. The role of miR-122, miR-146a, miR-21, miR-29a, and miR-34a in these pathological processes are explored (Figure 2.1).

Inflammation is a hallmark of metabolic disease pathophysiology. miRNAs are involved in monocyte/macrophage differentiation and immune cell inflammatory potential. This chapter looks specifically at the roles of miR-155, miR-126, and miR-223 in obesity-related inflammation and metabolic disease/T2DM. This chapter takes a systematic approach by using PubMed searches and focusing on miRNAs ordered by top five returns in order to be as unbiased as possible and address the most relevant miRNAs. Table 2.1 delineates the search results and Tables 2.2 and 2.3 contain most of the relevant human studies for the top 10 miRNAs across all search terms. Table 2.4 contains animal and *in vitro* studies. (Tables and text are ordered by conventional miRNA ordering scheme, not by importance.)

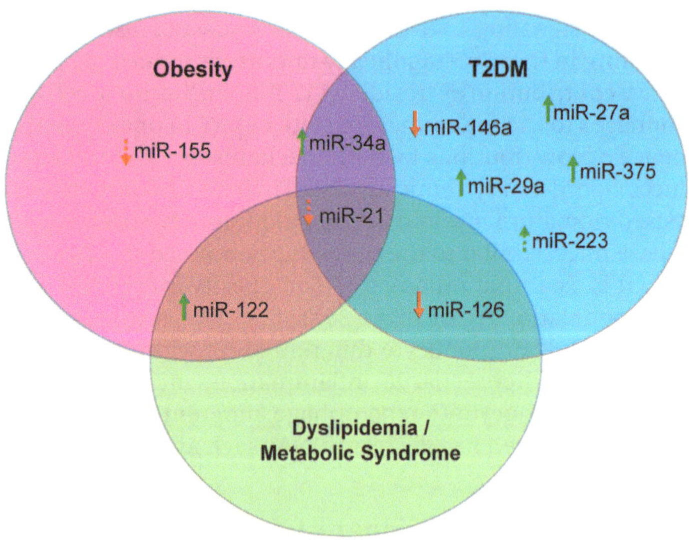

Figure 2.1 Top-cited miRNAs returned from PubMed search terms obesity, type 2 diabetes mellitus (T2DM), and dyslipidemia (see Table 2.1). miR-21 is the only species highly reported in all three search categories. miR-34a, miR-122, and miR-126 appear as highly reported in two of the three search categories. A dashed arrow indicates conflicting data about the directionality of differential expression in human metabolic disease.

Table 2.1 PubMed database search term returns.

Search term[a]	Date	Return 1 (N)	miRNA species (top 5)	Return 2[b] (N)
("Diabetes mellitus, type 2"[mesh]) AND "microRNAs"[mesh]	April 10, 2018	315	miR-126	29
			miR-146a	20
			miR-375	15
			miR-21	19
			miR-223	18
"Type 2 diabetes" AND "microRNAs"	April 28, 2018	530	miR-126	44
			miR-146a	29
			miR-375	26
			miR-223	22
			miR-34a	21
"T2DM" AND "microRNAs"	April 26, 2018	118	miR-146a	11
			miR-126	10
			miR-27a	8
			miR-29a	7
			miR-375	4
"Obesity" AND "microRNAs"	September 21, 2018	609	miR-122	31
			miR-21	30
			miR-143	23
			miR-155	23
			miR-34a	23
("Metabolic syndrome" [mesh]) AND "microRNAs"[mesh]	April 26, 2018	48	miR-21	4
			miR-130a	3
			miR-122	3
			miR-126	2
			miR-197	2

[a]Review articles and commentaries were excluded from search term return count.
[b]Return 2 is for the search term in the first column with "microRNA(s)" replaced with miRNA species name, *e.g.*, "Type 2 diabetes" AND "miR-126".

2.2 miR-122 is Elevated in Obesity and Human Metabolic Disease

The *hsa-mir-122* gene (previously referred to as *hsa-mir*-122a; HGNC : MIR122) is located on chromosome 18q21.31 (58451074–58451158) and transcribes hsa-miR-122-5p and hsa-miR-122-3p. hsa-miR-122-5p MIMAT0000421 (5′-UGGAGUGUGACAAUGGUGUUUG-3′) is the most abundant strand, whereas hsa-miR-122-3p MIMAT0004590 (5′-AACGCCAUUAUCACACUAAAUA-3′) is significantly less abundant. There is one other miRNA (*hsa-mir-122b*, MI0017383) that resides at 58451080–58451152 on the opposite strand from *hsa-mir-122*.

miR-122 is highly expressed in the liver and is the second most abundant miRNA in the plasma of humans by small RNA sequencing.[28] Findings of miR-122 being differentially expressed in obesity, dyslipidemia, and metabolic syndrome may be due in large part to the effects of these conditions on hepatic tissue, *e.g.*, fatty liver, liver fibrosis, liver inflammation, and hepatic insulin resistance. Reports with the most statistical

Table 2.2 Differential expression of microRNA in tissue from patients with cardiometabolic diseases.[a]

miRNA	Patient population	Tissue source	Findings	Reference
miR-122 miR-146a miR-146b	$N = 60$; all following bariatric surgery; obese with NAFLD ($n = 17$), obese with borderline NAFLD ($n = 24$), normal obese ($n = 19$)	Liver tissue	↓ miR-122 and ↑ miR-146b-5p with NAFLD; ↓ miR-122-5p associated with impaired fatty acid use	123
miR-122-5p miR-34a-5p	$N = 95$; NASH ($n = 68$) *vs.* normal control ($n = 27$)	Liver tissue	↓ miR-122-5p with NASH *vs.* non-NASH control ↓ miR-122-5p with increasing BMI ↓ miR-29c-3p with increasing BMI ↓ miR-29c-3p with NASH ↑ miR-34a-5p with NASH	124
miR-126 miR-21 miR-27a	$N = 30$; T2DM and NGT; Chinese population	EPCs	↓ miRNA levels in EPCs from T2DM *vs.* NGT	48
miR-126 miR-132	$N = 28$; T2DM (without history of CVD), non-diabetic CVD, healthy controls (cadavers, control only)	Cardiac tissue, plasma	↓ miR-126 and miR-132 in T2DM plasma and cardiac tissue	125
miR-146a	T1DM and T2DM with nephropathy and/or diabetic retinopathy	Genomic DNA	SNP of miR-146a associated with diabetic macular edema and microvascular complications in diabetics	50
miR-146a	Southwestern American Indians with early T2DM as part of clinical trial NCT00340678 examining the renal protective potential of losartan; sample sizes not clear	Kidney (biopsies)	Diabetic kidney ↓ miR-146a, ↓ miR-146a associated with impaired kidney function; miR-146a highly expresses in podocytes in humans and mice	53
miR-146a	$N = 1348$; Han Chinese population	Genomic DNA	C allele of rs2910164 in miR-146a associated with T2DM	52
miR-146a	$N = 17$; NGT ($n = 7$), obese ($n = 5$), obese with T2DM ($n = 5$)	Abdominal adipose tissue	↑ miR-146a with obesity/body fat percentage	126
miR-146a	$N = 975$; Chinese population, T2DM and CVD	Genomic DNA	rs2910164 may be associated with advanced CVD in T2DM	51
miR-146a	$N = 375$; Iranian population	Genomic DNA	miR-146a variant rs2910164 has a higher prevalence in T2DM *vs.* control	49

miR-155	$N = 25$; BMI<25 kg m^{-2} ($n = 11$), BMI>30 kg m^{-2} ($n = 14$)	Subcutaneous WAT *via* needle biopsy	↑ miR-155 in WAT of obese *vs.* lean humans	64
miR-223	$N = 12$; NGT ($n = 6$), T2DM ($n = 6$)	Left ventricular heart biopsies	↑ miR-223 in T2DM heart muscle *vs.* non-T2DM	127
miR-223-3p	$N = 50$; obese ($n = 25$) *vs.* normal-BMI control ($n = 25$)	VAT (whole tissue) and SVF isolated from VAT	↑ miR-223-3p in obesity; miR-223 suppressed macrophage activation *via* iNOS	85
miR-27a	$N = 413$; T2DM ($n = 204$), sex and age-matched non-T2DM controls ($n = 209$); Iranian population	Genomic DNA	Variant rs895819 positively correlated with T2DM	128
miR-29	$N = 22$; treatment-naïve males with T2DM ($n = 12$) *vs.* healthy age-/sex-matched controls ($n = 10$)	Skeletal muscle	↑ miR-29a and miR-29c in skeletal muscle	91
miR-29a miR-29b	Arteriole biopsies from T2DM	Intraluminal delivery of miR-29a and -29b mimics	Mimics restored endothelial-dependent vasodilation in *ex vivo* arterioles from T2DM	96
miR-29a miR-29b	$N = 83$; age ~60 years, T2DM stratified by carotid intimal thickness *via* ultrasound	Urine spiked with cel-miR-39 (non-human small RNA)	miR-29a positively correlated with urinary albuminuria and retinopathy, no differential expression of miR-29b or c; miR-29b correlated with carotid intimal thickness	93
miR-37124a5	$N = 15$; NGT ($n = 10$), T2DM ($n = 5$)	Primary human pancreatic islets from multi-organ donors	↑ miR-124a (~30-fold) *vs.* NGT	129
miR-375	GWAS meta-analysis	Genomic DNA	Found 37 genomic variations (SNP) in miRNA binding sites specific to cardiometabolic genes; rs11735092 (HSD17B13:miR-375)	111
miR-375	$N = 151$; NGT, IGT, and T2DM	Genomic DNA	IGT ↑ methylation of CpG sites in promoter of miR-375 *vs.* NGT and T2DM	108

[a]BMI: body mass index; CpG: cytosine phosphate guanine; CVD: cardiovascular disease; EPC: endothelial progenitor cell; GWAS: genome-wide association study; IGT: impaired glucose tolerance; iNOS: inducible nitric oxide synthase; NASH: non-alcoholic hepatosteatosis; NAFLD: non-alcoholic fatty liver disease; NGT: normal glucose tolerance; N: total study sample size; *n*: group-specific sample size; SNP: single nucleotide polymorphism; SVF: stromal vascular fraction; T2DM: type 2 diabetes mellitus; VAT: visceral adipose tissue; WAT: white adipose tissue.

Table 2.3 Differential expression of circulating microRNAs in patients with cardiometabolic diseases.[a]

miRNA	Patient population	Tissue source	Findings	Reference
miR-122	$N = 128$; Asian Indian population without known CVD or anti-diabetic meds over 2.5 years	Plasma	miR-122 negatively correlates with hyperglycemic progression over 2.5 years	130
miR-122 miR-222 miR-130b	$N = 80$; obese men and pre-/post-bariatric surgery	Plasma	↓miR-122 after bariatric surgery ↑miR-222 in obesity ↓miR-130b in obesity	131
miR-122	$N = 703$; age 34–49 years, Young Finns Study	Plasma	↑ miR-122 with fatty liver, increased with severity; large sample size	132
miR-122	$N = 90$; age 2–25 years, obese or overweight youth based on data from screening of Framingham Heart Study	Plasma/"extracellular RNAs"	↑ miR-122 with IR, miR-122 levels positively correlated with insulin levels when grouped by quartile	133
miR-122	$N = 112$; obese and control adults	Serum	↑ miR-122 in obese *vs.* control; Positive correlation to BMI and TG, as well as markers of NASH	134
miR-122	$N = 810$; Bruneck study and Anglo-scandinavian Cardiac Outcomes Trial (ASCOT)	Serum and plasma	↑ miR-122 with obesity and dyslipidemia; Positive correlation with markers of T2DM; atorvastatin (12 months) decreased miR-122 levels	29
miR-122 miR-21 miR-34a	$N = 403$; correlated plasma miRNA expression with clinical measures of NASH/NAFLD	Plasma	↑ miR-122 ↑ miR-21 ↑ miR-34a in NASH/NAFLD	135
miR-122 miR-375	$N = 70$; females with pre-gestational obesity, gestational obesity, or normal pregnancy	Plasma	↓ miR-122 and miR-375 in both obese groups *vs.* normal	136
miR-126	$N = 152$; NGT ($n = 50$), T2DM without nephropathy ($n = 52$), with nephropathy ($n = 50$)	Whole blood, peripheral	Negatively correlated with markers of metabolic syndrome and albuminuria	137
miR-126	$N = 120$; NGT, healthy ($n = 20$); T2DM with CAD ($n = 46$), without CAD ($n = 54$)	Plasma	↓ miR-126 in T2DM with CAD *vs.* T2DM without; miR-126 negatively correlated with CVD	40
miR-126	$N = 455$; T1DM and complications ($n = 312$) *vs.* T1DM without complications ($n = 143$)	Serum	↓ miR-126 negatively associated with all diabetic complications	37

miR-126 miR-222	$N = 40$; pre-, post-metformin + sitagliptin intervention; elderly population	Plasma	↑ miR-126 (3.3-fold), ↓ miR-222 (0.26-fold) in group responsive to sitagliptin; ↑ miR-222 (1.85-fold) in group non-responsive to sitagliptin	39
miR-126	$N = 40$; T2DM without CAD/CVD	Platelets *ex vivo* and plasma	Platelets are a major source of plasma miR-126; use of aspirin reduced platelet transfer of miR-126 to plasma	35
miR-126	$N = 160$; pre-diabetic, diabetic, and non-diabetic groups	miR-126 was measured in microparticles from blood using qPCR	miR-126 was negatively correlated with fasting glucose and positively correlated with plasma antioxidant capacity	43
miR-126	$N = 135$; adults with T2DM ($n = 55$), NGT controls ($n = 80$)	Endothelial vesicles Plasma and circulating endothelial microparticles	↓ miR-126 in T2DM and associated with increased CAD risk	42
miR-126	All male, very small sample sizes that vary by assay	$CD34^+$ PBMCs	miR-126 was enriched in $CD34^+$ cells	138
miR-126 miR-146a miR-155	$N = 23$; age >60 years, $n = 10$ non-diabetic, $n = 13$ diabetic. Both groups underwent intervention: strength and cardiovascular training	Whole blood including blood cells and plasma	Strength training ↑ miR-146a; cardiovascular training had no effect	139
miR-126 miR-21	$N = 300$; NGT, T2DM (stratified to with or without complication groups)	Plasma	↓ miR-126 ↓ miR-21 with T2DM (both with and without complications)	36
miR-126 miR-142-3p miR-222-3p	$N = 93$; T2DM *vs.* NGT controls	Plasma	↓ miR-126 in T2DM; ↑ miR-142-3p and miR-222-3p *vs.* NGT	140
miR-126 miR-146a miR-21 miR-375 miR-29a	$N = 92$; pre-diabetic, T2DM, LADA, T1DM, non-diabetic	Plasma	↓ miR-126 and miR-146a in prediabetic *vs.* nondiabetic; oddly ↑ in T1DM and T2DM *vs.* non-diabetic ↑ miR-21 in T1 and T2DM *vs.* non-diabetic	141
miR-126	$N = 40$; T2DM ($n = 20$) *vs.* normal controls ($n = 20$)	Plasma (obtained 2 years before the onset of T2DM)	↓ miR-126 prior to manifestation of T2DM	142

(*continued*)

Table 2.3 (*continued*)

miRNA	Patient population	Tissue source	Findings	Reference
miR-126 miR-223 miR-29a miR-375 miR-34a	Meta-analysis of 38 profiling studies between 1993 and 2014 in humans with T2DM; only studies with NGT controls	Circulating/blood (excluded blood and urine miRNAs)	40 miRNAs dysregulated; miR-126a was the most downregulated in T2DM *vs.* NGT	143
miR-146a	$N = 40$; T2DM ($n = 20$) *vs.* NGT controls ($n = 20$)	PBMCs	↓ miR-146a in T2DM	144
miR-146a	$N = 108$; NGT ($n = 44$), T2DM ($n = 64$); Ecuadoran population	Serum, PBMCs	miR-146a in monocytes from blood did not correlation with serum levels of miR-146a	54
miR-146a miR-155	$N = 96$; T2DM ($n = 56$) *vs.* NGT ($n = 40$); Ecuadorian population	Serum	↓ miR-146a in T2DM *vs.* NGT; miR-146a was negatively correlated with IL-8; miR-155 no differences	145
miR-146a miR-155	$N = 40$; non-insulin-dependent T2DM ($n = 20$), healthy control ($n = 20$); Mexican population, both genders	PBMCs	↓ miR-146a and miR-155 in T2DM *vs.* control	55
miR-146a	$N = 70$; T2DM ($n = 35$) *vs.* NGT ($n = 35$); Indian population	PBMCs	↓ miR-146a in T2DM	56
miR-146a	$N = 180$; healthy ($n = 56$), T1DM ($n = 78$), T2DM ($n = 46$)	PBMCs	↓ miR-146a in T1DM *vs.* other groups	146
miR-155	$N = 100$; NAFLD ($n = 50$), healthy ($n = 50$); Chinese population, Nanjing	Whole blood RNA, liver	↓ miR-155 in blood and liver of NAFLD *vs.* control	63
miR-21 miR-122 miR-223 miR-192	$N = 198$; NAFLD/NASH *vs.* "healthy"	Serum	↑ miR-122 ↑ miR-21 ↑ miR-223 ↑ miR-192	147
miR-21 miR-29a	$N = 50$; T2DM Taiwanese population with varying levels of kidney disease	Serum	↑ miR-21 in T2DM with overt proteinuria *vs.* normal or micro-albuminuria; miR-21 negative correlated with eGFR ↑ miR-29a in overt proteinuria *vs.* normal only	92
miR-21 miR-126 miR-146a	$N = 87$; T2DM ($n = 45$), NGT ($n = 42$)	Serum	↓ miR-21 obesity, but did not correlate to T2DM status; no difference for other miRNAs	66
miR-21	$N = 40$; T2DM ($n = 20$), non-diabetic ($n = 20$)	PBMCs and *ex vivo* PBMCs	↓ miR-21 in obesity; miR-21 negatively correlated with IL-6 and TNFα	67

miR-21	N = 97; obese and "lean" heart failure	Plasma	No change	148
miR-21 miR-34a miR-155	N = 35; T2DM males with treated HTN and stable CAD; randomized to placebo, GSE, or GSE + resveratrol	PBMCs	↑ miR-21; ↓ miR-34a and miR-155 with resveratrol treatment	149
miR-21 miR-222 miR-27a miR-29a miR-375	Meta-analysis, 26 studies included	Cell-free circulating miRNAs (serum or plasma)	↓ miR-21 and ↑ miR-222 in obese; ↑ miR-375 and miR-27a in T2DM; Good analysis of how to use miRNAs as biomarkers	68
miR-21 miR-126 miR-223	N = 160; T2DM (n = 80), normal controls (n = 80)	Plasma	↓ miR-21, miR-126, miR-223 in T2DM; miR-223 predictors of T2DM manifestation miR-126 negatively associated with severity of impaired glucose tolerance	65
miR-21-3p (less abundant form) miR-155-5p	N = 116; GDM (n = 36), controls (n = 80)	Plasma	↑ miR-21 ↑ miR-155 with GDM; differential expression of the less abundant form brings into question the relevance of these findings	150
miR-21-5p miR-375	N = 93; IGT (n = 75) *vs.* NGT (n = 18)	Plasma	↓ miR-375 in IGT *vs.* NGT; ↓ miR-21 in TZD responders *vs.* non-responders	151
miR-223 miR-143 miR-155	N = 44; age 25–60 years withT2DM, treated with insulin or metformin (n = 22) *vs.* healthy age-matched controls (n = 22)	Platelets	↓ miRNAs in T2DM	152
miR-223	N = 60; T2DM ± cerebral infarction *vs.* NGT	PBMCs	↓ miR-223 in T2DM without cerebral infarction *vs.* NGT	153
miR-223	N = 47; males undergoing 12 week weight-loss study with high (n = 20) or normal protein diet (n = 27)	HDL	↓ miR-223 after high-protein diet; miR-223 levels did not correlate with HDL composition or size	154
miR-223	N = 20 female T2DM before and after training interventions over 8 weeks	Platelets	Found no intervention-associated change in miR-223 with exercise	155
miR-223	N = 121; obese (n = 40), overweight (n = 40), NGT (n = 41); all aged 40–60 years	Plasma	↓ miR-223 with overweight and obesity	156
miR-223	Healthy NGT (n = 30), T2DM (n = 56), T2DM with acute ischemic stroke (n = 58)	Platelet-depleted plasma	T2DM and T2DM with ischemic stroke had ↓ miR-223 in platelets and plasma, no difference in leukocytes	157

(continued)

Table 2.3 (*continued*)

miRNA	Patient population	Tissue source	Findings	Reference
miR-223 miR-143	$N = 83$; morbidly obese ($n = 23$), obese ($n = 30$), normal BMI and overweight (BMI 20–29.9 kg m^{-2}) controls ($n = 30$)	Plasma	↓ miR-223 and miR-143 in obesity and morbid obesity *vs.* control	158
miR-27a	$N = 526$; Japanese population	Circulating	miR-27a associated with subQ AT, VAT, and BMI	159
miR-27a miR-375	$N = 265$; healthy ($n = 46$), metabolic syndrome ($n = 50$), T2DM ($n = 50$); Singaporean population	Whole blood, blood exosomes	↑ miR-27a and miR-375 in metabolic syndrome and T2DM *vs.* healthy controls	86
miR-27a	$N = 90$; obese ($n = 45$), normal weight ($n = 45$); children aged 6–15 years; Chinese population	Serum	↑ miR-27a in obese group	87
miR-29	$N = 182$; T2DM with elevated TGs ($n = 114$) *vs.* T2DM with normal TGs ($n = 68$)	Plasma	LPL is regulated by miR-29	160
miR-29a let-7b	$N = 278$; T2DM ($n = 112$), IGT ($n = 72$), NGT ($n = 94$)	Plasma	↑ miR-29a in IGT and T2DM *vs.* NGT	130
miR-29b miR-126 miR-155 miR-21 miR-34a miR-27a miR-146a miR-223	T2DM ($n = 17$) and pre-diabetes ($n = 21$) *vs.* NGT ($n = 20$); Obese ($n = 9$) *vs.* lean ($n = 23$)	Plasma	↓ miR-29b, miR-126, and miR-155, in pre-diabetes; ↑ miR-21, miR-24, and miR-34a in T2DM; ↓ miR-27a, miR-126, miR-146a, and miR-223 in obesity; ↑ miR-34a with obesity	99
miR-29b	$N = 152$, Iraqis and Swedes diagnosed with T2DM and controls	Plasma	↓ miR-29b; no difference in miR-223, miR-21, or miR-126	161
miR-34a miR-126 miR-146a	$N = 45$; sedentary middle-aged adults	Plasma	↑ miR-34a (2-fold) in obese *vs.* normal weight; ↓ miR-126 and miR-146a were lower in obese subjects	100
miR-34a miR-146a miR-375	$N = 56$; NGT, IGT, T2DM	Serum	↑ miR-34a, miR-146a and miR-375 in T2DM *vs.* NGT and IGT; no difference IGT *vs.* NGT	162

miR-34a miR-122 miR-7	$N = 22$; post-RYGB surgery followed for 6 months, randomized to exercise intervention or none	Plasma	↓ miR-122 with RYGB and exercise; small sample size	101
miR-34a	$N = 57$; women with or without insulin resistance and obesity	Plasma	↑ miR-34a in T2DM *vs.* control; small study	163
miR-34a	$N = 125$; NGT ($n = 52$), T2DM ($n = 73$)	PBMCs	miR-34a did not correlate with metabolic measures but did with LDL/HDL ratio	164
miR-375	$N = 200$; T2DM of different ethnic Chinese backgrounds: Han *vs.* Kazak	Plasma, genomic DNA	↑ miR-375 (3-fold) Han T2DM *vs.* Kazak T2DM; increased methylation in Han *vs.* Kazak	165
miR-375	$N = 46$; prospective intervention with T2DM patients randomized to dietary inulin supplement *vs.* control; Iranian population; 6 week duration	Plasma	Inulin ↑ miR-375 ~4-fold and decreased fasting glucose; one of the first reports to use a circulating miRNA as a measure of efficacy of anti-diabetic agent	109
miR-375	$N = 204$; T2DM *vs.* NGT	Serum	↑ miR-375 in T2DM *vs.* NGT	106
miR-375 miR-33b	$N \approx 50$; newborns with low, normal, and high birthweight	Dried blood spots	↑ miR-33b and miR-375 in macrosomia (high birthweight)	110
miR-375	$N = 200$; T2DM *vs.* NGT; Chinese population	Plasma, cell-free genomic DNA isolated from plasma	No differential CpG methylation status of miR-375 gene, ↑ miR-375 in plasma T2DM *vs.* NGT	107

[a] AT: adipose tissue; BMI: body mass index; CAD: coronary artery disease; CVD: cardiovascular disease; eGFR: estimated glomerular filtration rate; GDM: gestational diabetes mellitus; GSE: grapeseed extract; HDL: high-density lipoprotein; HTN: hypertension; IGT: impaired glucose tolerance; IL: interleukin; IR: insulin resistance; LADA: latent autoimmune diabetes in adults; LDL: low-density lipoprotein; LPL: lipoprotein lipase; NASH: non-alcoholic hepatosteatosis; NAFLD: non-alcoholic fatty liver disease; NGT: normal glucose tolerance; N: total study sample size; *n*: group-specific sample size; PBMCs: peripheral blood mononuclear cells; qPCR: quantitative PCR; RYGB: Roux-en-Y gastric bypass; subQ: subcutaneous; T1/2DM: type 1/2 diabetes mellitus; TG: triglycerides; TNFα: tumor necrosis factor α; VAT: visceral adipose tissue; WAT: white adipose tissue.

Table 2.4 *In vivo* murine studies and *in vitro* studies.[a]

miRNA	Model	Tissue source	Findings	Reference
miR-122	*In vitro*	Human Huh7 hepatocarcinoma cell line	Negatively regulates SIRT6, fatty acid oxidation genes	32
miR-126	Adipocytes *in vitro*		miR-126 decreased CCL2 expression in adipocytes	166
miR-126	EPCs treated with high glucose and AGEs		↓ miR-126 by high glucose and AGEs; miR-126 OVEX reduced IL-6 *in vitro*	47
miR-126	Pancreatic beta-cells *in vitro*		miR-126 inhibited beta-cell proliferation	167
miR-146a	*db/db vs. db/m* miR-146a loaded exosomes injected into brain (ventricles)		miR-146a ↑ PrP^c production and improved short-term memory measures	168
miR-146a miR-29a	Wistar rat SCD and HFD with low-dose streptozotocin injection	Insulin-sensitive tissues and blood	Poor model of T2DM	169
miR-146a	*db/db* ± miR-146a mimic injections	n/a	Mimic reduced diabetic peripheral neuropathy	170
miR-146a	Streptozotocin-induced diabetic nephropathy	Kidney and HK-2 cells	↓ miR-146a *in vivo* miR-146a OVEX decreased ROS generation and inflammation	171
miR-155	LDLR KO mice LDLR, miR-155 double-KO mice	miR-155/MAFB target site blocker	Double KO had increased atherosclerotic lesion size	172
miR-200 family		Beta cell-specific OVEX of miR-200	miR-200 OVEX induced beta cell apoptosis and death; propose Trp53 and Bax are involved	173
miR-21	Sprague Dawley rats, HFD; L6-SMCs		↑ miR-21 in rats fed HFD, metformin ↓ miR-21; miR-21 represses smad7 in SMCs	174
miR-21	*db/db vs.* C57BL/6 (SCD, HFD)	Hepatic adenoviral miR-21 OVEX	miR-21 OVEX decreased hyperglycemia	175
miR-21	*db/db*, aged	miR-21 inhibition *via* LNA-antagomiR injection	Decreased body weight and reduced adipocyte size	176
miR-21 miR-222 miR-27a	MKP1-muscle KO mouse model fed HFD	Skeletal muscle-specific KO of MKP1	MKP1-muscle KO increased miR-21, which repressed PTEN and increased insulin sensitivity in mice	71
miR-29a	Multiple rat models	Adipose, liver, skeletal muscle	Upregulated in adipose, various parameters[b]	90
miR-222	C57BL/6; SCD and HFHSD, 24 weeks	Liver; primary hepatocytes	↑ miR-222 in livers of HFHSD *vs.* SCD; *in vitro* OVEX of miR-222 resulted in ↓ IRS-1	177

miR-29a	INS-1E pancreatic cell line, and human and rat pancreatic islets	Synthetic miR-29a mimic and inhibitor	miR-29a is induced by glucose, however miR-29a OVEX results in reduced glucose stimulated insulin secretion	178
miR-29a miR-29b miR-29c	miR-29a/b-1, miR-29b-2/c KO mice	Basic metabolic phenotyping (GTT and ITT)	No strong effects on pancreatic function, 29a KO had marginally decreased insulin production; both KO models had increased lean body mass and lower adipose mass *vs.* WT; full miR-29a KO mice died at 7 months; double loci KO had 100% mortality before 100 days	97
miR-29a miR-29b	NONcNZO10/LtJ polygenic mouse model of diabetes; have decreased pancreatic insulin release with age; control mouse (NON/ShiLtJ)	Pancreas at 10, 13, and 22 weeks *vs.* baseline at 7 weeks of age	↑ miR-29a and miR-29b over progression of pancreatic failure	98
miR-34a	BALB/c mice SCD or HFD (BALB/c not a good model for diet-induced obesity)		miR-34a targets SIRT1 and NAMPT resulting in reduced NAD + biosynthesis and increased acetylation of SIRT1 in obese mice	103
miR-34a	Cholangiocytes (epithelial cells that line the bile duct); C57BL/6J mice on HFHS diet *vs.* SCD		HFHS-fed mice with NASH had 4-fold higher hepatic miR-34a *vs.* control; PA induces miR-34a expression and apoptosis, anti-miR-34a reduced PA-induced apoptosis	102
miR-34a	miR-34a$^{-/-}$ mice	Epididymal WAT	miR-34a$^{-/-}$ mice were heavier than WT mice when both fed a HFD, epididymal fat was heavier in KO *vs.* WT	104
miR-34a	Wistar rat and human	Primary beta-cells	miR-34a increased in islets with time, positive correlation with age	179
miR-375	MIN6 mouse pancreatic beta-cell line		miR-375 OVEX reduced insulin secretion	113

[a]AGEs: advanced glycation end products; EPCs: endothelial progenitor cells; GTT: glucose tolerance test; HFD: high-fat diet; HFHSD: high-fat high-sugar diet; IL: interleukin; IRS-1: insulin receptor substrate 1; ITT: insulin tolerance test; KO: knockout; LDLR: low-density lipoprotein receptor; LNA: locked nucleic acid; MKP-1: mitogen-activated protein kinase phosphatase-1; n/a: not applicable; NAMPT: nicotinamide phosphoribosyltransferase; NASH: non-alcoholic hepatosteatosis; OVEX: overexpression; PA: palmitic acid; PrP^{c}: cellular prion protein; PTEN: phosphatase and tensin homolog; ROS: reactive oxygen species; SCD: standard chow diet; SIRT: sirtuin; SMC: smooth muscle cells; T2DM: type 2 diabetes mellitus; WT: wild type.
[b]see primary literature for details.

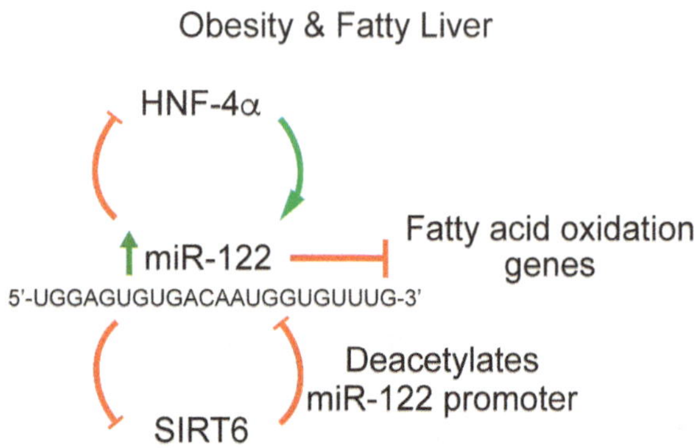

Figure 2.2 miR-122 is elevated in hepatic disease (fatty liver) related to obesity. miR-122 and the transcription factor HNF-4α (hepatocyte nuclear factor-4 alpha, mutated in maturity-onset diabetes of the young) exist in a feedback loop. miR-122 levels directly alter expression of fatty acid oxidation genes and indirectly alter glucose metabolism *via* effects of HNF-4α. miR-122 exists within a negative feedback loop with sirtuin 6 (SIRT6). SIRT6 is a protein deacetylase which regulates the expression of many genes *via* effects on chromatin structure.

power suggest that miR-122 is higher in the circulation (cell-free RNAs) in obesity *vs.* controls,[29] although there are conflicting findings in the literature. miR-122 was decreased substantially in humans on statin intervention (atorvastatin).[29] Interestingly, there are very few reports of miR-122 in human cardiovascular disease (CVD), but an abundance relating to lipid metabolism, especially in non-alcoholic steatohepatitis (NASH).[30] Circulating miR-122 is being developed as a clinical biomarker of nonalcoholic fatty liver disease (NAFLD) or NASH.[31] miR-122 targets sirtuin (SIRT) 6 and negatively regulates fatty acid oxidation genes, and is thus a potential mechanism by which elevated miR-122 potentiates hepatic lipid accumulation.[32] miR-122 expression is at least partially regulated by hepatocyte nuclear factor-4α (HNF-4α), which exists in a feedback loop with miR-122 and impinges upon lipid and glucose metabolism within the liver (Figure 2.2).[33] Farnesoid X receptor (FXR) is also a potential regulator of miR-122 expression.[34] Knowledge of the full functionality of miR-122 in the pathophysiology of obesity and metabolic disease is still evolving, but a substantial body of evidence exists to show that miR-122-5p plays a role in multiple steps of hepatic comorbidities in obesity.

2.3 miR-126 is Downregulated in T2DM and Human Metabolic Disease

The *mir-126* gene (HGNC:MIR126) is located on chromosome 9q34.3 (136670602–136670686) and transcribes hsa-miR-126-5p and hsa-miR-126-3p. hsa-miR-126-5p MIMAT0000444 (5′-CAUUAUUACUUUUGGUACGCG-3) is the most abundant strand, whereas hsa-miR-126-3p MIMAT0000445 (5′-UCGUACCGUGAGUAAUAAUGCG-3′) is less abundant, but well represented in RNA sequencing data in tissue. In human plasma, the 3p strand was #28 most abundant in plasma *vs.* #161 for the 5p strand.[28] *hsa-mir-126* is intronic within the epidermal growth factor-like domain multiple 7 (EGFL7) gene of humans and mice.

miR-126 is the most-reported miRNA species in T2DM and metabolic disease, based on PubMed search returns (Table 2.1). miR-126 is consistently differentially downregulated in circulating miRNAs in humans with cardiovascular and metabolic disease. Platelets and the endothelium are major sources of circulating miR-126. Aspirin has been shown to decrease the transfer of platelet miR-126 to plasma albumin,[35] which has implications for confounding effects in studies of patients on aspirin therapy. The largest human study of T2DM (N = 300) looked at normal glucose tolerance (NGT) subjects *vs.* T2DM patients with and without complications.[36] The authors found that miR-126 and miR-21 were significantly lower in the plasma of both T2DM groups. Although T1DM is not systematically covered in this chapter, it should be noted that miR-126 was found to be lower in T1DM patients with complications *vs.* T1DM patients without complications in a relatively large study (N = 455).[37] These findings imply that vascular damage created by T1DM leads to higher miR-126 levels, as miR-126 is the most abundant microRNA found in apoptotic bodies released by the endothelium.[38] Insulin-sensitizing treatment of elderly diabetics (pre- *vs.* post-paired analysis) resulted in a substantial increase in plasma miR-126 levels;[39] based on the study design, it is unclear whether this is due to improved endothelial function, due to increased insulin sensitivity, or both.

Studies examining miR-126 in diabetes focus heavily on circulating microparticles (derived from the endothelium). This is probably due to the strong link between miR-126 and CVD. Of the miRNAs examined here, miR-126 has the most studies looking at cardiovascular and cardiometabolic disease, with 192 returns for "miR-126 AND CVD". There is evidence that circulating microparticle-associated miR-126 may be useful in predicting cardiovascular events in patients diagnosed with T2DM/coronary artery disease (CAD)[40] and in classifying pre-diabetes in susceptible populations.[41–43]

Whole-body miR-126 knockout (KO) is ~50% embryonically lethal in mice,[44,45] making murine studies of obesity and insulin resistance more challenging and currently unreported. However, murine experiments indicate that miR-126 has high expression in plasmacytoid dendritic cells (pDCs) and regulates the innate immune response *via* genes in the TLR and NF-κB pathway.[46] miR-126 knockout mice had an impaired immune response to TLR7 (CpG-A) and TLR9 (R-848) ligands and lack of miR-126 resulted in loss of pDC populations in bone marrow.[46] miR-126 may be broadly anti-inflammatory, thus any decrease may be permissive to inflammation, generally (Figure 2.3).[47,48]

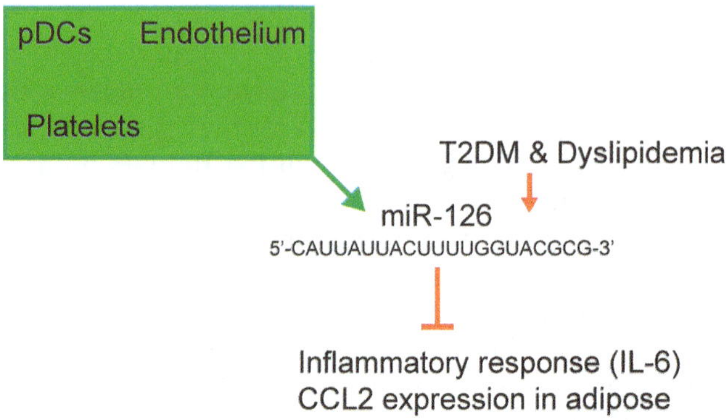

Figure 2.3 Measures of type 2 diabetes mellitus (T2DM) and dyslipidemia are negatively correlated with circulating miR-126 levels in humans. miR-126 is an abundant miRNA in plasmacytoid dendritic cells (pDCs), endothelial cells, and platelets, which contribute to circulating albumin-associated miR-126. miR-126 has anti-inflammatory actions which include repression of interleukin 6 (IL-6) and CCL2 [chemokine (C–C motif) ligand 2; receptor for MCP1, monocyte chemoattractant protein 1].

2.4 miR-146a is Downregulated in Human Obesity and Diabetic Kidney Disease

The *mir-146a* gene (HGNC : MIR146A) is located on chromosome 5q33.3 (160485352–160485450) and transcribes hsa-miR-146a-5p and hsa-miR-146a-3p. hsa-miR-146a-5p MIMAT0000449 (5′-UGAGAACUGAAUUCCAUGGGUU-3′) is the most abundant strand, whereas hsa-miR-146a-3p MIMAT0004608 (5′-CCUCUGAAAUUCAGUUCUUCAG-3′) is significantly less abundant, but well represented in RNA sequencing data. miR-146a-5p is the #46 most abundant in human plasma, while the 3p strand is almost undetectable at #851.[28]

miR-146a-5p has differentially lower expression in T2DM, as measured in multiple RNA sources, including peripheral blood mononuclear cells (PBMCs), plasma/serum, kidney, and adipose tissue, compared to controls. Furthermore, miR-146a has genomic variability, with the variant rs2910164 occurring with higher prevalence in T2DM *vs.* healthy controls.[49,50] The variant rs2910164 was found to be associated with CVD in T2DM in a Chinese population ($N = 975$).[51] A larger study in China ($N = 1348$) found a similar correlation between T2DM and rs2910164.[52]

Podocytes in humans and mice contain high levels of miR-146a. It was found that diabetic kidney tissue had lower levels of miR-146a in a population of Southwestern Native Americans and that low miR-146a was associated with poor kidney function in this (early) diabetic population.[53] Lee *et al.* suggest that hyperglycemia results in a downregulation of miR-146a in podocytes, causing de-repression of miR-146a targets notch homolog 1 (NOTCH1) and erb-b2 receptor tyrosine kinase 4 (ERBB4), resulting in proteinuria (diabetic

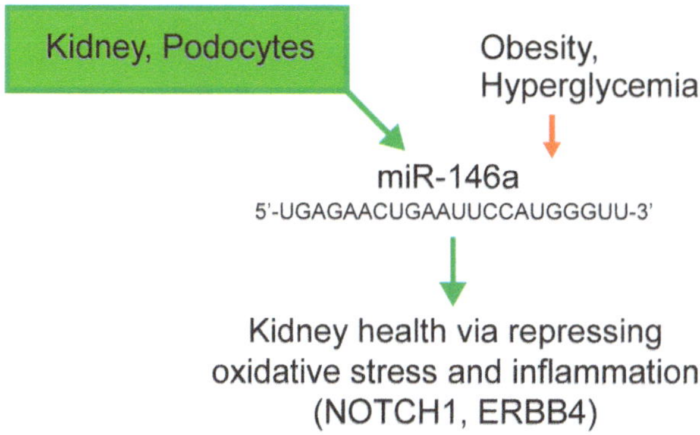

Figure 2.4 Obesity and hyperglycemia in humans negatively correlate with miR-146a levels in various tissues and in the circulation. miR-146a plays a prominent beneficial role in kidney health *via* repression of NOTCH1 (notch homolog 1) and ERBB4 (erb-b2 receptor tyrosine kinase 4), mediators of oxidative stress and inflammation. Obesity-related loss of miR-146a probably contributes to diabetic kidney disease such as diabetic glomerulopathy *via* apoptosis of kidney podocytes. Podocytes wrap around capillaries of the glomerulus to form filtration slits necessary for normal kidney function.

glomerulopathy).[53] If this is true, downregulation of miR-146a may not be an etiological change potentiating diabetes, but a mechanism by which hyperglycemia in diabetes negatively affects the health of the kidney (Figure 2.4). As mentioned earlier, miR-146a is downregulated in many human tissues in T2DM, including PBMCs. This may be of particular use as a marker of hyperglycemia, as miRNA quantification from PBMCs has fewer technical issues (fewer inhibitors, greater RNA yield) than from biofluids (plasma, serum) and currently has a lower cost. However, it was found that PMBC and plasma levels do not correlate.[54] miR-146a was found to be lower in PBMCs from T2DM Mexican patients *vs.* healthy controls,[55] and in a population of South Indian T2DM patients *vs.* NGT.[56]

2.5 miR-155 is Downregulated in Human Obesity and NAFLD

The *mir-155* gene (HGNC:MIR155) is located on chromosome 21q21.3 (25573980–25574044) and transcribes hsa-miR-155-5p and hsa-miR-155-3p. hsa-miR-155-5p MIMAT0000646 (5′- UUAAUGCUAAUCGUGAUAGGGGUU-3′) is the most abundant strand, whereas hsa-miR-155-3p MIMAT0004658 (5′-CUCCUACAUAUUAGCAUUAACA-3′) is significantly less abundant in RNA sequencing data. miR-155-5p is the #195 most abundant in human plasma, while the 3p strand was not reported.[28] miR-155 is an exonic miRNA within the non-coding transcription unit that contains the long non-coding (B-cell integration cluster) BIC gene (HGNC: 35 460; ENSG00000234883).[57]

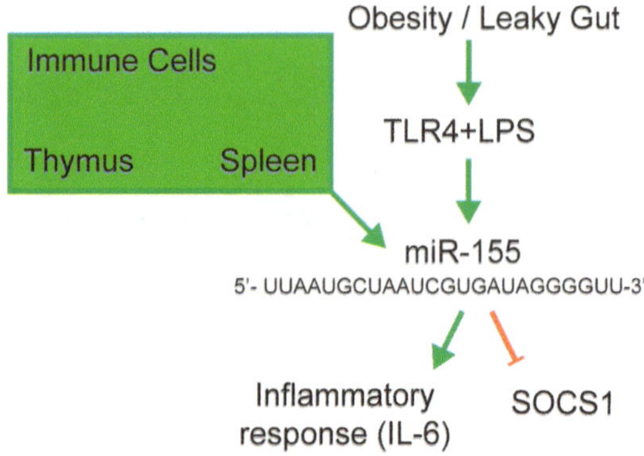

Figure 2.5 Upregulation of miR-155 in obesity potentiates immune cell-mediated inflammation. Obesity involves an increase in TLR4 (toll-like receptor 4) signaling within immune cells *via* multiple mechanisms including a leaky gut and increased circulating palmitic acid levels (a ligand for TLR4) which dramatically induces miR-155 expression. miR-155 potentiates an inflammatory response *via* multiple pathways including repression of SOCS1 (suppressor of cytokine signaling 1). LPS: lipopolysaccharide.

miR-155 is the prototypical immuno-miRNA; a comprehensive review of its functions has been described by Mashima.[58] Briefly, miR-155 is most highly expressed in the spleen and thymus, and plays roles in most immune cell types, including adaptive and innate cells; miR-155 acts in the macrophage as a positive regulator of pro-inflammatory cytokine production. miR-155 regulates and is regulated by common immune stimuli and pathways, including lipopolysaccharide (LPS)-TLR4-interleukin (IL) -1,[59] interferon gamma,[60] IL-2, IL-17,[61] T-cell receptor (TCR), and transforming growth factor beta (TGFβ). LPS ligation of TLR4 increases miR-155 levels, resulting in repression of SOCS1 (suppressor of cytokine signaling 1) (Figure 2.5).[62] Accordingly, obesity-mediated changes in immune cell populations and activation states are expected to impact miR-155 regulated pathways. In an inflammatory state, it would be expected that miR-155 levels would be elevated, but Wang *et al.* reported that miR-155 levels were dramatically lower in the peripheral blood and liver biopsies of NAFLD patients in a Chinese population compared to healthy controls.[63] Conversely, subcutaneous adipose tissue from obese humans had dramatically higher levels of miR-155 compared to controls with body mass index (BMI) <25 kg m^{-2}.[64]

Rodent and *in vitro* models demonstrate more of a role for miR-155 in obesity and insulin resistance pathways than do reports involving human tissues. This may be due to the preponderance of human studies examining plasma or serum miRNA levels in human disease. Based on the etiological role played by immune cells in metabolic disease, the authors believe that miR-155 is likely involved in the inflammatory changes associated with human cardiovascular and metabolic disease.

2.6 miR-21 is Differentially Expressed in Obesity, T2DM, and Human Metabolic Syndrome

The *mir-21* gene (HGNC:MIR21) is located on chromosome 17q23.1 (59841266–59841337) and transcribes hsa-miR-21-5p and hsa-miR-21-3p. hsa-miR-21-5p MIMAT0000076 (5′-UAGCUUAUCAGACUGAUGUUGA-3′) is the most abundant strand, whereas hsa-miR-21-3p MIMAT0004494 (5′-CAACACCAGUCGAUGGGCUGU-3′) is significantly less abundant in RNA sequencing data. miR-21-5p is the #10 most abundant in human plasma, while the 3p strand is #269.[28]

As previously mentioned, Olivieri *et al.* found that miR-21 (along with miR-126) was decreased in the plasma of T2DM with and without complications *vs.* NGT controls (in 2015),[36] but Zampetaki *et al.* were the first to report a plasma biomarker for T2DM in 2010 using a robust sample of 822 individuals from Bolzano Province, Italy. 80 T2DM subjects were identified from this cohort (50% male/50% female; mean ± SD age 66.3 ± 8.9 years; mean ± SD BMI 28 ± 4.4 kg m^{-2}).[65] Initially, two pools of plasma from five overnight-fasted T2DM individuals and six pools of plasma from controls (identical age, sex, and risk factors) were profiled using quantitative (q)PCR. They found that miR-21, miR-126, and miR-223 were downregulated in the T2DM group. During this study (1995–2005), 19 individuals who were initially non-diabetic manifested T2DM.[65] Analysis showed that these individuals had significantly lower baseline levels of four miRNAs, including miR-223 and miR-126, suggesting an etiological role.[65]

Differential plasma and PBMC miR-21-5p levels in humans more closely correlated with BMI (obesity) than measures of insulin resistance.[66–68] miR-21 is also widely reported in atherosclerosis and hypertension, with far more publications on these topics than in metabolic disease. Mechanical stress to the vascular endothelium likely results in increased circulating miR-21 levels.[69] Mechanistically, miR-21 has been shown in various models to enhance Akt signaling. In murine adipocytes, miR-21 mimics reversed an *in vitro* model of T2DM *via* targeting (repressing) phosphatase and tensin homolog (PTEN; a negative regulator of Akt signaling). miR-21 mimics also increased glycogen synthase kinase-3 beta (GSK3β) signaling (Figure 2.6).[70] Knocking out mitogen-activated protein kinase (MAPK) phosphatase-1 (MKP-1), which dephosphorylates JNK and p38, in the skeletal muscle resulted in increased p-JNK and p-p38.[71] Increased p-JNK and p38 would be expected to increase skeletal muscle insulin resistance, but the knockout enhanced miR-21 levels, which increased insulin sensitivity *via* repression of PTEN. It is likely that miR-21 is a key regulator of insulin sensitivity in muscle and adipose tissue, but this importance may not be evident in the majority of plasma studies, which show conflicting directionality of expression and correlation to metabolic disease.

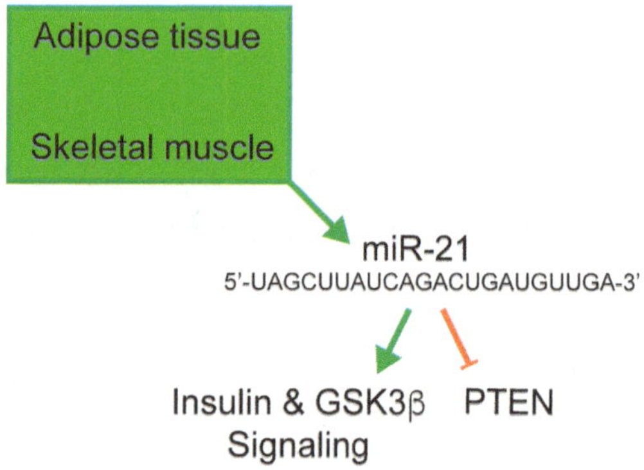

Figure 2.6 miR-21 is a positive regulator of insulin sensitivity and glucose metabolism in the adipose tissue and skeletal muscle. One mechanism for this is repression of PTEN (phosphatase and tensin homolog, a negative regulator of Akt/insulin signaling). GSK3β: glycogen synthase kinase-3 beta.

2.7 miR-223 is Important in T2DM, Although Reports Conflict on Direction of Differential Expression

The *mir-223* gene (HGNC : MIR21) is located on the Xq12 chromosome (66018870–66018979) and transcribes hsa-miR-223-5p and hsa-miR-223-3p. hsa-miR-223-5p MIMAT0004570 (5′-CGUGUAUUUGACAAGCUGAGUU-3′) is the less abundant of the two strands, whereas hsa-miR-223-3p MIMAT0000280 (5′-UGUCAGUUUGUCAAAUACCCCA-3′) is more abundant in RNA sequencing data based on miRbase.org data compilation. However, miR-223-3p is the #121 most abundant in human plasma, while the 5p strand is #105.[28]

miR-223 has been shown to affect the NF-κB signaling pathway in mouse and human macrophages. An extensive review of the NF-κB pathway is beyond the scope of this review; briefly, the NF-κB signaling pathway is categorized into classical (canonical) and alternative (non-canonical) pathways. The classical pathway involves NF-κB-p65 and an IκB kinase (IKK) complex consisting of IKKα, IKKβ and IKKγ, while the alternative pathway involves NF-κB-p52, NF-κB-inducing kinase (NIK) and IKKα homodimer. NF-κB, when activated and translocated to the nucleus, is a master transcriptional regulator of inflammatory gene expression. Li *et al.* showed that miR-223 overexpression, concomitantly with two other miRNAs (miR-15a and miR-16), resulted in decreased levels of IKKα protein in M1-activated human PBMCs.[72] mRNA degradation of IKKα resulted in reduced levels of NF-κB-p52. Therefore, researchers suggested that reduced levels of miR-223 may contribute to increased activation of the alternative NF-κB pathway.[72] miR-223 does not affect protein levels of IKKβ and IKKγ;[72] however, a study reported that the alteration of miR-223 levels inversely correlated with levels of activated phosphorylated NF-κB-p65 protein during M1-activation of mouse bone

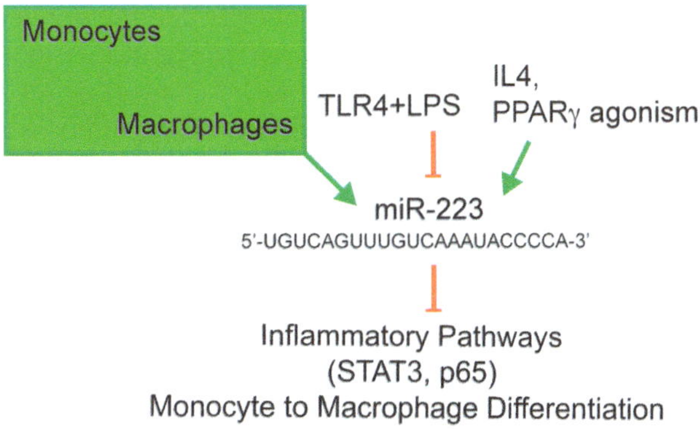

Figure 2.7 miR-223 is highly expressed in monocytes and monocyte-derived cells. Pro-inflammatory stimuli such as lipopolysaccharide (LPS) downregulate miR-223 expression and anti-inflammatory stimuli such as interleukin-4 (IL-4) upregulate miR-223 levels. miR-223 represses many inflammatory pathways as well as monocyte to macrophage differentiation. p65 (also known as RelA) is part of the nuclear factor kappaB (NFκB) complex. TLR4: toll-like receptor-4; PPAR-γ: peroxisome proliferator-activated receptor gamma.

marrow-derived macrophages (BMMs).[73] Thus, downregulation of miR-223 during inflammatory activation may contribute to the activation of both NF-κB pathways (Figure 2.7).

Obesity and metabolic disease are associated with chronic inflammation, mediated in part by increased infiltration and pro-inflammatory activation of adipose tissue macrophages. In murine models of obesity (*ob*/*ob* mice), the visceral adipose tissue has elevated expression of numerous inflammatory molecules, such as nucleotide-binding domain leucine-rich-containing family pyrin-domain-containing-3 (NLRP3), tumor necrosis factor alpha (TNF-α) and IL-1 β, along with hyperactivation of their respective signaling pathways.[4] miR-223 has been shown to be a critical regulator of macrophage polarization and plays a role in numerous pathways involved in inflammatory signaling. Generally, miR-223 levels decrease during pro-inflammatory activation and increase during anti-inflammatory activation.[74,75]

miR-223 levels in macrophages decrease in a time and dose-dependent manner with LPS treatment.[73] Multiple studies have reported that miR-223 levels decrease (five- to 10-fold) due to TLR4 ligation by LPS (M1, classical activation of macrophages) and/or IL-6, although the exact mechanism responsible for this transcriptional increase has yet to be determined.[73,75,76] Conversely, IL-4 treatment increases miR-223 levels and induces an alternative M2 activation *via* peroxisome proliferator-activated receptor gamma (PPARγ).[77] Treatment of mouse BMMs with PPARγ agonist (pioglitazone) and IL-4 concomitantly increased miR-223 approximately four-fold above IL-4 alone.[77] miR-223 represses important molecules involved in pro-inflammatory pathways, including NF-κB-p65, IL-1β, IL-6, STAT3, TNF-α, NLRP3, TLR4 and PBX/knotted 1 homeobox (Pknox1). Perhaps most relevant to macrophage function is the role of miR-223 in regulating TLR4 signaling. Elevated TLR4 signaling is a strong and possible etiologic link between obesity and metabolic disease[4,78,79] TLR4 activation is known to cause insulin resistance in diet-induced obesity (DIO) mice, *via* NF-κB-mediated inflammation.[78] Conversely, TLR4 knockout mice fed a high-fat diet displayed no evidence of impaired insulin responsiveness (or vascular inflammation) in DIO.[78] A reduction in miR-223 levels in visceral adipose tissue macrophages would upregulate TLR4 expression, increasing their sensitivity to an inflammatory stimulus or TLR4 ligands such as palmitic acid (elevated in obesity).[80,81]

Pescador *et al.* characterized the presence of miRNAs in serum from non-obese/NGT, obese/NGT, non-obese/T2DM, and obese/T2DM individuals with the intent of identifying reliable biomarkers of diabetes and obesity. Fasting serum samples from 69 individuals were pooled by group (one assay per group).[82] miR-223 was significantly lower in the three disease cohorts (obese/NGT, non-obese/T2DM, and obese/T2DM), compared with the healthy (non-obese/NGT) cohort.[82] Wen *et al.* examined the relationship between serum miR-223 and obesity in 121 subjects.[83] Baseline ($t = 0$, post-fast) serum miR-223 levels were found to be lower in overweight (1.65-fold) and obese (7.11-fold) individuals, compared to normal-weight age–sex matched controls using qPCR measurements.[83] A 3-month reduced-calorie diet/physical activity intervention program increased miR-223 levels.[83] A correlation analysis between miR-223 and obesity was performed by dividing the subjects into three groups based on miR-223 serum levels (low, middle, and high).[83] The highest odds ratio (OR) was observed between the low miR-223 group and obesity (OR 2.12, 95% CI 1.62–8.74), after adjusting for age, sex, and BMI.[83] The middle miR-223 group/obesity OR was 1.43 (95% CI 0.85–2.91), while the high miR-223 group/obesity OR was 1.00 (95% CI).[83] Milagro *et al.* studied PBMC-derived miRNAs in 10 obese Caucasian females enrolled in an energy-restricted weight-loss trial.[84] Baseline ($t = 0$) PBMC miRNA levels were measured in responders (>5% weight loss) and non-responders (<5% weight loss) by next-generation sequencing; miR-223 levels were 2.7-fold higher in the non-responder group ($t = 8$ weeks).[84] Note that the article shows that miR-223 was downregulated; this is an error and was confirmed *via* personal communication. However, the authors did not vet the use of PBMC miR-223 expression *vs.* standard biomarkers of metabolic disease (such as HbA1c, fasting glucose, and insulin).

The senior author of this chapter, Deiuliis *et al.*, found that miR-223 was elevated in the stromal vascular fraction of visceral adipose tissue macrophages from obese patients *vs.* normal BMI controls.[85] We believe this to be a compensatory upregulation.

2.8 miR-27a is Elevated in T2DM

The *mir-27a* gene (HGNC:MIR27A) is located on chromosome 19p13.12 (13836440–13836517) and transcribes hsa-miR-27a-5p and hsa-miR-27a-3p. hsa-miR-27a-5p MIMAT0004501 (5′- AGGGCUUAGCUGCUUGUGAGCA-3′) is the less abundant of the strands, whereas hsa-miR-27a-3p MIMAT0000084

(5′- UUCACAGUGGCUAAGUUCCGC-3′) is more abundant in RNA sequencing data based on miRbase.org data compilation. miR-27a-3p is the #87 most abundant in human plasma, while the 5p strand is #249.[28] *hsa-mir-27a* is part of a miRNA cluster including *mir-23a* and *mir-24-2*.

Whole-blood- and exosome-derived miR-27a was found to correlate highly with fasting glucose levels in humans and to be higher in patients differentially classified by the authors to T2DM or metabolic syndrome.[86] In serum from obese and normal-weight children in China, miR-27a was elevated in the obese group and positively correlated with BMI.[87] Animal models of obesity and insulin resistance show a similar elevation of circulating miR-27a.[87] Mechanistically, elevated miR-27a may be affecting adipose health by repressing adipogenesis, potentiating metabolic disease.[87–89] Herrera *et al.* reported in a rat model of T2DM that miR-27a was increased in the adipose tissue, and culture of mouse 3T3-L1 adipocyte cell line in high glucose increased miR-27a expression.[90] The authors reported similar findings for miR-29a, discussed in the next section.

2.9 miR-29a is Elevated in T2DM

The *mir-29a* gene (HGNC:MIR29A) is located on chromosome 7q32.3 (130876747–130876810) and transcribes hsa-miR-29a-5p and hsa-miR-29a-3p. hsa-miR-29a-5p MIMAT0004503 (5′- ACUGAUUUCUUUUGGUGUUCAG-3′) is the less abundant of the strands, whereas hsa-miR-29a-3p MIMAT0000086 (5′-UAGCACCAUCUGAAAUCGGUUA-3′) is more abundant in RNA sequencing data based on miRbase.org data compilation. miR-29a-3p is the #136 most abundant in human plasma, while the 5p strand is almost undetectable at #880.[28] *hsa-mir-29a* is part of a cluster including *hsa-mir-29b-1*. The miR-29 family member *hsa-mir-29c* is located on chromosome 1q32.2 in humans and within close proximity to *hsa-mir-29b-2*. Most studies examine miR-29 family expression, despite different chromosomal locations and transcriptional regulation.

Massart *et al.* found higher miR-29a and miR-29c levels in skeletal muscle biopsies from anti-diabetic treatment-naïve T2DM men ($n = 12$) compared to healthy age-matched male controls.[91] Chien *et al.* measured serum miR-29a in T2DM patients ($N = 50$) with diabetic nephropathy or retinopathy in Taiwan.[92] Patients with progression of nephropathy over the study had greater miR-29a levels than patients with disease that was stable.[92] Urinary miR-29a (but not miR-29b or miR-29c) was found to be positively correlated with urinary albuminuria and diabetic retinopathy.[93] However, miR-29b did correlate positively with carotid artery intimal thickness in the T2DM patients. The role of miR-29 in diabetic kidney disease has not been made clear by mechanistic studies.

It has been well described that T2DM dramatically increases risk of major cardiac events and mortality; miR-29 is a cardiometabolic disease-associated miRNA that is reported in human hypertension and atherosclerosis, in addition to T2DM.[94,95] Widlansky *et al.* found that miR-29a-3p

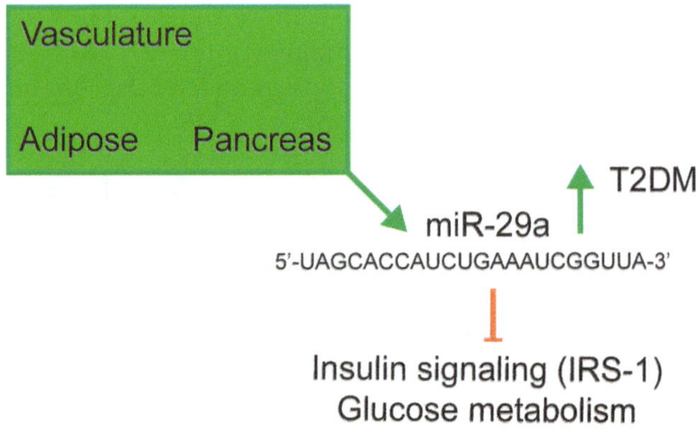

Figure 2.8 miR-29a, found to be upregulated in human diabetes, suppresses insulin signaling by multiple mechanisms including repression of IRS-1 (insulin receptor substrate 1). Tissues with high expression of miR-29a relative to other miRNAs include adipose tissue, pancreas, and the vasculature.

and miR-29b-3p were elevated in arterioles isolated from patients with T2DM.[96] Interestingly, when the same T2DM patient-derived vessels where cannulated and delivered mimics of miR-29, vessel dilation was improved as measured by pressure myography. miR-29 probably plays a role in regulating nitric oxide release by the endothelium, affecting vasodilation and vascular tone.

miR-29a/b-1 and miR-29b-2/c whole-body knockout mice provide mechanistic data as to the role of this family in metabolic disease. miR-29a is one of the most abundant microRNAs in the pancreas;[97] thus the pancreata from these knockout mice were well characterized; miR-29a/b-1 knockout reduced pancreatic islet glucose-stimulated insulin secretion. Lean body mass was higher in knockout mice, while total body weight and adipose mass (increasing serum leptin) was lower in knockout *vs.* wildtype mice. Knockout mice had enhanced insulin signaling and glucose uptake (Figure 2.8). Interestingly, knockout of both loci (miR-29a/b-1 and miR-29b-2c) resulted in 100% death rate before 100 days on a standard chow diet. Individual knockout mice had reduced survival compared to wildtype. This strong effect on mortality is not common for miRNA knockout mouse models. While the cause of the increased mortality of miR-29 knockout is unknown, it is hypothesized to be due to the complex effects on glucose handling in vital organs.

A role for miR-29 in the pancreas during murine diabetes has been reported by other groups. Hirata *et al.* reported that pancreatic miR-29 increased over the progression of T2DM in the recombinant congenic NONcNZO10/LtJ model that shows progressive pancreatic islet atrophy (without drug ablation) while developing mild obesity and dyslipidemia on a chow diet.[98] At 22 weeks, miR-29a-3p and miR-27a-3p in the pancreas were two-fold higher than in 7 week-old mice. This matches human studies that showed increased miR-29 expression with cardiometabolic disease; however, mechanisms within the islet were not pursued.

2.10 miR-34a is Elevated in Obesity and T2DM

The *mir-34a* gene (HGNC:MIR34A) is located on chromosome 1p36.22 (9151668–9151777) and transcribes hsa-miR-34a-5p and hsa-miR-34a-3p. hsa-miR-34a-5p MIMAT0000255 (5′- UGGCAGUGUCUUAGCUGGUUGU-3′) is the more abundant strand, whereas hsa-miR-34a-3p MIMAT0004557 (5′-CAAUCAGCAAGUAUACUGCCCU-3′) is less abundant in RNA sequencing data based on miRbase.org data compilation. miR-34a-5p and -3p were reported to be undetectable in human plasma by RNASeq.[28] miR-34b is often examined at the same time as miR-34a. *hsa-mir-34b* is located on chromosome 11q23.1 at 111512938–111513021, on the opposite strand.

miR-34a is another miRNA that is differentially regulated in both cardiovascular and metabolic diseases. miR-34a is consistently reported as being elevated in various human cohorts involving obesity and metabolic disease. Nunez Lopez *et al.* and Hijmans *et al.* found that miR-34a levels (positively) correlated best with BMI in normal weight and obese individuals with a wide range of metabolic disease, from pre-diabetes to diagnosed T2DM.[99,100] Exercise and weight loss by roux-en-Y gastric bypass (RYGB) surgery decreased plasma miR-34a *vs.* control groups.[101] Human and murine studies show a clear elevation of miR-34a in cardiometabolic disease.[102,103]

Mechanistic studies suggest that miR-34a targets nicotinamide phosphoribosyltransferase (NAMPT) and SIRT1 directly. NAMPT is required for biogenesis of NAD+, upon which SIRT1 is dependent. Thus, miR-34a may be downregulating SIRT1 activity *via* two mechanisms. miR-34a overexpression decreased metabolic genes: peroxisome proliferator-activated receptor-gamma coactivator (PGC-1α), sterol-responsive element-binding protein 1c (SREBP-1c), and FXR *via* NAMPT and SIRT1 effects on histone acetylation.[103] Anti-miR-34a dramatically reduced high fat diet induced hepatic lipid accumulation in BALB/c mice.[103] A high fat and high sugar diet in C57BL/6J mice also resulted in a dramatic increase in hepatic miR-34a, although no rescue experiment was shown.[103] Natarajan *et al.* did show that the transcription factor FOXO3 (forkhead box O3) is upstream of miR-34a, that palmitic acid treatment results in miR-34a expression, and that miR-34a partially mediates apoptosis in this model (Figure 2.9).[102] A whole-body knockout of miR-34a resulted in increased body weight and body fat. The authors of this study suggested that the knockout was not protective; however, the authors did not clearly address glucose clearance post-bolus or insulin-stimulated glucose disposal.[104] This is one of the few conflicting findings (anti-miR-34 rescued diabetic phenotype) and raises the question of whether a hepatic-specific miR-34a knockout would yield different results to whole-body knockout.

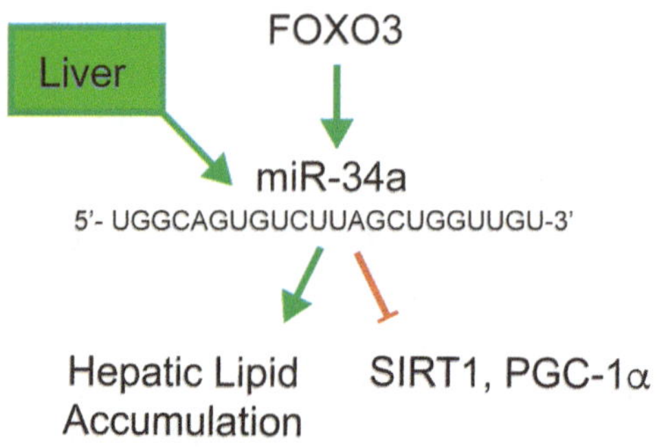

Figure 2.9 miR-34a is upregulated in human obesity and T2DM. The transcription factor FOXO3 (forkhead box O3), which is considered a master regulator of metabolism, mediates some of these effects by controlling miR-34a expression. miR-34a contributes to the modulation of insulin-responsive genes, metabolic substrate utilization, and lipid metabolism within the liver. More specifically, increases in miR-34a in obesity may drive hepatic lipid accumulation (fatty liver). SIRT1: sirtuin 1; PGC-1α: peroxisome proliferator-activated receptor-γ coactivator.

2.11 miR-375 is Elevated in T2DM

The *mir-375* gene (HGNC:MIR375) is located on chromosome 2q35 (219001645–219001708) and transcribes hsa-miR-375-5p and hsa-miR-375-3p. hsa-miR-375-5p MIMAT0037313 (5′- GCGACGAGCCCCUCGCACAAACC-3′) is the less abundant strand, whereas hsa-miR-34a-3p MIMAT0000728 (5′- UUUGUUCGUUCGGCUCGCGUGA-3′) is the more abundant in RNA sequencing data based on miRbase.org data compilation. miR-375 was the #40 most detectible in human plasma by RNASeq; strand specificity for miR-375 was not clear in this report.[28] miR-375 is highly expressed in pancreatic tissue and in organs with high epithelial content, such as lung, breast, and gastrointestinal tract.[105]

Circulating miR-375 has a robust body of literature reporting elevated levels in T2DM patients. Higuchi *et al.* found, in a population of 204 patients in Japan, that serum miR-375 was higher in T2DM *vs.* NGT individuals.[106] Similarly, in a Chinese population, plasma miR-375 was found to be elevated in T2DM patients compared to NGT controls.[107] Methylation status of miR-375 has been examined by multiple groups with conflicting findings.[107,108] Ghavami *et al.* led a prospective clinical trial enrolling T2DM patients randomized to the dietary fiber inulin or control and found that patients receiving inulin had a four-fold increase in miR-375, a significant increase in fasting insulin, and a decrease in fasting glucose.[109] This is one of the few clinical trials using a miRNA as a primary marker of efficacy. Macrosomia (high birthweight infants) had higher miR-375 and miR-33b levels derived from dried blood spots banked at birth, compared to normal or low birthweight infants.[110] An allelic variation that reduces miR-375 binding efficiency for the 3′-UTR of HSD17B13 was identified by genome-wide association studies (GWAS) to be associated with cardiometabolic disease in humans.[111] Hydroxysteroid 17-beta dehydrogenase 13 (HSD17B13) is a lipid droplet-associated protein that is upregulated in the liver of patients with NAFLD and has been shown to be etiologically relevant in disease progression.[112] These data suggest that upregulation of miR-375 in obesity is a beneficial compensatory change in obesity.

Obesity requires pancreatic islet expansion in order to produce more insulin, capable of driving larger amounts of glucose into increasing tissue mass. miR-375 became a prominent player in pancreatic biology in 2004 following a publication by Poy *et al.* which showed that adenoviral overexpression of miR-375 reduced insulin secretion by mouse pancreatic beta cells (MIN6 line) compared to control vector after *in vitro* stimulation, and that this effect did not result from impaired intracellular Ca^{2+} signaling (depolarization) post-stimuli.[113] Poy *et al.* followed this work by showing that miR-375 knockout mice are hyperglycemic due to impaired beta-cell proliferation.[114] Alpha-cell mass in knockout mice was increased compared to control, demonstrating an important dual role for miR-375 in pancreatic maintenance and alpha- to beta-cell ratio, and may play a role in pancreatic carcinogenesis. Interestingly, *ob*/*ob* mice exhibit elevated miR-375 and elevated beta-cell mass, and are overtly hyperinsulinemic at an early age.[114] However, the role of miR-375 in the pancreas is not fully clarified, and it should be noted that miR-184 and ARGONAUTE 2 (*Ago2*) also play a role in beta-cell proliferation in *ob*/*ob* mice.[115] Latreille *et al.* concluded that beta-cell-derived miR-375 did not represent the majority of circulating miR-375, and that miR-375 was not a good circulating biomarker for beta-cell function in mice;[116] it is unclear if this is also true in humans. Although murine data are conflicting, it is likely that miR-375 within the pancreas is required for glucose homeostasis *via* maintenance of endogenous insulin production, with higher miR-375 transcription being vital for pancreatic adaptation to increased body mass in overnutrition and hyperphagia.

2.12 miRNAs as Biomarkers in Metabolic Disease

A biomarker is any biological measurement with the ability to assess disease risk, diagnosis, severity, progression, and response to therapy.[117] To have diagnostic value in a clinical setting, a biomarker must be significantly changed in a diseased patient, when compared to a reference range established from normal control patients.[117] Establishing a reference range for a biomarker with Gaussian distribution requires at least 120 healthy subjects.[118,119] If distributions differ by subgroups (*i.e.*, age, sex, BMI), then separate reference ranges must be established using 120 healthy subjects for each subgroup.[118,119] A biomarker measurement falling outside the reference range (central 95% of measurements from healthy reference subjects) is considered abnormal or critically changed. In addition, a biomarker must have sensitivity, defined as the ability to diagnose a disease in patients who truly have the disease (true positives), and specificity, the ability to determine if a person is disease-free (true negative).[117] Successful implementation of plasma miRNAs as clinical biomarkers (miRmarkers) of metabolic disease will be requisite upon substantial evidence showing greater sensitivity and specificity than currently used diagnostic biomarkers such as HbA1c and overnight fasting blood glucose. In addition, important to clinical implementation of miRmarkers is ease of sample collection, small volume requirements, straightforward detection methods, and cost effectiveness.[120]

The study of miRNAs in biological fluids such as plasma, saliva, urine, bronchial lavage, and cerebrospinal fluid and their use as biomarkers of disease is rapidly expanding. qPCR-based methods utilizing <200 μL of plasma provide more than sufficient miRNA cDNA template for amplification.[121] In addition, advances in cDNA synthesis methods allow for a single cDNA reaction for all miRNA species, *vs.* miRNA-specific cDNA synthesis. Experiments by Weber *et al.* measured (using qPCR) the 20 most abundant miRNAs in six minimally invasive fluids (amniotic fluid, bronchial lavage, cerebrospinal fluid, peritoneal fluid, plasma, pleural fluid) and six non-invasive fluids [breast milk, colostrum (one donor), saliva, seminal fluid, tears, urine] from five healthy donors.[120] These biofluids were rich in miRNAs, with an average of 361 detectable miRNAs (range 204–458). More specifically, 600 different miRNAs were detected in at least one of these body fluids and 61 of the 706 human miRNAs assayed were detected in all biofluids tested.[120] Human plasma was found to contain the most miRNA species, followed by saliva. Interestingly, miR-223 was one of the top 20 most abundant miRNAs in tears, urine, amniotic fluid, bronchial lavage, cerebrospinal fluid, pleural fluid, colostrum, and peritoneal fluid.[120] miRmarkers in plasma/serum are of primary focus because obtaining blood is a standard procedure for many patients. In addition, large banks of plasma samples exist from various populations, which can be analyzed retrospectively. Circulating miRNAs are relatively simple to detect and amplify compared to other biomolecules due to their relative stability in specimens, even at room temperature (~48 h) and after multiple freeze–thaw cycles (due to their association with Ago2).[120,121] Circulating miRNAs are also protected from RNase degradation due to enrichment within membranous vesicles and association of mature miRNAs with Ago2.[121,122]

However, miRNA biomarkers have limitations; they are susceptible to pre-analytical variations (*i.e.*, sampling method/amplification bias, storage contamination, and hemolysis) and analytical variations (*i.e.*, co-purification of qPCR inhibitors).[121] These limitations have been examined by Exiqon (Vedbæk, Denmark; currently owned by Qiagen) in >1500 serum and plasma samples from a variety of sources, making plasma miRNA measures accurate when incorporating multilevel quality checks.[121] The miRNA assay is 95% valid when using only 10 copies (~4000 copies per 20 μL in a cDNA synthesis reaction) of a target miRNA in the PCR reaction, thereby achieving accuracy and adjusting for pre-analytical variations.[121] Importantly, normal miRNA reference ranges for 119 plasma miRNAs (most commonly present miRNAs in serum and plasma) have been determined from 381 high-quality samples, and were compared to a contaminated sample. However, precision of miRNA assay values was not addressed in the article.[121] Although miRNAs are challenging to measure accurately, their potential as diagnostic biomarkers needs further exploration.

Advanced analyses of circulating miRNA as biomarkers for the severity of metabolic disease or to predict risk of developing insulin resistance are largely absent in the current literature, with the exception of Zampetaki *et al.*,[65] Pescador *et al.*,[82] and Wen *et al.*.[83] Ray *et al.* state that "The studies of biomarkers, either as diagnostic or prognostic variables, are often presented with poor biostatistics and methodologic flaws that often preclude them from providing a reliable and reproducible scientific message".[117] To thoroughly conceptualize the utility of a miRNA biomarker, a study should perform statistical analyses appropriate for a clinical biomarker such as odds ratios, relative risks, risk differences, receiver operating characteristics, and likelihood ratios,[117] which are absent from the vast majority of circulating miRNA studies. Ray *et al.* further propose a Bayesian approach to determining the clinical usefulness of a biomarker, in addition to multiple other statistical analyses.[117] Bayes' theory warrants using the probability of likelihood of disease (pre-test probability) and predictive power of biomarker (positive likelihood ratio), to determine an adjusted probability of likelihood (post-test probability).[117] In addition, Ray *et al.* suggest that positive and negative predictive values provide clinical information, rather than just sensitivity and specificity.[117] A comprehensive review of the Bayesian approach is beyond the scope of this text, but interested readers are referred to the biomarker article by Ray *et al.*[117] In addition, techniques for miRNA quantification and sample handling before RNA isolation are heterogeneous, and these non-standardized methods probably contribute heavily to the variable results and findings.

Acknowledgements

This work was supported by the National Institutes of Health (NIH), National Institute of Diabetes and Digestive and Kidney Diseases (NIDDK, USA) with grant DK099475 to JAD. The content is solely the responsibility of the authors and does not necessarily represent the official views of the NIH, USA.

References

1. ASMBS T2D and Obesity, Twin epidemics Factsheet, https://asmbs.org/resources/weight-and-type-2-diabetes-after-bariatric-surgery-fact-sheet.
2. National Diabetes Statistics Report, 2017, https://www.niddk.nih.gov/health-information/health-statistics/diabetes-statistics.
3. D. L. Morris, K. E. Oatmen, T. Wang, J. L. DelProposto and C. N. Lumeng, *Obesity*, 2012, **20**, 1189–1199.

4. P. R. Nagareddy, M. Kraakman, S. L. Masters, R. A. Stirzaker, D. J. Gorman, R. W. Grant, D. Dragoljevic, E. S. Hong, A. Abdel-Latif, S. S. Smyth, S. H. Choi, J. Korner, K. E. Bornfeldt, E. A. Fisher, V. D. Dixit, A. R. Tall, I. J. Goldberg and A. J. Murphy, *Cell Metab.*, 2014, **19**, 821–835.
5. V. Elgazar-Carmon, A. Rudich, N. Hadad and R. Levy, *J. Lipid Res.*, 2008, **49**, 1894–1903.
6. H. Nascimento, S. Rocha, C. Rego, H. F. Mansilha, A. Quintanilha, A. Santos-Silva and L. Belo, *Open Biochem. J.*, 2010, **4**, 72–76.
7. Y. Herishanu, O. Rogowski, A. Polliack and R. Marilus, *Eur. J. Haematol.*, 2006, **76**, 516–520.
8. R. Feinstein, H. Kanety, M. Z. Papa, B. Lunenfeld and A. Karasik, *J. Biol. Chem.*, 1993, **268**, 26055–26058.
9. S. P. Weisberg, D. McCann, M. Desai, M. Rosenbaum, R. L. Leibel and A. W. Ferrante Jr, *J. Clin. Invest.*, 2003, **112**, 1796–1808.
10. S. E. Shoelson, J. Lee and A. B. Goldfine, *J. Clin. Invest.*, 2006, **116**, 1793–1801.
11. S. Cinti, G. Mitchell, G. Barbatelli, I. Murano, E. Ceresi, E. Faloia, S. Wang, M. Fortier, A. S. Greenberg and M. S. Obin, *J. Lipid Res.*, 2005, **46**, 2347–2355.
12. K. S. Ravichandran, *Immunity*, 2011, **35**, 445–455.
13. A. Kosteli, E. Sugaru, G. Haemmerle, J. F. Martin, J. Lei, R. Zechner and A. W. Ferrante, *J. Clin. Invest.*, 2010, **120**, 3466–3479.
14. J. I. Odegaard and A. Chawla, *Nat. Clin. Pract. Endocrinol. Metab.*, 2008, **4**, 619–626.
15. T. Jourdan, G. Godlewski, R. Cinar, A. Bertola, G. Szanda, J. Liu, J. Tam, T. Han, B. Mukhopadhyay, M. C. Skarulis, C. Ju, M. Aouadi, M. P. Czech and G. Kunos, *Nat. Med.*, 2013, **19**, 1132–1140.
16. W. Guo, S. Wong, W. Xie, T. Lei and Z. Luo, *Am. J. Physiol.: Endocrinol. Metab.*, 2007, **293**, E576–E586.
17. M. S. Han, D. Y. Jung, C. Morel, S. A. Lakhani, J. K. Kim, R. A. Flavell and R. J. Davis, *Science*, 2013, **339**, 218–222.
18. S. U. Amano, J. L. Cohen, P. Vangala, M. Tencerova, S. M. Nicoloro, J. C. Yawe, Y. Shen, M. P. Czech and M. Aouadi, *Cell Metab.*, 2014, **19**, 162–171.
19. S. P. Weisberg, D. Hunter, R. Huber, J. Lemieux, S. Slaymaker, K. Vaddi, I. Charo, R. L. Leibel, A. W. Ferrante and A. W. Ferrante Jr, *J. Clin. Invest.*, 2006, **116**, 115–124.
20. D. Patsouris, P.-P. Li, D. Thapar, J. Chapman, J. M. Olefsky and J. G. Neels, *Cell Metab.*, 2008, **8**, 301–309.
21. M. C. Arkan, A. L. Hevener, F. R. Greten, S. Maeda, Z.-W. Li, J. M. Long, A. Wynshaw-Boris, G. Poli, J. Olefsky and M. Karin, *Nat. Med.*, 2005, **11**, 191–198.
22. G. Solinas, C. Vilcu, J. G. Neels, G. K. Bandyopadhyay, J.-L. Luo, W. Naugler, S. Grivennikov, A. Wynshaw-Boris, M. Scadeng, J. M. Olefsky and M. Karin, *Cell Metab.*, 2007, **6**, 386–397.

23. D. Cai, M. Yuan, D. F. Frantz, P. A. Melendez, L. Hansen, J. Lee and S. E. Shoelson, *Nat. Med.*, 2005, **11**, 183–190.
24. D. M. Ferreira, A. L. Simao, C. M. Rodrigues and R. E. Castro, *FEBS J.*, 2014, **281**(11), 2503–2524.
25. Z. B. Deng, Y. Liu, C. Liu, X. Xiang, J. Wang, Z. Cheng, S. V. Shah, S. Zhang, L. Zhang, X. Zhuang, S. Michalek, W. E. Grizzle and H. G. Zhang, *Hepatology*, 2009, **50**, 1412–1420.
26. S. Xia, H. Sha, L. Yang, Y. Ji, S. Ostrand-Rosenberg and L. Qi, *J. Biol. Chem.*, 2011, **286**, 23591–23599.
27. J. A. Deiuliis, S. Oghumu, D. Duggineni, J. Zhong, J. Rutsky, A. Banerjee, B. Needleman, D. Mikami, V. Narula, J. Hazey, A. R. Satoskar and S. Rajagopalan, *Obesity*, 2014, **22**, 1264–1274.
28. M. Rubio, M. Bustamante, C. Hernandez-Ferrer, D. Fernandez-Orth, L. Pantano, Y. Sarria, M. Piqué-Borras, K. Vellve, S. Agramunt, R. Carreras, X. Estivill, J. R. Gonzalez and A. Mayor, *PLoS One*, 2018, **13**, e0193527.
29. P. Willeit, P. Skroblin, A. R. Moschen, X. Yin, D. Kaudewitz, A. Zampetaki, T. Barwari, M. Whitehead, C. M. Ramírez, L. Goedeke, N. Rotllan, E. Bonora, A. D. Hughes, P. Santer, C. Fernández-Hernando, H. Tilg, J. Willeit, S. Kiechl and M. Mayr, *Diabetes*, 2017, **66**, 347–357.
30. H. Sendi, I. Mead, M. Wan, M. Mehrab-Mohseni, K. Koch, A. Atala, H. L. Bonkovsky and C. E. Bishop, *PLoS One*, 2018, **13**, e0200847.
31. K. Jampoka, P. Muangpaisarn, K. Khongnomnan, S. Treeprasertsuk, P. Tangkijvanich and S. Payungporn, *MicroRNA*, 2018, **7**, 215–222.
32. S. Elhanati, R. Ben-Hamo, Y. Kanfi, A. Varvak, R. Glazz, B. Lerrer, S. Efroni and H. Y. Cohen, *Cell Rep.*, 2016, **14**, 234–242.
33. S. Wei, M. Zhang, Y. Yu, H. Xue, X. Lan, S. Liu, G. Hatch and L. Chen, *Eur. J. Pharmacol.*, 2016, **791**, 254–263.
34. J. He, K. Zhao, L. Zheng, Z. Xu, W. Gong, S. Chen, X. Shen, G. Huang, M. Gao, Y. Zeng, Y. Zhang and F. He, *Mol. Cancer*, 2015, **14**, 163.
35. H. C. de Boer, C. van Solingen, J. Prins, J. M. G. J. Duijs, M. V. Huisman, T. J. Rabelink and A. J. van Zonneveld, *Eur. Heart J.*, 2013, **34**, 3451–3457.
36. F. Olivieri, L. Spazzafumo, M. Bonafè, R. Recchioni, F. Prattichizzo, F. Marcheselli, L. Micolucci, E. Mensà, A. Giuliani, G. Santini, M. Gobbi, R. Lazzarini, M. Boemi, R. Testa, R. Antonicelli, A. D. Procopio and A. R. Bonfigli, *Oncotarget*, 2015, **6**, 35372–35382.
37. F. Barutta, G. Bruno, G. Matullo, N. Chaturvedi, S. Grimaldi, C. Schalkwijk, C. D. Stehouwer, J. H. Fuller and G. Gruden, *Acta Diabetol.*, 2017, **54**, 133–139.
38. X. Sun, N. Belkin and M. W. Feinberg, *Curr. Atheroscler. Rep.*, 2013, **15**, 372.
39. G. Catanzaro, Z. M. Besharat, M. Chiacchiarini, L. Abballe, C. Sabato, A. Vacca, P. Borgiani, F. Dotta, M. Tesauro, A. Po and E. Ferretti, *Int. J. Endocrinol.*, 2018, **2018**, 6872635.
40. K. S. Amr, H. Abdelmawgoud, Z. Y. Ali, S. Shehata and H. M. Raslan, *Br. J. Biomed. Sci.*, 2018, **75**, 82–87.

41. F. Jansen, X. Yang, S. Proebsting, M. Hoelscher, D. Przybilla, K. Baumann, T. Schmitz, A. Dolf, E. Endl, B. S. Franklin, J.-M. Sinning, M. Vasa-Nicotera, G. Nickenig and N. Werner, *J. Am. Heart Assoc.*, 2014, **3**, e001249.
42. F. Jansen, H. Wang, D. Przybilla, B. S. Franklin, A. Dolf, P. Pfeifer, T. Schmitz, A. Flender, E. Endl, G. Nickenig and N. Werner, *Cardiovasc. Diabetol.*, 2016, **15**, 49.
43. A. Giannella, C. M. Radu, L. Franco, E. Campello, P. Simioni, A. Avogaro, S. V. de Kreutzenberg and G. Ceolotto, *Cardiovasc. Diabetol.*, 2017, **16**, 118.
44. F. Kuhnert, M. R. Mancuso, J. Hampton, K. Stankunas, T. Asano, C.-Z. Chen and C. J. Kuo, *Development*, 2008, **135**, 3989–3993.
45. S. Wang, A. B. Aurora, B. A. Johnson, X. Qi, J. McAnally, J. A. Hill, J. A. Richardson, R. Bassel-Duby and E. N. Olson, *Dev. Cell*, 2008, **15**, 261–271.
46. J. Agudo, A. Ruzo, N. Tung, H. Salmon, M. Leboeuf, D. Hashimoto, C. Becker, L.-A. Garrett-Sinha, A. Baccarini, M. Merad and B. D. Brown, *Nat. Immunol.*, 2014, **15**, 54–62.
47. Y. Li, Q. Zhou, C. Pei, B. Liu, M. Li, L. Fang, Y. Sun, Y. Li and S. Meng, *J. Vasc. Res.*, 2016, **53**, 94–104.
48. S. Meng, J.-T. Cao, B. Zhang, Q. Zhou, C.-X. Shen and C.-Q. Wang, *J. Mol. Cell. Cardiol.*, 2012, **53**, 64–72.
49. B. Alipoor, R. Meshkani, H. Ghaedi, Z. Sharifi, G. Panahi and T. Golmohammadi, *Clin. Lab.*, 2016, **62**, 1553–1561.
50. G. Kaidonis, M. C. Gillies, S. Abhary, E. Liu, R. W. Essex, J. H. Chang, B. Pal, S. Sivaprasad, M. Pefkianaki, M. Daniell, S. Lake, N. Petrovsky, A. W. Hewitt, A. Jenkins, E. L. Lamoureux, J. M. Gleadle, J. E. Craig and K. P. Burdon, *Acta Diabetol.*, 2016, **53**, 643–650.
51. J. Shen, M. Zhang, M. Sun, K. Tang and B. Zhou, *Thromb. Res.*, 2015, **136**, 1149–1155.
52. Y. Li, Y. Zhang, X. Li, L. Shi, W. Tao, L. Shi, M. Yang, X. Wang, Y. Yang and Y. Yao, *Int. J. Med. Sci.*, 2015, **12**, 875–880.
53. H. W. Lee, S. Q. Khan, S. Khaliqdina, M. M. Altintas, F. Grahammer, J. L. Zhao, K. H. Koh, N. J. Tardi, M. H. Faridi, T. Geraghty, D. J. Cimbaluk, K. Susztak, L. F. Moita, D. Baltimore, P.-L. Tharaux, T. B. Huber, M. Kretzler, M. Bitzer, J. Reiser and V. Gupta, *J. Biol. Chem.*, 2017, **292**, 732–747.
54. L. Baldeón Rojas, K. Weigelt, H. de Wit, B. Ozcan, A. van Oudenaren, F. Sempértegui, E. Sijbrands, L. Grosse, A.-J. van Zonneveld, H. A. Drexhage and P. J. M. Leenen, *Diabetol. Metab. Syndr.*, 2016, **8**, 6.
55. N. E. Corral-Fernández, M. Salgado-Bustamante, M. E. Martínez-Leija, N. Cortez-Espinosa, M. H. García-Hernández, E. Reynaga-Hernández, R. Quezada-Calvillo and D. P. Portales-Pérez, *Exp. Clin. Endocrinol. Diabetes*, 2013, **121**, 347–353.
56. R. Lenin, A. Sankaramoorthy, V. Mohan and M. Balasubramanyam, *J. Leukocyte Biol.*, 2015, **98**, 615–622.
57. P. S. Eis, W. Tam, L. Sun, A. Chadburn, Z. Li, M. F. Gomez, E. Lund and J. E. Dahlberg, *Proc. Natl. Acad. Sci. U. S. A.*, 2005, **102**, 3627–3632.

58. R. Mashima, *Immunology*, 2015, **145**, 323–333.
59. M. Ceppi, P. M. Pereira, I. Dunand-Sauthier, E. Barras, W. Reith, M. A. Santos and P. Pierre, *Proc. Natl. Acad. Sci. U. S. A.*, 2009, **106**, 2735–2740.
60. H. Zhou, X. Huang, H. Cui, X. Luo, Y. Tang, S. Chen, L. Wu and N. Shen, *Blood*, 2010, **116**, 5885–5894.
61. R. M. O'Connell, D. Kahn, W. S. J. Gibson, J. L. Round, R. L. Scholz, A. A. Chaudhuri, M. E. Kahn, D. S. Rao and D. Baltimore, *Immunity*, 2010, **33**, 607–619.
62. A. Androulidaki, D. Iliopoulos, A. Arranz, C. Doxaki, S. Schworer, V. Zacharioudaki, A. N. Margioris, P. N. Tsichlis and C. Tsatsanis, *Immunity*, 2009, **31**, 220–231.
63. L. Wang, N. Zhang, Z. Wang, D.-M. Ai, Z.-Y. Cao and H.-P. Pan, *Cell. Physiol. Biochem.*, 2016, **39**, 2239–2248.
64. E. Karkeni, J. Astier, F. Tourniaire, M. El Abed, B. Romier, E. Gouranton, L. Wan, P. Borel, J. Salles, S. Walrand, J. Ye and J.-F. Landrier, *J. Clin. Endocrinol. Metab.*, 2016, **101**, 1615–1626.
65. A. Zampetaki, S. Kiechl, I. Drozdov, P. Willeit, U. Mayr, M. Prokopi, A. Mayr, S. Weger, F. Oberhollenzer, E. Bonora, A. Shah, J. Willeit and M. Mayr, *Circ. Res.*, 2010, **107**, 810–817.
66. S. Ghorbani, R. Mahdavi, B. Alipoor, G. Panahi, E. Nasli Esfahani, F. Razi, M. Taghikhani and R. Meshkani, *Arch. Physiol. Biochem.*, 2017, 1–6.
67. H. Mazloom, S. Alizadeh, E. N. Esfahani, F. Razi and R. Meshkani, *Mol. Cell. Biochem.*, 2016, **419**, 11–17.
68. A. Villard, L. Marchand, C. Thivolet and S. Rome, *J. Mol. Biomarkers Diagn.*, 2015, **6**(6), 251.
69. M. Weber, M. B. Baker, J. P. Moore and C. D. Searles, *Biochem. Biophys. Res. Commun.*, 2010, **393**, 643–648.
70. H. Y. Ling, B. Hu, X. B. Hu, J. Zhong, S. D. Feng, L. Qin, G. Liu, G. B. Wen and D. F. Liao, *Exp. Clin. Endocrinol. Diabetes*, 2012, **120**, 553–559.
71. A. Lawan, K. Min, L. Zhang, A. Canfran-Duque, M. J. Jurczak, J. P. G. Camporez, Y. Nie, T. P. Gavin, G. I. Shulman, C. Fernandez-Hernando and A. M. Bennett, *Diabetes*, 2018, **67**, 624–635.
72. T. Li, M. J. Morgan, S. Choksi, Y. Zhang, Y.-S. S. Kim and Z. G. Liu, *Nat. Immunol.*, 2010, **11**, 799–805.
73. J. Wang, X. Bai, Q. Song, F. Fan, Z. Hu, G. Cheng and Y. Zhang, *Int. J. Mol. Sci.*, 2015, **16**, 24965–24982.
74. M. Yang, J. Chen, F. Su, B. Yu, L. Lin, Y. Liu, J. D. Huang and E. Song, *Mol. Cancer*, 2011, **10**, 117.
75. Q. Chen, H. Wang, Y. Liu, Y. Song, L. Lai, Q. Han, X. Cao and Q. Wang, *PLoS One*, 2012, **7**, e42971.
76. G. Zhuang, C. Meng, X. Guo, P. S. Cheruku, L. Shi, H. Xu, H. Li, G. Wang, A. R. Evans, S. Safe, C. Wu and B. Zhou, *Circulation*, 2012, **125**, 2892–2903.
77. W. Ying, A. Tseng, R. C. A. Chang, A. Morin, T. Brehm, K. Triff, V. Nair, G. Zhuang, H. Song, S. Kanameni, H. Wang, M. C. Golding, F. W. Bazer, R. S. Chapkin, S. Safe and B. Zhou, *J. Clin. Invest.*, 2015, **125**, 4149–4159.

78. F. Kim, M. Pham, I. Luttrell, D. D. Bannerman, J. Tupper, J. Thaler, T. R. Hawn, E. W. Raines and M. W. Schwartz, *Circ. Res.*, 2007, **100**, 1589–1596.
79. D. J. Samuvel, J. Jin, K. P. Sundararaj, Y. Li, X. Zhang, M. F. Lopes-Virella and Y. Huang, *Endocrinology*, 2011, **152**, 4662–4671.
80. T. Suganami, K. Tanimoto-Koyama, J. Nishida, M. Itoh, X. Yuan, S. Mizuarai, H. Kotani, S. Yamaoka, K. Miyake, S. Aoe, Y. Kamei and Y. Ogawa, *Arterioscler., Thromb., Vasc. Biol.*, 2007, **27**, 84–91.
81. T. Suganami, X. Yuan, Y. Shimoda, K. Uchio-Yamada, N. Nakagawa, I. Shirakawa, T. Usami, T. Tsukahara, K. Nakayama, Y. Miyamoto, K. Yasuda, J. Matsuda, Y. Kamei, S. Kitajima and Y. Ogawa, *Circ. Res.*, 2009, **105**, 25–32.
82. N. Pescador, M. Perez-Barba, J. M. Ibarra, A. Corbaton, M. T. Martinez-Larrad, M. Serrano-Rios, M. Pérez-Barba, J. M. Ibarra, A. Corbatón, M. T. Martínez-Larrad, M. Serrano-Ríos, M. Perez-Barba, J. M. Ibarra, A. Corbaton, M. T. Martinez-Larrad and M. Serrano-Rios, *PLoS One*, 2013, **8**, e77251.
83. D. Wen, P. Qiao and L. Wang, *Obes. Res. Clin. Pract.*, 2015, 1–7.
84. F. I. Milagro, J. Miranda, M. P. Portillo, A. Fernandez-Quintela, J. Campión and J. A. Martínez, *PLoS One*, 2013, **8**, e54319.
85. J. A. Deiuliis, R. Syed, D. Duggineni, J. Rutsky, P. Rengasamy, J. Zhang, K. Huang, B. Needleman, D. Mikami, K. Perry, J. Hazey and S. Rajagopalan, *PLoS One*, 2016, **11**, e0165962.
86. D. S. Karolina, S. Tavintharan, A. Armugam, S. Sepramaniam, S. L. T. Pek, M. T. K. Wong, S. C. Lim, C. F. Sum and K. Jeyaseelan, *J. Clin. Endocrinol. Metab.*, 2012, **97**, E2271–E2276.
87. Y. Yu, H. Du, S. Wei, L. Feng, J. Li, F. Yao, M. Zhang, G. M. Hatch and L. Chen, *Theranostics*, 2018, **8**, 2171–2188.
88. T. Kang, W. Lu, W. Xu, L. Anderson, M. Bacanamwo, W. Thompson, Y. E. Chen and D. Liu, *J. Biol. Chem.*, 2013, **288**, 34394–34402.
89. S. Y. Kim, A. Y. Kim, H. W. Lee, Y. H. Son, G. Y. Lee, J.-W. Lee, Y. S. Lee and J. B. Kim, *Biochem. Biophys. Res. Commun.*, 2010, **392**, 323–328.
90. B. M. Herrera, H. E. Lockstone, J. M. Taylor, M. Ria, A. Barrett, S. Collins, P. Kaisaki, K. Argoud, C. Fernandez, M. E. Travers, J. P. Grew, J. C. Randall, A. L. Gloyn, D. Gauguier, M. I. McCarthy and C. M. Lindgren, *Diabetologia*, 2010, **53**, 1099–1109.
91. J. Massart, R. J. O. Sjögren, L. S. Lundell, J. M. Mudry, N. Franck, D. J. O'Gorman, B. Egan, J. R. Zierath and A. Krook, *Diabetes*, 2017, **66**, 1807–1818.
92. H.-Y. Chien, C.-Y. Chen, Y.-H. Chiu, Y.-C. Lin and W.-C. Li, *Int. J. Med. Sci.*, 2016, **13**, 457–465.
93. H. Peng, M. Zhong, W. Zhao, C. Wang, J. Zhang, X. Liu, Y. Li, S. D. Paudel, Q. Wang and T. Lou, *PLoS One*, 2013, **8**, e82607.
94. Y.-Q. Huang, A.-P. Cai, J.-Y. Chen, C. Huang, J. Li and Y.-Q. Feng, *Cell. Physiol. Biochem.*, 2016, **40**, 1521–1528.
95. C.-Z. Liu, Q. Zhong and Y.-Q. Huang, *Tohoku J. Exp. Med.*, 2017, **241**, 183–188.

96. M. E. Widlansky, D. M. Jensen, J. Wang, Y. Liu, A. M. Geurts, A. J. Kriegel, P. Liu, R. Ying, G. Zhang, M. Casati, C. Chu, M. Malik, A. Branum, M. J. Tanner, S. Tyagi, K. Usa and M. Liang, *EMBO Mol. Med.*, 2018, **10**, e8046.
97. J. Dooley, J. E. Garcia-Perez, J. Sreenivasan, S. M. Schlenner, R. Vangoitsenhoven, A. S. Papadopoulou, L. Tian, S. Schonefeldt, L. Serneels, C. Deroose, K. A. Staats, B. Van der Schueren, B. De Strooper, O. P. McGuinness, C. Mathieu and A. Liston, *Diabetes*, 2016, **65**, 53–61.
98. T. Hirata, T. Yoshitomi, M. Inoue, Y. Iigo, K. Matsumoto, K. Kubota and A. Shinagawa, *Gene*, 2017, **629**, 52–58.
99. Y. O. Nunez Lopez, G. Garufi and A. A. Seyhan, *Mol. BioSyst.*, 2016, **13**, 106–121.
100. J. G. Hijmans, K. J. Diehl, T. D. Bammert, P. J. Kavlich, G. M. Lincenberg, J. J. Greiner, B. L. Stauffer and C. A. DeSouza, *MicroRNA*, 2018, **7**, 148–154.
101. Y. O. Nunez Lopez, P. M. Coen, B. H. Goodpaster and A. A. Seyhan, *Int. J. Obes.*, 2017, **41**, 1121–1130.
102. S. K. Natarajan, B. A. Stringham, A. M. Mohr, C. J. Wehrkamp, S. Lu, M. A. Phillippi, D. Harrison-Findik and J. L. Mott, *J. Lipid Res.*, 2017, **58**, 866–875.
103. S.-E. Choi, T. Fu, S. Seok, D.-H. Kim, E. Yu, K.-W. Lee, Y. Kang, X. Li, B. Kemper and J. K. Kemper, *Aging Cell*, 2013, **12**, 1062–1072.
104. C. A. Lavery, M. Kurowska-Stolarska, W. M. Holmes, I. Donnelly, M. Caslake, A. Collier, A. H. Baker and A. M. Miller, *Obesity*, 2016, **24**, 1741–1751.
105. Y. Liang, D. Ridzon, L. Wong and C. Chen, *BMC Genomics*, 2007, **8**, 166.
106. C. Higuchi, A. Nakatsuka, J. Eguchi, S. Teshigawara, M. Kanzaki, A. Katayama, S. Yamaguchi, N. Takahashi, K. Murakami, D. Ogawa, S. Sasaki, H. Makino and J. Wada, *Metabolism*, 2015, **64**, 489–497.
107. K. Sun, X. Chang, L. Yin, J. Li, T. Zhou, C. Zhang and X. Chen, *Mol. Med. Rep.*, 2014, **9**, 967–972.
108. X. Wang, X. Chang, J. Li, L. Yin and K. Sun, *Exp. Ther. Med.*, 2014, **8**, 775–780.
109. A. Ghavami, N. Roshanravan, S. Alipour, M. Barati, B. Mansoori, F. Ghalichi, E. Nattagh-Eshtivan and A. Ostadrahimi, *Adv. Pharm. Bull.*, 2018, **8**, 39–47.
110. P. Rodil-Garcia, E. D. C. Arellanes-Licea, A. Montoya-Contreras and L. A. Salazar-Olivo, *Int. J. Mol. Sci.*, 2017, **18**, 2552.
111. M. Ghanbari, O. H. Franco, H. W. J. de Looper, A. Hofman, S. J. Erkeland and A. Dehghan, *Circ.: Cardiovasc. Genet.*, 2015, **8**, 473–486.
112. W. Su, Y. Wang, X. Jia, W. Wu, L. Li, X. Tian, S. Li, C. Wang, H. Xu, J. Cao, Q. Han, S. Xu, Y. Chen, Y. Zhong, X. Zhang, P. Liu, J.-Å. Gustafsson and Y. Guan, *Proc. Natl. Acad. Sci. U. S. A.*, 2014, **111**, 11437–11442.
113. M. N. Poy, L. Eliasson, J. Krutzfeldt, S. Kuwajima, X. Ma, P. E. Macdonald, S. Pfeffer, T. Tuschl, N. Rajewsky, P. Rorsman and M. Stoffel, *Nature*, 2004, **432**, 226–230.

114. M. N. Poy, J. Hausser, M. Trajkovski, M. Braun, S. Collins, P. Rorsman, M. Zavolan and M. Stoffel, *Proc. Natl. Acad. Sci. U. S. A.*, 2009, **106**, 5813–5818.
115. S. G. Tattikota, T. Rathjen, S. J. McAnulty, H.-H. Wessels, I. Akerman, M. van de Bunt, J. Hausser, J. L. S. Esguerra, A. Musahl, A. K. Pandey, X. You, W. Chen, P. L. Herrera, P. R. Johnson, D. O'Carroll, L. Eliasson, M. Zavolan, A. L. Gloyn, J. Ferrer, R. Shalom-Feuerstein, D. Aberdam and M. N. Poy, *Cell Metab.*, 2014, **19**, 122–134.
116. M. Latreille, K. Herrmanns, N. Renwick, T. Tuschl, M. T. Malecki, M. I. McCarthy, K. R. Owen, T. Rülicke and M. Stoffel, *J. Mol. Med.*, 2015, **93**, 1159–1169.
117. P. Ray, Y. Le Manach, B. Riou and T. T. Houle, *Anesthesiology*, 2010, **112**, 1023–1040.
118. A. Katayev, C. Balciza and D. W. Seccombe, *Am. J. Clin. Pathol.*, 2010, **133**, 180–186.
119. G. Jones and A. Barker, *Clin. Biochem. Rev.*, 2008, **29**(Suppl 1), S93–S97.
120. J. A. Weber, D. H. Baxter, S. Zhang, D. Y. Huang, K. H. Huang, M. J. Lee, D. J. Galas and K. Wang, *Clin. Chem.*, 2010, **56**, 1733–1741.
121. T. Blondal, S. Jensby Nielsen, A. Baker, D. Andreasen, P. Mouritzen, M. Wrang Teilum and I. K. Dahlsveen, *Methods*, 2013, **59**, S1–S6.
122. J. D. Arroyo, J. R. Chevillet, E. M. Kroh, I. K. Ruf, C. C. Pritchard, D. F. Gibson, P. S. Mitchell, C. F. Bennett, E. L. Pogosova-Agadjanyan, D. L. Stirewalt, J. F. Tait and M. Tewari, *Proc. Natl. Acad. Sci. U. S. A.*, 2011, **108**, 5003–5008.
123. J. Latorre, J. M. Moreno-Navarrete, J. M. Mercader, M. Sabater, Ò. Rovira, J. Gironès, W. Ricart, J. M. Fernández-Real and F. J. Ortega, *Int. J. Obes.*, 2017, **41**, 620–630.
124. A. Braza-Boïls, J. Marí-Alexandre, P. Molina, M. A. Arnau, M. Barceló-Molina, D. Domingo, J. Girbes, J. Giner, L. Martínez-Dolz and E. Zorio, *Liver Int.*, 2016, **36**, 1221–1229.
125. S. Rawal, P. E. Munasinghe, A. Shindikar, J. Paulin, V. Cameron, P. Manning, M. J. A. Williams, G. T. Jones, R. Bunton, I. Galvin and R. Katare, *Cardiovasc. Res.*, 2017, **113**, 90–101.
126. Y. O. Nunez Lopez, G. Garufi, M. Pasarica and A. A. Seyhan, *Int. J. Endocrinol.*, 2018, **2018**, 7351902.
127. H. Lu, R. J. Buchan and S. A. Cook, *Cardiovasc. Res.*, 2010, **86**, 410–420.
128. H. Ghaedi, M. Tabasinezhad, B. Alipoor, F. Shokri, A. Movafagh, R. Mirfakhraie, M. D. Omrani and A. Masotti, *J. Endocrinol. Invest.*, 2016, **39**, 1187–1193.
129. G. Sebastiani, A. Po, E. Miele, G. Ventriglia, E. Ceccarelli, M. Bugliani, L. Marselli, P. Marchetti, A. Gulino, E. Ferretti and F. Dotta, *Acta Diabetol.*, 2015, **52**, 523–530.
130. E. Flowers, M. Gadgil, B. E. Aouizerat and A. M. Kanaya, *Biomarker Res.*, 2015, **3**, 22.
131. F. J. Ortega, J. M. Mercader, V. Catalán, J. M. Moreno-Navarrete, N. Pueyo, M. Sabater, J. Gómez-Ambrosi, R. Anglada, J. A. Fernández-Formoso, W.

Ricart, G. Frühbeck and J. M. Fernández-Real, *Clin. Chem.*, 2013, **59**, 781–792.
132. E. Raitoharju, I. Seppälä, L.-P. Lyytikäinen, J. Viikari, M. Ala-Korpela, P. Soininen, A. J. Kangas, M. Waldenberger, N. Klopp, T. Illig, J. Leiviskä, B.-M. Loo, N. Oksala, M. Kähönen, N. Hutri-Kähönen, R. Laaksonen, O. Raitakari and T. Lehtimäki, *Sci. Rep.*, 2016, **6**, 38262.
133. R. Shah, V. Murthy, M. Pacold, K. Danielson, K. Tanriverdi, M. G. Larson, K. Hanspers, A. Pico, E. Mick, J. Reis, S. de Ferranti, E. Freinkman, D. Levy, U. Hoffmann, S. Osganian, S. Das and J. E. Freedman, *Diabetes Care*, 2017, **40**, 546–553.
134. R. Wang, J. Hong, Y. Cao, J. Shi, W. Gu, G. Ning, Y. Zhang and W. Wang, *Eur. J. Endocrinol.*, 2015, **172**, 291–300.
135. H. Yamada, K. Suzuki, N. Ichino, Y. Ando, A. Sawada, K. Osakabe, K. Sugimoto, K. Ohashi, R. Teradaira, T. Inoue, N. Hamajima and S. Hashimoto, *Clin. Chim. Acta*, 2013, **424**, 99–103.
136. G. Carreras-Badosa, A. Bonmatí, F.-J. Ortega, J.-M. Mercader, M. Guindo-Martínez, D. Torrents, A. Prats-Puig, J.-M. Martinez-Calcerrada, E. Platero-Gutierrez, F. De Zegher, L. Ibáñez, J.-M. Fernandez-Real, A. Lopez-Bermejo and J. Bassols, *J. Clin. Endocrinol. Metab.*, 2015, **100**, E1446–E1456.
137. G. Al-Kafaji, G. Al-Mahroos, H. A. Al-Muhtaresh, C. Skrypnyk, M. A. Sabry and A. R. Ramadan, *Exp. Ther. Med.*, 2016, **12**, 815–822.
138. P. Mocharla, S. Briand, G. Giannotti, C. Dörries, P. Jakob, F. Paneni, T. Lüscher and U. Landmesser, *Blood*, 2013, **121**, 226–236.
139. G. S. Morais Junior, V. C. Souza, W. Machado-Silva, A. D. Henriques, A. Melo Alves, D. Barbosa Morais, O. T. Nóbrega, C. J. Brito and R. J. Dos Santos Silva, *Clin. Interventions Aging*, 2017, **12**, 1443–1450.
140. F. J. Ortega, J. M. Mercader, J. M. Moreno-Navarrete, O. Rovira, E. Guerra, E. Esteve, G. Xifra, C. Martínez, W. Ricart, J. Rieusset, S. Rome, M. Karczewska-Kupczewska, M. Straczkowski and J. M. Fernández-Real, *Diabetes Care*, 2014, **37**, 1375–1383.
141. A. A. Seyhan, Y. O. Nunez Lopez, H. Xie, F. Yi, C. Mathews, M. Pasarica and R. E. Pratley, *Sci. Rep.*, 2016, **6**, 31479.
142. T. Zhang, L. Li, Q. Shang, C. Lv, C. Wang and B. Su, *Biochem. Biophys. Res. Commun.*, 2015, **463**, 60–63.
143. H. Zhu and S. W. Leung, *Diabetologia*, 2015, **58**, 900–911.
144. M. Balasubramanyam, S. Aravind, K. Gokulakrishnan, P. Prabu, C. Sathishkumar, H. Ranjani and V. Mohan, *Mol. Cell. Biochem.*, 2011, **351**, 197–205.
145. R. L. Baldeon, K. Weigelt, H. de Wit, B. Ozcan, A. van Oudenaren, F. Sempertegui, E. Sijbrands, L. Grosse, W. Freire, H. A. Drexhage and P. J. M. Leenen, *PLoS One*, 2014, **9**, e115209.
146. G. Wang, Y. Gu, N. Xu, M. Zhang and T. Yang, *Biochem. Biophys. Res. Commun.*, 2018, **498**, 382–387.
147. P. P. Becker, M. Rau, J. Schmitt, C. Malsch, C. Hammer, H. Bantel, B. Müllhaupt and A. Geier, *PLoS One*, 2015, **10**, e0142661.

148. J. G. Thomé, M. R. Mendoza, A. V. Cheuiche, V. L. La Porta, D. Silvello, K. G. Dos Santos, M. E. Andrades, N. Clausell, L. E. Rohde and A. Biolo, *Gene*, 2015, **574**, 1–10.
149. J. Tomé-Carneiro, M. Larrosa, M. J. Yáñez-Gascón, A. Dávalos, J. Gil-Zamorano, M. Gonzálvez, F. J. García-Almagro, J. A. Ruiz Ros, F. A. Tomás-Barberán, J. C. Espín and M.-T. García-Conesa, *Pharmacol. Res.*, 2013, **72**, 69–82.
150. P. L. Wander, E. J. Boyko, K. Hevner, V. J. Parikh, M. G. Tadesse, T. K. Sorensen, M. A. Williams and D. A. Enquobahrie, *Diabetes Res. Clin. Pract.*, 2017, **132**, 1–9.
151. E. Flowers, B. E. Aouizerat, F. Abbasi, C. Lamendola, K. M. Grove, Y. Fukuoka and G. M. Reaven, *Metabolism*, 2015, **64**, 1051–1059.
152. A. Elgheznawy, L. Shi, J. Hu, I. Wittig, H. Laban, J. Pircher, A. Mann, P. Provost, V. Randriamboavonjy and I. Fleming, *Circ. Res.*, 2015, **117**, 157–165.
153. Y. Long, Q. Zhan, M. Yuan, X. Duan, J. Zhou, J. Lu, Z. Li, F. Yu, X. Zhou, Q. Yang and J. Xia, *Cardiovasc. Toxicol.*, 2017, **17**, 42–48.
154. F. Tabet, L. F. Cuesta Torres, K. L. Ong, S. Shrestha, S. A. Choteau, P. J. Barter, P. Clifton and K.-A. Rye, *PLoS One*, 2016, **11**, e0151061.
155. M. Taghizadeh, S. Ahmadizad and M. Naderi, *Clin. Hemorheol. Microcirc.*, 2018, **68**, 391–399.
156. D. Wen, P. Qiao and L. Wang, *Obes. Res. Clin. Pract.*, 2015, **9**, 398–404.
157. S. Yang, J. Zhao, Y. Chen and M. Lei, *Cardiovasc. Toxicol.*, 2016, **16**, 213–222.
158. I. D. Kilic, Y. Dodurga, B. Uludag, Y. I. Alihanoglu, B. S. Yildiz, Y. Enli, M. Secme and H. E. Bostancı, *Gene*, 2015, **560**, 140–142.
159. E. Munetsuna, H. Yamada, Y. Ando, M. Yamazaki, Y. Tsuboi, M. Kondo, G. Mizuno, H. Ishikawa, K. Sugimoto, K. Osakabe, N. Ichino, K. Ohashi, N. Hamajima and K. Suzuki, *Ann. Clin. Biochem.*, 2018, **55**, 437–445.
160. C. Caussy, S. Charrière, A. Meirhaeghe, J. Dallongeville, E. Lefai, S. Rome, C. Cuerq, V. Euthine, M. Delay, O. Marmontel, M. Di Filippo, M. Lagarde, P. Moulin and C. Marçais, *Atherosclerosis*, 2016, **246**, 280–286.
161. X. Wang, J. Sundquist, B. Zöller, A. A. Memon, K. Palmér, K. Sundquist and L. Bennet, *PLoS One*, 2014, **9**, e86792.
162. L. Kong, J. Zhu, W. Han, X. Jiang, M. Xu, Y. Zhao, Q. Dong, Z. Pang, Q. Guan, L. Gao, J. Zhao and L. Zhao, *Acta Diabetol.*, 2011, **48**, 61–69.
163. A. Jones, K. M. Danielson, M. C. Benton, O. Ziegler, R. Shah, R. S. Stubbs, S. Das and D. Macartney-Coxson, *Obesity*, 2017, **25**, 1734–1744.
164. Y. Shen, H. Xu, X. Pan, W. Wu, H. Wang, L. Yan, M. Zhang, X. Liu, S. Xia and Q. Shao, *Exp. Ther. Med.*, 2017, **14**, 5589–5596.
165. X. Chang, S. Li, J. Li, L. Yin, T. Zhou, C. Zhang, X. Chen and K. Sun, *J. Diabetes Res.*, 2014, **2014**, 761938.
166. A. Kulyte, Y. Belarbi, S. Lorente-Cebrian, C. Bambace, E. Arner, C. O. Daub, P. Heden, M. Ryden, N. Mejhert and P. Arner, *Diabetes*, 2014, **63**, 1248–1258.

167. H. Tao, M. Wang, M. Zhang, S. Zhang, C. Wang, W. Yuan, T. Sun, L. He and Q. Hu, *PLoS One*, 2016, **11**, e0149954.
168. A. Kalani, P. Chaturvedi, C. Maldonado, P. Bauer, I. G. Joshua, S. C. Tyagi and N. Tyagi, *Mol. Cell. Neurosci.*, 2017, **80**, 58–65.
169. D. S. Karolina, A. Armugam, S. Tavintharan, M. T. K. Wong, S. C. Lim, C. F. Sum and K. Jeyaseelan, *PLoS One*, 2011, **6**, e22839.
170. X. S. Liu, B. Fan, A. Szalad, L. Jia, L. Wang, X. Wang, W. Pan, L. Zhang, R. Zhang, J. Hu, X. M. Zhang, M. Chopp and Z. G. Zhang, *Diabetes*, 2017, **66**, 3111–3121.
171. R. J. Wan and Y. H. Li, *Mol. Med. Rep.*, 2018, **17**, 4759–4766.
172. M. Zhu, Y. Wei, C. Geißler, K. Abschlag, J. C. Campos, M. Hristov, J. Möllmann, M. Lehrke, E. Karshovska and A. Schober, *Diabetes*, 2017, **66**, 3072–3084.
173. B.-F. Belgardt, K. Ahmed, M. Spranger, M. Latreille, R. Denzler, N. Kondratiuk, F. von Meyenn, F. N. Villena, K. Herrmanns, D. Bosco, J. Kerr-Conte, F. Pattou, T. Rülicke and M. Stoffel, *Nat. Med.*, 2015, **21**, 619–627.
174. J. Wang, Y. Gao, L. Duan, S. Wei, J. Liu, L. Tian, J. Quan, Q. Zhang, J. Liu and J. Yang, *Oncotarget*, 2017, **8**, 98029–98039.
175. A. Luo, H. Yan, J. Liang, C. Du, X. Zhao, L. Sun and Y. Chen, *Gene*, 2017, **627**, 194–201.
176. T. Seeger, A. Fischer, M. Muhly-Reinholz, A. M. Zeiher and S. Dimmeler, *Obesity*, 2014, **22**, 2352–2360.
177. K. Ono, M. Igata, T. Kondo, S. Kitano, Y. Takaki, S. Hanatani, M. Sakaguchi, R. Goto, T. Senokuchi, J. Kawashima, N. Furukawa, H. Motoshima and E. Araki, *PLoS One*, 2018, **13**, e0191553.
178. A. Bagge, T. R. Clausen, S. Larsen, M. Ladefoged, M. W. Rosenstierne, L. Larsen, O. Vang, J. H. Nielsen and L. T. Dalgaard, *Biochem. Biophys. Res. Commun.*, 2012, **426**, 266–272.
179. K. Tugay, C. Guay, A. C. Marques, F. Allagnat, J. M. Locke, L. W. Harries, G. A. Rutter and R. Regazzi, *Diabetologia*, 2016, **59**, 161–169.

CHAPTER 3

Circulating MicroRNAs as Biomarkers of Celiac Disease and Other Intestinal Pediatric Diseases: Experimental and Bioinformatics Challenges From Bench to Bedside

CRISTINA FELLI[†a], ANTONELLA BALDASSARRE[†a], FRANCESCA FERRETTI[b], MARIANGELA CARUSO[c] AND ANDREA MASOTTI*[a]

[a]Bambino Gesù Children's Hospital-IRCCS, Research Laboratories, V.le di San Paolo 15, 00146, Rome, Italy; [b]Bambino Gesù Children's Hospital-IRCCS, Hepatology, Gastroenterology and Nutrition, V.le di San Paolo 15, 00146, Rome, Italy; [c]Sapienza University of Rome, Department of Pediatrics, V.le Regina Elena 324, 00161, Rome, Italy
*E-mail: andrea.masotti@opbg.net

3.1 Introduction

3.1.1 Biogenesis and Regulatory Functions of miRNAs

MicroRNAs (miRNAs) are short non-coding RNAs that are widely recognized as fundamental regulators of several biological and cellular functions.[1,2] In the early 1990s, miRNAs were first discovered in *Caenorhabditis elegans* as

[†]These two authors contributed equally.

Drug Discovery Series No. 69
MicroRNAs in Diseases and Disorders: Emerging Therapeutic Targets
Edited by Philip V. Peplow, Bridget Martinez, George A. Calin and Aurora Esquela-Kerscher

Published by the Royal Society of Chemistry, www.rsc.org

small RNA molecules that regulate the developmental fates of these organisms.[3] Since then, many miRNAs have been discovered and ascertained to regulate not only the developmental stages, but also many other cellular functions, such as apoptosis, proliferation and differentiation. To date, 1917 precursors and 2654 mature miRNAs (miRBase Rel.22, March 2018) have been annotated in humans.

Long primary miRNA transcripts (pri-miRNA) are initially transcribed by RNA polymerase II in the nucleus, then cleaved to precursor hairpin (pre-miRNA) by the microprocessor complex Drosha and DGCR8. However, although this is the most common transcription mechanism where miRNAs are generated as intron byproducts from splicing reactions, other miRNAs are found to be transcribed by Pol III.[4] Pre-miRNAs are exported into the cytoplasm (by Exportin 5 and other cofactors) and further cleaved to mature miRNAs by the RNase III endonuclease Dicer. One of the two RNA strands of the mature miRNA is incorporated into the RNA-induced silencing complex (RISC) and guided to bind the 3′ untranslated regions (UTR) of the "target" mRNA. Although 3′ UTR-mediated regulation by miRNAs is the most common mechanism of regulation, miRNAs are found to also regulate target gene expression *via* binding to the open reading frame or 5′ UTR.[5] Moreover, miRNAs have been reported to act in the nucleus.[6] After recognition, this interaction leads to a translational repression or mRNA degradation and, as a consequence, protein and gene expression is reduced or abolished.[7] This complex system has been demonstrated to be a fine-tuned regulatory machine, conserved in many species and evolutionarily designed to control gene expression efficiently. To understand the effects of post-transcriptional gene regulation by miRNAs, many studies have focused on bioinformatics predictions of their target genes and the molecular functions and biological pathways in which these genes are involved.[8] A large number of bioinformatics tools such as TargetScan,[9] MiRanda,[10] MiRò,[11] PITA,[12] miRWalk2,[13] TransmiR[14] and miRecords[15] are available to investigate the biological role of the circulating miRNAs of interest.

3.1.2 Biogenesis, Stability and Cellular Recognition of Circulating miRNAs

In recent years, the discovery of miRNAs in biological fluids has generated a great interest for their potential use as biomarkers. Circulating biomarkers play a significant role in clinical applications especially for the diagnosis of specific diseases, to monitor the therapeutic effect of a drug or to predict the tumor recurrence in chemotherapy-treated patients.[16] As an example, Chim and coworkers were the first to identify circulating placental miRNAs in the plasma of pregnant women.[17] Circulating miRNAs have many of the essential characteristics of good biomarkers: they are stable in the circulation and resistant to RNAses digestion, extreme pH, high temperatures, extended storage and multiple freeze–thaw cycles.[18,19] In many cases, changes in circulating miRNA expression levels have been associated with different diseases

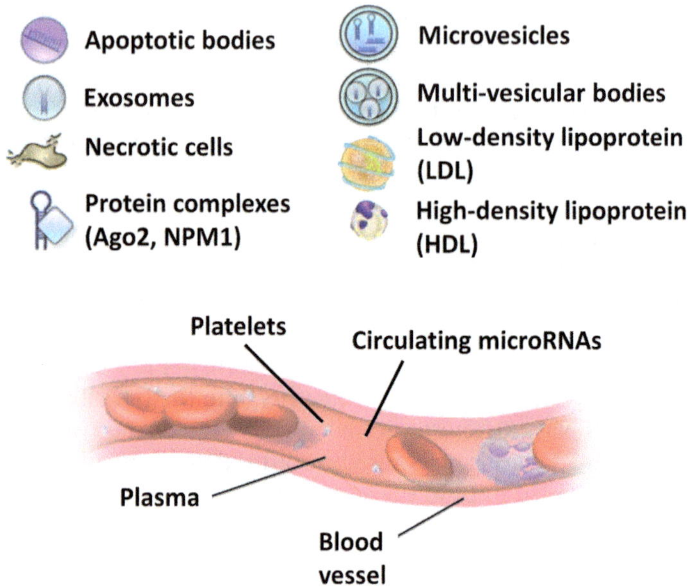

Figure 3.1 Different species of circulating miRNAs in the bloodstream.

or certain biological/pathological stages. Circulating miRNAs are released into the bloodstream in many forms (Figure 3.1), although their origin and the mechanism of their release have not been completely elucidated.

The reasons for the high stability of circulating miRNAs remain largely unknown, although several hypotheses have been suggested.[20] First, circulating miRNAs might have unique modifications, such as methylation, adenylation and uridylation that modulate their stability and thus protect them against RNAses.[21,22] Second, circulating miRNAs are protected by the encapsulation in the cell-derived microvesicles[23] or by specific RNA-binding proteins.[24]

Despite the sparse information about the biogenesis and stability of circulating miRNAs, more is known about their functional role in determining a biological effect even in distant sites of the body. In fact, three different hypotheses explaining the origin and stability of circulating miRNAs in body fluids have been suggested.[20] The first suggests a passive release of miRNAs from apoptotic/necrotic cells and injured tissues, chronic inflammation, or from cells with a short half-life, such as platelets. The second suggests an active secretion *via* cell-derived microvesicles, including exosomes and shedding vesicles, which are membrane-enclosed cell fragments released by cells under both normal and pathological conditions.[25] Exosomes are formed *via* inward budding of early endosomal membranes, giving rise to intracellular multivesicular bodies that later fuse with the plasma membrane and are released to the extracellular environment. Shedding vesicles are larger vesicles that are generated by outward budding and fission of the plasma membrane. The loading of miRNAs into microvesicles seems to be controlled by specific proteins of the RISC.[26] The last hypothesis involves the active secretion by cells of miRNAs and RNA-binding protein complexes. Some of these RNA-binding proteins are represented by nucleophosmin (*i.e.*, the nucleolar phosphoprotein NMP1)[20] high-/low-density lipoproteins[27] and Argonaute 2 (AGO2).[23] In one study, the release of miRNAs in the bloodstream was shown to be controlled by neutral sphingomyelinase 2 and through ceramide-dependent secretory machinery.[27] These results suggest the existence of a specific miRNA export system by which certain miRNAs are exported by some cells, but are recognized, taken up and employed by others.

3.1.3 Circulating miRNAs and Toll-like Receptors

There is debate how circulating miRNAs are taken up by recipient cells and whether this process occurs randomly (passive transport) or involves targeting of specific cells (active transport).[20] It has been postulated that miRNAs within extracellular vesicles could be internalized by recipient cells through endocytosis, phagocytosis or direct fusion with the plasma membranes, whereas RNA binding protein-associated miRNAs could be taken up *via* specific receptors on the recipient cell surface. Alternatively, specific antigen molecules on the microvesicle surface (*i.e.*, clusters of differentiation CD9, CD63 and CD81) could be recognized by specific cell receptors. Recent studies highlight the importance of extracellular miRNAs in cell-to-cell communication and indicate a possible mechanism of intercellular transfer of genetic information.[23,28] The identification of circulating miRNAs released by cells within exosomes and the demonstration of their ability to be transferred from one cell to another suggested a functional comparison of circulating miRNAs with hormones. Following this hypothesis, "hormone-like" miRNAs require a specific receptor for cellular internalization. The toll-like receptor (TLR) family is a particularly appealing candidate for several reasons. First, TLRs are a family of receptors through which the mammalian innate immune system recognizes the presence of invading pathogens.[29] It has been demonstrated that members of the TLR family (*i.e.*, murine TLR7 and human TLR8) on dendritic cells and B-lymphocytes can recognize and bind viral single-stranded RNA (ssRNA) sequences, leading to cell activation and production of cytokines.[30] Both murine TLR7 and human TLR8 are activated by the binding to a 20-nucleotide ssRNAs, a physiologic ligand for these two receptors. Therefore, as circulating miRNAs have a comparable length, they could represent endogenous agonists of these TLRs. Fabbri and colleagues observed that the pro-inflammatory and pro-metastatic responses in human cancers are triggered by two circulating miRNAs, miR-21 and miR-29a, and that this response is mediated only by TLR8.[31] This is a finding that prompts further studies aimed at understanding the selective role of TLRs, a necessary background to properly develop selective and specific TLR7 (or TLR8)-binding anticancer drugs.[31] Interestingly, the implications of circulating miRNAs binding to TLRs are not disease-specific and go beyond cancer. In fact, Lehmann and colleagues showed that *let-7b*, a key regulator of gene expression in the central nervous system, is expressed at higher concentration in the cerebrospinal fluid of patients with Alzheimer's disease.[32] The exogenous intrathecal injection of *let-7b* into the cerebrospinal fluid of wild-type mice resulted in neurodegeneration, whereas mice lacking TLR7 were resistant to this neurodegenerative effect. However, the susceptibility to *let-7b* was restored in neurons of TLR7−/− fetuses transfected with TLR7 through intrauterine electroporation.[32] These findings emphasize that in addition to miR-21 and miR-29a, *let-7b* can also bind to TLR7 and indicate that other circulating miRNAs may have a role in mediating the activation of different processes (*i.e.*, pro-metastatic inflammatory response) leading to different diseases.[33]

3.1.4 The Role of miRNAs in Intestinal Development and Homeostasis

The intestinal barrier separates the lumen from mucosal tissues and is formed by a monolayer of epithelial cells, including columnar, Paneth, endocrine and goblet cells. Inflammatory disorders, such as Crohn's disease, inflammatory bowel disease or celiac disease, are characterized by a functional impairment of the intestinal epithelium (Figure 3.2).

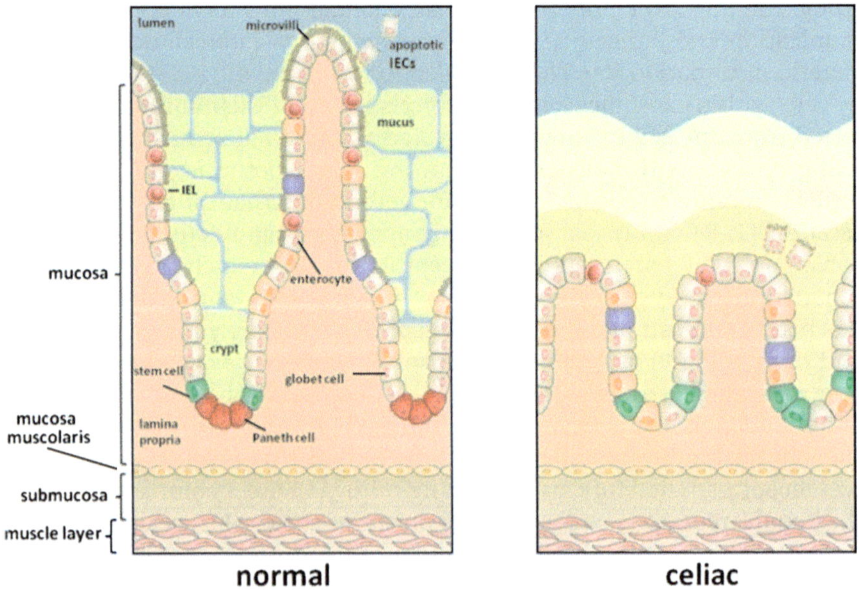

Figure 3.2 The small intestinal mucosa of a healthy individual (left) compared to that of celiac patients (right). The gluten exposure induces a damage of the intestinal mucosa characterized by a complete loss of villi and crypt hyperplasia. IEL: intraepithelial lymphocyte; IEC: intestinal epithelial cell.

Many studies have revealed that the mouse intestinal barrier function is dysregulated by the presence of miRNAs. McKenna and collaborators performed an expression profile of the total miRNAs present in mouse intestinal epithelium and determined their contribution to intestinal homeostasis.[34] The study identified 453 families of miRNAs which included mmu-miR-192, mmu-miR-215 and mmu-*let-7* that were highly expressed in both the small and large intestine. The intestinal epithelium of Dicer1-ablated mice was highly impaired in architecture and barrier function.[34] The effective deletion was confirmed by TaqMan quantitative reverse transcription-polymerase chain reaction (qRT-PCR) measuring the amount of intestinal mmu-miR-21 and mmu-*let-7b*. Dicer-1 mutants had a disorganized epithelium in both small and large intestine, more extended crypts in the small intestine and a decrease of goblet cells in the large intestine.[34]

Epithelial cells, which constitute the intestinal barrier, are sealed by the apical junctional complexes formed by adherens and tight junctions (TJ). The alteration of the TJ barrier induces an increase of permeability, followed by inflammation. The activation of immune response is stimulated by the release of chemokines and cytokines, such as tumor necrosis factor (TNF)-α. Its administration at physiological concentration, both *in vivo* using the mouse small-intestine model and *in vitro* employing intestinal epithelial Caco-2 cells model, induced depletion of the TJ protein occludin.[35] Ye and collaborators demonstrated that mmu-miR-122a may act as a modulator of intestinal TJ permeability. In fact, TNF-α caused an increase of mmu-miR-122a, which, in turn, by recognizing the occludin mRNA 3′ UTR induced its degradation and the consequent depletion of occludin.[35]

3.2 Celiac Disease

Celiac disease is an autoimmune enteropathy triggered by the interaction of environmental and genetic factors. In genetically susceptible individuals, the ingestion of dietary gluten induces an inflammatory process that results in the damage of the bowel mucosa. This is followed by villous atrophy and crypt hyperplasia in addition to infiltration of a consistent number of lymphocytes and plasma cells in the lamina propria (Figure 3.2). In addition to environmental factors, the susceptibility to celiac disease depends on genetic background. In fact, celiac disease is associated with specific major histocompatibility complex class II alleles that encode for the human leukocyte antigens HLA-DQ2 or HLA-DQ8 heterodimers (the first occurs in at least 90–95% of patients with celiac disease). The DQ2–DQ8 heterodimer characterizes the remaining 5–10% of celiac individuals. Presently, the only available treatment for celiac disease is a lifelong adherence to a gluten-free diet (GFD). Gluten is the storage protein fraction present in grains of wheat, rye and barley and it includes two major protein types, gliadins and glutenins (hordeins and secalins in barley and rye, respectively). Gliadins and glutenins have a high proline and glutamine content and are resistant to gastrointestinal enzymatic proteolysis. Therefore, the presence of large and potentially immunogenic peptides at the intestinal mucosal surface is quite high.

After food ingestion and initial degradation, gliadin peptides undergo deamidation by tissue transglutaminase (TG2). Thus, peptides acquire negative charges able to bind efficiently to HLA-DQ2 or HLA-DQ8 molecules on the surface of antigen presenting cells, which present them to $CD4^+$ T-lymphocytes. After activation by gliadin peptides, T-cells secrete high levels of pro-inflammatory cytokines which, in turn, stimulate natural killer cells and $CD8^+$ intraepithelial T-lymphocytes, ultimately leading to the typical celiac mucosa injury.[36–39] Primarily, celiac disease affects the gastrointestinal tract and the most common symptoms are malabsorption, diarrhea, weight loss and nutritional deficiency. However, individuals with celiac disease can have extraintestinal symptoms, including anemia, osteoporosis or arthritis. The gold standard for the diagnosis of celiac disease is a combination of serological screening tests often accompanied by intestinal mucosal biopsy to confirm diagnosis. Serological assays determine the presence of autoantibodies produced during the immunological response to gluten, a phenomenon that characterizes the disease. The two major autoantibodies tested in individuals with celiac disease symptoms are anti-tTG2 and anti-endomysium IgA (EMA).[40] High serum levels of anti-tTG2 IgA and anti-EMA IgA identify individuals who need an intestinal mucosa biopsy to confirm celiac disease diagnosis. According to the European Society for Paediatric Gastroenterology, Hepatology and Nutrition (ESPGHAN) guidelines,[41] in adolescents and children with symptoms or signs suggestive of celiac disease and with high anti-TG2 levels (>10 times the normal upper limit) as well as positivity for anti-EMA antibodies and HLA-DQ2 and/or HLA-DQ8 heterodimer the duodenal biopsy can be avoided because the likelihood of villous atrophy (Marsh 3 classification) is high.[41,42] To date, a screening test that totally avoids an invasive approach is not available.

Numerous researchers have tried to unravel the mechanisms underlying the pathogenesis of celiac disease and to find novel molecular biomarkers to employ for the diagnosis or follow-up of the intestinal damage, or to find alternative methods to avoid invasive biopsies. In the past two decades, miRNAs have been widely studied and discovered to regulate many biological processes such as differentiation, development and cell death (to cite only a few). Despite the growing number of studies about the role of miRNAs in autoimmune disorders, data about miRNAs and celiac disease are scarce. A few studies have mainly focused on intestinal epithelium, in which the regulation of gene expression is complex and controlled by different signaling pathways that regulate the balance between proliferation and differentiation of epithelial cells, processes that are altered in pathologies such as celiac disease.[43] In patients affected by celiac disease, these processes and the expression of specific miRNAs (*i.e.*, miR-212, miR-192 and miR-122a) are altered, suggesting that miRNAs could be involved in the pathogenesis of intestinal barrier dysfunction and be associated with particular clinical features.[44] These molecules may have a role in disease characterization, for celiac disease diagnosis or as predictors of GFD outcome. Moreover, starting from these dysregulated miRNAs, we investigated the putative miRNA targets by using simple computational approaches in order to classify them by gene ontology and pathway analysis. We suggest that this approach may allow researchers to identify and explore miRNA-mediated molecular mechanisms that could be potentially involved in the pathogenesis of celiac disease.

3.3 Intestinal miRNAs of Celiac Disease Patients

As mentioned previously, in the past few years many studies have focused attention on the expression profiles of miRNAs in human small intestine of patients affected by celiac disease, in an attempt to understand the mechanisms implicated in celiac disease through the investigation of their mRNA targets.[45–49] The expression patterns of miRNAs in the small intestine of a cohort of 20 children with active celiac disease (nine on GFD, and 11 controls) have been investigated.[45] Using TaqMan low-density arrays, authors identified nine upregulated and 20 downregulated miRNAs in children with celiac disease compared to controls. Significant miRNAs (upregulated and downregulated) reported in Capuano's paper are listed in Table 3.1. The authors further validated the upregulation of miR-449a and the downregulation of miR-124a in patients with celiac disease and on GFD.[45]

To investigate whether the modulation of miRNAs is dependent on the clinical symptoms, Vaira *et al.* examined the deregulation of miRNAs in the duodenal mucosa of adult celiac disease patients with different clinical manifestations.[46] Therefore, they separated untreated adult celiac patients with classical clinical symptoms (CC), iron-deficiency anemia (CA), patients on GFD (NT-C) and non-celiac disease subjects with normal duodenal mucosa. Using microarray technology, authors characterized the miRNA expression profiles of untreated CC, CA and NT-C patients, compared to control subjects. Seven miRNAs (miR-31-5p, miR-192-3p, miR-194-5p, miR-551a, miR-551b-5p, miR-638 and miR-1290) were further validated using real-time qPCR. Both CC and CA patients showed a significant downregulation of miR-31-5p and miR-192-3p, and upregulation of miR-1290, miR-638 and miR-551b-5p, compared to control subjects (Table 3.1). In addition, CC patients presented reduced levels of miR-551a, while miR-194-5p was downregulated in CA patients, compared to non-celiac individuals. Notably, the downregulation of miR-194-5p or the overexpression of miR-638 appeared to be typical of CA patients compared with CC patients. Conversely, the downregulation of miR-31-5p and miR-192-3p and overexpression of miR-1290 appeared to be related to celiac disease independently of the clinical presentations (Table 3.1). The downregulation of miR-192-3p was also confirmed in fibroblasts of patients with celiac disease, after incubation with gliadin peptides (13- and 33-mer).[46]

Table 3.1 List of upregulated and downregulated miRNAs in intestinal biopsies of patients affected by celiac disease (CD) (active CD, on gluten free-diet (GFD), with or without anemia) compared to control subjects. The list of miRNAs has been obtained from the studies indicated in the last column.

Patient status	Upregulated miRNAs	Downregulated miRNAs	Reference
Biopsies of CD patients (active and on GFD)	miR-182, miR-196a, miR-330, miR-449a, miR-492, miR-500, miR-503, miR-504	miR-105, miR-124a, miR-135a, miR-189, miR-202, miR-219, miR-299-5p, miR-323, miR-379, miR-380-5p, miR-409-5p	45
	miR-644	miR-412, miR-512-3p, miR-566, miR-576, miR-600, miR-614, miR-616, miR-618, miR-631, miR-659	
GFD	miR-422a, miR-551a, miR-1285-3p miR-3681-5p	N.A.	46
CD patients	miR-24-2-5p, miR-2113, miR-4300	miR-215, miR-451a, miR-192-5p	
	miR-551b-5p, miR-519d, miR-642b-3p, miR-523-3p, miR-491-3p	miR-31-5p, miR-194-5p, miR-338-3p	
CD patients with anemia	miR-491-3p, miR-337-3p, miR-2355-3p, miR-3148, miR-24-2-5p, miR-920	miR-215, miR-31-5p, miR-193a-5p	
	miR-638, miR-146b-3p, miR-1304-5p		
	miR-498, miR-490-3p, miR-1285-3p		
	miR-3183, miR-1290, miR-1299, miR-1270, miR-3663-5p, miR-618, miR-3135a	miR-194-5p, miR-192-5p, miR-451a	
	miR-4268, miR-4300, miR-4324, miR-300, miR-519d, miR-422a, miR-302a-3p		
	miR-3654, miR-3611, miR-4329	miR-192-3p, miR-138-1-3p, miR-30b-5p, miR-664-5p	
	miR-3681-5p, miR-551b-5p, miR-4303, miR-642a-5p/miR-642b-5p, miR-550b-3p, miR-593-3p, miR-146a-5p, miR-1273e, miR-432-5p		
Human biopsies of CD patients	N/A	miR-192-5p, miR-194-5p, miR-197 miR-31-5p, miR-338-3p	47
	miR-21-5p, miR-21-3p, miR-486-5p	miR-192-5p, miR-31-5p, miR-338-3p	48

Since the profile of miRNAs may change as a function of the severity of intestinal damage, the miRNA expression patterns in duodenal biopsies of adult celiac disease patients with a classification of histological findings (Marsh) 3C, was evaluated by microarray analysis.[47] The authors showed that the expression of miR-192-5p, miR-194-5p, miR-31-5p, miR-338-3p and miR-197 was significantly reduced in biopsies obtained from celiac disease patients at diagnosis, especially with Marsh 3C lesions. miR-338-3p and miR-197 were significantly decreased in celiac patients independently of the severity of the mucosal impairment (Table 3.1).[47] Changes in intestinal morphology and the immune system were found to be closely associated with age, and this may explain the different clinical manifestations among pediatric and adult patients.

In a recent study, Buoli Comani and colleagues evaluated a panel of miRNAs and their target genes in duodenal biopsies of Marsh 3AB and 3C in a pediatric cohort of celiac patients compared to controls, to assess whether the miRNAs previously identified in adults as differentially expressed, presented the same pattern in children.[48] Similar to their previous study, Buoli Comani *et al.* coauthors detected a significant downregulation of miR-31-5p and miR-338-3p, whereas other miRNAs were involved in the regulation of the immune response in celiac duodenum (*i.e.*, miR-21-5p). In Marsh 3C children, miR-21-5p showed a significant upregulation in the duodenal biopsies, but this upregulation was not observed in adult celiac disease patients. miR-486-5p was upregulated in pediatric celiac disease patients as detected in biopsies, but it was not significant.

The intestinal epithelium is regulated by the balance between proliferation and differentiation of epithelial cells. Thus, perturbations of the homeostasis in the autophagic process could be the culprit for the beginning of celiac disease. In this regard, Comincini and collaborators investigated the expression of ATG7 and BECN1, two genes involved in the autophagy process and their negative regulators miR-17 and miR-30, respectively. Quantitative PCR was performed on blood samples and intestinal biopsies derived from pediatric celiac disease patients, and results compared with controls. The analysis highlighted the association of autophagy-related genes and miRNAs with celiac disease condition outlining the importance and involvement of autophagy processes in celiac disease pathogenesis.[49]

3.4 Circulating miRNAs of Celiac Disease Patients

Past research focused on miRNA modulation/profiling and examined the expression levels of miRNAs in the intestinal mucosa of individuals with celiac disease (tissue miRNAs). However, the role of cell-free miRNAs in celiac disease has been underestimated and poorly investigated. Only in recent years has scientific evidence shown that miRNAs circulating in different body fluids (*i.e.*, serum and plasma) are protected by protein binding or enclosed in vesicles and released into the extracellular space.[50,51] Many of these studies have emphasized their importance as potential biomarkers.[52,53] Moreover, miRNAs may act as intercellular communication "actors" by conveying their "message" to other distant cells. Since circulating miRNAs are highly stable in the circulation and under extreme laboratory conditions (*i.e.*, pH and temperature, repeated freeze–thaw cycles, *etc.*) and their expression is associated with specific conditions, circulating miRNAs represent reliable and promising diagnostic, prognostic and therapeutic biomarkers.[54–56] In order to investigate whether some of the miRNAs found to be dysregulated in tissue were also significantly dysregulated in the circulation and to demonstrate their effective role as biomarkers, Buoli Comani and colleagues assessed the circulating levels of the miRNAs previously identified by their group in duodenal biopsies of Marsh 3AB and 3C pediatric celiac disease patients. The authors found that in GFD patients, the expression levels of miR-192-5p and 486-5p remained significantly downregulated, whereas the expression levels of miR-31-5p, miR-21-5p and miR-21-3p tended to return to control levels.[48] However, this study focused on a limited number of patients and controls (*i.e.*, 12 controls, 17 celiac disease patients at the first diagnosis and an additional seven celiac disease patients on GFD for at least 1 year). This study was performed on plasma, which may contain a higher amount of PCR contaminants, and without adequate quality controls (*i.e.*, hemolysis that may release red blood cell miRNAs). Moreover, this study did not report the overall number of miRNAs studied in their examination or the normalization techniques used. Finally, the authors concluded that the use of a wider panel of miRNAs would be required to obtain more conclusive results.

3.5 Regulation of Post-transcriptional Gene Expression by miRNAs in Celiac Disease

A miRNA transcriptome study of the small intestine of Dicer-1 knockout mice allowed an investigation of miRNA–mRNA relationships.[34] Microarray analysis showed that mutation of Dicer-1 affected principally the expression of genes belonging to the immune pathway, also supported by the increase of neutrophils in the lamina propria of the small and large intestine. Among all the differentially expressed genes, McKenna and collaborators identified some upregulated genes that were predicted as targets of miRNAs detected in the intestinal epithelium. In particular, miR-10a was proposed to target bone morphogenetic protein 2 (BMP2), which inhibits colonic epithelial cell growth *in vitro*, promotes apoptosis and differentiation and inhibits proliferation. Three miRNAs, miR-22, miR-375 and miR-145* were predicted to regulate BMP7, which is abundantly expressed in the developing intestine and regulates the anti-inflammatory response in the gut tissue. Moreover, miR-215 and miR-93 were reported to potentially regulate kruppel-like factor 9 (KLF9), which is downregulated in the epithelium in human colon cancer. All these data suggest that miRNAs play multiple important roles in the differentiation and function of mouse intestinal epithelium.[34]

Among the miRNAs differentially expressed in the human small intestine, Capuano *et al.* showed that miR-449a is upregulated in children with active celiac disease and in children on a GFD.[45] Using miRecords, which integrates predicted miRNA targets produced by 11 established miRNA target prediction programs, the authors identified mRNA targets of miR-449a and found that some of them are involved in the NOTCH signaling pathway (*i.e.*, NOTCH1, KLF4, DLL1, LEF1 and NUMBL). The binding of miR-449a to the 3′ UTR of NOTCH1 and KLF4 has been validated by a luciferase reporter assay, and a reduction of mRNA expression levels was observed *in vitro* in HEK293 cells. A decrease of protein levels has been detected by immunochemistry in small intestine biopsies of children with celiac disease and on a GFD, compared to healthy controls, suggesting a reduced activity of the Notch pathway, which has a key role in maintaining intestinal homeostasis. In addition, celiac disease patients with a lower number of goblet cells, exhibited more nuclear beta-catenin (an indicator of Wnt pathway activation) and an increased proliferation in the crypts (as indicated by Ki67 levels) were observed. The overexpression of miR-449a may therefore appear as a distinctive feature of celiac disease patients. However, in celiac disease individuals, clinical parameters are highly heterogeneous and the miRNA dysregulation that has been observed in children cannot explain all of this variability.[45]

Vaira and collaborators distinguished CA from CC patients and found that the underexpression of miR-194-5p as well as the overexpression of miR-638 appeared to be a feature of CA patients compared with CC patients.[46] In fact, miR-638 is upregulated exclusively in patients with iron-deficiency anemia. Computational methods (although not supported by functional analysis) showed that a putative target of miR-638 is TGM2, the gene that encodes the autoantigen TG2 implicated in celiac disease. Conversely, downregulation of miR-31-5p and miR-192-3p appeared to be related to celiac disease independently of the clinical presentation. The authors investigated the targets of these selected miRNAs using three different prediction tools: DIANA-microT,[57] TargetScan[9] and PicTar.[58] Bioinformatics analyses suggested that miR-31-5p and miR192/194 cluster are involved in the mitogen-associated protein kinase pathway, in cytoskeletal remodeling or in Wnt signaling.

miRNA expression is also significantly altered in duodenal mucosa of celiac disease patients and is dependent on the severity of the damage. This alteration can modulate the expression of factors involved in innate and adaptive immunity. Bioinformatics analyses of possible targets of miRNAs identified (miR-192-5p, miR-194-5p and miR-31-5p, miR-338-3p and miR-197) have been used also by Magni *et al.* to identify several genes involved in the immune response, including C-X-C motif ligand 2 (CXCL2), nucleotide-binding oligomerization domain-containing protein 2 (NOD2), interleukin 18 (IL-18) and Forkhead Box P3 (FOXP3).[47] CXCL2 and NOD2 are putative targets of miR-192-5p. For both targets, the authors detected significant increases in mRNA and protein expression in Marsh 3C biopsies.[47] A direct interaction between miR-192-5p and NOD2 was shown by luciferase reporter assays using constructs that contain the 3′ UTR of NOD2 mRNA and co-transfection of miR-192. In Marsh 3C patients, the analysis using qPCR showed a significant increase in the expression of FOXP3 a target of the miR-31-5p, and RUNX1, target of miR-338-3p. IL-18 is a target of miR-197 and its mRNA level was significantly upregulated in Marsh 3C patients. Furthermore, in celiac disease patients the expression level of CXCL2, NOD2, FOXP3 and miR-192-5p and miR-31-5p were altered also by gliadin exposure, suggesting the multifactorial nature of this disease. For all of these miRNA targets, the mRNA expression has been shown to be increased in Marsh 3C patients by qRT-PCR and Western blots (Figure 3.3).[47]

Among the miRNAs differentially expressed in adults, miR-192-5p was also significantly reduced in children.[48] Surprisingly, Buoli Comani and coauthors noticed a reduction of its two mRNA targets, CXCL2 and NOD2. However, bioinformatics tools revealed that CXCL2 and NOD2 are also targets of the upregulated miR-486-5p, suggesting that these two genes might be targeted with a different kinetic by miR-486-5p and miR-192-5p. Accordingly, mRNA expression of the mitotic arrest deficient-like 1 (MAD2L1), another target of miR-192-5p, was shown to be upregulated in Marsh 3C biopsies of children

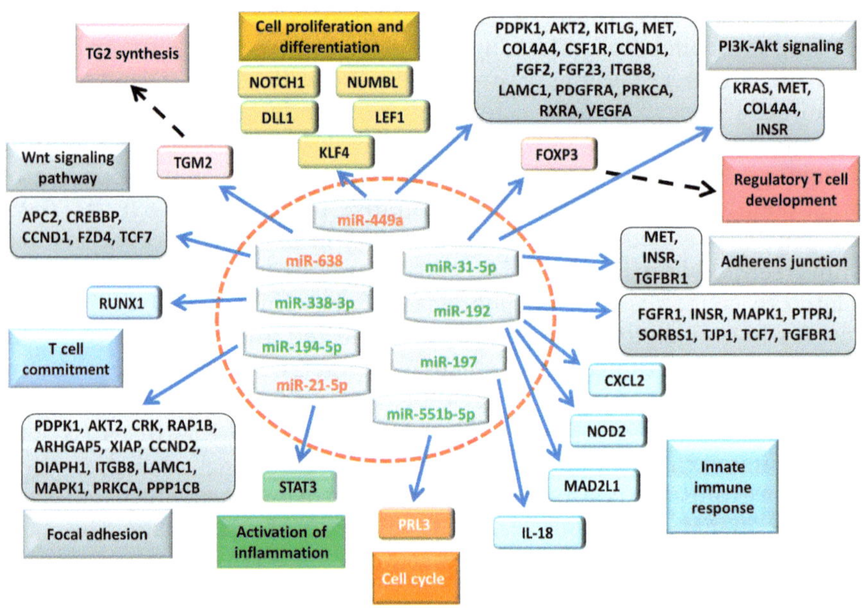

Figure 3.3 Schematic representation of miRNAs found to be deregulated in human small intestine of patients with celiac disease and their target genes and reported in the literature (colored boxes). The genes targeted by upregulated (red) or downregulated (green) miRNAs are directly implicated in celiac disease, as discussed in this review. Gray circles and boxes report the predicted genes and pathways obtained by bioinformatics predictions. Although not comprehensive, these genes extend the picture of the biological processes involved in celiac disease.

with celiac disease, but not in adult celiac disease subjects. Interestingly, a significant downregulation of miR-31-5p and miR-338-3p was accompanied by an upregulation of their mRNA targets, FOXP3 and RUNX1, respectively. In addition, the increase of miR-21-5p in duodenal biopsies of children with celiac disease was accompanied by an enhancement of STAT3 mRNA expression. All these mRNAs are involved in the intestinal development processes and in the immune response at mucosal level.[48] Celiac disease is a severe intestinal autoimmune disease that has been widely investigated for its increasing prevalence and important social and economic implications. So far, little is known about the presence of reliable circulating miRNA biomarkers to diagnose the disease or to monitor the progression or remission as a result of a GFD. Our group aimed at investigating intestinal and circulating miRNAs in celiac disease, and is concluding a large cohort study where the circulating miRNAs of celiac disease patients (at the diagnosis and after a 6-month GFD) were examined and compared to controls (unpublished results). We are aware that further studies are needed in this field and to recapitulate the efforts already done by other groups, and we collected and discussed the papers about intestinal and circulating miRNAs and celiac disease in a recent review paper.[51]

3.6 Circulating miRNAs in Inflammatory Bowel Disease

Inflammatory bowel disease (IBD) is a multi-factorial inflammatory disorder of the gastrointestinal tract and includes other diseases such as Crohn's disease and ulcerative colitis. Both Crohn's disease and ulcerative colitis are characterized by chronic inflammation of the gastrointestinal tract that may arise as a consequence of an aberrant immune response to gut microbiota in genetically predisposed individuals. The specific role of miRNAs in intestinal diseases are not well understood. In mouse models, the loss of intestinal miRNAs has been demonstrated to impair the epithelial barrier function, resulting in acute inflammation.[59] Moreover, little is known about the interplay between the gut microbiota and regulation of gene expression by miRNAs,[60] although a recent paper has suggested a potential mechanism of host–guest inter-kingdom communication.[61] In Crohn's disease, the alteration of disease-specific circulating miRNA levels correlates with epithelial damage and inflammation.[62] In particular, Zahm and collaborators found that circulating miR-192 is elevated in Crohn's disease and this miRNA is greatly expressed in intestinal epithelia.[62] Moreover, these authors examined the diagnostic properties of circulating miRNAs compared to standard serological markers (*i.e.*, C-reactive protein, anti-*Saccharomyces cerevisiae* antibodies IgG and albumin) and found that miR-484 and *let-7b* displayed a sensitivity of 80%, whereas the use of three miRNAs (miR-30e, miR-160a and miR-195) showed a sensitivity >90%. To exclude the possibility that these miRNAs could represent unselective indicators of intestinal inflammation conditions, the expression of circulating miRNAs in patients with celiac disease was compared with that of IBD. In this latter case, the expression level of these miRNAs remained unaltered, thus confirming their specificity for IBD or celiac disease. Moreover, the expression level of circulating miR-484 and miR-195 were reduced after 6 months of treatment [systemic steroids 16 (67%), methotrexate 1 (4%), 6-mercaptopurine or azathioprine 10 (42%) and infliximab 2 (8%), suggesting that these miRNAs could be employed not only as diagnostic biomarkers, but also as prognostic indicators of the outcome of therapeutic interventions]. For the above-mentioned diseases, circulating miRNAs have been considered more versatile biomarkers compared to serological ones.

Duttagupta and collaborators have investigated the expression of circulating miRNAs within microvesicles, peripheral blood mononuclear cells and platelets in a cohort of 20 ulcerative colitis patients and 20 normal individuals and they found a distinct diagnostic signature of 31 miRNAs.[63] The platelet miRNA profile can stratify ulcerative colitis patients from normal individuals with 92.8% accuracy, 96.2% specificity and 89.5% sensitivity.[63] These authors observed a significant overlap between the miRNA profiles from the platelets-only fraction and the combined platelets and microvesicles fraction.[63] Therefore, Duttagupta and collaborators concluded that the origin of the majority of circulating miRNAs in ulcerative colitis derived from the anucleate platelet fraction.

Zahm and collaborators suggested that circulating miRNAs could be used as potential biomarkers of colitis in adolescents with IBD.[64] The authors compared the expression of miRNAs in 50 mucosal biopsies with the corresponding serum miRNA profiles. Several miRNAs have altered expression in biopsies of IBD patients and controls, but only miR-142-3p and miR-21 were among the most abundant, together with miR-192 and miR-194. miR-192 and miR-21, two circulating miRNAs previously associated with pediatric celiac disease, displayed high expression level in IBD samples compared to controls, but were not significantly different between the two disease subgroups (Crohn's disease and ulcerative colitis). miR-192, miR-142-3p and miR-21 correctly classified 78.72%, 72.34% and 72.34% of patients, respectively, but serum miRNA levels did not significantly correlate with tissue miRNAs. However, this study revealed that these miRNAs failed to be significantly altered between the clinical groups and the authors concluded that serum miRNAs are not useful biomarkers to distinguish ulcerative colitis from Crohn's disease in adolescents.[64]

3.7 Experimental and Bioinformatics Challenges for miRNA Analysis

3.7.1 Quantification and Analysis of Circulating miRNAs

The expression of miRNAs can be easily determined by various methods (*i.e.*, qPCR, microarrays and next-generation sequencing).[65–70] One of the most powerful methods for the analysis of circulating miRNAs is qPCR, which in some applications involves the use of a RT-PCR stem–loop primer.[71] In principle, this method can also be employed for the detection of poorly expressed circulating miRNAs, although its practical use is debated.[72] Microarray profiling of circulating miRNAs is a technique largely employed in those studies that use an amount of starting material greater than is required for qRT-PCR. Moreover, the need to develop updated probes and dedicated hybridization conditions for the detection of different miRNAs has somehow concurrently hindered the use of this technique.[16] Next-generation sequencing (deep sequencing) appeared to be a very promising technique for the identification of novel miRNA biomarkers.[73–75] This technology has been used to identify tissue- and stage-specific mRNA and miRNA expression, and to compare miRNA profiles in different diseases.[73] In addition, this method seems to be suited for the quantification of circulating miRNAs, provided that particular care is given to sample preparation. Significant examples are described here, but the list of applications for profiling miRNA expression is growing continuously. Williams *et al.* profiled cell-free miRNAs in serum and plasma samples from human volunteers by introducing synthetic oligonucleotides as calibrators during library preparation, thus enabling the calculation of the total as well as the specific concentration of circulating miRNAs.[76] These authors studied trios of samples (*i.e.*, newborn babies and their parents) and were able to detect placental-specific miRNAs in both maternal and newborn circulations and to quantify the relative contribution of placental miRNAs to the circulating pool of miRNAs. Therefore, this work can be considered a model to be employed not only in the field of prenatal diagnosis, but also to detect other diseases. Circulating exosomes contain miRNAs that can be sequenced. Bellingham and coworkers performed the first small RNA sequencing (small RNA-Seq) experiment to investigate the profile of miRNAs contained in exosomes released from prion-infected neuronal cells.[77] They demonstrated that neuronal exosomes contained a wide range of RNA species, which included retroviral RNA repeat regions, mRNA fragments, tRNAs, non-coding RNAs, small nuclear RNAs, small nucleolar RNA, other RNAs and novel candidate miRNAs. These results clearly demonstrated that it is possible to obtain a "miRNA signature" from exosomes and that this signature can be potentially exploited, not only to better understand the pathogenesis of these diseases, but also for their diagnosis. The sequencing of circulating miRNAs has been exploited to identify novel markers predicting clinical outcome of locally advanced breast cancer.[78] In fact, Wu and coworkers used a comprehensive *de novo* sequencing approach and identified specific sets of miRNAs that were associated with various histopathological parameters. Two of these, miR-375 and miR-122, were associated with clinical outcomes (higher levels of these miRNAs are associated with metastatic recurrences) in breast cancer patients (stage II–III). These results are noteworthy and may allow optimized chemotherapy treatments and preventive anti-metastasis interventions in clinical applications.

However, one of the main issues when dealing with the quantification of circulating miRNAs is reproducibility across various platforms,[79] methodologies (*i.e.*, microarrays or RT-PCR arrays)[18] and different protocols for sample preparation[80] and miRNA isolation procedures.[81] Moreover, inevitably, statistical data analysis is different from the classic approach. Several normalization strategies have been used for the analysis of circulating miRNAs.[82] Initially, scientists employed calibrators classically used for mRNA normalization of qPCR data (*i.e.*, GAPDH, RNU6B, 18S and 5S),[83–85] but their variability or absent expression in the circulation and their possible degradation prevented their further use.[86] Other strategies involved the use of synthetic *C. elegans* miRNAs, such as cel-miR-39 and cel-miR-54, two species without homology with human miRNAs, as spike-in and normalization controls.[87,88] However, since they are not endogenous miRNAs, their use is not recommended methodologically. To overcome this limitation, a constitutively expressed miRNA (*i.e.*, miR-16) has been used as an internal normalization control.[89,90] Although quite stable and highly expressed in many conditions, normalization miRNAs such as miR-16 are not suitable for this purpose since they are often themselves modulated in many diseases and in tumors[91] and expression can be hemolysis-dependent.[92] To circumvent all of these problems, Mestdagh *et al.* introduced the global mean normalization method to normalize qPCR miRNA profiling data.[93] In fact, in these studies a large number of miRNAs are evaluated at the same time and the use of this normalization strategy has been demonstrated to outperform other normalization strategies.[93] Other normalization strategies can be adopted (*i.e.*, the use of other internal normalizers such as miR-103a-3p or miR-191-5p) or even other techniques,[94] although there is no consensus for the optimal normalization strategy, especially in the pediatric setting.

3.8 Bioinformatics Analysis Tools and Public Databases

3.8.1 Bioinformatics Tools and Databases for Circulating Extracellular miRNAs

The growing evidence about the presence of circulating miRNAs and their role in cell–cell communication and use as disease biomarkers has outlined the need for suitable tools for collecting and rationalizing these experimental data. To this end, the first database, miRandola (http://mirandola.iit.cnr.it/) for the classification of different extracellular/circulating miRNAs has been developed (Table 3.2).[95]

miRNAs are classified into four categories, based on their extracellular forms: miRNA-Ago2, miRNA-exosome, miRNA-HDL and miRNA-circulating. The database offers numerous information details including isolation and experimental methods, type of source samples and associated diseases. Moreover, a direct link to the knowledge base "miRò" is provided to explore the biological functions of circulating miRNAs and their relationships with the phenotype.[11] Different tools embedded in miRandola allow users to search by mature miRNAs, miRNA family, sample, diseases and malignant cell lines, and potential biomarker role. Results are directly linked to miRò

Table 3.2 List of analysis tools and database for circulating extracellular miRNAs, web links and year of publication.

Bioinformatics resources	Year	Web link	Reference
miRandola	2012	http://mirandola.iit.cnr.it/	95
miRò 2	2014	http://microrna.osumc.edu/miro/	11
Evpedia	2012	http://evpedia.info	96
ExoCarta	2009	http://www.exocarta.org	97
FunRich	2015	http://www.funrich.org	98
Vesiclepedia	2011	http://www.microvesicles.org	99
exRNA Research portal	2014	http://exrna.org	100

for further functional annotations. Noteworthy, users can also download the comprehensive database miRandola, which contains all the information about extracellular/circulating miRNAs or contribute to update the database through an online data submission form. The current version of miRandola (last update February 2017) is focused on human circulating miRNAs, long non-coding and circular RNAs, integrates Vesiclepedia miRNA data and is linked to the National Institutes of Health exRNA.org Research Portal. Moreover, miRandola can help to retrieve specific information about miRNAs complexed with exosomes taken from the ExoCarta database, an exosome database that provides links to proteins, RNAs and lipids identified in exosomes isolated from multiple organisms.[96] The new version of ExoCarta database (version 5; 29 July 2015) has more than two-fold data compared to the previous release in 2012 and has new additional features including the annotation of the International Society for Extracellular Vesicles standards and dynamic protein–protein interaction networks. An integrated and comprehensive proteome, transcriptome and lipidome database of extracellular vesicles derived from archaea, bacteria and eukarya, including human, is collected in EVpedia (update 6 February 2015), which provides a search and browse tool for vesicular proteins, comparison of vesicular datasets by orthologue identification, Gene Ontology enrichment analyses and network analyses of vesicular proteins. Furthermore, EVpedia provides databases of vesicular mRNAs, miRNAs, and lipids.[97] FunRich, which provides functional enrichment and network analysis integrated with gene ontology, biological pathways, protein interactions and domains or associated diseases offers the opportunity to compare proteomic data with ExoCarta.[98] In addition, FunRich has an integrated plugin pertaining to extracellular vesicles and a convenient tool to perform proteomic analysis in a quick and reliable manner. FunRich is an alternative to currently available software that allows the generation of graphical representations (Venn diagrams, pie charts, bar graphs or heatmaps) as well as protein–protein interaction network visualization. Vesiclepedia (version 3.1 updated 9 January 2015) is a manually curated compendium of molecular data (lipid, RNA and protein) identified in different classes of extracellular vesicles. Currently, Vesiclepedia comprises 92 897 protein, 27 642 mRNAs, 4934 miRNAs and 584 lipid entries from 538 independent studies that were published over the past several years.[99] The exRNA tool is a research portal aimed at discovering fundamental biological principles about the mechanisms of extracellular RNA generation, secretion, transport and to investigate the potential for using extracellular RNAs in the clinic as therapeutic molecules or biomarkers of disease.[100]

3.9 Conclusions

The discovery that miRNAs circulate in body fluids has generated a great interest for their potential use as disease biomarkers. Circulating miRNAs are not cell-associated and generally escape degradation by endogenous ribonucleases owing to their localization within membrane-structured bodies as well as protein and lipid complexes. The functional roles that these circulating miRNAs might play in the human body even at distant sites will continue to be the focus of future studies. In fact, several highly abundant miRNAs can be present at the same time in multiple fluids, whereas some miRNAs are enriched only in a specific fluid source. Extracellular miRNAs can potentially interact with recipient cells *via* a number of different processes, including direct fusion, passive uptake and receptor-mediated interactions.[54] Over the past several years, the presence of circulating miRNAs has been associated with a variety of medical conditions and linked to many pediatric diseases.

Specific miRNAs are modulated in duodenal mucosa of patients affected by celiac disease, suggesting a role in the pathogenesis and their potential use for celiac disease diagnosis or as predictors of GFD outcome. All of these studies suggest the crucial role of miRNAs in different pathways involved in the development of celiac disease.

In fact, miRNAs are important factors in the differentiation and function of the intestinal epithelium and have a relevant role in regulating gene expression in physiological and pathological conditions, including inflammatory and autoimmune disorders. However, the molecular determinants underlying the pathogenesis of celiac disease still remain unclear. Therefore, the profiling of miRNAs and the study of their function is of paramount importance. An alteration of miRNA expression pattern in biopsies of celiac disease patients when compared to healthy controls has been observed and suggested the use of miRNAs as non-invasive and innovative diagnostic biomarkers.

Mucosal epithelium has a pathogenic role in diseases that have an intestinal inflammation component. Generally, epithelial cells derived from stem cells, proliferate and migrate from the base of the crypt onto the villi, and then to the villus tips, finally differentiating into absorptive enterocytes. The homeostasis and function of the intestinal epithelium are regulated by a balance between cell proliferation and cell death. In individuals with celiac disease, the ingestion of gluten causes the damage of intestinal mucosa that is mediated both by adaptive and innate immune activation, and by proliferation of enterocytes. The celiac intestine is characterized, in fact, by decreased differentiation and increased proliferation of epithelial cells that lead to villi atrophy and crypt hyperplasia, respectively.[101,102] A pathway involved in the development of the intestine and in maintaining adult tissue homeostasis is represented by Notch signaling, which regulates cell proliferation and cellular differentiation. In fact, in the small intestine of pediatric celiac disease patients, the downregulation of Notch1 reduces the number of goblet cells.[45] The importance of cell fate has been also demonstrated by investigating the role of Wnt signaling in the homeostasis of adult tissue.[103] In mouse models, the inactivation of beta-catenin, the key component of Wnt cascade, showed a loss of intestinal crypts.[104] In the small intestine of children with celiac disease, an increased expression of beta-catenin has been observed, which suggests an enhanced cellular proliferation.[45] In digestive and autoimmune-related disorder as celiac disease, the regulation of the autophagy process might also be involved. In fact, the decrease in the expression of miR-17 and miR-30a may be the result of a compensatory mechanism triggered by pathological loss of homeostasis in the autophagic process.[49] Accordingly, bioinformatics analyses show an involvement of miRNAs both in the transcription of Wnt target genes and in positive regulation of cell proliferation.

The gut is a tissue continuously renewing, and in untreated celiac disease patients the apoptosis of enterocytes is greatly increased.[105] Another deeply investigated pathway is the Fas cell surface death receptor (FAS) signaling. In *in vitro* culture of duodenal cells (of celiac disease patients) treated with gliadin an intense FAS expression has been observed compared to controls.[106]

IL-15, member of the γc family of cytokines, which also includes IL-2, IL-4, IL-9 and IL-21, is the major mediator of the innate immune system. IL-15 and IL-2 receptors share a β-chain, which combines with γc to form a common signaling module. In active celiac disease, gluten-activated $CD4^+$ T-cells secrete IL-2 and IL-21, while epithelium and dendritic cells produce high levels of IL-15. After the receptor engagement, IL-15 signaling activates both STAT3 and STAT5 transcription factor to exert its regulatory function in immune cells.[36] Again, bioinformatics predictions indicated that the upregulated and downregulated miRNAs that we have considered might interact with STAT5, which is involved in the response to the cytokine stimulus.

Many studies have shown that an enhancement of intestinal permeability is a feature of celiac disease subjects.[44,107] Nevertheless, the exposure of rat intestinal epithelial cells to gliadin led to the release of TJ protein zonulin from enterocytes and to cytoskeleton reorganization with a redistribution of actin filaments.[108] By regulating the degradation of mRNAs at the post-transcriptional level, miRNAs may provide novel insights into understanding the pathological and physiological process of intestinal barrier dysfunction (Figure 3.2).

The expression level of circulating miRNAs can be easily assessed using various methods, mainly qPCR, which allows signal amplification, and next-generation sequencing.

However, although computational methods are useful to gain information on diseases such as celiac disease, the algorithms behind them may be prone to errors for many reasons, giving rise to many false positive predictions. Therefore, we are aware that the predictions obtained need to be carefully validated experimentally.[8]

Although the commonest serological tests are highly accurate to diagnose pediatric celiac disease patients, new biomarkers have to be discovered and used to avoid the need for biopsies, especially in those cases where the antibody levels are lower than the limits suggested by ESPGHAN guidelines.[41,42]

Finding informative biomarkers from biological fluids will not only be key to the understanding of the physio-pathological processes of diseases, but will also be the next crucial challenge for the development of new therapeutic strategies.

Circulating miRNAs applied to pediatric diseases might reveal all of their potential as diagnostic and prognostic biomarkers as already demonstrated successfully for other intestinal (*i.e.*, Crohn's disease)[62] and non-intestinal diseases[52,109] and might be used routinely as powerful non-invasive biomarkers in the diagnosis of pediatric celiac disease.

References

1. D. P. Bartel, MicroRNAs: target recognition and regulatory functions, *Cell*, 2009, **136**, 215–233.
2. R. W. Carthew and E. J. Sontheimer, Origins and Mechanisms of miRNAs and siRNAs, *Cell*, 2009, **136**, 642–655.
3. R. C. Lee, R. L. Feinbaum and V. Ambros, The *C. elegans* heterochronic gene lin-4 encodes small RNAs with antisense complementarity to lin-14, *Cell*, 1993, **75**, 843–854.
4. G. M. Borchert, W. Lanier and B. L. Davidson, RNA polymerase III transcribes human microRNAs, *Nat. Struct. Mol. Biol.*, 2006, **13**, 1097–1101.
5. L. Da Sacco and A. Masotti, Recent Insights and Novel Bioinformatics Tools to Understand the Role of MicroRNAs Binding to 5′ Untranslated Region, *Int. J. Mol. Sci.*, 2012, **14**, 480–495.
6. H. Liu, C. Lei, Q. He, Z. Pan, D. Xiao and Y. Tao, Nuclear functions of mammalian MicroRNAs in gene regulation, immunity and cancer, *Mol. Cancer*, 2018, **17**, 64.
7. B. Wightman, I. Ha and G. Ruvkun, Posttranscriptional regulation of the heterochronic gene lin-14 by lin-4 mediates temporal pattern formation in *C. elegans*, *Cell*, 1993, **75**, 855–862.
8. S. Ekimler and K. Sahin, Computational Methods for MicroRNA Target Prediction, *Genes*, 2014, **5**, 671–683.

9. B. P. Lewis, C. B. Burge and D. P. Bartel, Conserved seed pairing, often flanked by adenosines, indicates that thousands of human genes are microRNA targets, *Cell*, 2005, **120**, 15–20.
10. A. J. Enright, B. John, U. Gaul, T. Tuschl, C. Sander and D. S. Marks, MicroRNA targets in Drosophila, *Genome Biol.*, 2003, **5**, R1.
11. A. Lagana, S. Forte, A. Giudice, M. R. Arena, P. L. Puglisi, R. Giugno, A. Pulvirenti, D. Shasha and A. Ferro, miRo: a miRNA knowledge base, *Database*, 2009, **2009**, bap008.
12. M. Kertesz, N. Iovino, U. Unnerstall, U. Gaul and E. Segal, The role of site accessibility in microRNA target recognition, *Nat. Genet.*, 2007, **39**, 1278–1284.
13. H. Dweep, C. Sticht, P. Pandey and N. Gretz, miRWalk–database: prediction of possible miRNA binding sites by "walking" the genes of three genomes, *J. Biomed. Inf.*, 2011, **44**, 839–847.
14. J. Wang, M. Lu, C. Qiu and Q. Cui, TransmiR: a transcription factor-microRNA regulation database, *Nucleic Acids Res.*, 2010, **38**, D119–D122.
15. F. Xiao, Z. Zuo, G. Cai, S. Kang, X. Gao and T. Li, miRecords: an integrated resource for microRNA-target interactions, *Nucleic Acids Res.*, 2009, **37**, D105–D110.
16. N. Kosaka, H. Iguchi and T. Ochiya, Circulating microRNA in body fluid: a new potential biomarker for cancer diagnosis and prognosis, *Cancer Sci.*, 2010, **101**, 2087–2092.
17. S. S. Chim, T. K. Shing, E. C. Hung, T. Y. Leung, T. K. Lau, R. W. Chiu and Y. M. Lo, Detection and characterization of placental microRNAs in maternal plasma, *Clin. Chem.*, 2008, **54**, 482–490.
18. Y. Chen, J. A. Gelfond, L. M. McManus and P. K. Shireman, Reproducibility of quantitative RT-PCR array in miRNA expression profiling and comparison with microarray analysis, *BMC Genomics*, 2009, **10**, 407.
19. P. S. Mitchell, R. K. Parkin, E. M. Kroh, B. R. Fritz, S. K. Wyman, E. L. Pogosova-Agadjanyan, A. Peterson, J. Noteboom, K. C. O'Briant, A. Allen, D. W. Lin, N. Urban, C. W. Drescher, B. S. Knudsen, D. L. Stirewalt, R. Gentleman, R. L. Vessella, P. S. Nelson, D. B. Martin and M. Tewari, Circulating microRNAs as stable blood-based markers for cancer detection, *Proc. Natl. Acad. Sci. U. S. A.*, 2008, **105**, 10513–10518.
20. M. A. Cortez, C. Bueso-Ramos, J. Ferdin, G. Lopez-Berestein, A. K. Sood and G. A. Calin, MicroRNAs in body fluids–the mix of hormones and biomarkers, *Nat. Rev. Clin. Oncol.*, 2011, **8**, 467–477.
21. M. R. Jones, L. J. Quinton, M. T. Blahna, J. R. Neilson, S. Fu, A. R. Ivanov, D. A. Wolf and J. P. Mizgerd, Zcchc11-dependent uridylation of microRNA directs cytokine expression, *Nat. Cell Biol.*, 2009, **11**, 1157–1163.
22. T. Katoh, Y. Sakaguchi, K. Miyauchi, T. Suzuki, S. Kashiwabara, T. Baba and T. Suzuki, Selective stabilization of mammalian microRNAs by 3′ adenylation mediated by the cytoplasmic poly(A) polymerase GLD-2, *Genes Dev.*, 2009, **23**, 433–438.

23. H. Valadi, K. Ekstrom, A. Bossios, M. Sjostrand, J. J. Lee and J. O. Lotvall, Exosome-mediated transfer of mRNAs and microRNAs is a novel mechanism of genetic exchange between cells, *Nat. Cell Biol.*, 2007, **9**, 654–659.
24. N. Kosaka, H. Iguchi, Y. Yoshioka, F. Takeshita, Y. Matsuki and T. Ochiya, Secretory mechanisms and intercellular transfer of microRNAs in living cells, *J. Biol. Chem.*, 2010, **285**, 17442–17452.
25. Z. Hu, X. Chen, Y. Zhao, T. Tian, G. Jin, Y. Shu, Y. Chen, L. Xu, K. Zen, C. Zhang and H. Shen, Serum microRNA signatures identified in a genome-wide serum microRNA expression profiling predict survival of non-small-cell lung cancer, *J. Clin. Oncol.*, 2010, **28**, 1721–1726.
26. D. J. Gibbings, C. Ciaudo, M. Erhardt and O. Voinnet, Multivesicular bodies associate with components of miRNA effector complexes and modulate miRNA activity, *Nat. Cell Biol.*, 2009, **11**, 1143–1149.
27. K. C. Vickers, B. T. Palmisano, B. M. Shoucri, R. D. Shamburek and A. T. Remaley, MicroRNAs are transported in plasma and delivered to recipient cells by high-density lipoproteins, *Nat. Cell Biol.*, 2011, **13**, 423–433.
28. S. Bala, J. Petrasek, S. Mundkur, D. Catalano, I. Levin, J. Ward, H. Alao, K. Kodys and G. Szabo, Circulating microRNAs in exosomes indicate hepatocyte injury and inflammation in alcoholic, drug-induced, and inflammatory liver diseases, *Hepatology*, 2012, **56**, 1946–1957.
29. R. Medzhitov and C. Janeway Jr, Innate immune recognition: mechanisms and pathways, *Immunol. Rev.*, 2000, **173**, 89–97.
30. F. Heil, H. Hemmi, H. Hochrein, F. Ampenberger, C. Kirschning, S. Akira, G. Lipford, H. Wagner and S. Bauer, Species-specific recognition of single-stranded RNA *via* toll-like receptor 7 and 8, *Science*, 2004, **303**, 1526–1529.
31. M. Fabbri, TLRs as miRNA receptors, *Cancer Res.*, 2012, **72**, 6333–6337.
32. S. M. Lehmann, C. Kruger, B. Park, K. Derkow, K. Rosenberger, J. Baumgart, T. Trimbuch, G. Eom, M. Hinz, D. Kaul, P. Habbel, R. Kalin, E. Franzoni, A. Rybak, D. Nguyen, R. Veh, O. Ninnemann, O. Peters, R. Nitsch, F. L. Heppner, D. Golenbock, E. Schott, H. L. Ploegh, F. G. Wulczyn and S. Lehnardt, An unconventional role for miRNA: *let-7* activates Toll-like receptor 7 and causes neurodegeneration, *Nat. Neurosci.*, 2012, **15**, 827–835.
33. M. Fabbri, A. Paone, F. Calore, R. Galli, E. Gaudio, R. Santhanam, F. Lovat, P. Fadda, C. Mao, G. J. Nuovo, N. Zanesi, M. Crawford, G. H. Ozer, D. Wernicke, H. Alder, M. A. Caligiuri, P. Nana-Sinkam, D. Perrotti and C. M. Croce, MicroRNAs bind to Toll-like receptors to induce prometastatic inflammatory response, *Proc. Natl. Acad. Sci. U. S. A.*, 2012, **109**, E2110–E2116.
34. L. B. McKenna, J. Schug, A. Vourekas, J. B. McKenna, N. C. Bramswig, J. R. Friedman and K. H. Kaestner, MicroRNAs control intestinal epithelial differentiation, architecture, and barrier function, *Gastroenterology*, 2010, **139**, 1654–1664.e1.
35. D. Ye, S. Guo, R. Al-Sadi and T. Y. Ma, MicroRNA regulation of intestinal epithelial tight junction permeability, *Gastroenterology*, 2011, **141**, 1323–1333.

36. B. Meresse, J. Ripoche, M. Heyman and N. Cerf-Bensussan, Celiac disease: from oral tolerance to intestinal inflammation, autoimmunity and lymphomagenesis, *Mucosal Immunol.*, 2009, **2**, 8–23.
37. L. Giordani, T. Del Pinto, O. Vincentini, C. Felli, M. Silano and M. Viora, Two wheat decapeptides prevent gliadin-dependent maturation of human dendritic cells, *Exp. Cell Res.*, 2014, **321**, 248–254.
38. L. M. Sollid, Intraepithelial lymphocytes in celiac disease: license to kill revealed, *Immunity*, 2004, **21**, 303–304.
39. D. Schuppan, Y. Junker and D. Barisani, Celiac disease: from pathogenesis to novel therapies, *Gastroenterology*, 2009, **137**, 1912–1933.
40. S. Guandalini and A. Assiri, Celiac disease: a review, *JAMA Pediatr.*, 2014, **168**, 272–278.
41. S. Husby, S. Koletzko, I. R. Korponay-Szabo, M. L. Mearin, A. Phillips, R. Shamir, R. Troncone, K. Giersiepen, D. Branski, C. Catassi, M. Lelgeman, M. Maki, C. Ribes-Koninckx, A. Ventura and K. P. Zimmer, ESPGHAN Working Group on Coeliac Disease Diagnosis; ESPGHAN Gastroenterology Committee; European Society for Pediatric Gastroenterology, Hepatology, and Nutrition European Society for Pediatric Gastroenterology, Hepatology, and Nutrition guidelines for the diagnosis of coeliac disease, *J. Pediatr. Gastroenterol. Nutr.*, 2012, **54**, 136–160.
42. K. Giersiepen, M. Lelgemann, N. Stuhldreher, L. Ronfani, S. Husby, S. Koletzko and I. R. Korponay-Szabo, ESPGHAN Working Group on Coeliac Disease Diagnosis Accuracy of diagnostic antibody tests for coeliac disease in children: summary of an evidence report, *J. Pediatr. Gastroenterol. Nutr.*, 2012, **54**, 229–241.
43. K. A. Bascunan-Gamboa, M. Araya-Quezada and F. Perez-Bravo, MicroRNAs: an epigenetic tool to study celiac disease, *Rev. Esp. Enferm. Dig.*, 2014, **106**, 325–333.
44. L. Zhang, J. Cheng and X. M. Fan, MicroRNAs: New therapeutic targets for intestinal barrier dysfunction, *World J. Gastroenterol.*, 2014, **20**, 5818–5825.
45. M. Capuano, L. Iaffaldano, N. Tinto, D. Montanaro, V. Capobianco, V. Izzo, F. Tucci, G. Troncone, L. Greco and L. Sacchetti, MicroRNA-449a overexpression, reduced NOTCH1 signals and scarce goblet cells characterize the small intestine of celiac patients, *PLoS One*, 2011, **6**, e29094.
46. V. Vaira, L. Roncoroni, D. Barisani, G. Gaudioso, S. Bosari, G. Bulfamante, L. Doneda, D. Conte, C. Tomba, M. T. Bardella, S. Ferrero, M. Locatelli and L. Elli, microRNA profiles in coeliac patients distinguish different clinical phenotypes and are modulated by gliadin peptides in primary duodenal fibroblasts, *Clin. Sci.*, 2014, **126**, 417–423.
47. S. Magni, G. Buoli Comani, L. Elli, S. Vanessi, E. Ballarini, G. Nicolini, M. Rusconi, M. Castoldi, R. Meneveri, M. U. Muckenthaler, M. T. Bardella and D. Barisani, miRNAs affect the expression of innate and adaptive immunity proteins in celiac disease, *Am. J. Gastroenterol.*, 2014, **109**, 1662–1674.

48. G. Buoli Comani, R. Panceri, M. Dinelli, A. Biondi, C. Mancuso, R. Meneveri and D. Barisani, miRNA-regulated gene expression differs in celiac disease patients according to the age of presentation, *Genes Nutr.*, 2015, **10**, 482.
49. S. Comincini, F. Manai, C. Meazza, S. Pagani, C. Martinelli, N. Pasqua, G. Pelizzo, M. Biggiogera and M. Bozzola, Identification of Autophagy-Related Genes and Their Regulatory miRNAs Associated with Celiac Disease in Children, *Int. J. Mol. Sci.*, 2017, **18**, 391.
50. A. Baldassarre, C. Felli, G. Prantera and A. Masotti, Circulating microRNAs and Bioinformatics Tools to Discover Novel Diagnostic Biomarkers of Pediatric Diseases, *Genes*, 2017, **8**, 234.
51. C. Felli, A. Baldassarre and A. Masotti, Intestinal and Circulating MicroRNAs in Coeliac Disease, *Int. J. Mol. Sci.*, 2017, **18**(9), 1907.
52. A. Masotti, A. Baldassarre, M. Fabrizi, G. Olivero, M. C. Loreti, P. Giammaria, P. Veronelli, M. P. Graziani and M. Manco, Oral glucose tolerance test unravels circulating miRNAs associated with insulin resistance in obese preschoolers, *Pediatr. Obes.*, 2017, **12**, 229–238.
53. A. Masotti, A. Baldassarre, M. P. Guzzo, C. Iannuccelli, C. Barbato and M. Di Franco, Circulating microRNA Profiles as Liquid Biopsies for the Characterization and Diagnosis of Fibromyalgia Syndrome, *Mol. Neurobiol.*, 2017, **54**, 7129–7136.
54. X. Chen, H. Liang, J. Zhang, K. Zen and C. Y. Zhang, Secreted microRNAs: a new form of intercellular communication, *Trends Cell Biol.*, 2012, **22**, 125–132.
55. I. Salido-Guadarrama, S. Romero-Cordoba, O. Peralta-Zaragoza, A. Hidalgo-Miranda and M. Rodriguez-Dorantes, MicroRNAs transported by exosomes in body fluids as mediators of intercellular communication in cancer, *OncoTargets Ther.*, 2014, **7**, 1327–1338.
56. E. Bell and M. A. Taylor, Functional Roles for Exosomal MicroRNAs in the Tumour Microenvironment, *Comput. Struct. Biotechnol. J.*, 2016, **15**, 8–13.
57. M. D. Paraskevopoulou, G. Georgakilas, N. Kostoulas, I. S. Vlachos, T. Vergoulis, M. Reczko, C. Filippidis, T. Dalamagas and A. G. Hatzigeorgiou, DIANA-microT web server v5.0: service integration into miRNA functional analysis workflows, *Nucleic Acids Res.*, 2013, **41**, W169–W173.
58. A. Krek, D. Grun, M. N. Poy, R. Wolf, L. Rosenberg, E. J. Epstein, P. MacMenamin, I. da Piedade, K. C. Gunsalus, M. Stoffel and N. Rajewsky, Combinatorial microRNA target predictions, *Nat. Genet.*, 2005, **37**, 495–500.
59. L. B. McKenna, J. Schug, A. Vourekas, J. B. McKenna, N. C. Bramswig, J. R. Friedman and K. H. Kaestner, MicroRNAs control intestinal epithelial differentiation, architecture, and barrier function, *Gastroenterology*, 2010, **139**, 1654–1664.e1.
60. A. Masotti, Interplays between gut microbiota and gene expression regulation by miRNAs, *Front. Cell. Infect. Microbiol.*, 2012, **2**, 137.
61. A. Celluzzi and A. Masotti, How Our Other Genome Controls Our Epi-Genome, *Trends Microbiol.*, 2016, **24**, 777–787.

62. A. M. Zahm, M. Thayu, N. J. Hand, A. Horner, M. B. Leonard and J. R. Friedman, Circulating microRNA is a biomarker of pediatric Crohn disease, *J. Pediatr. Gastroenterol. Nutr.*, 2011, **53**, 26–33.
63. R. Duttagupta, S. DiRienzo, R. Jiang, J. Bowers, J. Gollub, J. Kao, K. Kearney, D. Rudolph, N. B. Dawany, M. K. Showe, T. Stamato, R. C. Getts and K. W. Jones, Genome-wide maps of circulating miRNA biomarkers for ulcerative colitis, *PLoS One*, 2012, **7**, e31241.
64. A. M. Zahm, N. J. Hand, D. M. Tsoucas, C. L. L. Guen, R. N. Baldassano and J. R. Friedman, Rectal microRNAs are perturbed in pediatric inflammatory bowel disease of the colon, *J. Crohn's Colitis*, 2014, **8**, 1108–1117.
65. A. Etheridge, I. Lee, L. Hood, D. Galas and K. Wang, Extracellular microRNA: a new source of biomarkers, *Mutat. Res.*, 2011, **717**, 85–90.
66. Z. Huang, D. Huang, S. Ni, Z. Peng, W. Sheng and X. Du, Plasma microRNAs are promising novel biomarkers for early detection of colorectal cancer, *Int. J. Cancer*, 2010, **127**, 118–126.
67. N. J. Park, H. Zhou, D. Elashoff, B. S. Henson, D. A. Kastratovic, E. Abemayor and D. T. Wong, Salivary microRNA: discovery, characterization, and clinical utility for oral cancer detection, *Clin. Cancer Res.*, 2009, **15**, 5473–5477.
68. M. F. Corsten, R. Dennert, S. Jochems, T. Kuznetsova, Y. Devaux, L. Hofstra, D. R. Wagner, J. A. Staessen, S. Heymans and B. Schroen, Circulating MicroRNA-208b and MicroRNA-499 reflect myocardial damage in cardiovascular disease, *Circ.: Cardiovasc. Genet.*, 2010, **3**, 499–506.
69. M. J. Lodes, M. Caraballo, D. Suciu, S. Munro, A. Kumar and B. Anderson, Detection of cancer with serum miRNAs on an oligonucleotide microarray, *PLoS One*, 2009, **4**, e6229.
70. K. E. Resnick, H. Alder, J. P. Hagan, D. L. Richardson, C. M. Croce and D. E. Cohn, The detection of differentially expressed microRNAs from the serum of ovarian cancer patients using a novel real-time PCR platform, *Gynecol. Oncol.*, 2009, **112**, 55–59.
71. C. Chen, D. A. Ridzon, A. J. Broomer, Z. Zhou, D. H. Lee, J. T. Nguyen, M. Barbisin, N. L. Xu, V. R. Mahuvakar, M. R. Andersen, K. Q. Lao, K. J. Livak and K. J. Guegler, Real-time quantification of microRNAs by stem-loop RT-PCR, *Nucleic Acids Res.*, 2005, **33**, e179.
72. A. Schamberger and T. I. Orban, 3′ IsomiR species and DNA contamination influence reliable quantification of microRNAs by stem-loop quantitative PCR, *PLoS One*, 2014, **9**, e106315.
73. S. C. Schuster, Next-generation sequencing transforms today's biology, *Nat. Methods*, 2008, **5**, 16–18.
74. S. Wach, E. Nolte, J. Szczyrba, R. Stohr, A. Hartmann, T. Orntoft, L. Dyrskjot, E. Eltze, W. Wieland, B. Keck, A. B. Ekici, F. Grasser and B. Wullich, MicroRNA profiles of prostate carcinoma detected by multiplatform microRNA screening, *Int. J. Cancer*, 2012, **130**, 611–621.
75. Y. Han, J. Chen, X. Zhao, C. Liang, Y. Wang, L. Sun, Z. Jiang, Z. Zhang, R. Yang, J. Chen, Z. Li, A. Tang, X. Li, J. Ye, Z. Guan, Y. Gui and Z. Cai, MicroRNA expression signatures of bladder cancer revealed by deep sequencing, *PLoS One*, 2011, **6**, e18286.

76. Z. Williams, I. Z. Ben-Dov, R. Elias, A. Mihailovic, M. Brown, Z. Rosenwaks and T. Tuschl, Comprehensive profiling of circulating microRNA *via* small RNA sequencing of cDNA libraries reveals biomarker potential and limitations, *Proc. Natl. Acad. Sci. U. S. A.*, 2013, **110**, 4255–4260.
77. S. A. Bellingham, B. M. Coleman and A. F. Hill, Small RNA deep sequencing reveals a distinct miRNA signature released in exosomes from prion-infected neuronal cells, *Nucleic Acids Res.*, 2012, **40**, 10937–10949.
78. X. Wu, G. Somlo, Y. Yu, M. R. Palomares, A. X. Li, W. Zhou, A. Chow, Y. Yen, J. J. Rossi, H. Gao, J. Wang, Y. C. Yuan, P. Frankel, S. Li, K. T. Ashing-Giwa, G. Sun, Y. Wang, R. Smith, K. Robinson, X. Ren and S. E. Wang, De novo sequencing of circulating miRNAs identifies novel markers predicting clinical outcome of locally advanced breast cancer, *J. Transl. Med.*, 2012, **10**, 42.
79. P. Mestdagh, N. Hartmann, L. Baeriswyl, D. Andreasen, N. Bernard, C. Chen, D. Cheo, P. D'Andrade, M. DeMayo, L. Dennis, S. Derveaux, Y. Feng, S. Fulmer-Smentek, B. Gerstmayer, J. Gouffon, C. Grimley, E. Lader, K. Y. Lee, S. Luo, P. Mouritzen, A. Narayanan, S. Patel, S. Peiffer, S. Ruberg, G. Schroth, D. Schuster, J. M. Shaffer, E. J. Shelton, S. Silveria, U. Ulmanella, V. Veeramachaneni, F. Staedtler, T. Peters, T. Guettouche, L. Wong and J. Vandesompele, Evaluation of quantitative miRNA expression platforms in the microRNA quality control (miRQC) study, *Nat. Methods*, 2014, **11**, 809–815.
80. H. H. Cheng, H. S. Yi, Y. Kim, E. M. Kroh, J. W. Chien, K. D. Eaton, M. T. Goodman, J. F. Tait, M. Tewari and C. C. Pritchard, Plasma processing conditions substantially influence circulating microRNA biomarker levels, *PLoS One*, 2013, **8**, e64795.
81. M. A. McAlexander, M. J. Phillips and K. W. Witwer, Comparison of Methods for miRNA Extraction from Plasma and Quantitative Recovery of RNA from Cerebrospinal Fluid, *Front. Genet.*, 2013, **4**, 83.
82. B. D'haene, P. Mestdagh, J. Hellemans and J. Vandesompele, miRNA expression profiling: from reference genes to global mean normalization, *Methods Mol. Biol.*, 2012, **822**, 261–272.
83. S. S. Luo, O. Ishibashi, G. Ishikawa, T. Ishikawa, A. Katayama, T. Mishima, T. Takizawa, T. Shigihara, T. Goto, A. Izumi, A. Ohkuchi, S. Matsubara, T. Takeshita and T. Takizawa, Human villous trophoblasts express and secrete placenta-specific microRNAs into maternal circulation *via* exosomes, *Biol. Reprod.*, 2009, **81**, 717–729.
84. M. N. Lardizabal, A. L. Nocito, S. M. Daniele, L. A. Ornella, J. F. Palatnik and L. M. Veggi, Reference genes for real-time PCR quantification of microRNAs and messenger RNAs in rat models of hepatotoxicity, *PLoS One*, 2012, **7**, e36323.
85. R. Wang, N. Li, Y. Zhang, Y. Ran and J. Pu, Circulating microRNAs are promising novel biomarkers of acute myocardial infarction, *Intern. Med.*, 2011, **50**, 1789–1795.
86. M. P. Hunter, N. Ismail, X. Zhang, B. D. Aguda, E. J. Lee, L. Yu, T. Xiao, J. Schafer, M. L. Lee, T. D. Schmittgen, S. P. Nana-Sinkam, D. Jarjoura and

C. B. Marsh, Detection of microRNA expression in human peripheral blood microvesicles, *PLoS One*, 2008, **3**, e3694.

87. A. S. Ho, X. Huang, H. Cao, C. Christman-Skieller, K. Bennewith, Q. T. Le and A. C. Koong, Circulating miR-210 as a Novel Hypoxia Marker in Pancreatic Cancer, *Transl. Oncol.*, 2010, **3**, 109–113.
88. W. Zhu, W. Qin, U. Atasoy and E. R. Sauter, Circulating microRNAs in breast cancer and healthy subjects, *BMC Res. Notes*, 2009, **2**, 89.
89. H. Zhao, J. Shen, L. Medico, D. Wang, C. B. Ambrosone and S. Liu, A pilot study of circulating miRNAs as potential biomarkers of early stage breast cancer, *PLoS One*, 2010, **5**, e13735.
90. C. Roth, B. Rack, V. Muller, W. Janni, K. Pantel and H. Schwarzenbach, Circulating microRNAs as blood-based markers for patients with primary and metastatic breast cancer, *Breast Cancer Res.*, 2010, **12**, R90.
91. K. Z. Qu, K. Zhang, H. Li, N. H. Afdhal and M. Albitar, Circulating microRNAs as biomarkers for hepatocellular carcinoma, *J. Clin. Gastroenterol.*, 2011, **45**, 355–360.
92. A. Yamada, M. A. Cox, K. A. Gaffney, A. Moreland, C. R. Boland and A. Goel, Technical factors involved in the measurement of circulating microRNA biomarkers for the detection of colorectal neoplasia, *PLoS One*, 2014, **9**, e112481.
93. P. Mestdagh, P. Van Vlierberghe, A. De Weer, D. Muth, F. Westermann, F. Speleman and J. Vandesompele, A novel and universal method for microRNA RT-qPCR data normalization, *Genome Biol.*, 2009, **10**, R64.
94. N. Vigneron, M. Meryet-Figuiere, A. Guttin, J. P. Issartel, B. Lambert, M. Briand, M. H. Louis, M. Vernon, P. Lebailly, Y. Lecluse, F. Joly, S. Krieger, S. Lheureux, B. Clarisse, A. Leconte, P. Gauduchon, L. Poulain and C. Denoyelle, Towards a new standardized method for circulating miRNAs profiling in clinical studies: Interest of the exogenous normalization to improve miRNA signature accuracy, *Mol. Oncol.*, 2016, **10**, 981–992.
95. F. Russo, S. D. Bella, G. Nigita, V. Macca, A. Lagana, R. Giugno, A. Pulvirenti and A. Ferro, miRandola: extracellular circulating microRNAs database, *PLoS One*, 2012, **7**, e47786.
96. S. Mathivanan, C. J. Fahner, G. E. Reid and R. J. Simpson, ExoCarta 2012: database of exosomal proteins, RNA and lipids, *Nucleic Acids Res.*, 2012, **40**, D1241–D1244.
97. D. K. Kim, J. Lee, R. J. Simpson, J. Lotvall and Y. S. Gho, EVpedia: A community web resource for prokaryotic and eukaryotic extracellular vesicles research, *Semin. Cell Dev. Biol.*, 2015, **40**, 4–7.
98. A. Benito-Martin and H. Peinado, FunRich proteomics software analysis, let the fun begin, *Proteomics*, 2015, **15**, 2555–2556.
99. H. Kalra, R. J. Simpson, H. Ji, E. Aikawa, P. Altevogt, P. Askenase, V. C. Bond, F. E. Borras, X. Breakefield, V. Budnik, E. Buzas, G. Camussi, A. Clayton, E. Cocucci, J. M. Falcon-Perez, S. Gabrielsson, Y. S. Gho, D. Gupta, H. C. Harsha, A. Hendrix, A. F. Hill, J. M. Inal, G. Jenster, E. M. Kramer-Albers, S. K. Lim, A. Llorente, J. Lotvall, A. Marcilla, L. Mincheva-Nilsson, I. Nazarenko, R. Nieuwland, E. N. N. Hoen, A. Pandey, T. Patel,

M. G. Piper, S. Pluchino, T. S. Prasad, L. Rajendran, G. Raposo, M. Record, G. E. Reid, F. Sanchez-Madrid, R. M. Schiffelers, P. Siljander, A. Stensballe, W. Stoorvogel, D. Taylor, C. Thery, H. Valadi, B. W. v. Balkom, J. Vazquez, M. Vidal, M. H. Wauben, M. Yanez-Mo, M. Zoeller and S. Mathivanan, Vesiclepedia: a compendium for extracellular vesicles with continuous community annotation, *PLoS Biol.*, 2012, **10**, e1001450.

100. F. Russo, S. D. Bella, V. Bonnici, A. Lagana, G. Rainaldi, M. Pellegrini, A. Pulvirenti, R. Giugno and A. Ferro, A knowledge base for the discovery of function, diagnostic potential and drug effects on cellular and extracellular miRNAs, *BMC Genomics*, 2014, **15**(Suppl 3), S4.
101. M. V. Barone, D. Zanzi, M. Maglio, M. Nanayakkara, S. Santagata, G. Lania, E. Miele, M. T. Ribecco, F. Maurano, R. Auricchio, C. Gianfrani, S. Ferrini, R. Troncone and S. Auricchio, Gliadin-mediated proliferation and innate immune activation in celiac disease are due to alterations in vesicular trafficking, *PLoS One*, 2011, **6**, e17039.
102. K. Juuti-Uusitalo, M. Maki, H. Kainulainen, J. Isola and K. Kaukinen, Gluten affects epithelial differentiation-associated genes in small intestinal mucosa of coeliac patients, *Clin. Exp. Immunol.*, 2007, **150**, 294–305.
103. A. Gregorieff and H. Clevers, Wnt signaling in the intestinal epithelium: from endoderm to cancer, *Genes Dev.*, 2005, **19**, 877–890.
104. T. Fevr, S. Robine, D. Louvard and J. Huelsken, Wnt/beta-catenin is essential for intestinal homeostasis and maintenance of intestinal stem cells, *Mol. Cell. Biol.*, 2007, **27**, 7551–7559.
105. S. F. Moss, L. Attia, J. V. Scholes, J. R. Walters and P. R. Holt, Increased small intestinal apoptosis in coeliac disease, *Gut*, 1996, **39**, 811–817.
106. L. Maiuri, C. Ciacci, V. Raia, L. Vacca, I. Ricciardelli, F. Raimondi, S. Auricchio, S. Quaratino and M. Londei, FAS engagement drives apoptosis of enterocytes of coeliac patients, *Gut*, 2001, **48**, 418–424.
107. S. C. Bischoff, G. Barbara, W. Buurman, T. Ockhuizen, J. D. Schulzke, M. Serino, H. Tilg, A. Watson and J. M. Wells, Intestinal permeability–a new target for disease prevention and therapy, *BMC Gastroenterol.*, 2014, **14**, 189.
108. M. G. Clemente, S. De Virgiliis, J. S. Kang, R. Macatagney, M. P. Musu, M. R. Di Pierro, S. Drago, M. Congia and A. Fasano, Early effects of gliadin on enterocyte intracellular signalling involved in intestinal barrier function, *Gut*, 2003, **52**, 218–223.
109. A. Masotti, A. Baldassarre, M. P. Guzzo, C. Iannuccelli, C. Barbato and M. Di Franco, Circulating microRNA Profiles as Liquid Biopsies for the Characterization and Diagnosis of Fibromyalgia Syndrome, *Mol. Neurobiol.*, 2016, **54**(9), 7129–7136.

CHAPTER 4

MicroRNAs and Mycobacterial Infections in Humans and Domestic Animals

SARAH L. FAHERTY O'DONNELL[a], CAROLINA N. CORREIA[a] AND DAVID E. MACHUGH*[a,b]

[a]Animal Genomics Laboratory, UCD School of Agriculture and Food Science, UCD College of Health and Agricultural Sciences, University College Dublin, Dublin D04 V1W8, Ireland; [b]UCD Conway Institute of Biomolecular and Biomedical Research, University College Dublin, Dublin D04 V1W8, Ireland
*E-mail: david.machugh@ucd.ie

4.1 Introduction: Many Mycobacteria, Many Diseases

Tuberculosis (TB) is a disease associated with poverty that has afflicted humans and animals for millennia. Responsible for the deaths of approximately 1.7 million people in 2016,[1] TB develops following infection with different lineages (strains) of mycobacteria from the *Mycobacterium tuberculosis* complex (MTBC). The MTBC comprises 10 Linnaean binomial species with very similar genome sequences, but different host adaptation ranges: *Mycobacterium tuberculosis*, *M. africanum*, *M. canetti*, *M. bovis*, *M. microti*, *M. orygis*,

Drug Discovery Series No. 69
MicroRNAs in Diseases and Disorders: Emerging Therapeutic Targets
Edited by Philip V. Peplow, Bridget Martinez, George A. Calin and Aurora Esquela-Kerscher

Published by the Royal Society of Chemistry, www.rsc.org

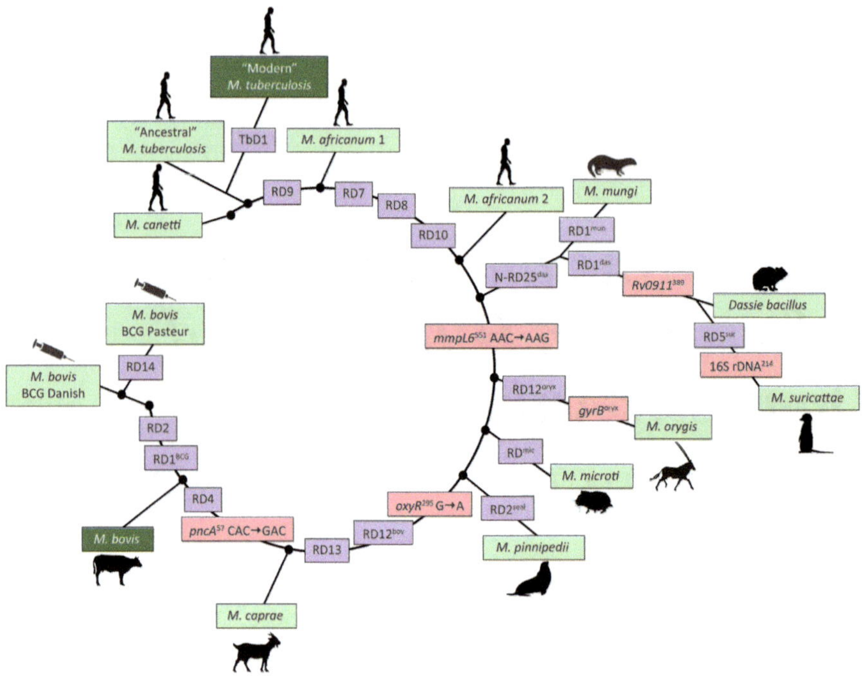

Figure 4.1 The genomic variations and regions of deletion that distinguish the animal-adapted species of the *Mycobacterium tuberculosis* complex (MTBC). This diagram depicts the regions of deletion (RD; light purple) and single nucleotide polymorphisms (SNPs; pink) that can be used to reconstruct the evolution of the MTBC phylogeny, from an *M. canetti*-like ancestor to *M. bovis* bacille Calmette–Guérin (BCG) Pasteur. The strains that represent the largest burdens of TB disease in humans (*M. tuberculosis*) and cattle (*M. bovis*) are shown in dark green. Adapted from ref. 12 with permission from Springer Nature, Copyright 2017.

M. caprae, *M. pinnipedii*, *M. suricattae* and *M. mungi* (Figure 4.1). Without therapeutic intervention, infection in humans with any one of these strains can result in fatal pulmonary and extra-pulmonary TB disease.

MicroRNAs (miRNAs) are short, non-coding regulators of gene expression that play a key role in maintaining cellular homeostasis within an organism by targeting and blocking or inhibiting translation of messenger RNA (mRNA) molecules. Generally 18–22 nucleotides in length, miRNAs act by binding with perfect or partial complementarity to the 3′ untranslated region (UTR) of mRNA molecules.[2,3] Due to their role in many biological processes, including the innate and adaptive immune responses, miRNAs are fundamental to mounting an effective immune response to bacterial infections, such as those caused by mycobacteria.[4–6]

In 1882, Robert Koch identified the tubercle bacillus as the causative agent of TB and 16 years later, Theobald Smith demonstrated a difference in host specificity between *M. tuberculosis* and *M. bovis*.[7,8] These discoveries provided the groundwork for the development of the bacille Calmette–Guérin (BCG) vaccine, TB diagnostics, and ultimately, effective therapeutic intervention.[9]

There has been a dramatic reduction in rates of human mortality caused by TB since the introduction of the BCG vaccine in the early 20th century[10] and the implementation of combination therapy for TB during the 1950s.[11] However, TB remains one of the top 10 causes of death worldwide. In 2016 alone, 10.4 million people fell ill with TB disease and 1.7 million of these cases resulted in death, with >95% of these deaths occurring in developing countries.[1] Consequently, human TB is an infectious disease which still poses significant health, social, and economic challenges.

M. tuberculosis is a slow-growing, acid-fast bacteria that preferentially establishes infection in humans leading to pulmonary and extra-pulmonary TB. *M. tuberculosis* is transmitted *via* aerosolised droplets that are excreted from an infected host and inhaled into the lungs of another, non-infected host. If not successfully cleared by the host immune response, *M. tuberculosis* infection can develop into active or latent TB disease. One-third of the global population are thought to be latently infected and with one in 10 latent TB cases leading to reactivation of infection, *M. tuberculosis* poses a significant threat to global human health.[9,13] *M. africanum*, another member of the MTBC, is clinically identical to *M. tuberculosis* and endemic in West Africa.[8,14] The treatment regimen for TB patients lasts up to 9 months and requires administration of a combination of drugs. The emergence of multidrug resistant strains of *M. tuberculosis* increases the need for novel diagnostics and therapies.[15,16]

M. bovis is a strain of the MTBC that primarily infects cattle, but can also infect a wider host range including humans, badgers, deer, and antelopes.[17,18] Prior to pasteurisation of milk, extra-pulmonary TB often developed in humans following consumption of milk from *M. bovis*-infected cattle. In developing countries, occupational exposure to *M. bovis*-infected cattle and consumption of unpasteurised milk results in *M. bovis*-derived TB disease in humans that can prove fatal if left untreated.[19,20]

BCG is a live attenuated version of *M. bovis* that was developed by two French bacteriologists, Albert Calmette and Camille Guérin, between 1908 and 1919. Calmette and Guérin generated the attenuated strain of *M. bovis* by repeatedly sub-culturing the virulent bacteria on a medium containing ox bile, potato, and glycerine every 3 weeks for 11 years, continuing their work throughout World War I.[7,21] First administered to a human child in 1921, the BCG vaccine was found to be safe and widespread vaccination commenced.[10]

Hansen's disease, more commonly known as leprosy, is caused by *M. leprae*, an evolutionarily related (but phylogenetically distinct) species to the MTBC strains.[22] *M. leprae* has been shown to primarily target the skin, nervous system, and upper mucosa of the respiratory tract in humans and animals.[23] Despite being completely curable using a combination of antimicrobial drugs, leprosy remains a global health problem, with 214 783 new cases reported to the World Health Organization during 2016 alone.[24]

In a similar fashion to human TB, the number of reported leprosy cases indicates a persistent level of transmission despite therapeutic intervention. Compared to the members of the MTBC, *M. leprae* is an extremely slow-growing bacteria with an incubation period estimated at ~5 years, making *M. leprae* a difficult pathogen to detect prior to clinical presentation.[25] Recently, a second strain, *M. lepromatosis*, was isolated from red squirrels (*Sciurus vulgaris*) exhibiting leprosy-like lesions in England, Scotland and Ireland signifying a wildlife reservoir of infection in these animals.[26]

Johne's disease is a chronic inflammatory bowel disease of ruminant animals caused by *M. avium* subspecies *paratuberculosis* (MAP). Exhibiting similar symptoms to that of Crohn's disease in humans, MAP predominantly infects the intestinal tract of animals and domestic livestock.[27] Johne's disease is a slow-progressing condition that varies in severity, with some MAP-infected animals presenting no clinical symptoms whatsoever while others develop chronic enteric inflammation leading to rapid weight loss and significantly decreased production yields.[28] As well as posing a threat to animal welfare and the global farming sector, MAP has been isolated from humans and has been controversially hypothesised to be a causative agent of Crohn's disease.[29,30]

Non-tuberculous mycobacteria (NTM) are strains that do not cause TB or leprosy disease. *M. smegmatis* is a rapid-growing environmental NTM that was first discovered in 1885.[31] Although not usually considered a pathogenic threat to humans, on occasion, *M. smegmatis* has presented as an opportunistic infection, with several cases being reported as fatal.[32–34] The opportunistic nature of NTM is most often seen in immunocompromised individuals, but can also present in immunocompetent patients as established lung infection, which can be mistaken for TB. Clinical isolates allow for the differentiation of NTMs from *M. tuberculosis* and other members of the MTBC, with the most common NTM strains in Europe being reported as members of the *Mycobacterium avium* complex (MAC), *M. gordonae*, and *M. xenopi*.[35]

4.2 Host–Pathogen Interaction, Interplay, and the Immunobiology of Human and Bovine Tuberculosis

Human TB occurs following infection with *M. tuberculosis*, but can also develop following zoonotic infection with the cattle-adapted *M. bovis* strain. For both *M. tuberculosis* and *M. bovis*, infectious bacilli are transmitted *via* aerosolised droplets that have been excreted or exhaled from a clinically infected host and inhaled into the lungs of a non-infected individual.[9,36,37] Once inside the lungs, invading bacilli travel down the bronchi to the alveoli where they are actively engulfed (phagocytosed) by resident alveolar macrophages.[38,39] Initial detection of *M. tuberculosis* by the macrophage is facilitated by cell surface receptors that recognise bacterial antigens.[40,41] This leads to phagocytosis of the bacteria into phagosomes that fuse with lysosomes to form phagolysosomes.[38,42] The intracellular bacteria can then be enzymatically degraded within the acidic environment of the phagolysosome and presented on the surface of the macrophage by the major histocompatibility complexes (MHCs) class I and class II, for recognition by T-cells as part of the adaptive immune response.[43,44]

A second response to bacterial antigen detection by the macrophage is an intracellular cascade of signalling events that leads to the expression and secretion of signalling proteins known as cytokines and chemokines.[45,46] Cytokines are signalling molecules that activate, differentiate, and effectively control immune cells by binding to their corresponding cell surface receptors. Chemokines are chemotactic cytokines that attract other immune cells to the site of infection, heightening the potency of the immune response. Interferon-gamma (IFN-γ) is one of the most well characterised cytokines due to its involvement in driving inflammation, while interleukin-8 (IL-8) is a well-known chemoattractant cytokine that attracts neutrophils to the site of infection.

In the context of intracellular mycobacterial pathogens, it is important to note that macrophages are said to be classically activated (M1 macrophages) when stimulated by IFN-γ and lipopolysaccharides (LPS) or tumour necrosis factor (TNF, TNF-α), and lead to the activation of T-helper I (T_H1) cell activation and inflammatory responses.[47,48] M1 macrophages are associated with intracellular killing of pathogens, tumour resistance, and secretion of the cytokines IL-1, IL-6, IL-12, IL-23 and TNF. They are also characterised by intracellular inducible nitric oxide synthase production, and secretion of reactive nitrogen and oxygen intermediates.[49,50] Furthermore, macrophages activated *via* alternative processes (M2 macrophages) are associated with immune regulation, T-helper II (T_H2) cell activation, tissue remodelling, and allergy.[47,48] Depending on their stimulating molecules, M2 macrophages produce IL-10, matrix proteins, polyamines, transforming growth factor beta (TGF-β), and low levels of IL-12. The plasticity of macrophages allows for rapid phenotype switching between pro-inflammatory and anti-inflammatory responses, which are key to maintaining homeostasis and preventing immune-mediated damage to host cells and tissues.[49]

Dysregulation of macrophage polarisation is detrimental to maintaining homeostasis and can significantly affect the health status of an organism. Therefore, a key observation regarding intracellular mycobacterial pathogens is that macrophage cells infected with *M. tuberculosis* and *M. bovis* can adopt an M2 phenotype despite type I immunity being crucial to host control and clearance.[51,52] Ultimately, this interaction is critical to the formation of granulomas in the lung, which are complex aggregates of macrophages containing mycobacteria in a dormant state, surrounded by layers of immune cells—neutrophils, dendritic cells, B- and T-cells and natural killer (NK) cells—that work together to contain infection.[53] For the most part, this granulomatous structure can maintain the mycobacterial infection in a dormant state for the lifespan of the host, but for individuals that become immunocompromised or, for example, are co-infected with the human immunodeficiency virus (HIV), the granuloma can become unstable and disintegrate, allowing for re-activation of infection to occur.[9,13,36]

Mycobacterial pathogens have evolved a bewildering array of molecular strategies to avoid destruction by the microbicidal processes inside host macrophages. These intracellular evasion and subversion strategies are best understood for MTBC strains that cause TB disease in humans and other animals, particularly for *M. tuberculosis*.[39,42,54–57] Indeed, it has been proposed that *M. tuberculosis* represents a paradigm of host–pathogen adaptation with a distinct "pathogenic personality".[56] Consequently, most of what follows summarises examples of macrophage interactions with *M. tuberculosis*, although it is well understood that much of what has been discovered also applies to the closely related *M. bovis* strain.

The macrophage is a highly efficient phagocytic cell that plays a crucial role in activating the adaptive immune response. Infecting *M. tuberculosis* bacilli are detected by macrophages and engulfed into phagosomes that are destined for fusion with low-pH lysosomes containing degradative enzymes. The hostile environment of a phagolysosome is usually sufficient for bacterial degradation; however, *M. tuberculosis* has evolved numerous immunoevasion strategies to survive within macrophage phagosomes and prevent phagolysosomal fusion.

To promote intracellular survival, *M. tuberculosis* employs several mechanisms to inhibit phagolysosome maturation. First, *M. tuberculosis* is thought to prevent phagolysosome maturation by secreting a phosphatase enzyme, phosphatidylinositol-3-phosphatase (SapM), which interferes with the synthesis and accumulation of an essential host membrane component, phosphatidylinositol 3-phosphate (PI3P).[58] A second *M. tuberculosis*-secreted phosphatase, phosphotyrosine protein phosphatase (PtpA), inhibits phagolysosomal fusion in the presence of a host vesicular sorting protein, vacuolar protein sorting-associated protein 33B (VPS33B).[59] In addition, chemical inhibition of serine/threonine-protein kinase (PknG), a soluble kinase protein conserved across all pathogenic mycobacteria, unblocks phagolysosome fusion in *M. tuberculosis*-infected macrophages, thereby supporting the hypothesis that mycobacterial PknG has evolved to prevent phagolysosome fusion and promote intracellular bacterial survival.[60] In addition to secreting phosphatases and kinases to promote intracellular survival, pathogenic mycobacteria also exploit and subvert components of the host macrophage cellular machinery. For example, it has been shown that *M. tuberculosis* recruits a host factor, the leukocyte-specific protein coronin-1A, to regulate calcium-dependent signalling and prevent lysosomal delivery of the bacteria by the phagosome.[61]

The stimulation of macrophages by IFN-γ and TNF induces a more potent microbicidal response within these cells and is associated with successful lysosomal destruction of intracellular mycobacteria; however, it has been shown that the cell signalling processes governing stimulation by IFN-γ can vary between mammalian host species.[62] Oxygen and nitrogen intermediates also play a central role in intracellular antimicrobial responses of macrophages, but have on occasion been found to be ineffective against *M. tuberculosis*.[63,64] Therefore, it is unsurprising that mycobacteria have evolved to survive within these toxic environments by secreting enzymes to neutralise reactive oxygen species and by synthesising protective proteases against nitric-oxide stress.[65,66]

Lipoarabinomannan (LAM) is an abundant *M. tuberculosis* surface glycolipid and a well-characterised virulence factor important for infection of macrophages. LAM mediates uptake of *M. tuberculosis* into macrophages *via* interaction with host mannose and complement receptors, but is then localised to the macrophage membrane to prevent phagosome maturation.[67] LAM has also been shown to shift the cytokine secretion profiles of T-cells from a T_H1 to a T_H2 response.[68] In a study by Knutson *et al.*,[69] LAM was shown to activate the protein tyrosine phosphatase, non-receptor type 6 (PTPN6) (previously known as src homology region 2 domain-containing phosphatase-1) which leads to dephosphorylation of mitogen-activated protein kinase, thereby attenuating downstream signalling events and preventing macrophage activation and microbicidal activity.

In addition to intra-phagosomal mechanisms of survival, it has also been proposed that *M. tuberculosis* can employ phagosomal escape mechanisms to replicate in the cytosol. Transmission electron microscopy has been used to show that different *M. tuberculosis* strains show varying capacities to escape the phagosome,[70] and various mycobacterial factors—including phthiodiolone dimycocerosate and early secreted antigenic target of 6 kDa (ESAT-6), which is secreted by the homeobox protein ESX-1 system—have been implicated in phagosomal escape and cell lysis.[71–73]

4.3 Mammalian miRNAs

First discovered in 1993, miRNAs have since been shown to play a vital role in biochemical and physiological processes in eukaryotic cells and tissues; in particular, through maintenance of intracellular and systemic homeostasis, by post-transcriptionally regulating gene expression.[74,75] In addition, miRNAs are critical to mounting, maintaining, and controlling host responses to microbial infections, including those caused by mycobacterial pathogens, through regulation and fine-tuning of the expression of key immune genes.[4,6,76,77] miRNAs are short, non-coding, single-stranded, ribonucleic acids that range from ~18 to 22 nucleotides in length. These highly abundant and evolutionarily conserved molecules bind to the 3′ UTR of a target mRNA to prevent translation, thereby regulating gene expression.[2,3] Due to their short length and ability to bind genomic targets with perfect and imperfect complementarity, miRNAs have low complexity and are often highly redundant in their regulatory function. For this reason, miRNAs play an integral role in development, proliferation, differentiation, apoptosis, autophagy, quorum sensing, and immune responses.[5,6,76,78,79]

Mammalian miRNA biogenesis is usually classified into canonical and non-canonical pathways, depending on how miRNAs are processed within the cell nucleus. The canonical pathway comprises the transcription of miRNA genes by either RNA polymerase II (Pol II) or RNA polymerase III (Pol III) into long primary-miRNA (pri-miRNA) molecules, which are further processed by the RNase III enzyme Drosha and its co-factor DiGeorge critical region 8 (DGCR8) to form ~70 nucleotide precursor-miRNA (pre-miRNA) hairpin products.[80–83] Pre-miRNAs are then exported to the cytoplasm by Exportin V and RanGTP,[84] where they are cleaved by Dicer, an RNA polymerase III, generating a transient (~22 nucleotide) mature miRNA : miRNA duplex.[85] The duplex is then incorporated into an Argonaute-containing RNA induced silencing complex (RISC) that retains a single-stranded mature miRNA known as the guide strand. The function of the guide strand is to post-transcriptionally regulate a target mRNA that shares sequence complementarity to its seed region.[86] Conversely, non-canonical miRNA biogenesis encompasses a range of different processing routes that are usually further grouped into Drosha/DGCR8-independent and Dicer-independent pathways.[87] These include mature miRNAs originated from introns of mRNAs during splicing, which are called mirtrons. In a Dicer-independent manner, the cleavage of endogenous small hairpin RNA (shRNA) molecules by Drosha/DGCR8 also generates functional mature miRNAs.[87]

Seed regions are short, conserved nucleotide sequences consisting of 2–8 bp at the 5′ end of a mature miRNA that must bind to a region of perfect or partial complementarity on the 3′ UTR of a target mRNA in order to regulate gene expression. There has been debate in the scientific literature concerning which mode of miRNA action is most important in metazoan animals, with experimental evidence supporting both mRNA destabilisation[75,88] and translational repression.[89,90] miRNAs are capable of repressing mRNA translation by preventing the formation of the ribosome–tRNA–mRNA complex at the initiation of translation and associating with polysomes and polyribosomes at the post-initiation and elongation phases of translation.[2,91–93] The scope and precision of current miRNA target detection protocols has been critically reviewed, suggesting that the dominant mechanism of miRNA action is translational repression.[94,95] Finally, and somewhat counterintuitively, under specific conditions miRNAs have also been observed to promote translation of target mRNA molecules.[96]

4.3.1 miRNAs and Development of Immune Cell Types

It is now well understood that differentiation of immune cells into their various lineages is influenced by microRNA transcriptional regulatory networks.[4,97–99] For example, myeloid and lymphoid cells are associated with the expression of miR-125b, miR-126, and miR-196b, while B-cell development and differentiation are modulated by expression of miR-17-92, miR-181a, miR-150, and miR-34a.[100–107] By regulating notch signalling, miR-181 was shown to be involved in the development of NK cells,[108] while miR-181a has been proposed to play a unique role in T-cell receptor signalling.[109] Conversely, Dicer, the ribonuclease required in the biosynthesis of miRNAs, is conditionally depleted in T-cell progenitors, which suggests that miRNAs may not play a significant role in T-cell differentiation.[110,111]

In a study performed by Bezman *et al.*,[112] a greater frequency of mature NK cells was seen in mice constitutively expressing miR-150, while mice with targeted depletion of miR-150 were unable to generate mature NK cells. In addition, the miRNA let-7 has been shown to modulate the cytokine secretion profile of murine natural killer T (NKT)-cells by targeting the zinc finger and BTB domain containing 16-protein (ZBTB16; previously known as promyelocytic leukaemia zinc finger).[113] When increased in expression, let-7 was observed to downregulate synthesis of *Zbtb16* mRNA and the NKT cells were found to produce high levels of IFN-γ. Conversely, depleted levels of let-7 resulted in an increase in ZBTB16 protein synthesis and an accumulation of IL-4- and IL-17-producing NKT cells.[113]

An example of indirect regulation of cell differentiation by a single miRNA was demonstrated by Hager *et al.*[114] who identified two miR-130a binding sites in the 3′ UTR of the human SMAD family member 4 gene (*SMAD4*) mRNA transcript. When investigated further, it was discovered that increased expression of miR-130a led to repression of this mRNA in HEK293, A549, and 32Dcl3 cells.

It has been demonstrated that B-cells lacking miR-155 failed to generate immunoglobulin G antibodies, implicating miR-155 as a critical regulator of the terminal differentiation programme of B-cells.[115] A central role for miRNAs encoded by the *mir144/mir451A* gene cluster in erythroblast maturation has also been demonstrated in the mouse.[116] Although mature erythrocytes cannot synthesise miRNAs, it has been observed that these cells can retain these molecules following enucleation.[117,118] Interestingly, a recent study has demonstrated a potential immunological role for red blood cells through isolation of Argonaute 2 (Ago2) and miRNA-containing microvesicles from red blood cells infected with *Plasmodium falciparum*.[119] Moreover, when transferred into the infecting parasites, the miRNA : Ago2 complexes downregulated the expression of *P. falciparum* erythrocyte membrane protein 1 (PfEMP1), a key malaria antigen.[119]

4.3.2 miRNAs and Immune Cell Functions

Immune cell functions are determined by transcriptional regulatory networks that generate specific proteomic profiles under various conditions and a large body of evidence now supports a central role for miRNAs in regulating these processes.[4–6,76] In particular, cytokine expression and signalling are fine-tuned by miRNAs that regulate translation of suppressor of cytokine signalling (SOCS) proteins, and *via* direct targeting of mRNA molecules transcribed by transcription factors. For example, miR-19a represses *SOCS3* mRNA transcripts to drive cytokine signalling, whereas an increase in expression of miR-146a attenuated innate immune responses by targeting

TNF receptor associated factor 6 gene (*TRAF6*) and interleukin 1 receptor associated kinase 1 gene (*IRAK1*) mRNA transcripts.[120,121] miR-29 and miR-99b were found to regulate IFN-γ and TNF signalling in macrophage and T-cells,[122–125] while miR-146a-5p is induced by IL-1β to subsequently repress IRAK signalling.[126] Additionally, miR-132 has been implicated in regulating innate immune signalling by targeting IRAK4 in macrophage cells stimulated with peptidoglycan,[127] and miR-21 has been shown to negatively regulate the murine toll-like receptor (TLR) 4 signalling pathway, nuclear factor kappa-light-chain-enhancer of activated B cells (NF-κB) activity, and type I IFN production by targeting the programmed cell death 4 (*Pdcd4*) mRNA transcript.[128]

MiR-146a and miR-155 are both examples of miRNAs that create a balance between inflammation and repair and illustrate miRNA modulation of complex cellular processes through negative feedback loops. For instance, miR-146a, although activated by the pro-inflammatory cytokine NF-κB, targets *TRAF6* and *IRAK1* transcripts, thereby negatively regulating TLR signalling.[126] Similarly, miR-99a expression driven by NF-κB regulates endothelial cell inflammation through targeted inhibition of NF-κB signalling.[129] MiR-155 drives myeloid cell differentiation and amplifies inflammatory signals by decreasing expression of negative regulators of inflammation in some cells.[130,131] However, in LPS-stimulated macrophages, IL-10 downregulates miR-155 expression *via* the transcription factor signal transducer and activator of transcription 3 (STAT3) and the enzyme phosphatidylinositol-3,4,5-trisphosphate 5-phosphatase 1 (INPP5D or SHIP1), thereby attenuating inflammation.[132] Conversely, it has also been shown that IL-10 can upregulate miR-155, which in turn targets suppressor of cytokine signalling 1 gene (*SOCS1*) mRNA transcripts, leading to enhanced mast cell responses.[132]

4.3.3 miRNA Binding Site Variation, Immunobiology and Mycobacterial Infections

An important and perhaps underappreciated aspect of miRNA-directed transcriptional regulation is the evolutionary role of genomic sequence variation at 3′ UTR target sites for these regulatory RNA molecules.[133–137] The importance of this phenomenon was first demonstrated more than a decade ago in domestic sheep bred for meat production. Clop *et al.*[136] showed that an initially rare myostatin gene (*MSTN*) variant had rapidly become almost fixed in the Texel sheep breed, which has been subject to intense artificial selection for increased musculature over the past 200 years. The phenotypic effect of this *MSTN* allele became apparent when Clop *et al.* showed that a 3′ UTR G→A change characterised the selected variant, which generated a *de novo* target site for miR-1 and miR-206, thereby leading to downregulation of the myostatin protein, a negative regulator of muscle growth.[136]

The role of miRNA binding site variation in immunobiology is becoming more widely recognised, although to date, there have been relatively few reports of 3′ UTR sequence variation being associated with resistance or susceptibility to infectious diseases. However, it has recently been shown that miR-4717 binding site variation at the programmed cell death 1 gene (*PDCD1*) is associated with host responses to hepatitis B virus infection.[138,139] At a broader level, it has also been shown that the expression of mammalian MHC genes can be profoundly influenced by 3′ UTR miRNA target site polymorphisms.[140] Finally, miRNA binding site polymorphism has been shown to be key to post-transcriptional regulation of the immunomodulatory MHC class I G gene (*HLA-G*).[141]

Taking these observations into account, it is likely that the importance of miRNA binding site variation in the gene regulatory networks that underpin vertebrate immunobiology will become increasingly apparent during the coming years. In particular, it is likely that miRNA binding site polymorphisms account for a portion of the genomic variation underlying intra- and interpopulation differences in susceptibility or resistance to mycobacterial infections in humans and animals.[142–145] For example, it has been shown that 3′ UTR polymorphisms probably influencing miRNA binding sites in the caprine solute carrier family 11 member 1 gene (*SLC11A1*) are associated with outcomes of MAP infection in domestic goats.[146] Similarly, recent work has demonstrated that a miR-125b binding site polymorphism in the 3′ UTR of the *IFNG* gene significantly affects *IFNG* expression and is associated with susceptibility to bovine tuberculosis in Mediterranean water buffalo (*Bubalus bubalis*).[147]

4.4 miRNAs and Mycobacterial Infections

Dynamic and orchestrated control of miRNA and mRNA expression profiles are critical in mounting an effective host immune response to infection. However, perturbations of transcriptional regulatory networks, including those involving miRNAs, can also benefit infecting pathogens and contribute to pathogenesis and disease transmission. During the past 5 years, this subversion of host miRNA regulatory networks has been a significant focus of infection biology research on mycobacterial diseases.[77,148–158] Outlined below are some examples of *in vitro* and *in vivo* studies that demonstrate how mycobacterial infection alters host miRNA expression profiles to evade immune responses and promote survival.

It has been shown that infection of murine macrophages with *M. tuberculosis* leads to downregulation of the miRNA let-7f and upregulation of TNF-α-induced protein 3 (TNFAIP3), an inhibitor of NF-κB activity.[152] In addition, it has been demonstrated that IFN-γ signalling is inhibited in *M. tuberculosis*-infected human macrophages due to the upregulation of two histone acetyltransferase p300-targeting miRNAs, miR-132 and miR-26a.[159]

Upregulation of miR-155 promoted intracellular survival of *M. tuberculosis* in experimentally infected murine RAW264.7 cells and bone marrow-derived macrophages (BMDMs) by targeting BTB domain and CNC homolog gene (*Bach1*) and inositol polyphosphate-5-phosphatase D gene (*Inpp5d*) mRNAs.[160] MiR-155 was also shown to inhibit apoptosis in $CD14^+$ monocytes—derived from peripheral blood of human patients with active TB—by targeting and regulating the expression of the forkhead box O3 gene (*FOXO3*) mRNA, which encodes a transcriptional activator of apoptosis-related genes.[161] Using a systems biology approach, Rothchild *et al.*[155] identified a dual role for miR-155 during *M. tuberculosis* infection, in which miR-155 protects the bacteria within macrophages, but also increases the immune response of T-cells to *M. tuberculosis*. Moreover, infection of murine wild-type and miR-$155^{-/-}$ knockout BMDMs with *M. tuberculosis* showed that macrophage cells lacking miR-155 display a reduction in growth, and an increase in cellular apoptosis in response to infection. Additionally, miR-$155^{-/-}$ T-cells produced significantly less IFN-γ than the wild-type cells. Intrinsic expression of miR-155 in T-cells was found to regulate cytokine production, thereby controlling the potency of these immune effector cells. Rothchild *et al.* also demonstrated that *Inpp5d* mRNA is a target for miR-155 due to an observed increase of INPP5D protein levels in miR-$155^{-/-}$ cells.[155]

Lipomannan, a surface molecule of *M. tuberculosis*, induces high expression of miR-125b which directly targets and inhibits the biosynthesis of TNF.[162] Cathepsin S encoded by the *CTSS* gene is a cysteine protease involved in lysosomal proteolysis that was recently found to be modulated by miR-106b-5p in *M. tuberculosis*-infected human macrophages, thereby promoting intracellular survival of the pathogen.[163] *M. tuberculosis* has also been shown to regulate host autophagy in murine macrophages *via* upregulation of miR-23a-5p and through targeted modulation in dendritic cells by miR-155 of the autophagy related 3 gene (*Atg3*) mRNA transcript.[149,164]

Infection with *M. bovis* BCG has been shown to increase the production of IFN-γ by inhibiting the expression of miR-29 in $CD4^+$ and $CD8^+$ T-cells.[125] In addition, infection of macrophages with *M. bovis* BCG reduces inducible nitric oxide synthase synthesis and nitric oxide production and regulates inflammation by driving overexpression of miR-146a and miR-142-3p and inhibiting the translation of TRAF6 and IRAK1 proteins.[153,165] In addition, the presence of *M. bovis* BCG was observed to increase expression of miR-21, leading to apoptosis in murine dendritic cells *via* targeting of B-cell lymphoma 2 gene (*Bcl2*) mRNA, and modulation of T-cell anti-mycobacterial activity through repressing translation of *Il12a* transcripts.[166] Interestingly, while *M. bovis* BCG was found to increase reactive oxygen species production in macrophage cells by upregulating miR-155,[167] an earlier study implicated miR-155 in *M. bovis* BCG-mediated apoptosis of macrophages.[168] In addition, it has been demonstrated that *M. bovis* BCG regulates cathepsin S, thereby promoting intracellular survival of bacteria by modulating miR-3619-59.[169]

In a similar fashion to strains of the MTBC, microRNA-sequencing (miRNA-seq) of whole blood from MAP-infected cattle revealed changes in miRNA expression that led to alterations in immune responses.[170] For instance, miR-301, an indirect activator of NF-κB, and miR-32, which targets phosphatase and tensin homolog gene (*PTEN*) mRNA transcripts, both exhibited decreased expression in MAP-infected cattle compared to MAP-negative control cattle.

In addition, infection of human macrophages with *M. avium* subspecies *hominissuis* resulted in decreased expression of miR-20a, miR-191, miR-378, and miR-185, but an increase in expression of miR-29a, miR-146a, miR-146b, miR-155, miR-886-5p, and let-7e.[171]

To further our understanding of bovine tuberculosis immunobiology and pathogenesis and to elucidate the role of miRNAs in these processes, transcriptomics studies focusing on expression of host miRNAs in *M. bovis*-infected cattle macrophages have been performed. For example, we and our collaborators have characterised the miRNA expression profile in *M. bovis*-infected alveolar macrophages that is associated with enhanced pathogen survival.[157] In this study, one, six, and 40 miRNAs were observed to be differentially expressed at 2, 24, and 48 h post-infection, respectively. Subsequent target binding site prediction analysis indicated involvement of these miRNAs in a range of molecular pathways, including IL-1 signalling and lysosomal trafficking. Moreover, using miRNA mimics, miR-146a and miR-146b were overexpressed in a bovine macrophage cell line resulting in suppression of *IRAK1* and TGF-β receptor 1 gene (*TGFBR1*) transcripts.[157] These results support the hypothesis that *M. bovis* modulates expression of host miRNAs to manipulate immune responses to infection.

4.4.1 Mycobacteria-encoded Regulatory Small RNAs

As outlined in the previous section, most miRNA-based transcriptomics studies of host–pathogen interactions for mycobacterial infections have focused on host-encoded miRNAs, particularly as "smoking guns" of mycobacterial manipulation and subversion of host innate immune processes. However, there has also been notable work dissecting and cataloguing non-coding regulatory small RNAs (sRNAs) expressed by mycobacterial pathogens and it is becoming clear that RNA biology plays a central role in regulation of mycobacterial cell responses and adaptation to a myriad of host environments.[172,173]

A decade ago, Arnvig and Young[174] were the first to experimentally demonstrate the expression of sRNAs from the *M. tuberculosis* genome; they also showed that overexpression of *trans*-encoded sRNAs impacted the growth of both *M. tuberculosis* and *M. smegmatis*. In addition, they observed that the expression profiles of sRNAs differed between stationary and exponential growth phases, and that two of the putative sRNAs profiled were associated with lipid metabolism. Another early study by DiChiara *et al.* used a combination of experimental and computational approaches to identify 34 sRNAs expressed by *M. bovis* BCG that had putative homologs in the *M. tuberculosis* and/or *M. smegmatis* genomes.[175] A number of years later, Tsai *et al.*[176] used computational algorithms to predict 17 novel sRNAs in *M. smegmatis* and 23 in *M. bovis* BCG.

More recently, two *M. tuberculosis* sRNAs have been shown to exhibit extensive secondary RNA folding capabilities, similar to those of eukaryotic pre-miRNAs. These sRNAs, designated *Mtb*-mR-1 and *Mtb*-mR-6, were observed to be transcribed by MAC strains and *M. tuberculosis*, but not *M. smegmatis*, suggesting an association of these small RNAs with pathogenic strains of mycobacteria.[177] Finally, specific regulatory roles for mycobacterial-encoded sRNAs are beginning to come to light. For example, an important study described large-scale sRNA discovery and expression in *M. tuberculosis* under five different stress conditions relevant to pathogenesis.[178] This work identified a subset of sRNAs that are consistently upregulated across a range of stress conditions. It also showed that one of these sRNAs, mycobacterial regulatory sRNA in iron (MrsI), is key to inducing an appropriate response to limited iron availability in the host macrophage, thereby enhancing mycobacterial survival.[178]

4.5 Clinical Applications of miRNAs in Mycobacterial Infections

With an increasing body of evidence highlighting miRNAs as key modulators of host immune responses to mycobacterial infections,[77,148,150,179] research is under way to identify diagnostic and therapeutic applications of these small regulatory molecules in preventing and treating diseases.

4.5.1 Diagnostic Potential of Circulating miRNAs

A point-of-care diagnostic modality that can accurately differentiate healthy individuals from those who are infected is urgently needed for human TB, since current diagnostics are limited in their ability to distinguish active infection from pre-exposure and latent infection.[9,180–182] Unlike mRNAs, circulating miRNAs in peripheral blood are protected from RNases because they are associated with proteins and lipids, and in many cases, encapsulated in endosomal vesicles.[183] Consequently, the stability of miRNA molecules, in combination with their critical role in fine-tuning immune responses,[4–6,76] suggests that circulating miRNAs in peripheral blood are promising biomarker candidates for diagnosing infections in humans and other animals.[184,185] Figure 4.2 broadly illustrates how a specific circulating miRNA profile may arise as a consequence of TB disease in humans or cattle caused by infection with *M. tuberculosis* or *M. bovis*, respectively.

Detection of differentially expressed miRNAs in plasma or serum from human TB patients has become a widely used strategy for evaluating the suitability of circulating miRNAs as biomarkers for TB disease. For example, in a study published in 2011, Fu *et al.* identified 92 differentially expressed circulating miRNAs in serum from TB patients, compared to healthy controls.[186] Following on from this, Miotto *et al.* defined circulating serum miRNA biosignatures for distinct human TB disease categories by examination of circulating miRNA profiles from individuals that were classified as (1) healthy; (2) exhibiting active pulmonary TB; (3) exhibiting active pulmonary TB with HIV co-infection; (4) exhibiting latent TB; or (5) exhibiting another pulmonary infection.[187] Using a multivariate logistic model and a machine learning method (relevance vector machine), a 15-miRNA signature was identified that could discriminate healthy individuals from those with active pulmonary TB with 77% and 82% accuracy, respectively.[187] In a similar fashion, Zhang *et al.* used miRNA-seq to compare miRNA expression profiles of active TB and asymptomatic latent TB in patients with and without BCG vaccination.[188] They observed that 33 miRNAs were differentially expressed between patients with active TB and healthy controls, with 24 upregulated and nine downregulated in the infected group. Interestingly, 134 miRNAs were found to be differentially expressed in patients that had been previously vaccinated with BCG. Functional network analysis of the gene targets for these miRNAs also revealed that the biological pathways likely to be affected by the differences in these miRNA expression profiles were immune related.[188]

In 2016, miRNAs purified from peripheral blood mononuclear cells (PBMCs) were investigated as diagnostic markers of early stage *M. tuberculosis* infection in young children. A total of 29 miRNAs were observed to be differentially expressed in children with TB, and 14 of these were considered critical to the disease process.[189] Receiver operating characteristic (ROC) curve analysis to evaluate diagnostic sensitivity was performed on all 29 differentially expressed miRNAs, with miR-150, miR-146a, miR-125b, miR-10a, and miR-31 all yielding area under the curve (AUC) values of ≥0.95. Subsequent combination of these miRNAs yielded a very high AUC value of 0.996.[189] More recently, comparative analysis of circulating plasma miRNA expression profiles of TB patients and healthy controls identified a panel of three miRNAs, miR-769-5p, miR-320a, and miR-22-3p, which could be used as a differential diagnostic marker.[190]

As early diagnosis of *M. leprae* infection can improve recovery rates and limit transmission, skin tissue miRNAs have been evaluated as potential early-stage leprosy biomarkers.[191] Low-density TaqMan™ arrays were used to analyse the expression of 377 miRNAs in skin lesions from leprosy patients and skin specimens from healthy controls. A selection of differentially expressed miRNAs were subsequently validated by reverse transcription-quantitative polymerase chain reaction and further investigated using principal component analysis, logistic regression modelling, and ROC analysis to generate a panel of four potential diagnostic miRNA biomarkers, miR-101, miR-196b, miR-27b, and miR-29c, with a sensitivity score of 80% and a specificity of 91%.[191]

Similar work has been performed to evaluate the potential of miRNAs as biomarkers of bovine TB. For example, Golby *et al.* have investigated the potential of miRNAs as disease biomarkers by isolating PBMCs from BCG-vaccinated and -unvaccinated cattle that had been experimentally infected with *M. bovis*. Microarray analysis of PBMCs from the two experimental animal groups subsequently indicated a diagnostic and prognostic potential for miR-155.[192]

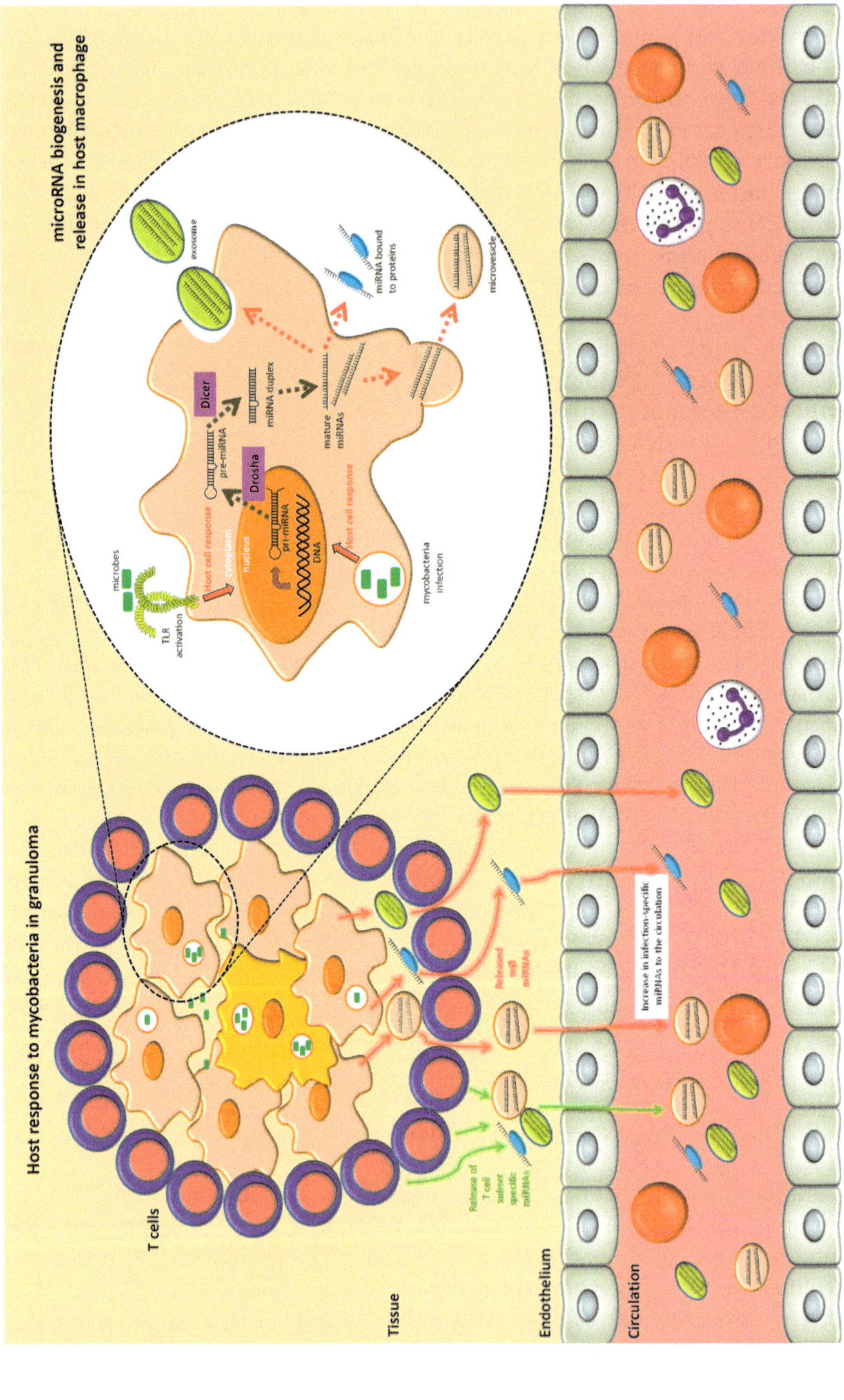
Host response to mycobacteria in granuloma
microRNA biogenesis and release in host macrophage
T cells
Tissue
Endothelium
Circulation
Release of T cell subset specific miRNAs
Released mΦ miRNAs
Increase in infection-specific miRNAs to the circulation
TLR activation
microbes
Host cell response
cytoplasm
nucleus
DNA
pri-miRNA
Drosha
pre-miRNA
Dicer
miRNA duplex
mature miRNAs
Host cell response
mycobacteria infection
exosome
miRNA bound to proteins
microvesicle

Comparable research has also been undertaken to assess circulating miRNAs as biomarkers for Johne's disease in cattle caused by infection with MAP. Shaughnessy *et al.* have demonstrated the long-term stability of circulating miRNAs in stored serum samples from cattle infected with MAP.[193] In addition, using miRNA-seq of serum RNA isolates, the same group evaluated the potential of circulating serum miRNAs as biomarkers of MAP infection in cattle.[194] However, although a panel of novel circulating miRNAs was identified in sera from MAP-challenged and control cattle, no MAP-specific miRNA biomarkers were detected. More recent work by an Italian group has reported panels of differentially expressed miRNAs in whole blood from MAP-positive, MAP-exposed and unexposed MAP-negative cattle.[170] Nine differentially expressed miRNAs were observed between MAP-positive and MAP-negative cattle, eight between MAP-exposed and MAP-negative cattle, and two between MAP-positive and MAP-exposed cattle.[170]

4.5.2 Therapeutic Potential of miRNA Modulation

During the past decade, a range of therapeutic strategies have been developed to harness the immunoregulatory features of miRNAs in modulating host responses to infection and disease. While some of these attempts have been successful in reaching clinical trials, others have been hindered by limitations in specificity, sensitivity, and safety.[195,196] What follows is a summary of these therapeutic applications and where they may be translatable to treatment of diseases caused by mycobacterial infections.

Figure 4.2 A tuberculosis lung granuloma demonstrates how specific circulating miRNAs may arise during an infection process. Mycobacterial pathogen-associated molecular patterns are recognised by Toll-like receptors (TLRs) and other pattern recognition receptors, which results in the upregulation of primary miRNAs in macrophages. These transcripts are then cleaved in the nucleus and cytoplasm by Drosha and Dicer respectively, resulting in 21–25-nucleotide mature miRNAs that act to fine-tune intracellular immune processes. Specific pathways and components of the immune response may be regulated by different miRNA subsets. Concurrently, the surrounding T-lymphocytes involved in granuloma formation/maintenance upregulate T-cell subset-specific miRNAs as a means of modulating the type of adaptive immune response. Mature miRNAs generated in macrophages and T-cells may also be released into the extracellular environment within exosomes, heterogeneous microvesicles or in association with high-density lipoproteins, low-density lipoproteins, or other protein complexes. Subsequently, by means not yet fully understood, these extracellular miRNAs move from local sites of infection to the circulatory system. This process can therefore give rise to infection-specific circulating miRNA expression signatures that are readily accessed from multiple biological fluids (*e.g.*, serum, plasma, or sputum). Reproduced from ref. 184 [https://doi.org/10.3389/fimmu.2017.00118] under the terms of a CC BY 4.0 license [https://creativecommons.org/licenses/by/4.0/].

The capacity of natural miRNAs to direct immune responses by targeting key immunological regulators has led to the development of synthetic miRNA mimics that can temporarily downregulate the expression of target genes.[197–199] miRNA mimics are small artificial RNA duplexes that behave similarly to natural endogenous miRNAs encoded by metazoan genomes. These synthetic double-stranded RNA molecules are also normally chemically modified to improve stability and cellular uptake; for example, methylation of the passenger strand can increase stability and conjugation with cholesterol can enhance cellular uptake.[196] Because miRNAs can normally affect the expression of many target genes simultaneously, miRNA mimics are considered powerful new tools to modulate and direct complex molecular pathways within cells and tissues.[195,198] However, it is important to note that double-stranded miRNA mimics can induce a non-specific interferon response by activating Toll-like receptors on host cells.[200] There are also major hurdles to be overcome in developing efficient systems and technologies for delivery of miRNA mimics to target cells and tissues.[199]

An alternative therapeutic strategy to take advantage of the gene regulatory properties of miRNAs is to inhibit the function of individual miRNAs or groups of miRNAs using antisense oligonucleotides termed antimiRs. AntimiRs are synthetic molecules that have been designed to contain sequences complementary to mature miRNAs and will therefore prevent silencing of the specific genes targeted by a particular miRNA or set of miRNAs.[201,202] AntimiRs with a particular chemical modification (2′-*O*-methoxyethyl) are also called antagomiRs.[199] Again, as with miRNA mimics, the problems with stability and delivery represent significant obstacles to deployment of antimiRs in clinical settings. However, with regard to infectious disease, it is notable that an engineered locked nucleic acid antimiR therapy for hepatitis C infection is in phase II clinical trials.[185] Miravirsen is an antimiR that targets miR-122, which has been subverted by hepatitis C virus (HCV) to stabilise its RNA genome during infection.[203] Consequently, repression of miR-122 activity using the miravirsen antimiR-122 drug has been shown to reduce viral titres and liver damage in animal models and clinical trials with HCV patients.[196,199]

Another route to leveraging the immunomodulatory features of miRNAs to tackle infectious disease—albeit one currently restricted to domestic animals—would be to use germline genome editing tools such as clustered regularly interspaced short palindromic repeats and CRISPR-associated protein 9 (CRISPR-Cas9)[204] to abrogate existing, or introduce new 3′ miRNA target binding sites at key immune genes. In cattle this strategy could be used to modulate host immune responses and enhance resistance to infection with *M. bovis*; for example, to strongly induce a pro-inflammatory and microbicidal M1 phenotype[47,48] in alveolar macrophages when they initially encounter the pathogen in the lung. Existing natural sequence variants at specific miRNA binding sites in domestic cattle or closely related taxa could also be exploited for this purpose.[146,147,205]

4.6 Concluding Remarks

At present, almost two decades into the 21st century, mycobacterial infections still represent a major burden on human and animal health, particularly in the context of human and bovine TB caused by infection with *M. tuberculosis* and *M. bovis*, respectively. These mycobacterial pathogens have evolved myriad strategies for evading and subverting host immune responses, including manipulation and modulation of the internal cellular machinery of the phagocytic alveolar macrophage, the most important host effector cell during the early stages of infection. In this regard, research work during the past 10 years has shown that host- and pathogen-encoded miRNAs have evolved to play key roles in the interplay between the macrophage and the intracellular mycobacteria. Dissecting and reconstructing these miRNA-directed host–pathogen interactions will be key to developing a better scientific understanding of disease pathogenesis for both human and animal TB, and for other diseases caused by intracellular mycobacterial infections such as leprosy in humans and Johne's disease in ruminants, which are caused by *M. leprae* and *M. avium* subspecies *paratuberculosis*, respectively. Circulating serum or plasma miRNAs are also emerging as important new sources of biomarkers for mycobacterial infections, which can augment existing diagnostic tools currently used in human and animal populations. Finally, new technologies such as miRNA-based therapies and genome editing have opened up exciting new research avenues for developing novel strategies to tackle mycobacterial pathogens and the diseases they cause.

References

1. World Health Organization, *Global Tuberculosis Report 2017*, World Health Organization, Geneva, Switzerland, 2017.
2. D. P. Bartel, *Cell*, 2018, **173**, 20–51.
3. E. Berezikov, *Nat. Rev. Genet.*, 2011, **12**, 846–860.
4. A. Mehta and D. Baltimore, *Nat. Rev. Immunol.*, 2016, **16**, 279–294.
5. R. M. O'Connell, D. S. Rao and D. Baltimore, *Annu. Rev. Immunol.*, 2012, **30**, 295–312.
6. R. M. O'Connell, D. S. Rao, A. A. Chaudhuri and D. Baltimore, *Nat. Rev. Immunol.*, 2010, **10**, 111–122.
7. R. J. Dubos and J. Dubos, *The White Plague: Tuberculosis, Man, and Society*, Rutgers University Press, New Brunswick, 1987.
8. J. Esteban and M. C. Munoz-Egea, *Microbiol. Spectrum*, 2016, **4**(6), DOI: 10.1128/microbiolspec.TNMI7-0021-2016.
9. M. Pai, M. A. Behr, D. Dowdy, K. Dheda, M. Divangahi, C. C. Boehme, A. Ginsberg, S. Swaminathan, M. Spigelman, H. Getahun, D. Menzies and M. Raviglione, *Nat. Rev. Dis. Primers*, 2016, **2**, 16076.
10. V. Tran, J. Liu and M. A. Behr, *Microbiol. Spectrum*, 2014, **2**, Mgm2-0028-2013.
11. C. A. Kerantzas and W. R. Jacobs Jr, *mBio*, 2017, **8**(2), e01586–16.
12. K. M. Malone and S. V. Gordon, *Adv. Exp. Med. Biol.*, 2017, **1019**, 135–154.

13. A. O'Garra, P. S. Redford, F. W. McNab, C. I. Bloom, R. J. Wilkinson and M. P. Berry, *Annu. Rev. Immunol.*, 2013, **31**, 475–527.
14. B. C. de Jong, M. Antonio and S. Gagneux, *PLoS Neglected Trop. Dis.*, 2010, **4**, e744.
15. A. Koch, H. Cox and V. Mizrahi, *Curr. Opin. Pharmacol.*, 2018, **42**, 7–15.
16. R. Mishra, P. Shukla, W. Huang and N. Hu, *Tuberculosis*, 2015, **95**, 1–5.
17. S. D. Fitzgerald and J. B. Kaneene, *Vet. Pathol.*, 2013, **50**, 488–499.
18. F. Olea-Popelka, A. Muwonge, A. Perera, A. S. Dean, E. Mumford, E. Erlacher-Vindel, S. Forcella, B. J. Silk, L. Ditiu, A. El Idrissi, M. Raviglione, O. Cosivi, P. LoBue and P. I. Fujiwara, *Lancet Infect. Dis.*, 2017, **17**, e21–e25.
19. B. Muller, S. Durr, S. Alonso, J. Hattendorf, C. J. Laisse, S. D. Parsons, P. D. van Helden and J. Zinsstag, *Emerging Infect. Dis.*, 2013, **19**, 899–908.
20. W. Y. Ayele, S. D. Neill, J. Zinsstag, M. G. Weiss and I. Pavlik, *Int. J. Tuberc. Lung Dis.*, 2004, **8**, 924–937.
21. A. Calmette, *Proc. R. Soc. Med.*, 1931, **24**, 1481–1490.
22. A. Benjak, C. Avanzi, P. Singh, C. Loiseau, S. Girma, P. Busso, A. N. B. Fontes, Y. Miyamoto, M. Namisato, K. Bobosha, C. G. Salgado, M. B. da Silva, R. C. Bouth, M. A. C. Frade, F. B. Filho, J. G. Barreto, J. A. C. Nery, S. Buhrer-Sekula, A. Lupien, A. R. Al-Samie, Y. Al-Qubati, A. S. Alkubati, G. Bretzel, L. Vera-Cabrera, F. Sakho, C. R. Johnson, M. Kodio, A. Fomba, S. O. Sow, M. Gado, O. Konate, M. M. A. Stefani, G. O. Penna, P. N. Suffys, E. N. Sarno, M. O. Moraes, P. S. Rosa, I. Baptista, J. S. Spencer, A. Aseffa, M. Matsuoka, M. Kai and S. T. Cole, *Nat. Commun.*, 2018, **9**, 352.
23. L. C. Rodrigues and D. Lockwood, *Lancet Infect. Dis.*, 2011, **11**, 464–470.
24. World Health Organization, *Weekly Epidemiological Record*, 2017, vol. 92, pp. 501–519.
25. World Health Organization, *Leprosy Fact Sheet*, 2018, available at www.who.int/news-room/fact-sheets/detail/leprosy.
26. C. Avanzi, J. Del-Pozo, A. Benjak, K. Stevenson, V. R. Simpson, P. Busso, J. McLuckie, C. Loiseau, C. Lawton, J. Schoening, D. J. Shaw, J. Piton, L. Vera-Cabrera, J. S. Velarde-Felix, F. McDermott, S. V. Gordon, S. T. Cole and A. L. Meredith, *Science*, 2016, **354**, 744–747.
27. G. Rathnaiah, D. K. Zinniel, J. P. Bannantine, J. R. Stabel, Y. T. Grohn, M. T. Collins and R. G. Barletta, *Front. Vet. Sci.*, 2017, **4**, 187.
28. A. B. Garcia and L. Shalloo, *J. Dairy Sci.*, 2015, **98**, 5019–5039.
29. W. C. Davis, *World J. Gastroenterol.*, 2015, **21**, 13411–13417.
30. A. L. McNees, D. Markesich, N. R. Zayyani and D. Y. Graham, *Expert Rev. Gastroenterol. Hepatol.*, 2015, **9**, 1523–1534.
31. B. A. Brown-Elliott and R. J. Wallace Jr, *Clin. Microbiol. Rev.*, 2002, **15**, 716–746.
32. C. Pierre-Audigier, E. Jouanguy, S. Lamhamedi, F. Altare, J. Rauzier, V. Vincent, D. Canioni, J. F. Emile, A. Fischer, S. Blanche, J. L. Gaillard and J. L. Casanova, *Clin. Infect. Dis.*, 1997, **24**, 982–984.
33. M. Driks, F. Weinhold and Q. Cokingtin, *BMJ case rep.*, 2011, **2011**, 3281.
34. B. Ergan, L. Coplu, A. Alp and M. Artvinli, *Respirology*, 2004, **9**, 283–285.

35. W. Hoefsloot, J. van Ingen, C. Andrejak, K. Angeby, R. Bauriaud, P. Bemer, N. Beylis, M. J. Boeree, J. Cacho, V. Chihota, E. Chimara, G. Churchyard, R. Cias, R. Daza, C. L. Daley, P. N. Dekhuijzen, D. Domingo, F. Drobniewski, J. Esteban, M. Fauville-Dufaux, D. B. Folkvardsen, N. Gibbons, E. Gomez-Mampaso, R. Gonzalez, H. Hoffmann, P. R. Hsueh, A. Indra, T. Jagielski, F. Jamieson, M. Jankovic, E. Jong, J. Keane, W. J. Koh, B. Lange, S. Leao, R. Macedo, T. Mannsaker, T. K. Marras, J. Maugein, H. J. Milburn, T. Mlinko, N. Morcillo, K. Morimoto, D. Papaventsis, E. Palenque, M. Paez-Pena, C. Piersimoni, M. Polanova, N. Rastogi, E. Richter, M. J. Ruiz-Serrano, A. Silva, M. P. da Silva, H. Simsek, D. van Soolingen, N. Szabo, R. Thomson, T. Tortola Fernandez, E. Tortoli, S. E. Totten, G. Tyrrell, T. Vasankari, M. Villar, R. Walkiewicz, K. L. Winthrop and D. Wagner, *Eur. Respir. J.*, 2013, **42**, 1604–1613.
36. K. Dheda, C. E. Barry 3rd and G. Maartens, *Lancet*, 2016, **387**, 1211–1226.
37. J. P. Cassidy and A. R. Martineau, *J. Comp. Pathol.*, 2014, **151**, 291–308.
38. G. Weiss and U. E. Schaible, *Immunol. Rev.*, 2015, **264**, 182–203.
39. B. R. Bloom and R. L. Modlin, *Microbiol. Spectrum*, 2016, **4**(3), DOI: 10.1128/microbiolspec.MCHD-0006-2015.
40. K. E. Killick, C. Ni Cheallaigh, C. O'Farrelly, K. Hokamp, D. E. MacHugh and J. Harris, *Cell. Microbiol.*, 2013, **15**, 1484–1495.
41. A. Mishra, S. Akhtar, C. Jagannath and A. Khan, *Mol. Immunol.*, 2017, **87**, 240–248.
42. C. J. Queval, R. Brosch and R. Simeone, *Front. Microbiol.*, 2017, **8**, 2284.
43. S. M. Behar, S. M. Carpenter, M. G. Booty, D. L. Barber and P. Jayaraman, *Semin. Immunol.*, 2014, **26**, 559–577.
44. C. V. Harding and W. H. Boom, *Nat. Rev. Microbiol.*, 2010, **8**, 296–307.
45. R. Domingo-Gonzalez, O. Prince, A. Cooper and S. A. Khader, *Microbiol. Spectrum*, 2016, **4**(5), DOI: 10.1128/microbiolspec.TBTB2-0018-2016.
46. L. Monin and S. A. Khader, *Semin. Immunol.*, 2014, **26**, 552–558.
47. A. Sica, M. Erreni, P. Allavena and C. Porta, *Cell. Mol. Life Sci.*, 2015, **72**, 4111–4126.
48. P. J. Murray, *Annu. Rev. Physiol.*, 2017, **79**, 541–566.
49. F. O. Martinez and S. Gordon, *F1000Prime Rep.*, 2014, **6**, 13.
50. P. Italiani and D. Boraschi, *Front. Immunol.*, 2014, **5**, 514.
51. Z. Huang, Q. Luo, Y. Guo, J. Chen, G. Xiong, Y. Peng, J. Ye and J. Li, *PLoS One*, 2015, **10**, e0129744.
52. C. M. McClean and D. M. Tobin, *Pathog. Dis.*, 2016, **74**, ftw068.
53. L. Ramakrishnan, *Nat. Rev. Immunol.*, 2012, **12**, 352–366.
54. E. Guirado, L. S. Schlesinger and G. Kaplan, *Semin. Immunopathol.*, 2013, **35**, 563–583.
55. S. A. Stanley and J. S. Cox, *Curr. Top. Microbiol. Immunol.*, 2013, **374**, 211–241.
56. C. J. Cambier, S. Falkow and L. Ramakrishnan, *Cell*, 2014, **159**, 1497–1509.
57. M. D. Stutz, M. P. Clark, M. Doerflinger and M. Pellegrini, *J. Leukocyte Biol.*, 2018, **103**, 259–268.
58. I. Vergne, J. Chua, H. H. Lee, M. Lucas, J. Belisle and V. Deretic, *Proc. Natl. Acad. Sci. U. S. A.*, 2005, **102**, 4033–4038.

59. H. Bach, K. G. Papavinasasundaram, D. Wong, Z. Hmama and Y. Av-Gay, *Cell Host Microbe*, 2008, **3**, 316–322.
60. A. Walburger, A. Koul, G. Ferrari, L. Nguyen, C. Prescianotto-Baschong, K. Huygen, B. Klebl, C. Thompson, G. Bacher and J. Pieters, *Science*, 2004, **304**, 1800–1804.
61. R. Jayachandran, V. Sundaramurthy, B. Combaluzier, P. Mueller, H. Korf, K. Huygen, T. Miyazaki, I. Albrecht, J. Massner and J. Pieters, *Cell*, 2007, **130**, 37–50.
62. M. Fabri, S. Stenger, D. M. Shin, J. M. Yuk, P. T. Liu, S. Realegeno, H. M. Lee, S. R. Krutzik, M. Schenk, P. A. Sieling, R. Teles, D. Montoya, S. S. Iyer, H. Bruns, D. M. Lewinsohn, B. W. Hollis, M. Hewison, J. S. Adams, A. Steinmeyer, U. Zugel, G. Cheng, E. K. Jo, B. R. Bloom and R. L. Modlin, *Sci. Transl. Med.*, 2011, **3**, 104ra102.
63. M. I. Voskuil, I. L. Bartek, K. Visconti and G. K. Schoolnik, *Front. Microbiol.*, 2011, **2**, 105.
64. N. Fazal, *Biochem. Mol. Biol. Int.*, 1997, **43**, 399–408.
65. Z. Li, C. Kelley, F. Collins, D. Rouse and S. Morris, *J. Infect. Dis.*, 1998, **177**, 1030–1035.
66. K. H. Darwin, S. Ehrt, J. C. Gutierrez-Ramos, N. Weich and C. F. Nathan, *Science*, 2003, **302**, 1963–1966.
67. A. Welin, M. E. Winberg, H. Abdalla, E. Sarndahl, B. Rasmusson, O. Stendahl and M. Lerm, *Infect. Immun.*, 2008, **76**, 2882–2887.
68. A. K. Shabaana, K. Kulangara, I. Semac, Y. Parel, S. Ilangumaran, K. Dharmalingam, C. Chizzolini and D. C. Hoessli, *Cell. Mol. Life Sci.*, 2005, **62**, 179–187.
69. K. L. Knutson, Z. Hmama, P. Herrera-Velit, R. Rochford and N. E. Reiner, *J. Biol. Chem.*, 1998, **273**, 645–652.
70. S. V. Jamwal, P. Mehrotra, A. Singh, Z. Siddiqui, A. Basu and K. V. Rao, *Sci. Rep.*, 2016, **6**, 23089.
71. R. Simeone, A. Bobard, J. Lippmann, W. Bitter, L. Majlessi, R. Brosch and J. Enninga, *PLoS Pathog.*, 2012, **8**, e1002507.
72. J. Augenstreich, A. Arbues, R. Simeone, E. Haanappel, A. Wegener, F. Sayes, F. Le Chevalier, C. Chalut, W. Malaga, C. Guilhot, R. Brosch and C. Astarie-Dequeker, *Cell. Microbiol.*, 2017, **19**, e12726.
73. T. R. Lerner, C. J. Queval, A. Fearns, U. Repnik, G. Griffiths and M. G. Gutierrez, *BMC Biol.*, 2018, **16**, 1.
74. R. C. Lee, R. L. Feinbaum and V. Ambros, *Cell*, 1993, **75**, 843–854.
75. H. Guo, N. T. Ingolia, J. S. Weissman and D. P. Bartel, *Nature*, 2010, **466**, 835–840.
76. L. A. O'Neill, F. J. Sheedy and C. E. McCoy, *Nat. Rev. Immunol.*, 2011, **11**, 163–175.
77. P. Bettencourt, D. Pires and E. Anes, *Tuberculosis*, 2016, **97**, 1–7.
78. A. Shenoy and R. H. Blelloch, *Nat. Rev. Mol. Cell Biol.*, 2014, **15**, 565–576.
79. C. Alberti and L. Cochella, *Development*, 2017, **144**, 2548–2559.
80. J. Han, Y. Lee, K. H. Yeom, Y. K. Kim, H. Jin and V. N. Kim, *Genes Dev.*, 2004, **18**, 3016–3027.

81. Y. Lee, M. Kim, J. Han, K. H. Yeom, S. Lee, S. H. Baek and V. N. Kim, *EMBO J.*, 2004, **23**, 4051–4060.
82. J. Han, Y. Lee, K. H. Yeom, J. W. Nam, I. Heo, J. K. Rhee, S. Y. Sohn, Y. Cho, B. T. Zhang and V. N. Kim, *Cell*, 2006, **125**, 887–901.
83. J. O. Westholm and E. C. Lai, *Biochimie*, 2011, **93**, 1897–1904.
84. M. T. Bohnsack, K. Czaplinski and D. Gorlich, *RNA*, 2004, **10**, 185–191.
85. Y. K. Kim, B. Kim and V. N. Kim, *Proc. Natl. Acad. Sci. U. S. A.*, 2016, **113**, E1881–E1889.
86. R. I. Gregory, T. P. Chendrimada, N. Cooch and R. Shiekhattar, *Cell*, 2005, **123**, 631–640.
87. J. O'Brien, H. Hayder, Y. Zayed and C. Peng, *Front. Endocrinol.*, 2018, **9**, 402.
88. S. W. Eichhorn, H. Guo, S. E. McGeary, R. A. Rodriguez-Mias, C. Shin, D. Baek, S. H. Hsu, K. Ghoshal, J. Villen and D. P. Bartel, *Mol. Cell*, 2014, **56**, 104–115.
89. H. A. Meijer, Y. W. Kong, W. T. Lu, A. Wilczynska, R. V. Spriggs, S. W. Robinson, J. D. Godfrey, A. E. Willis and M. Bushell, *Science*, 2013, **340**, 82–85.
90. J. L. Clancy, G. H. Wei, N. Echner, D. T. Humphreys, T. H. Beilharz and T. Preiss, *RNA*, 2011, **17**, 1025–1031.
91. L. A. Macfarlane and P. R. Murphy, *Curr. Genomics*, 2010, **11**, 537–561.
92. P. H. Olsen and V. Ambros, *Dev. Biol.*, 1999, **216**, 671–680.
93. N. Molotski and Y. Soen, *RNA*, 2012, **18**, 1612–1623.
94. N. Cloonan, *BioEssays*, 2015, **37**, 379–388.
95. A. Wilczynska and M. Bushell, *Cell Death Differ.*, 2015, **22**, 22–33.
96. S. Vasudevan, Y. Tong and J. A. Steitz, *Science*, 2007, **318**, 1931–1934.
97. R. M. O'Connell and D. Baltimore, *Curr. Top. Dev. Biol.*, 2012, **99**, 145–174.
98. S. S. Lazare, E. E. Wojtowicz, L. V. Bystrykh and G. de Haan, *Exp. Cell Res.*, 2014, **329**, 234–238.
99. D. Baltimore, M. P. Boldin, R. M. O'Connell, D. S. Rao and K. D. Taganov, *Nat. Immunol.*, 2008, **9**, 839–845.
100. K. Okuyama, T. Ikawa, B. Gentner, K. Hozumi, R. Harnprasopwat, J. Lu, R. Yamashita, D. Ha, T. Toyoshima, B. Chanda, T. Kawamata, K. Yokoyama, S. Wang, K. Ando, H. F. Lodish, A. Tojo, H. Kawamoto and A. Kotani, *Proc. Natl. Acad. Sci. U. S. A.*, 2013, **110**, 13410–13415.
101. R. Popovic, L. E. Riesbeck, C. S. Velu, A. Chaubey, J. Zhang, N. J. Achille, F. E. Erfurth, K. Eaton, J. Lu, H. L. Grimes, J. Chen, J. D. Rowley and N. J. Zeleznik-Le, *Blood*, 2009, **113**, 3314–3322.
102. R. M. O'Connell, A. A. Chaudhuri, D. S. Rao, W. S. Gibson, A. B. Balazs and D. Baltimore, *Proc. Natl. Acad. Sci. U. S. A.*, 2010, **107**, 14235–14240.
103. K. Y. Kong, K. S. Owens, J. H. Rogers, J. Mullenix, C. S. Velu, H. L. Grimes and R. Dahl, *Exp. Hematol.*, 2010, **38**, 629.e1–640.e1.
104. D. S. Rao, R. M. O'Connell, A. A. Chaudhuri, Y. Garcia-Flores, T. L. Geiger and D. Baltimore, *Immunity*, 2010, **33**, 48–59.
105. A. Ventura, A. G. Young, M. M. Winslow, L. Lintault, A. Meissner, S. J. Erkeland, J. Newman, R. T. Bronson, D. Crowley, J. R. Stone, R. Jaenisch, P. A. Sharp and T. Jacks, *Cell*, 2008, **132**, 875–886.

106. X. Li, J. Zhang, L. Gao, S. McClellan, M. A. Finan, T. W. Butler, L. B. Owen, G. A. Piazza and Y. Xi, *Cell Death Differ.*, 2012, **19**, 378–386.
107. C. Z. Chen, L. Li, H. F. Lodish and D. P. Bartel, *Science*, 2004, **303**, 83–86.
108. F. Cichocki, M. Felices, V. McCullar, S. R. Presnell, A. Al-Attar, C. T. Lutz and J. S. Miller, *J. Immunol.*, 2011, **187**, 6171–6175.
109. Q. J. Li, J. Chau, P. J. Ebert, G. Sylvester, H. Min, G. Liu, R. Braich, M. Manoharan, J. Soutschek, P. Skare, L. O. Klein, M. M. Davis and C. Z. Chen, *Cell*, 2007, **129**, 147–161.
110. B. S. Cobb, T. B. Nesterova, E. Thompson, A. Hertweck, E. O'Connor, J. Godwin, C. B. Wilson, N. Brockdorff, A. G. Fisher, S. T. Smale and M. Merkenschlager, *J. Exp. Med.*, 2005, **201**, 1367–1373.
111. S. A. Muljo, K. M. Ansel, C. Kanellopoulou, D. M. Livingston, A. Rao and K. Rajewsky, *J. Exp. Med.*, 2005, **202**, 261–269.
112. N. A. Bezman, T. Chakraborty, T. Bender and L. L. Lanier, *J. Exp. Med.*, 2011, **208**, 2717–2731.
113. L. A. Pobezinsky, R. Etzensperger, S. Jeurling, A. Alag, T. Kadakia, T. M. McCaughtry, M. Y. Kimura, S. O. Sharrow, T. I. Guinter, L. Feigenbaum and A. Singer, *Nat. Immunol.*, 2015, **16**, 517–524.
114. M. Hager, C. C. Pedersen, M. T. Larsen, M. K. Andersen, C. Hother, K. Gronbaek, H. Jarmer, N. Borregaard and J. B. Cowland, *Blood*, 2011, **118**, 6649–6659.
115. E. Vigorito, K. L. Perks, C. Abreu-Goodger, S. Bunting, Z. Xiang, S. Kohlhaas, P. P. Das, E. A. Miska, A. Rodriguez, A. Bradley, K. G. Smith, C. Rada, A. J. Enright, K. M. Toellner, I. C. Maclennan and M. Turner, *Immunity*, 2007, **27**, 847–859.
116. K. D. Rasmussen, S. Simmini, C. Abreu-Goodger, N. Bartonicek, M. Di Giacomo, D. Bilbao-Cortes, R. Horos, M. Von Lindern, A. J. Enright and D. O'Carroll, *J. Exp. Med.*, 2010, **207**, 1351–1358.
117. J.-T. A. Chi, C. Sangokoya and C. M. de Castro, *Blood*, 2007, **110**, 3675.
118. T. Umemura, Y. Tanaka, M. Fujisaki, S. Masaki, H. Shiotsu, K. Srinoun, S. Fucharoen, Y. Abe, H. Ichihara, H. Toh and H. Hisaeda, *Blood*, 2009, **114**, 4042.
119. Z. Wang, J. Xi, X. Hao, W. Deng, J. Liu, C. Wei, Y. Gao, L. Zhang and H. Wang, *Emerging Microbes Infect.*, 2017, **6**, e75.
120. C. Z. Zheng, Y. B. Shu, Y. L. Luo and J. Luo, *Eur. Rev. Med. Pharmacol. Sci.*, 2017, **21**, 1041–1048.
121. H. Park, X. Huang, C. Lu, M. S. Cairo and X. Zhou, *J. Biol. Chem.*, 2015, **290**, 2831–2841.
122. A. S. Collins, C. E. McCoy, A. T. Lloyd, C. O'Farrelly and N. J. Stevenson, *PLoS One*, 2013, **8**, e69090.
123. W. Shang, X. Chen, L. Nie, M. Xu, N. Chen, H. Zeng and Q. Zhou, *Int. J. Mol. Sci.*, 2013, **14**, 8422–8436.
124. D. F. Steiner, M. F. Thomas, J. K. Hu, Z. Yang, J. E. Babiarz, C. D. Allen, M. Matloubian, R. Blelloch and K. M. Ansel, *Immunity*, 2011, **35**, 169–181.

125. F. Ma, S. Xu, X. Liu, Q. Zhang, X. Xu, M. Liu, M. Hua, N. Li, H. Yao and X. Cao, *Nat. Immunol.*, 2011, **12**, 861–869.
126. K. D. Taganov, M. P. Boldin, K. J. Chang and D. Baltimore, *Proc. Natl. Acad. Sci. U. S. A.*, 2006, **103**, 12481–12486.
127. M. A. Nahid, B. Yao, P. R. Dominguez-Gutierrez, L. Kesavalu, M. Satoh and E. K. Chan, *J. Immunol.*, 2013, **190**, 1250–1263.
128. F. J. Sheedy, E. Palsson-McDermott, E. J. Hennessy, C. Martin, J. J. O'Leary, Q. Ruan, D. S. Johnson, Y. Chen and L. A. O'Neill, *Nat. Immunol.*, 2010, **11**, 141–147.
129. M. H. Bao, J. M. Li, H. Q. Luo, L. Tang, Q. L. Lv, G. Y. Li and H. H. Zhou, *Mediators Inflammation*, 2016, **2016**, 5308170.
130. E. Vigorito, S. Kohlhaas, D. Lu and R. Leyland, *Immunol. Rev.*, 2013, **253**, 146–157.
131. R. M. O'Connell, J. L. Zhao and D. S. Rao, *Blood*, 2011, **118**, 2960–2969.
132. S. T. Cheung, E. Y. So, D. Chang, A. Ming-Lum and A. L. Mui, *PLoS One*, 2013, **8**, e71336.
133. E. Mor and N. Shomron, *BioEssays*, 2013, **35**, 881–888.
134. K. Chen and N. Rajewsky, *Nat. Rev. Genet.*, 2007, **8**, 93–103.
135. R. Zhang and B. Su, *J. Genet. Genomics*, 2009, **36**, 1–6.
136. A. Clop, F. Marcq, H. Takeda, D. Pirottin, X. Tordoir, B. Bibe, J. Bouix, F. Caiment, J. M. Elsen, F. Eychenne, C. Larzul, E. Laville, F. Meish, D. Milenkovic, J. Tobin, C. Charlier and M. Georges, *Nat. Genet.*, 2006, **38**, 813–818.
137. M. Georges, A. Clop, F. Marcq, H. Takeda, D. Pirottin, S. Hiard, X. Tordoir, F. Caiment, F. Meish, B. Bibe, J. Bouix, J. M. Elsen, F. Eychenne, E. Laville, C. Larzul, D. Milenkovic, J. Tobin and A. C. Charlier, *Cold Spring Harbor Symp. Quant. Biol.*, 2006, **71**, 343–350.
138. H. Chihab, F. Z. Jadid, P. Foka, I. Zaidane, R. El Fihry, U. Georgopoulou, A. Marchio, A. Elhabazi, M. Chair, P. Pineau, S. Ezzikouri and S. Benjelloun, *J. Med. Virol.*, 2018, **90**, 1730–1738.
139. G. Zhang, N. Li, Z. Li, Q. Zhu, F. Li, C. Yang, Q. Han, Y. Lv, Z. Zhou and Z. Liu, *Oncotarget*, 2015, **6**, 18933–18944.
140. M. L. Endale Ahanda, E. R. Fritz, J. Estelle, Z. L. Hu, O. Madsen, M. A. Groenen, D. Beraldi, R. Kapetanovic, D. A. Hume, R. R. Rowland, J. K. Lunney, C. Rogel-Gaillard, J. M. Reecy and E. Giuffra, *PLoS One*, 2012, **7**, e48607.
141. E. C. Castelli, L. C. Veiga-Castelli, L. Yaghi, P. Moreau and E. A. Donadi, *J. Immunol. Res.*, 2014, **2014**, 734068.
142. L. Abel, J. El-Baghdadi, A. A. Bousfiha, J. L. Casanova and E. Schurr, *Philos. Trans. R. Soc., B*, 2014, **369**, 20130428.
143. A. R. Allen, G. Minozzi, E. J. Glass, R. A. Skuce, S. W. McDowell, J. A. Woolliams and S. C. Bishop, *Proc. Biol. Sci.*, 2010, **277**, 2737–2745.
144. N. le Roex, P. D. van Helden, A. P. Koets and E. G. Hoal, *Physiol. Genomics*, 2013, **45**, 631–637.
145. V. Naranbhai, *Microbiol. Spectrum*, 2016, **4**(5), DOI: 10.1128/microbiolspec.TBTB2-0011-2016.

146. S. Taka, M. Gazouli, K. Sotirakoglou, E. Liandris, M. Andreadou, K. Triantaphyllopoulos and J. Ikonomopoulos, *Vet. Immunol. Immunopathol.*, 2015, **167**, 75–79.
147. M. Iannaccone, G. Cosenza, A. Pauciullo, G. Martino and R. Capparelli, *Anim. Genet.*, 2018, **49**, 496–497.
148. M. D. Mehta and P. T. Liu, *Front. Genet.*, 2014, **5**, 231.
149. M. P. Etna, A. Sinigaglia, A. Grassi, E. Giacomini, A. Romagnoli, M. Pardini, M. Severa, M. Cruciani, F. Rizzo, E. Anastasiadou, B. Di Camillo, L. Barzon, G. M. Fimia, R. Manganelli and E. M. Coccia, *PLoS Pathog.*, 2018, **14**, e1006790.
150. J. K. Kim, T. S. Kim, J. Basu and E. K. Jo, *Cell. Microbiol.*, 2017, **19**, e12687.
151. J. K. Kim, J. M. Yuk, S. Y. Kim, T. S. Kim, H. S. Jin, C. S. Yang and E. K. Jo, *J. Immunol.*, 2015, **194**, 5355–5365.
152. M. Kumar, S. K. Sahu, R. Kumar, A. Subuddhi, R. K. Maji, K. Jana, P. Gupta, J. Raffetseder, M. Lerm, Z. Ghosh, G. van Loo, R. Beyaert, U. D. Gupta, M. Kundu and J. Basu, *Cell Host Microbe*, 2015, **17**, 345–356.
153. M. Li, J. Wang, Y. Fang, S. Gong, M. Li, M. Wu, X. Lai, G. Zeng, Y. Wang, K. Yang and X. Huang, *Sci. Rep.*, 2016, **6**, 23351.
154. M. Ouimet, S. Koster, E. Sakowski, B. Ramkhelawon, C. van Solingen, S. Oldebeken, D. Karunakaran, C. Portal-Celhay, F. J. Sheedy, T. D. Ray, K. Cecchini, P. D. Zamore, K. J. Rayner, Y. L. Marcel, J. A. Philips and K. J. Moore, *Nat. Immunol.*, 2016, **17**, 677–686.
155. A. C. Rothchild, J. R. Sissons, S. Shafiani, C. Plaisier, D. Min, D. Mai, M. Gilchrist, J. Peschon, R. P. Larson, A. Bergthaler, N. S. Baliga, K. B. Urdahl and A. Aderem, *Proc. Natl. Acad. Sci. U. S. A.*, 2016, **113**, E6172–E6181.
156. K. J. Siddle, L. Tailleux, M. Deschamps, Y. H. Loh, C. Deluen, B. Gicquel, C. Antoniewski, L. B. Barreiro, L. Farinelli and L. Quintana-Murci, *PLoS Genet.*, 2015, **11**, e1005064.
157. P. Vegh, D. A. Magee, N. C. Nalpas, K. Bryan, M. S. McCabe, J. A. Browne, K. M. Conlon, S. V. Gordon, D. G. Bradley, D. E. MacHugh and D. J. Lynn, *Tuberculosis*, 2015, **95**, 60–67.
158. U. von Both, M. Berk, P. M. Agapow, J. D. Wright, A. Git, M. S. Hamilton, G. Goldgof, N. Siddiqui, E. Bellos, V. J. Wright, L. J. Coin, S. M. Newton and M. Levin, *Sci. Rep.*, 2018, **8**, 661.
159. B. Ni, M. V. Rajaram, W. P. Lafuse, M. B. Landes and L. S. Schlesinger, *J. Immunol.*, 2014, **193**, 4537–4547.
160. R. Kumar, P. Halder, S. K. Sahu, M. Kumar, M. Kumari, K. Jana, Z. Ghosh, P. Sharma, M. Kundu and J. Basu, *Cell. Microbiol.*, 2012, **14**, 1620–1631.
161. J. Huang, J. Jiao, W. Xu, H. Zhao, C. Zhang, Y. Shi and Z. Xiao, *Mol. Med. Rep.*, 2015, **12**, 7102–7108.
162. M. V. Rajaram, B. Ni, J. D. Morris, M. N. Brooks, T. K. Carlson, B. Bakthavachalu, D. R. Schoenberg, J. B. Torrelles and L. S. Schlesinger, *Proc. Natl. Acad. Sci. U. S. A.*, 2011, **108**, 17408–17413.

163. D. Pires, E. M. Bernard, J. P. Pombo, N. Carmo, C. Fialho, M. G. Gutierrez, P. Bettencourt and E. Anes, *Front. Immunol.*, 2017, **8**, 1819.
164. X. Gu, Y. Gao, D. G. Mu and E. Q. Fu, *Exp. Cell Res.*, 2017, **354**, 71–77.
165. G. Xu, Z. Zhang, J. Wei, Y. Zhang, Y. Zhang, L. Guo and X. Liu, *Tuberculosis*, 2013, **93**, 606–611.
166. Z. Wu, H. Lu, J. Sheng and L. Li, *FEBS Lett.*, 2012, **586**, 2459–2467.
167. J. Wang, M. Wu, J. Wen, K. Yang, M. Li, X. Zhan, L. Feng, M. Li and X. Huang, *Mol. Immunol.*, 2014, **62**, 29–36.
168. D. S. Ghorpade, R. Leyland, M. Kurowska-Stolarska, S. A. Patil and K. N. Balaji, *Mol. Cell. Biol.*, 2012, **32**, 2239–2253.
169. K. Pawar, J. Sharbati, R. Einspanier and S. Sharbati, *Front. Cell. Infect. Microbiol.*, 2016, **6**, 27.
170. M. Malvisi, F. Palazzo, N. Morandi, B. Lazzari, J. L. Williams, G. Pagnacco and G. Minozzi, *PLoS One*, 2016, **11**, e0164461.
171. J. Sharbati, A. Lewin, B. Kutz-Lohroff, E. Kamal, R. Einspanier and S. Sharbati, *PLoS One*, 2011, **6**, e20258.
172. S. Schwenk and K. B. Arnvig, *Pathog. Dis.*, 2018, **76**, fty035.
173. K. Haning, S. H. Cho and L. M. Contreras, *Front. Cell. Infect. Microbiol.*, 2014, **4**, 96.
174. K. B. Arnvig and D. B. Young, *Mol. Microbiol.*, 2009, **73**, 397–408.
175. J. M. DiChiara, L. M. Contreras-Martinez, J. Livny, D. Smith, K. A. McDonough and M. Belfort, *Nucleic Acids Res.*, 2010, **38**, 4067–4078.
176. C. H. Tsai, C. Baranowski, J. Livny, K. A. McDonough, J. T. Wade and L. M. Contreras, *PLoS One*, 2013, **8**, e79411.
177. N. S. van Oers, S. Srivastava, P. Raj, I. Dozmorov, E. K. Wakeland and T. Gumbo, *J. Immunol.*, 2016, **196**, 65.13.
178. E. R. Gerrick, T. Barbier, M. R. Chase, R. Xu, J. Francois, V. H. Lin, M. J. Szucs, J. M. Rock, R. Ahmad, B. Tjaden, J. Livny and S. M. Fortune, *Proc. Natl. Acad. Sci. U. S. A.*, 2018, **115**, 6464–6469.
179. N. Sabir, T. Hussain, S. Z. A. Shah, A. Peramo, D. Zhao and X. Zhou, *Front. Microbiol.*, 2018, **9**, 602.
180. UNITAID, *Tuberculosis: Diagnostics Technology and Market Landscape*, World Health Organization (WHO), Geneva, Switzerland, 5th edn, 2017.
181. D. Goletti, M. R. Lee, J. Y. Wang, N. Walter and T. H. M. Ottenhoff, *Respirology*, 2018, **23**, 455–466.
182. J. Weiner and S. H. Kaufmann, *Int. J. Infect. Dis.*, 2017, **56**, 258–262.
183. A. Turchinovich, L. Weiz, A. Langheinz and B. Burwinkel, *Nucleic Acids Res.*, 2011, **39**, 7223–7233.
184. C. N. Correia, N. C. Nalpas, K. E. McLoughlin, J. A. Browne, S. V. Gordon, D. E. MacHugh and R. G. Shaughnessy, *Front. Immunol.*, 2017, **8**, 118.
185. R. E. Drury, D. O'Connor and A. J. Pollard, *Front. Immunol.*, 2017, **8**, 1182.
186. Y. Fu, Z. Yi, X. Wu, J. Li and F. Xu, *J. Clin. Microbiol.*, 2011, **49**, 4246–4251.
187. P. Miotto, G. Mwangoka, I. C. Valente, L. Norbis, G. Sotgiu, R. Bosu, A. Ambrosi, L. R. Codecasa, D. Goletti, A. Matteelli, E. N. Ntinginya,

F. Aloi, N. Heinrich, K. Reither and D. M. Cirillo, *PLoS One*, 2013, **8**, e80149.
188. H. Zhang, Z. Sun, W. Wei, Z. Liu, J. Fleming, S. Zhang, N. Lin, M. Wang, M. Chen, Y. Xu, J. Zhou, C. Li, L. Bi and G. Zhou, *PLoS One*, 2014, **9**, e88909.
189. M. Zhou, G. Yu, X. Yang, C. Zhu, Z. Zhang and X. Zhan, *Mol. Med. Rep.*, 2016, **13**, 4620–4626.
190. J. Y. Cui, H. W. Liang, X. L. Pan, D. Li, N. Jiao, Y. H. Liu, J. Fu, X. Y. He, G. X. Sun, C. L. Zhang, C. H. Zhao, D. H. Li, E. Y. Dai, K. Zen, F. M. Zhang, C. Y. Zhang, X. Chen and H. Ling, *PLoS One*, 2017, **12**, e0184113.
191. K. Jorge, R. P. Souza, M. T. A. Assis, M. G. Araujo, M. Locati, A. M. R. Jesus, I. M. F. Dias Baptista, C. X. Lima, A. L. Teixeira, M. M. Teixeira and F. M. Soriani, *J. Clin. Microbiol.*, 2017, **55**, 1516–1525.
192. P. Golby, B. Villarreal-Ramos, G. Dean, G. J. Jones and M. Vordermeier, *Vaccine*, 2014, **32**, 5839–5844.
193. R. G. Shaughnessy, D. Farrell, K. Riepema, D. Bakker and S. V. Gordon, *PLoS One*, 2015, **10**, e0145089.
194. D. Farrell, R. G. Shaughnessy, L. Britton, D. E. MacHugh, B. Markey and S. V. Gordon, *PLoS One*, 2015, **10**, e0134310.
195. W. Huang, *Methods Mol. Biol.*, 2017, **1617**, 57–67.
196. E. van Rooij and S. Kauppinen, *EMBO Mol. Med.*, 2014, **6**, 851–864.
197. N. W. Gibson, *J. R. Coll. Physicians Edinb.*, 2014, **44**, 196–200.
198. J. Krutzfeldt, *Best Pract. Res., Clin. Endocrinol. Metab.*, 2016, **30**, 551–561.
199. R. Rupaimoole and F. J. Slack, *Nat. Rev. Drug Discovery*, 2017, **16**, 203–222.
200. H. Peacock, R. V. Fucini, P. Jayalath, J. M. Ibarra-Soza, H. J. Haringsma, W. M. Flanagan, A. Willingham and P. A. Beal, *J. Am. Chem. Soc.*, 2011, **133**, 9200–9203.
201. J. Stenvang, A. Petri, M. Lindow, S. Obad and S. Kauppinen, *Silence*, 2012, **3**, 1.
202. K. A. Lennox and M. A. Behlke, *Gene Ther.*, 2011, **18**, 1111–1120.
203. C. L. Jopling, M. Yi, A. M. Lancaster, S. M. Lemon and P. Sarnow, *Science*, 2005, **309**, 1577–1581.
204. M. Jinek, K. Chylinski, I. Fonfara, M. Hauer, J. A. Doudna and E. Charpentier, *Science*, 2012, **337**, 816–821.
205. M. Braud, D. A. Magee, S. D. Park, T. S. Sonstegard, S. M. Waters, D. E. MacHugh and C. Spillane, *Front. Genet.*, 2017, **8**, 3.

CHAPTER 5

Regulation and Function of MicroRNAs in Kidney Diseases

P. TRIONFINI AND A. BENIGNI*

Istituto di Ricerche Farmacologiche Mario Negri IRCCS, Centro Anna Maria Astori, Science and Technology Park Kilometro Rosso, 24126 Bergamo, Italy
*E-mail: ariela.benigni@marionegri.it

5.1 Introduction

The kidney is a vital organ whose main functions are the reabsorption of vital nutrients, removal of wastes from the body, maintenance of plasma ion homeostasis and the regulation of blood volume, pressure, and pH. The functional and structural unit of the kidney is the nephron, composed of a filtering unit, the glomerulus, and renal tubules that reabsorb water and select metabolically valuable substrates from toxic wastes (Figure 5.1). Each kidney has approximately one million nephrons. The glomerulus is made of a specialized bundle of capillaries surrounded by the glomerular (Bowman's) capsule and encased by podocytes composed of long interdigitating processes interconnected by a thin membranous structure, the slit diaphragm. The glomerular tuft is structurally supported by mesangial cells, which secrete growth factors and matrix proteins. The renal tubule extends from the capsule and is made of four portions that differ anatomically and functionally: the proximal and distal convoluted tubules, the loop of Henle, and the collecting duct. The heterogeneity of cell types composing the kidney mirrors the heterogeneity of kidney diseases, which can involve several

Drug Discovery Series No. 69
MicroRNAs in Diseases and Disorders: Emerging Therapeutic Targets
Edited by Philip V. Peplow, Bridget Martinez, George A. Calin and Aurora Esquela-Kerscher

Published by the Royal Society of Chemistry, www.rsc.org

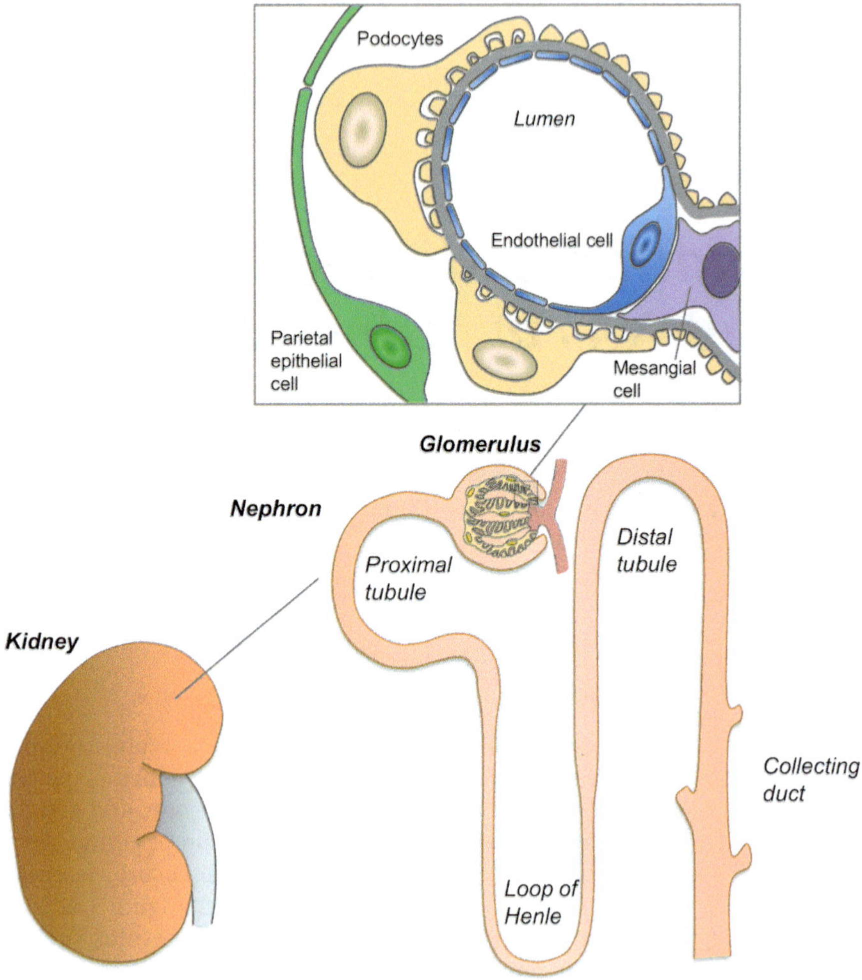

Figure 5.1 Schematic representation of the kidney and the nephron, the functional unit of the kidney, composed of a filter, the glomerulus, and of renal tubules extending the capsule (the proximal and distal convoluted tubules, the loop of Henle, and the collecting duct). Moreover, a schematic cross-section of the glomerulus is shown. The glomerular capillaries are composed of fenestrated endothelial cells surrounded by the basement membrane (GBM). The capillaries are wrapped by podocytes covering the outer face of GBM and the entire glomerular tuft is structurally supported by mesangial cells. Parietal epithelial cells line the inside of the Bowman's capsule.

types of pathological changes in glomerular, tubulointerstitial, and/or vascular compartments. Generally, kidney diseases are classified as either acute kidney injury or chronic kidney disease (CKD).

Recently, with the development of high-throughput technologies, several studies have identified perturbed expression of microRNAs (miRNAs) in the kidney during the progression of CKD.[1] miRNAs are short non-coding RNA (~21–24 nucleotides long) that function as negative regulators of target messenger RNA (mRNA) expression through mRNA degradation or translational repression.[2] Partial pairing with the target mRNA, instead of full complementarity, is required for the miRNA machinery to work, implying that a given miRNA can regulate several hundred transcripts belonging to different cellular pathways and networks. Moreover, one specific target gene can be regulated by many different miRNAs. To add to the complexity, miRNAs display tissue and cell-specific expression and this, combined with the complex morpho-functional organization of the mammalian kidney, makes it even more difficult and tricky to define the exact role of miRNAs in kidney homeostasis and disease. The cellular localization of miRNAs through *in situ* hybridization methods or miRNA expression analysis of laser-captured micro-dissected kidney tissues, in combination with *in vitro* models, is helpful for determining mRNA targets and their cell-specific effects.

Here we focus on miRNAs whose role in the pathophysiology of renal diseases has been ascertained.

5.2 Specific miRNA Signature in Kidney Homeostasis

Adult kidneys are characterized by a miRNA expression profile. Some miRNAs, such as miR-215, miR-146a and miR-886 are expressed mainly in the adult human kidney, while miR-192, miR-194, miR-21, miR-200a, miR-204 and *let-7a–g* are highly expressed in the kidney as well as in other organs.[3–5] The generation of mouse models with conditional deletion of two key miRNA processing enzymes (Drosha and Dicer) enabled the identification of the pivotal roles miRNAs have during renal cell homeostasis. Briefly, Drosha is a nuclear RNase III enzyme that turns primary miRNAs (thousand base pairs in length) into 70-nt stem–looped structures (precursor miRNAs, pre-miRNAs) soon after their transcription by RNA polymerase II. Pre-miRNAs are then shuttled by the exportin-5-RanGTP complex from the nucleus to the cytoplasm, where they are cleaved by Dicer, an RNase type III endonuclease that generates the mature miRNAs. Targeted deletion of Drosha or Dicer in podocytes, in the developing as well as in the adult kidney, causes disruption of the glomerular filtration barrier, leading to proteinuria and glomerulosclerosis,[6–9] while mice with inactivation of Dicer in post-natal proximal tubules displayed fully functional kidneys under normal conditions but were protected against renal ischemia-reperfusion injury.[10,11] Moreover, the suggestion that miRNAs play a role in controlling blood pressure derives from mice with specific inactivation of Dicer in renin-expressing juxtaglomerular cells, which are characterized by the loss of these cells, consequent low plasma renin levels, hypotension and kidney fibrosis.[12]

5.3 Chronic Kidney Disease

CKD is a very common syndrome characterized by gradual loss of kidney function over a period of months or years. More than 700 million people are affected worldwide, and in most cases, it is diagnosed only when the disease is well advanced, ultimately leading to end-stage renal disease requiring renal replacement therapy (dialysis or kidney transplantation).[13] CKD mainly develops secondary to major non-communicable diseases, such as diabetes mellitus and hypertension and, less frequently, hereditary diseases (polycystic kidney disease, infectious diseases and glomerulonephritis) or herbal and environmental toxins.[14] The most common pathological manifestation of CKD, regardless of disease etiology, is tubulointerstitial and glomerular fibrosis, accompanied by inflammatory cell infiltration, tubular cell atrophy, mesangial cell hypertrophy and podocyte apoptosis.[15] Current interventions targeting CKD centre on the blockade of the renin–angiotensin system, through angiotensin converting enzyme inhibitors or angiotensin II receptor blockers, but they are not always effective.[16]

Here, we describe miRNAs that, according to the available evidence, are involved in the development or progression of CKD, in diabetic and non-diabetic nephropathies.

5.3.1 Diabetic Nephropathy

Diabetic nephropathy is a severe vascular complication associated with both type 1 and 2 diabetes and it is the leading cause of kidney failure in the developed world, accounting for 45% of end-stage renal disease (ESRD) cases.[17,18] Diabetic nephropathy is a progressive kidney disease caused by alterations in the kidney architecture and function, induced by disturbed glucose homeostasis. The molecular patho-mechanisms are still poorly understood and there is increasing evidence to suggest that epigenetic mechanisms have an important role in the pathogenesis of diabetic nephropathy. The main pathophysiological hallmarks are glomerular hypertrophy, due to mesangial expansion, glomerular and tubulointerstitial fibrosis, caused by deposition of extracellular matrix proteins, and podocyte dysfunction.[19–21] Transforming growth factor-β1 (TGF-β) is one of the most important mediators of kidney fibrosis, cellular hypertrophy and survival. Its expression increases in glomerular cells and tubular cells of diabetic nephropathy patients because of aberrant glucose metabolism. The prosclerotic action of TGF-β is mainly mediated by the SMAD (mothers against decapentaplegic homolog) 2/3/4 complex and counteracted by SMAD7.[22]

Different experimental models that mimic diabetic nephropathy have been instrumental in dissecting the role of miRNAs, although they do not develop progressive loss of renal function leading to ESRD.[23] Type 1 diabetes mellitus in rodents is induced by streptozotocin, a chemical toxin that is toxic to pancreatic β-cells, whereas type 2 is mainly modeled by db/db mice with genetic defects in the leptin receptor (db/db mice).

Here, we discuss miRNAs that are consistently dysregulated in patients with diabetic nephropathy and in experimental models.

5.3.1.1 miR-21

miR-21 is a profibrotic miRNA, and as such its dysregulation accompanies several renal diseases whose outcome is scarring. Different studies have confirmed its upregulation in diabetic mice with diabetic nephropathy,[24,25] as well as in the kidneys of patients with both type 1 and 2 diabetes with different stages of diabetic nephropathy, and its expression correlates with the severity of fibrosis and rate of decline in renal function.[26–29] miR-21 expression is augmented in cultured proximal tubular epithelial cells, in mesangial cells and in podocytes exposed to TGF-β or high glucose.[26–28,30] miR-21 contributes to glomerular mesangial hypertrophy by activating the PI3K (phosphatidyl inositol 3 kinase)/AKT (protein kinase B) pathway through the direct downregulation of PTEN (phosphatase and tensin homolog)[26,27,30] and by impairing cell cycle progression, targeting Cdc25 (cell division cycle 25C) and Cdk6 (cyclin dependent kinase 6).[25] Moreover, it can further fuel TGF-β signaling in mesangial cells by suppressing the inhibitory SMAD7.[24] The coordinated targeting of tubular PTEN/SMAD7 by miR-21 enhances the TGF-β-mediated fibrotic response in proximal tubular cells, as well.[29] In addition, *in vitro* evidence suggests that a PTEN decrease downstream of the TGF-β-miR-21 axis in murine podocytes enhances podocyte motility.[25]

Notably, miR-21 silencing through antisense oligonucleotides designed against the miRNA mature sequence ameliorated albuminuria and renal fibrosis in murine models of type 1 and 2 diabetes.[24,25] Anti-miR-21 treatment also positively counteracted fibrosis in experimental models of chronic kidney disease.[31,32] Recently, an anti-miR (antagomir) against miR-21, RG-012, has been developed to treat Alport nephropathy and is planned to enter a phase II clinical trial. Alport syndrome is a genetic disorder characterized by glomerulonephritis, which progresses to ESRD starting in young adulthood. It is caused by mutations at the genetic locus encoding type IV collagen α3, α4 and α5 chains leading to structural abnormalities in the glomerular basement membrane.[33] Preclinical studies in Alport mouse models, where renal miR-21 expression increases as a consequence of the stress caused by defects in the glomerular basement membrane inducing inflammatory and profibrotic signaling, demonstrated that miR-21 silencing was able to improve survival and reduced albuminuria, glomerulosclerosis, interstitial fibrosis, tubular injury and inflammation.[34] In addition to Alport nephropathy, RG-012 also appears to be a promising drug for the cure of human diabetic nephropathy, but more studies are needed to fully understand the role of miR-21 in this disease. Indeed, genetic loss of miR-21 aggravated the progression of diabetic nephropathy in mice, suggesting a possible protective role of miR-21 in glomerular injury, although it is worth noting that knocking out miR-21 in the developing embryo could trigger phenotypic alterations independently to the onset of diabetes.[28]

5.3.1.2 *miR-29*

The miR-29 family, composed of miR-29a, miR-29b and miR-29c, is generally known as an antifibrotic family, since it is able to target collagen I and IV, counteracting extracellular matrix expansion.[35] It is highly conserved and expressed in the human kidney.[36] The members of the miR-29 family have a different cellular distribution and display functional differences, although they share an identical seed region (nucleotide positions 2–7), which is the key region determining the target mRNA.

Loss of miR-29a and miR-29b is associated with the progression of diabetic kidney disease in different murine models of diabetic nephropathy.[37–39] In particular, high levels of glucose and TGF-β decrease miR-29a and miR-29b expression in proximal tubular epithelial cells, mesangial cells and podocytes, through TGF-β-mediated binding of SMAD3 to the miR-29 promoter. Besides increasing collagen expression, low levels of miR-29a perturb the integrity of the podocyte ultrastructure and correlate with increased levels of its target histone deacetylase, which modulates the acetylation status of nephrin, a key functional component of the slit diaphragm. The overexpression of miR-29a attenuated nephrin ubiquitination and restored proper protein acetylation through the inhibition of histone deacetylase. Consistently, diabetic miR-29a transgenic mice had higher nephrin levels, podocyte viability and less glomerular fibrosis compared with diabetic wild-type mice. In addition, miR-29a counteracts apoptosis and fibrosis in mesangial cells by modulating the Dickkopf-1/Wnt/β-catenin signaling.[40]

Similarly to miR-29a, miR-29b exerts an antifibrotic role. Its expression is inhibited by the lncRNA Erbb4-IR, which in turn is upregulated in response to high levels of advanced glycosylation end products *via* a SMAD3-dependent mechanism in db/db mice.[41] Erbb4-IR functions as a molecular sponge, binding to miR-29 3′ UTR and suppressing its transcription. Kidney-specific silencing of Erbb4-IR protected the kidneys of db/db mice against progressive renal injury by increasing renoprotective miR-29b levels. Promising results in halting renal fibrosis were also obtained by directly overexpressing miR-29b in db/db mice.[37,39]

Unlike miR-29a and miR-29b, miR-29c is upregulated in glomeruli of diabetic db/db mice and increased expression of miR-29c in podocytes under hyperglycemic conditions promotes both cell apoptosis and fibronectin synthesis by directly inhibiting Spry (sprouty homolog 1) and consequently inducing Rho-kinase activation. Moreover, miR-29c can induce inflammation by targeting the anti-inflammatory protein tristetraprolin.[42,43] Pharmacological inhibition of miR-29c prevented the development of albuminuria in db/db mice.

5.3.1.3 *miR-25*

Different groups have demonstrated the protective role of miR-25. miR-25 was found to be downregulated in the sera of patients with diabetes with or without diabetic nephropathy[44] and its expression correlated inversely with proteinuria levels in diabetic nephropathy.[45] Similarly, miR-25 levels were significantly reduced in the kidney of type 1 diabetic rats and in mice with type 1 and 2 diabetes.[44–47] In diabetic mice and in high-glucose or TGF-β treated mesangial cells, the downregulation of miR-25 is a consequence of increased phosphorylation of the transcriptional repressor methyl-CpG binding protein (p-MeCP2). MeCP2 is phosphorylated by HIPK2 (homeodomain interacting protein kinase 2) whose expression increased in a diabetic setting as a consequence of reduced levels of SIAH1 (seven in absentias homolog 1), which mediates its proteosomal degradation.[47] In addition, it has been suggested that miR-25 has an anti-apoptotic effect on tubular cells by targeting PTEN and therefore activating the PTEN/AKT pathway. Notably, PTEN was upregulated in the kidneys of patients with diabetic nephropathy.[45] Another possible mechanism of miR-25 action is through the inhibition of oxidative stress, either by negatively regulating the Ras-related gene CDC42 (cell division cycle 42) or NOX4 (nicotinamide adenine dinucleotide phosphate oxidase 4), responsible for the production of superoxide.[44,46] However, further studies are needed to clarify how miR-25 reduction promotes kidney dysfunction. Notably, the overexpression of miR-25 in db/db mice induced by the systemic administration of a miR-25 agomir, suitably designed to mimic the mature miR-25, ameliorated glomerular fibrosis and reduced blood pressure, whereas its knockdown caused severe kidney dysfunction and hypertension with renin–angiotensin system activation.[47]

5.3.1.4 miR-192

miR-192 is highly expressed in the kidney and has been observed to have pleiotropic roles during kidney homeostasis and fibrosis.[48] There have been conflicting findings regarding its modulation in diabetic settings, which may be due to intrinsic differences between the disease stages when miR-192 expression analysis was performed. Indeed, in the early stages, diabetic nephropathy exhibited increased renal miR-192 expression,[49] whereas its expression was found to have decreased in renal biopsies from advanced diabetic nephropathy patients and negatively correlated with tubulointerstitial fibrosis and a reduction in estimated glomerular filtration rate.[50] miR-192 was also found to be up or downregulated in experimental diabetic nephropathy.[51–53] Moreover, it is well known that miR-192 displays cell-specific behaviors, acting as a fibrotic agent in mesangial cells or as an antifibrotic one in proximal tubular epithelial cells.[51–54]

In murine mesangial cells, TGF-β initially induces miR-192 expression through the SMAD pathway, followed by a secondary mechanism that relaxes the locked chromatin structure of the *mir-192* gene through epigenetic regulation *via* ETS1 (ETS proto-oncogene 1, transcription factor).[55] Similar findings were observed in glomeruli from diabetic db/db mice. Upregulation of miR-192 induces COL1A2 (collagen type I alpha 2 chain) and COL4A1 (collagen type IV alpha 1 chain) expression by suppressing the E-box repressors ZEB (zinc finger e-box binding homeobox) 1 and 2.[51,52] ZEB1 and ZEB2 expression consistently correlates inversely with miR-192 expression in Southwestern American Indians with type 2 diabetes.[49] In addition, miR-192 contributes to glomerular mesangial hypertrophy by activating the PI3K/AKT signalling pathway through the induction of four further miRNAs: miR-216a, miR-217, miR-200b and miR-200c. Indeed, miR-216a and miR-217[56,57] target the AKT inhibitor PTEN, while miR-200b and miR-200c repress the PI3K inhibitor FOG2 (friend of GATA 2).[58] In turn, miR-216 further increases collagen deposition in a diabetic nephropathy setting through TFE3 (transcription factor binding to IGHM enhancer 3) by targeting YBX1 (Y-box binding protein 1).[59] An increase in AKT activation downstream of the TGF-β/miR-192-driven cascade results in transcriptional repression of autophagic genes in cultured mesangial cells and in type 1 and 2 diabetic murine models. This results in the accumulation of damaged proteins and organelles, disturbing cellular homeostasis.[60] Moreover, emerging evidence suggests that autophagic gene expression could also be directly affected by miR-192 and miR-217. Notably, the inhibition of miR-192 with an anti-miR-192 oligonucleotide in diabetic mice increased the expression of renal Zeb1/2, decreased collagen, TGF-β and fibronectin, and restored the expression of autophagic genes. More importantly, the inhibition of miR-192 decreases renal fibrosis and improves proteinuria.[52,60] Finally, miR-192 has also been shown to sustain TGF-β signalling, which amplifies and creates a chronic state of profibrotic pathway activation in mesangial cells. Under diabetic conditions, miR-192 is co-regulated with p53 by a common pathway downstream of TGF-β signaling. In turn, miR-192 and p53 up-regulate each other's expression; in particular, p53 is upregulated by miR-192 *via* the targeting of ZEB2.[49]

In human proximal tubular cells miR-192 is expressed normally and TGF-β represses its transcription by decreasing HNF (hepatocyte nuclear factor) binding to the *mir-192* gene.[61] In the kidney, HNF expression is restricted to tubular cells, thus partly explaining cell-specific regulation of miR-192. TGF-β-dependent loss of miR-192 expression results in increased levels of the E-cadherin transcriptional repressors ZEB1 and ZEB2 and consequently decreased levels of E-cadherin, an epithelial marker involved in cell–cell adhesion, whose dysregulation induces epithelial cell phenotype changes. Notably, the miR-200 family[62,63] and miR-215[53] act in concert with miR-192 to protect proximal tubular epithelial cells from TGF-β-induced epithelial to mesenchymal transition by favouring E-cadherin expression.[53]

Overall, the available data on miRNAs represent promising results in the cure of experimental diabetic nephropathy. However, other miRNAs and their related pathways are noteworthy and promising therapeutic candidates that should still be tested in humans. These are the miR-30c-Snail1 (snail family transcriptional repressor 1)-TGF-β pathway,[64] the miR-137/Notch1 (translocation-associated notch protein TAN-1) pathway,[65] the miR-27a/PPARγ (peroxisome proliferator activated receptor gamma)/β-catenin pathway,[66,67] the miR-184/LPP3 (phospholipid phosphatase 3) pathway (Figure 5.2),[68] the miR-214/PTEN/mTORC1 (mammalian target of rapamycin complex 1) pathway,[69] *etc.*

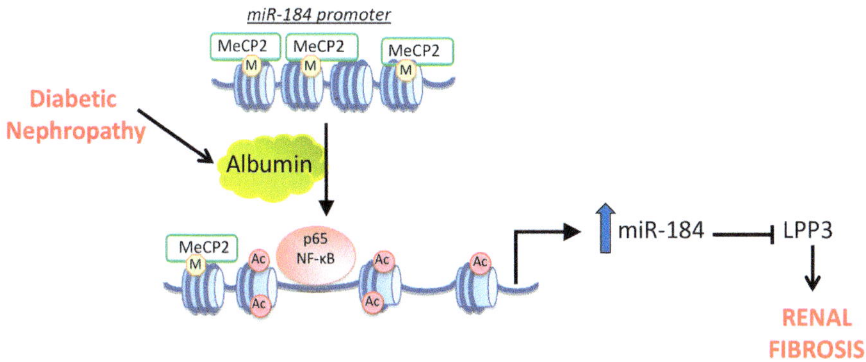

Figure 5.2 A miRNA regulatory network promoting tubulointerstitial fibrosis in Zucker diabetic fatty (ZDF) rats during overt nephropathy. Albuminuria is one of the best clinical indicators of diabetes-induced renal damage and a predictor of progression to ESRD. Besides transcription of proinflammatory and fibrogenic genes, albumin regulates the expression of miR-184 gene in tubular cells through epigenetic mechanisms. By reducing the transcriptional repressor MeCP2 (Methylcytosine-binding Protein 2) binding to methyl-CpG regions (M) and fostering histone lysine acetylation (Ac), albumin favours accessibility of NF-κB-p65 to its recognition sequence on the miR-184 promoter. In turn, miR-184 increased expression translates in reduced levels of its target lipid phosphate phosphatase 3 (LPP3), a kidney glycoprotein that catalyses dephosphorylation of lipid phosphates involved in multiple organ fibrosis. Loss of LPP3 leads to changes in bioactive lipid profile and consequent development of tubulointerstitial fibrosis.[67]

5.3.2 Immunoglobulin A Nephropathy

Immunoglobulin A nephropathy (IgAN) is the most prevalent type of primary glomerulonephritis, characterized by an increase in circulating levels of IgA1 with galactose-deficient *O*-glycans, followed by the synthesis of antibodies directed against galactose-deficient IgA1. IgA1–IgG immune complexes accumulate in the glomeruli, where they activate mesangial cells and cause their proliferation and excess matrix production, together with the secretion of cytokines and chemokines, finally resulting in glomerular injury.[70] IgA1 *O*-glycosylation is catalysed by GALNT2 (polypeptide *N*-acetylgalactosaminyltransferase 2) and C1GALT1 (core 1 synthase, glycoprotein-*N*-acetylgalactosamine 3-beta-galactosyltransferase 1), two enzymes that work in immune cells and whose expression is controlled by *let-7b* and miR-148b, respectively.[71,72] Both miRNAs are upregulated in peripheral blood mononuclear cells in patients with IgAN.[73] C1GALT1 acts with COSMC (C1GALT1-specific chaperone), whose activity is controlled by miR-347b, which is found to be overexpressed in the B-lymphocytes of IgAN patients.[74] An immune response against defective IgA1 is not triggered in all patients, and a second hit, such as mucosal infection with overproduction of secretory IgA is often necessary for the disease to develop. *In vitro* experiments have demonstrated that secretory IgA isolated from patients with IgAN induced mesangial cell proliferation and the production of the pro-inflammatory cytokines interleukin (IL)-8, IL-1β and IL-6 through the downregulation of miR-100-3p, miR-877-3p and miR-16, respectively.[75,76] Other miRNA networks are important in the development of inflammatory and profibrotic responses in the glomeruli and tubulointerstitium of IgAN patients. The miR-200b-c/429 cluster is downregulated in IgAN kidneys and is probably involved in the regulation of the inflammatory response, whereas renal miR-320 is upregulated and promotes B cell proliferation by suppressing PTEN expression.[77] In contrast, miR-21-5p, miR-214-3p and miR-199a-5p are well known fibrotic miRNAs and their involvement in interstitial fibrosis in the course of IgAN has been confirmed, with the first two being associated with an increased risk of renal failure.[78]

miRNAs are also involved in a more severe variant of IgAN, characterized by endocapillary hypercellularity. In particular, these patients are characterized by low levels of glomerular miR-223. More in-depth studies have revealed that the downregulation of miR-223 is associated with the increased expression of importin α4 and α5, both responsible for NF-κB (nuclear factor kappa B subunit 1)- and STAT3 (signal transducer and activator of transcription 3)-mediated cell proliferation and monocyte adhesion.[79]

5.3.3 Lupus Nephritis

Lupus nephritis is a major complication of systemic lupus erythematosus, a systemic disease characterized by dysregulated immune system activity with the formation of anti-double strand DNA IgG antibodies. The deposition of these immune complexes within glomeruli and tubulointerstitial compartments fosters kidney inflammation and fibrosis. Resident renal cells, notably mesangial cells, endothelial and proximal renal tubular epithelial cells and podocytes contribute differently to the pathogenesis of lupus nephritis inducing inflammatory and fibrotic processes in the kidney.[80]

The expression of several miRNAs is modulated in kidney biopsies from lupus nephritis patients compared to healthy controls, indicating that they are active players in the course of lupus nephritis.[81] Mesangial cell proliferation, apoptosis and activation are prominent features of lupus nephritis and the expression of several mesangial genes has been shown to be modulated by miRNAs during lupus nephritis. In patients and mice with lupus nephritis, mesangial expansion is mediated by the upregulation of miR-148a-3p through the post-transcriptional inhibition of PTEN with consequent AKT activation and cell proliferation.[82] The activation of human mesangial cells, in turn, induces the production of chemokines and cytokines, which cause both proliferation and intense inflammation. The renal activation of the type I interferon (IFN) signaling pathway contributes to the pathogenesis of lupus nephritis. Indeed, the expression level of IFN-inducible genes correlates with disease activity and severity.[83] Mesangial IFN signaling promotes the transcription of pro-inflammatory genes through miR-744, which targets phosphatase PTP1B (protein tyrosine phosphatase 1B).[84] Moreover, in mesangial cells, type I IFN indirectly switches off protective miRNAs, including miR-26a and miR-30b, which directly regulate the cell cycle in mesangial cells. Both miRNAs have been found to be downregulated in the kidneys and urine of lupus nephritis patients.[85] In addition, miRNAs play pivotal roles upstream of the type I IFN pathway that is negatively regulated by miR-130b.[86] Low miR-130b expression levels were observed in kidney tissues from patients and lupus-prone mice and correlated negatively with abnormal activation of the IFN response. Interestingly, treatment with the miR-130b agomir of a lupus-prone murine model reduced the IFN-accelerated progression of experimental lupus nephritis and ameliorated glomerular lesions. Recent evidence from cultured mesangial cells suggests that anti-double strand DNA IgG antibodies purified from lupus nephritis patients downregulate the expression of miR-10, which affects the expression of several genes involved in lupus nephritis pathogenesis, including IL-8.[87]

miR-150 is another crucial miRNA involved in the pathogenesis of lupus nephritis, whose renal expression differs in lupus nephritis patients with varying severity of renal dysfunction. miR-150 levels decreased in lupus nephritis renal biopsies from six female patients with class IV lupus nephritis (characterized by diffuse proliferative nephritis), compared to healthy controls.[88] Notably, it has been predicted recently that miR-150 is the target of circHLA-C, a circular RNA that may participate in lupus nephritis pathogenesis, functioning as a miR-150 sponge.[88] circHLA-C was consistently overexpressed in lupus nephritis kidney biopsies compared to healthy controls and showed a tendency towards a negative correlation with miR-150. Following disease onset, renal miR-150 levels increased in lupus nephritis patients with a high chronicity index (a morphologic index for the evaluation of renal biopsies in lupus nephritis), predominantly in proximal tubular cells and in podocytes, compared with patients with a low chronicity index.[89] A TGF-β increase during chronic inflammation induces miR-150 expression in proximal tubular cells and podocytes, which promotes the transcription of genes involved in cell proliferation, inflammation and fibrosis by directly targeting SOCS1 (suppressor of cytokine signaling 1), a negative regulator of the JAK (janus kinase)/STAT signaling pathway.[89]

Several additional recent projects have identified dysregulated miRNA-based mechanisms modulating inflammatory factor releases in lupus nephritis, from the miR-663a/miR-423-5p-TNIP2 (TNFAIP3 interacting protein 2)/NF-κB to the miR-146-TRAF6 (TNF receptor associated factor 6)/NF-κB axis, although more robust data are needed.[90,91]

5.3.4 Polycystic Kidney Disease

Polycystic kidney disease (PKD) is the most common monogenic renal disorder and is characterized by progressive enlargement of the kidneys, caused by sustained expansion of many fluid-filled cysts that originate from the tubule wall, leading to crowding of adjacent nephrons, injury of normal parenchyma, and eventually kidney failure.[92] PKD can be inherited as an autosomal recessive (ARPKD) or autosomal dominant (ADPKD) disorder. Mutations in either the PKD1 or PKD2 genes, which encode polycystin 1 and polycystin 2, are the underlying cause of ADPKD, whereas mutations in PKHD1, encoding for fibrocystin, are responsible for the less common ARPKD. Cyst expansion is mediated by cyclic AMP-driven mechanisms that stimulate both fluid secretion and cell proliferation.

miRNAs are closely associated with cystogenesis and are involved both in the induction and progression of PKD. A first piece of evidence derives from mice with conditional deletion of miRNAs in the ureteric bud derivatives that are characterized by dilatation of the renal pelvis and cortical cyst formation.[93] Similarly, spatial conditional deletion of miRNAs from maturing renal tubules and collecting ducts induces tubular and glomerular cyst growth in mice.[10]

The miR-17~92 cluster is upregulated in murine models of PKD. miR-17 inhibits the expression of PKD1 and PKD2,[94,95] while miR-92 represses PKHD1 expression through the transcription factor HNF-1β.[94,96] The repressive activity of the miR-17~92 cluster on PKD2 is antagonized by BICC1 (BicC family RNA binding protein 1).[97] BICC1 can also silence two key enzymes of the cAMP pathway through the modulation of miR-125a and miR-27a.[98] Kidney-specific overexpression of miR-17~92 produces kidney cysts and aggravates disease progression in a ciliopathy model of cystic kidney disease, while its inactivation slows kidney cyst growth, improves renal function and prolongs survival in a mouse model of PKD.[94] The Patel group analysed in depth the role of miR-17, a cMYC-regulated miRNA, that is induced in kidney cysts of mouse and human ADPKD.[99] miR-17 can enhance cyst proliferation by inhibiting of mitochondrial function. miR-17 targets PPARα, a key regulator of mitochondrial function, which activates the expression of many genes involved in oxidative phosphorylation and fatty acid oxidation. In the presence of low levels of PPARα, carbon is shuttled to produce DNA, proteins and lipids for cell membranes/organelles at the expense of ATP generation through oxidative phosphorylation.[99] Preclinical studies in different murine models of PKD have demonstrated the efficacy of anti-miR-17 treatment in attenuating cyst growth. Moreover, treatment with anti-miR-17 suppressed the proliferation and cyst growth of primary human ADPKD cultures. These findings have led to the development of a new drug, RGLS4326, which has just entered a phase I clinical trial. RGLS4326 is a specific anti-miR-17 inhibitor with a unique chemical formulation to preferentially target the kidney. Another potential therapeutic target for PKD is miR-21, for which a miRNA-based drug has already been developed to treat Alport syndrome, as described in Section 5.3.1.1. Indeed, murine and human ADPKD cysts are characterized by the upregulation of miR-21, which functions downstream of the cAMP pathway, inhibiting apoptosis of cyst epithelial cells—probably through the inhibition of *programmed cell death 4* in experimental PKD.[100] Recently, constitutive BCL2 (B-cell CLL/Lymphoma 2) downregulation by miR-181, causing spontaneous apoptosis of T-cells, was shown to aggravate the PKD phenotype in a patient with mutant PKD1.[101]

In addition to the aforementioned miRNAs, many other miRNAs (such as miRs-10a, miR-30a-5p, miR-182-5p, miR-200a, *etc.*) are involved in pathways linked to cystogenesis, including TGF-β, mitogen-activated protein kinase, calcium signaling and actin cytoskeleton pathways, as demonstrated in experimental models of autosomal dominant PKD.[102–104]

5.3.5 Focal Segmental Glomerulosclerosis

Focal segmental glomerulosclerosis (FSGS) is the typical histological renal lesion in familiar steroid-resistant nephrotic syndrome. It is characterized by proteinuria, due to progressive glomerular scarring, beginning with partial sclerosis (segmental) of a minority of glomeruli (focal).[105] The major cell targets of injury in FSGS are glomerular podocytes. Podocytes express high levels of miR-30 family members that have a protective effect on their integrity. Indeed, they prevent podocyte apoptosis by directly targeting Notch1 and p53 [106] and they contribute to actin fibre stability by suppressing several components of calcium/calcineurin signaling (TRPC6, PP3CA, PP3CB, PPP3R1 and NFATC3) whose dysregulation leads to podocyte cytoskeletal injury.[107] The loss of miR-30s occurs both in podocytes from FSGS patients and in the rat FSGS experimental model (rats treated with podocyte toxin puromycin aminoucleoside) with a TGF-β-triggered mechanism.[106,108] It is notable that glucocorticoids have a renoprotective effect in glomerular diseases by restoring miR-30 levels.[106] Moreover, the brain-derived neurotrophic factor repairs podocyte injury by increasing actin polymerization through miR-132 and miR-134 signaling.[109]

miRNAs also play a key role in podocyte–matrix crosstalk. Indeed, miR-378a-3p is overexpressed in biopsies from FSGS patients and it targets the glomerular basement membrane matrix component nephronectin, affecting podocyte matrix production and therefore basement membrane composition. In mice, overexpression of miR-378a-3p altered glomerular basement membrane protein composition and induced podocyte injury, finally resulting in albuminuria.[110]

miR-193 is another key miRNA involved in the pathogenesis of FSGS. miR-193 induces podocyte de-differentiation by targeting the transcription factor Wilms tumour protein (WT1), a master regulator of podocyte homeostasis.[111,112] Low levels of WT1 are associated with reduced expression of podocyte-relevant genes, such as podocalyxin and nephrin.[113] Notably, increased levels of miR-193a were detected in glomeruli of FSGS patients.[111] Moreover, miR-193-overexpressing mice develop FSGS with extensive podocyte foot process effacement. Besides holding great promise as a potential therapeutic target, recent findings also suggest that miR-193a could serve as a potential biomarker for the early diagnosis of primary FSGS. Indeed, levels of urinary exosomal miR-193a were higher in children with primary FSGS than in children with a different podocytopathy.[114]

5.3.6 Rapidly Progressive Glomerulonephritis

Crescentic rapidly progressive glomerulonephritis (RPGN) is a severe form of acquired glomerular disease characterized by extracapillary proliferation of podocytes and parietal epithelial cells (crescents).[115] Human extracapillary lesions from patients with RPGN are characterized by increased levels of miR-193a. In addition to being responsible for podocyte de-differentiation (see Section 5.3.5), miR-193a also has a key role in parietal epithelial cells (PECs), potential renal progenitors that contribute to glomerulosclerosis.[116] miR-193a is highly expressed in PECs and its downregulation causes PEC transdifferentiation towards podocytes. In contrast, its overexpression induces exacerbated activation of PECs, typical of crescent formation.[112] Notably, increased levels of miR-193a were detected both in human and murine extracapillary lesions, whereas pharmacological inhibition of miR-193a expression in mice with nephrotoxic serum-induced crescentic nephritis (NTN) reduced the number of crescents.[112]

Kidney biopsies from patients with RPGN also exhibited high abundance of miR-92a expression, especially in podocytes.[117] Similar findings were obtained in NTN mice. miR-92a is activated by STAT3, and it targets the cyclin-dependent kinase inhibitor $p57^{Kip2}$, which is involved in cell cycle regulation and control of the quiescent state of podocytes. The loss of $p57^{Kip2}$ translates into the removal of the break in the podocyte cell cycle and podocyte transition to a proliferative phenotype. Mice with podocyte-specific deletion of miR-92a had $p57^{Kip2}$-restored expression, reduced albuminuria and glomerular injury after NTN. Moreover, the administration of an anti-miR-92a in NTN mice prevented albuminuria and kidney failure. Therefore, anti-miR-92a seems to be a possible therapeutic strategy for RPGN.

5.4 miRNAs as Therapeutic Targets

Current strategies for halting CKD focus on reducing urinary protein excretion and controlling blood pressure through the blockade of the renin–angiotensin system, which slows, but does not invariably stop the progression of CKD.[118] Attempts made so far to find new therapies have yielded few therapeutic advances. Now several researchers and companies are moving towards investigating in depth miRNA-based mechanisms underlying both the disease and the fibrotic process. Hopefully, this will translate into finding new nephroprotective agents, in light of the fact that the kidney is a privileged site for miRNA treatment.

miRNA-based therapy foresees the use of antagomirs designed against pathogenic miRNAs, or agomirs that simulate protective miRNAs. So far, only two miRNA-based therapeutics to treat kidney disease have been tested in clinical trials: anti-miR-21 to treat Alport syndrome and anti-miR-17 to treat PKD. Following intravenous injection, anti-miRs—suitably chemically modified in order to increase their plasma stability (primarily with phosphorothioates or 2′*O*-methoxyethyl or locked nucleic acid backbones) and entrance into the cells[119]—are readily delivered to the kidney and the liver.[34,120] Although their half-life is only ~8 h in blood, they last for up to 3 weeks in the kidney, yielding significant inhibition of their targets in the kidneys of both mice[121] and non-human primates.[122] In the healthy kidney, anti-miRs are filtered through the glomerulus and accumulate in proximal tubules, while the cellular tropism of antagomirs in diseased kidney can be altered, as happens in cystic kidneys where anti-miRs are delivered to collecting duct-derived cysts.[99,123]

Contrary to expectations, major challenges significantly limit the translation of miRNA-based therapeutics into clinical practice, including delivery methods and safety concerns. Indeed, therapeutic miRNA should be delivered only to the injured cells of the kidney in order to avoid any potential adverse effects on other cells, tissues and organs and the target miRNA should affect only one pathway or at least only pathways with related functions.

5.5 The Role of miRNAs as Biomarkers

So far, several clinical studies have aimed to find reliable miRNA-based biomarkers able to provide information about kidney disease diagnosis and progression to improve the accuracy of common CKD diagnostic tools, such as urinary albumin, blood urea nitrogen and serum creatinine level. miRNAs are very promising candidates because they are stable in biological fluids. Indeed, they circulate either encapsulated into microvescicles, exosomes and apoptotic bodies, or complexed with RNA-binding proteins (argonaute 2 complexes) or lipoproteins and are consequently protected from degradation by ribonucleases.[124,125] Moreover, miRNAs can be detected with high accuracy by real-time quantitative PCR, microarrays and next-generation sequencing. However, in order to implement miRNA biomarkers in a clinical setting, researchers should remember that the serum miRNA profile can be affected by pre-analytical (such as haemolysis, lifestyle, hormonal activity, *etc.*) and analytical variables (quantification platform and normalization methods).[126] Moreover, serum miRNAs affect the urine miRNA profile, since they are filtered in the urine. Urine miRNAs could also derive from exosomes actively secreted from cells of the urinary tract. Considering these findings, the evaluation of urinary exosome miRNA content should be more informative than total urine miRNAs, although it requires specialized equipment and lengthy protocols.

Stability in the circulation and high sensitivity and specificity suggest that the applicability of miRNAs as biomarkers for kidney disease might even exceed that of protein-based biomarkers. However, to date, several clinical studies have identified urine miRNAs as being correlated with clinical and histological parameters in diabetic nephropathy, IgAN, lupus nephritis and FSGS, but this role has not yet been adopted in clinical practice. To this end, large-scale studies are needed for a miRNA to become a useful biomarker.

5.6 Conclusions

The emerging importance of miRNAs in the field of nephrology has opened several lines of research, encompassing investigations into issues ranging from their roles as disease biomarkers to the development of tailored therapies. They could be either pathogenic or mediators of a protective physiological response to kidney injury. However, despite the huge number of papers that stress the importance of miRNAs in a given kidney disease claiming their potential as therapeutic agents, only two miRNA-based therapeutics have entered clinical trials. Indeed, data are often inconsistent across different samples, reflecting heterogeneity in patient, species, disease models and evaluation time points.[127] Moreover, most results on miRNA expression studies derive from whole-kidney samples, underestimating the cell-specific effects of miRNAs. Instead, although this is admittedly more difficult, the miRNA profile of the cells that are primarily affected in the disease should be given preference when attempting to understand the kidney's response to different types of insults and in designing potential intervention strategies.[128] Another major concern is the limited power of most studies, often involving a small population, in the context of a complex regulatory network. Indeed, one miRNA simultaneously regulates different genes that work together, forming networks, and different miRNAs in the same cell can connect different cellular networks, including signaling, metabolism, protein interaction and gene regulatory networks. In light of this, researchers of the 'omics' era are moving towards system biology when trying to detect relevant miRNA regulatory networks, instead of studying a single candidate miRNA, by combining various high-throughput approaches, such as genome sequencing and microarray technologies.[129]

Acknowledgements

The authors would like to thank Kerstin Mierke for English language editing and Antonella Piccinelli for creating the artworks. Dr Piera Trionfini participated to the European STELLAR project (grant no. HEALTH-F4-2012-305436).

References

1. P. Trionfini, A. Benigni and G. Remuzzi, *Nat. Rev. Nephrol.*, 2015, **11**, 23–33.
2. M. Ha and V. N. Kim, *Nat. Rev. Mol. Cell Biol.*, 2014, **15**, 509–524.
3. S. Baskerville and D. P. Bartel, *RNA*, 2005, **11**, 241–247.
4. Y. Sun, S. Koo, N. White, E. Peralta, C. Esau, N. M. Dean and R. J. Perera, *Nucleic Acids Res.*, 2004, **32**, e188.
5. P. Landgraf, M. Rusu, R. Sheridan, A. Sewer, N. Iovino, A. Aravin, S. Pfeffer, A. Rice, A. O. Kamphorst, M. Landthaler, C. Lin, N. D. Socci, L. Hermida, V. Fulci, S. Chiaretti, R. Foà, J. Schliwka, U. Fuchs, A. Novosel,

R.-U. Müller, B. Schermer, U. Bissels, J. Inman, Q. Phan, M. Chien, D. B. Weir, R. Choksi, G. De Vita, D. Frezzetti, H.-I. Trompeter, V. Hornung, G. Teng, G. Hartmann, M. Palkovits, R. Di Lauro, P. Wernet, G. Macino, C. E. Rogler, J. W. Nagle, J. Ju, F. N. Papavasiliou, T. Benzing, P. Lichter, W. Tam, M. J. Brownstein, A. Bosio, A. Borkhardt, J. J. Russo, C. Sander, M. Zavolan and T. Tuschl, *Cell*, 2007, **129**, 1401–1414.

6. J. Ho, K. H. Ng, S. Rosen, A. Dostal, R. I. Gregory and J. A. Kreidberg, *J. Am. Soc. Nephrol.*, 2008, **19**, 2069–2075.
7. S. J. Harvey, G. Jarad, J. Cunningham, S. Goldberg, B. Schermer, B. D. Harfe, M. T. McManus, T. Benzing and J. H. Miner, *J. Am. Soc. Nephrol.*, 2008, **19**, 2150–2158.
8. S. Shi, L. Yu, C. Chiu, Y. Sun, J. Chen, G. Khitrov, M. Merkenschlager, L. B. Holzman, W. Zhang, P. Mundel and E. P. Bottinger, *J. Am. Soc. Nephrol.*, 2008, **19**, 2159–2169.
9. O. Zhdanova, S. Srivastava, L. Di, Z. Li, L. Tchelebi, S. Dworkin, D. B. Johnstone, J. Zavadil, M. M. Chong, D. R. Littman, L. B. Holzman, L. Barisoni and E. Y. Skolnik, *Kidney Int.*, 2011, **80**, 719–730.
10. V. Patel, S. Hajarnis, D. Williams, R. Hunter, D. Huynh and P. Igarashi, *J. Am. Soc. Nephrol.*, 2012, **23**, 1941–1948.
11. Q. Wei, K. Bhatt, H.-Z. He, Q.-S. Mi, V. H. Haase and Z. Dong, *J. Am. Soc. Nephrol.*, 2010, **21**, 756–761.
12. M. L. S. Sequeira-Lopez, E. T. Weatherford, G. R. Borges, M. C. Monteagudo, E. S. Pentz, B. D. Harfe, O. Carretero, C. D. Sigmund and R. A. Gomez, *J. Am. Soc. Nephrol.*, 2010, **21**, 460–467.
13. B. Bikbov, N. Perico, G. Remuzzi and on behalf of the GBD Genitourinary Diseases Expert Group, *Nephron*, 2018, 1–6.
14. P. Romagnani, G. Remuzzi, R. Glassock, A. Levin, K. J. Jager, M. Tonelli, Z. Massy, C. Wanner and H.-J. Anders, *Nat. Rev. Dis. Primers*, 2017, **3**, 17088.
15. Y. Liu, *Nat. Rev. Nephrol.*, 2011, **7**, 684–696.
16. N. Perico, P. Ruggenenti and G. Remuzzi, *Curr. Opin. Pharmacol.*, 2017, **33**, 34–40.
17. G. Remuzzi, A. Schieppati and P. Ruggenenti, *N. Engl. J. Med.*, 2002, **346**, 1145–1151.
18. K. R. Tuttle, G. L. Bakris, R. W. Bilous, J. L. Chiang, I. H. de Boer, J. Goldstein-Fuchs, I. B. Hirsch, K. Kalantar-Zadeh, A. S. Narva, S. D. Navaneethan, J. J. Neumiller, U. D. Patel, R. E. Ratner, A. T. Whaley-Connell and M. E. Molitch, *Diabetes Care*, 2014, **37**, 2864–2883.
19. B. Najafian, C. E. Alpers and A. B. Fogo, *Contrib. Nephrol.*, 2011, **170**, 36–47.
20. Y. Qian, E. Feldman, S. Pennathur, M. Kretzler and F. C. Brosius, *Diabetes*, 2008, **57**, 1439–1445.
21. C. Hu, L. Sun, L. Xiao, Y. Han, X. Fu, X. Xiong, X. Xu, Y. Liu, S. Yang, F. Liu and Y. S. Kanwar, *Curr. Med. Chem.*, 2015, **22**, 2858–2870.
22. G. J. Inman, F. J. Nicolás and C. S. Hill, *Mol. Cell*, 2002, **10**, 283–294.

23. K. Azushima, S. B. Gurley and T. M. Coffman, *Nat. Rev. Nephrol.*, 2018, **14**, 48–56.
24. X. Zhong, A. C. K. Chung, H. Y. Chen, Y. Dong, X. M. Meng, R. Li, W. Yang, F. F. Hou and H. Y. Lan, *Diabetologia*, 2013, **56**, 663–674.
25. M. Kölling, T. Kaucsar, C. Schauerte, A. Hübner, A. Dettling, J.-K. Park, M. Busch, X. Wulff, M. Meier, K. Scherf, N. Bukosza, G. Szénási, M. Godó, A. Sharma, M. Heuser, P. Hamar, C. Bang, H. Haller, T. Thum and J. M. Lorenzen, *Mol. Ther.*, 2017, **25**, 165–180.
26. Z. Zhang, H. Peng, J. Chen, X. Chen, F. Han, X. Xu, X. He and N. Yan, *FEBS Lett.*, 2009, **583**, 2009–2014.
27. X. Lu, Q. Fan, L. Xu, L. Li, Y. Yue, Y. Xu, Y. Su, D. Zhang and L. Wang, *PLoS One*, 2015, **10**, e0117400.
28. J. Y. Lai, J. Luo, C. O'Connor, X. Jing, V. Nair, W. Ju, A. Randolph, I. Z. Ben-Dov, R. N. Matar, D. Briskin, J. Zavadil, R. G. Nelson, T. Tuschl, F. C. Brosius, M. Kretzler and M. Bitzer, *J. Am. Soc. Nephrol.*, 2015, **26**, 805–816.
29. A. D. McClelland, M. Herman-Edelstein, R. Komers, J. C. Jha, C. E. Winbanks, S. Hagiwara, P. Gregorevic, P. Kantharidis and M. E. Cooper, *Clin. Sci.*, 2015, **129**, 1237–1249.
30. L. Denby, V. Ramdas, M. W. McBride, J. Wang, H. Robinson, J. McClure, W. Crawford, R. Lu, D. Z. Hillyard, R. Khanin, R. Agami, A. F. Dominiczak, C. C. Sharpe and A. H. Baker, *Am. J. Pathol.*, 2011, **179**, 661–672.
31. X. Zhong, A. C. K. Chung, H.-Y. Chen, X.-M. Meng and H. Y. Lan, *J. Am. Soc. Nephrol.*, 2011, **22**, 1668–1681.
32. B. N. Chau, C. Xin, J. Hartner, S. Ren, A. P. Castano, G. Linn, J. Li, P. T. Tran, V. Kaimal, X. Huang, A. N. Chang, S. Li, A. Kalra, M. Grafals, D. Portilla, D. A. MacKenna, S. H. Orkin and J. S. Duffield, *Sci. Transl. Med.*, 2012, **4**, 121ra18.
33. J. Kruegel, D. Rubel and O. Gross, *Nat. Rev. Nephrol.*, 2013, **9**, 170–178.
34. I. G. Gomez, D. A. MacKenna, B. G. Johnson, V. Kaimal, A. M. Roach, S. Ren, N. Nakagawa, C. Xin, R. Newitt, S. Pandya, T.-H. Xia, X. Liu, D.-B. Borza, M. Grafals, S. J. Shankland, J. Himmelfarb, D. Portilla, S. Liu, B. N. Chau and J. S. Duffield, *J. Clin. Invest.*, 2015, **125**, 141–156.
35. Z. Deng, Y. He, X. Yang, H. Shi, A. Shi, L. Lu and L. He, *Mol. Diagn. Ther.*, 2017, **21**, 285–294.
36. A. J. Kriegel, Y. Liu, Y. Fang, X. Ding and M. Liang, *Physiol. Genomics*, 2012, **44**, 237–244.
37. H.-Y. Chen, X. Zhong, X. R. Huang, X.-M. Meng, Y. You, A. C. Chung and H. Y. Lan, *Mol. Ther.*, 2014, **22**, 842–853.
38. B. Wang, R. Komers, R. Carew, C. E. Winbanks, B. Xu, M. Herman-Edelstein, P. Koh, M. Thomas, K. Jandeleit-Dahm, P. Gregorevic, M. E. Cooper and P. Kantharidis, *J. Am. Soc. Nephrol.*, 2012, **23**, 252–265.
39. W. Qin, A. C. K. Chung, X. R. Huang, X.-M. Meng, D. S. C. Hui, C.-M. Yu, J. J. Y. Sung and H. Y. Lan, *J. Am. Soc. Nephrol.*, 2011, **22**, 1462–1474.
40. Y.-C. Hsu, P.-J. Chang, C. Ho, Y.-T. Huang, Y.-H. Shih, C.-J. Wang and C.-L. Lin, *Sci. Rep.*, 2016, **6**, 30575.

41. S. F. Sun, P. M. K. Tang, M. Feng, J. Xiao, X. R. Huang, P. Li, R. C. W. Ma and H. Y. Lan, *Diabetes*, 2018, **67**, 731–744.
42. J. Long, Y. Wang, W. Wang, B. H. J. Chang and F. R. Danesh, *J. Biol. Chem.*, 2011, **286**, 11837–11848.
43. J. Guo, J. Li, J. Zhao, S. Yang, L. Wang, G. Cheng, D. Liu, J. Xiao, Z. Liu and Z. Zhao, *Sci. Rep.*, 2017, **7**, 2314.
44. Y. Liu, H. Li, J. Liu, P. Han, X. Li, H. Bai, C. Zhang, X. Sun, Y. Teng, Y. Zhang, X. Yuan, Y. Chu and B. Zhao, *J. Am. Soc. Nephrol.*, 2017, **28**, 3627–3638.
45. H. Li, X. Zhu, J. Zhang and J. Shi, *Biomed. Pharmacother.*, 2017, **96**, 471–479.
46. Y. Fu, Y. Zhang, Z. Wang, L. Wang, X. Wei, B. Zhang, Z. Wen, H. Fang, Q. Pang and F. Yi, *Am. J. Nephrol.*, 2010, **32**, 581–589.
47. H. J. Oh, M. Kato, S. Deshpande, E. Zhang, S. Das, L. Lanting, M. Wang and R. Natarajan, *Sci. Rep.*, 2016, **6**, 38789.
48. R. H. Jenkins, J. Martin, A. O. Phillips, T. Bowen and D. J. Fraser, *Biochem. Soc. Trans.*, 2012, **40**, 762–767.
49. S. D. Deshpande, S. Putta, M. Wang, J. Y. Lai, M. Bitzer, R. G. Nelson, L. L. Lanting, M. Kato and R. Natarajan, *Diabetes*, 2013, **62**, 3151–3162.
50. A. Krupa, R. Jenkins, D. D. Luo, A. Lewis, A. Phillips and D. Fraser, *J. Am. Soc. Nephrol.*, 2010, **21**, 438–447.
51. M. Kato, J. Zhang, M. Wang, L. Lanting, H. Yuan, J. J. Rossi and R. Natarajan, *Proc. Natl. Acad. Sci. U. S. A.*, 2007, **104**, 3432–3437.
52. S. Putta, L. Lanting, G. Sun, G. Lawson, M. Kato and R. Natarajan, *J. Am. Soc. Nephrol.*, 2012, **23**, 458–469.
53. B. Wang, M. Herman-Edelstein, P. Koh, W. Burns, K. Jandeleit-Dahm, A. Watson, M. Saleem, G. J. Goodall, S. M. Twigg, M. E. Cooper and P. Kantharidis, *Diabetes*, 2010, **59**, 1794–1802.
54. A. C. K. Chung, X. R. Huang, X. Meng and H. Y. Lan, *J. Am. Soc. Nephrol.*, 2010, **21**, 1317–1325.
55. M. Kato, V. Dang, M. Wang, J. T. Park, S. Deshpande, S. Kadam, A. Mardiros, Y. Zhan, P. Oettgen, S. Putta, H. Yuan, L. Lanting and R. Natarajan, *Sci. Signaling*, 2013, **6**, ra43.
56. M. Kato, S. Putta, M. Wang, H. Yuan, L. Lanting, I. Nair, A. Gunn, Y. Nakagawa, H. Shimano, I. Todorov, J. J. Rossi and R. Natarajan, *Nat. Cell Biol.*, 2009, **11**, 881–889.
57. L. Mahimainathan, F. Das, B. Venkatesan and G. G. Choudhury, *Diabetes*, 2006, **55**, 2115–2125.
58. J. T. Park, M. Kato, H. Yuan, N. Castro, L. Lanting, M. Wang and R. Natarajan, *J. Biol. Chem.*, 2013, **288**, 22469–22480.
59. M. Kato, L. Wang, S. Putta, M. Wang, H. Yuan, G. Sun, L. Lanting, I. Todorov, J. J. Rossi and R. Natarajan, *J. Biol. Chem.*, 2010, **285**, 34004–34015.
60. S. Deshpande, M. Abdollahi, M. Wang, L. Lanting, M. Kato and R. Natarajan, *Sci. Rep.*, 2018, **8**, 6954.

61. H. Butz, K. Rácz, L. Hunyady and A. Patócs, *Trends Pharmacol. Sci.*, 2012, **33**, 382–393.
62. O. Tang, X.-M. Chen, S. Shen, M. Hahn and C. A. Pollock, *Am. J. Physiol. Renal Physiol.*, 2013, **304**, F1266–F1273.
63. M. Xiong, L. Jiang, Y. Zhou, W. Qiu, L. Fang, R. Tan, P. Wen and J. Yang, *Am. J. Physiol. Renal Physiol.*, 2012, **302**, F369–F379.
64. Y. Zhao, Z. Yin, H. Li, J. Fan, S. Yang, C. Chen and D. W. Wang, *Aging Cell*, 2017, **16**, 387–400.
65. F. Han, S. Wang, Y. Chang, C. Li, J. Yang, Z. Han, B. Chang, B. Sun and L. Chen, *J. Cell. Physiol.*, 2018, **233**, 2225–2237.
66. Z. Zhou, J. Wan, X. Hou, J. Geng, X. Li and X. Bai, *Cell Death Dis.*, 2017, **8**, e2658.
67. L. Wu, Q. Wang, F. Guo, X. Ma, H. Ji, F. Liu, Y. Zhao and G. Qin, *Sci. Rep.*, 2016, **6**, 26072.
68. C. Zanchi, D. Macconi, P. Trionfini, S. Tomasoni, D. Rottoli, M. Locatelli, M. Rudnicki, J. Vandesompele, P. Mestdagh, G. Remuzzi, A. Benigni and C. Zoja, *Diabetologia*, 2017, **60**, 1114–1125.
69. A. Bera, F. Das, N. Ghosh-Choudhury, M. M. Mariappan, B. S. Kasinath and G. Ghosh Choudhury, *Am. J. Physiol. Cell Physiol.*, 2017, **313**, C430–C447.
70. H. Suzuki, K. Kiryluk, J. Novak, Z. Moldoveanu, A. B. Herr, M. B. Renfrow, R. J. Wyatt, F. Scolari, J. Mestecky, A. G. Gharavi and B. A. Julian, *J. Am. Soc. Nephrol.*, 2011, **22**, 1795–1803.
71. G. Serino, F. Sallustio, S. N. Cox, F. Pesce and F. P. Schena, *J. Am. Soc. Nephrol.*, 2012, **23**, 814–824.
72. G. Serino, F. Sallustio, C. Curci, S. N. Cox, F. Pesce, G. De Palma and F. P. Schena, *Nephrol., Dial., Transplant.*, 2015, **30**, 1132–1139.
73. G. Serino, F. Pesce, F. Sallustio, G. De Palma, S. N. Cox, C. Curci, G. Zaza, K. N. Lai, J. C. K. Leung, S. C. W. Tang, A. Papagianni, M. Stangou, D. Goumenos, M. Gerolymos, K. Takahashi, Y. Yuzawa, S. Maruyama, E. Imai and F. P. Schena, *Kidney Int.*, 2016, **89**, 683–692.
74. S. Hu, H. Bao, X. Xu, X. Zhou, W. Qin, C. Zeng and Z. Liu, *FEBS Lett.*, 2015, **589**, 4019–4025.
75. Y. Liang, J. Zhang, Y. Zhou, G. Xing, G. Zhao and Z. Liu, *Cell. Physiol. Biochem.*, 2015, **36**, 1793–1808.
76. Y. Liang, G. Zhao, L. Tang, J. Zhang, T. Li and Z. Liu, *Exp. Cell Res.*, 2016, **347**, 312–321.
77. C. Li, J. Shi and Y. Zhao, *J. Cell. Biochem.*, 2018, **119**, 4607–4614.
78. M.-F. Hennino, D. Buob, C. Van der Hauwaert, V. Gnemmi, Z. Jomaa, N. Pottier, G. Savary, E. Drumez, C. Noël, C. Cauffiez and F. Glowacki, *Sci. Rep.*, 2016, **6**, 27209.
79. H. Bao, H. Chen, X. Zhu, M. Zhang, G. Yao, Y. Yu, W. Qin, C. Zeng, K. Zen and Z. Liu, *Kidney Int.*, 2014, **85**, 624–635.
80. S. Yung and T. M. Chan, *Clin. Immunol.*, 2017, **185**, 40–50.
81. Y. Dai, W. Sui, H. Lan, Q. Yan, H. Huang and Y. Huang, *Rheumatol. Int.*, 2009, **29**, 749–754.

82. L. Qingjuan, F. Xiaojuan, Z. Wei, W. Chao, K. Pengpeng, L. Hongbo, Z. Sanbing, H. Jun, Y. Min and L. Shuxia, *Am. J. Physiol. Cell Physiol.*, 2016, **310**, C470–C478.
83. X. Feng, H. Wu, J. M. Grossman, P. Hanvivadhanakul, J. D. FitzGerald, G. S. Park, X. Dong, W. Chen, M. H. Kim, H. H. Weng, D. E. Furst, A. Gorn, M. McMahon, M. Taylor, E. Brahn, B. H. Hahn and B. P. Tsao, *Arthritis Rheumatol.*, 2006, **54**, 2951–2962.
84. X. Zhang, X. Han, Y. Tang, Y. Wu, B. Qu and N. Shen, *Sci. Rep.*, 2015, **5**, 12987.
85. P. Costa-Reis, P. A. Russo, Z. Zhang, L. Colonna, K. Maurer, S. Gallucci, S. W. Schulz, A. N. Kiani, M. Petri and K. E. Sullivan, *Arthritis Rheumatol.*, 2015, **67**, 2415–2426.
86. X. Han, Y. Wang, X. Zhang, Y. Qin, B. Qu, L. Wu, J. Ma, Z. Zhou, J. Qian, M. Dai, Y. Tang, E. K. L. Chan, J. B. Harley, S. Zhou and N. Shen, *Arthritis Rheumatol.*, 2016, **68**, 2232–2243.
87. P. Tangtanatakul, B. Thammasate, A. Jacquet, R. Reantragoon, T. Pisitkun, Y. Avihingsanon, A. Leelahavanichkul and N. Hirankarn, *Sci. Rep.*, 2017, **7**, 14517.
88. J. Luan, C. Jiao, W. Kong, J. Fu, W. Qu, Y. Chen, X. Zhu, Y. Zeng, G. Guo, H. Qi, L. Yao, J. Pi, L. Wang and H. Zhou, *Mol. Ther.--Nucleic Acids*, 2018, **10**, 245–253.
89. H. Zhou, S. A. Hasni, P. Perez, M. Tandon, S.-I. Jang, C. Zheng, J. B. Kopp, H. Austin, J. E. Balow, I. Alevizos and G. G. Illei, *J. Am. Soc. Nephrol.*, 2013, **24**, 1073–1087.
90. C.-Z. Zheng, Y.-B. Shu, Y.-L. Luo and J. Luo, *Eur. Rev. Med. Pharmacol. Sci.*, 2017, **21**, 1041–1048.
91. W. Wang, J. Gao and F. Wang, *Am. J. Transl. Res.*, 2017, **9**, 3796–3803.
92. J. J. Grantham, *N. Engl. J. Med.*, 2008, **359**, 1477–1485.
93. L. M. Pastorelli, S. Wells, M. Fray, A. Smith, T. Hough, B. D. Harfe, M. T. McManus, L. Smith, A. S. Woolf, M. Cheeseman and A. Greenfield, *Mamm. Genome*, 2009, **20**, 140–151.
94. V. Patel, D. Williams, S. Hajarnis, R. Hunter, M. Pontoglio, S. Somlo and P. Igarashi, *Proc. Natl. Acad. Sci. U. S. A.*, 2013, **110**, 10765–10770.
95. H. Sun, Q.-W. Li, X.-Y. Lv, J.-Z. Ai, Q.-T. Yang, J.-J. Duan, G.-H. Bian, Y. Xiao, Y.-D. Wang, Z. Zhang, Y.-H. Liu, R.-Z. Tan, Y. Yang, Y.-Q. Wei and Q. Zhou, *Mol. Biol. Rep.*, 2010, **37**, 2951–2958.
96. C. Bergmann, *et al.*, PKHD1 mutations in autosomal recessive polycystic kidney disease (ARPKD), - PubMed - NCBI, https://www.ncbi.nlm.nih.gov/pubmed/15108277, accessed 14 June 2018.
97. U. Tran, L. Zakin, A. Schweickert, R. Agrawal, R. Döger, M. Blum, E. M. De Robertis and O. Wessely, *Development*, 2010, **137**, 1107–1116.
98. N. Piazzon, C. Maisonneuve, I. Guilleret, S. Rotman and D. B. Constam, *J. Mol. Cell Biol.*, 2012, **4**, 398–408.
99. S. Hajarnis, R. Lakhia, M. Yheskel, D. Williams, M. Sorourian, X. Liu, K. Aboudehen, S. Zhang, K. Kersjes, R. Galasso, J. Li, V. Kaimal, S. Lockton, S. Davis, A. Flaten, J. A. Johnson, W. L. Holland, C. M. Kusminski, P. E.

Scherer, P. C. Harris, M. Trudel, D. P. Wallace, P. Igarashi, E. C. Lee, J. R. Androsavich and V. Patel, *Nat. Commun.*, 2017, **8**, 14395.
100. R. Lakhia, S. Hajarnis, D. Williams, K. Aboudehen, M. Yheskel, C. Xing, M. E. Hatley, V. E. Torres, D. P. Wallace and V. Patel, *J. Am. Soc. Nephrol.*, 2016, **27**, 2319–2330.
101. L. Duplomb, N. Droin, O. Bouchot, C. Thauvin-Robinet, A.-L. Bruel, J. Thevenon, P. Callier, G. Meurice, N. Pata-Merci, R. Loffroy, D. Vandroux, R. D. A. Costa, V. Carmignac, E. Solary and L. Faivre, *Hum. Mol. Genet.*, 2017, **26**, 4680–4688.
102. H. Dweep, C. Sticht, A. Kharkar, P. Pandey and N. Gretz, *PLoS One*, 2013, **8**, e53780.
103. P. Pandey, B. Brors, P. K. Srivastava, A. Bott, S. N. E. Boehn, H.-J. Groene and N. Gretz, *BMC Genomics*, 2008, **9**, 624.
104. P. Pandey, S. Qin, J. Ho, J. Zhou and J. A. Kreidberg, *BMC Syst. Biol.*, 2011, **5**, 56.
105. V. D. D'Agati, F. J. Kaskel and R. J. Falk, *N. Engl. J. Med.*, 2011, **365**, 2398–2411.
106. J. Wu, C. Zheng, Y. Fan, C. Zeng, Z. Chen, W. Qin, C. Zhang, W. Zhang, X. Wang, X. Zhu, M. Zhang, K. Zen and Z. Liu, *J. Am. Soc. Nephrol.*, 2014, **25**, 92–104.
107. J. Wu, C. Zheng, X. Wang, S. Yun, Y. Zhao, L. Liu, Y. Lu, Y. Ye, X. Zhu, C. Zhang, S. Shi and Z. Liu, *J. Clin. Invest.*, 2015, **125**, 4091–4106.
108. S. Shi, L. Yu, T. Zhang, H. Qi, S. Xavier, W. Ju and E. Bottinger, *PLoS One*, 2013, **8**, e75572.
109. M. Li, S. Armelloni, C. Zennaro, C. Wei, A. Corbelli, M. Ikehata, S. Berra, L. Giardino, D. Mattinzoli, S. Watanabe, C. Agostoni, A. Edefonti, J. Reiser, P. Messa and M. P. Rastaldi, *J. Pathol.*, 2015, **235**, 731–744.
110. J. Müller-Deile, J. Dannenberg, P. Schroder, M.-H. Lin, J. H. Miner, R. Chen, J.-H. Bräsen, T. Thum, J. Nyström, L. B. Staggs, H. Haller, J. Fiedler, J. M. Lorenzen and M. Schiffer, *Kidney Int.*, 2017, **92**, 836–849.
111. C. A. Gebeshuber, C. Kornauth, L. Dong, R. Sierig, J. Seibler, M. Reiss, S. Tauber, M. Bilban, S. Wang, R. Kain, G. A. Böhmig, M. J. Moeller, H.-J. Gröne, C. Englert, J. Martinez and D. Kerjaschki, *Nat. Med.*, 2013, **19**, 481–487.
112. L. Kietzmann, S. S. O. Guhr, T. N. Meyer, L. Ni, M. Sachs, U. Panzer, R. A. K. Stahl, M. A. Saleem, D. Kerjaschki, C. A. Gebeshuber and C. Meyer-Schwesinger, *J. Am. Soc. Nephrol.*, 2015, **26**, 1389–1401.
113. P. Hohenstein, K. Pritchard-Jones and J. Charlton, *Genes Dev.*, 2015, **29**, 467–482.
114. Z. Huang, Y. Zhang, J. Zhou and Y. Zhang, *BioMed Res. Int.*, 2017, **2017**, 7298160.
115. J. C. Jennette and D. B. Thomas, *Nephrol., Dial., Transplant.*, 2001, **16**(Suppl 6), 80–82.
116. S. J. Shankland, B. Smeets, J. W. Pippin and M. J. Moeller, *Nat. Rev. Nephrol.*, 2014, **10**, 158–173.

117. C. Henique, G. Bollée, X. Loyer, F. Grahammer, N. Dhaun, M. Camus, J. Vernerey, L. Guyonnet, F. Gaillard, H. Lazareth, C. Meyer, I. Bensaada, L. Legrès, T. Satoh, S. Akira, P. Bruneval, S. Dimmeler, A. Tedgui, A. Karras, E. Thervet, D. Nochy, T. B. Huber, L. Mesnard, O. Lenoir and P.-L. Tharaux, *Nat. Commun.*, 2017, **8**, 1829.
118. P. Ruggenenti, P. Cravedi and G. Remuzzi, *J. Am. Soc. Nephrol.*, 2012, **23**, 1917–1928.
119. B. Simonson and S. Das, *Mini-Rev. Med. Chem.*, 2015, **15**, 467–474.
120. S. Obad, C. O. dos Santos, A. Petri, M. Heidenblad, O. Broom, C. Ruse, C. Fu, M. Lindow, J. Stenvang, E. M. Straarup, H. F. Hansen, T. Koch, D. Pappin, G. J. Hannon and S. Kauppinen, *Nat. Genet.*, 2011, **43**, 371–378.
121. R. S. Geary, D. Norris, R. Yu and C. F. Bennett, *Adv. Drug Delivery Rev.*, 2015, **87**, 46–51.
122. E. M. Straarup, N. Fisker, M. Hedtjärn, M. W. Lindholm, C. Rosenbohm, V. Aarup, H. F. Hansen, H. Ørum, J. B. R. Hansen and T. Koch, *Nucleic Acids Res.*, 2010, **38**, 7100–7111.
123. M. Yheskel and V. Patel, *Curr. Opin. Nephrol. Hypertens.*, 2017, **26**, 282–289.
124. J. M. Lorenzen and T. Thum, *Clin. J. Am. Soc. Nephrol.*, 2012, **7**, 1528–1533.
125. T. Papadopoulos, J. Belliere, J.-L. Bascands, E. Neau, J. Klein and J. P. Schanstra, *Expert Rev. Mol. Diagn.*, 2015, **15**, 361–374.
126. S. A. MacLellan, C. MacAulay, S. Lam and C. Garnis, *BMC Clin. Pathol.*, 2014, **14**, 27.
127. A. Levin, W. Lancashire and R. G. Fassett, *Kidney Int.*, 2013, **83**, 1001–1009.
128. M. A. Baker, S. J. Davis, P. Liu, X. Pan, A. M. Williams, K. A. Iczkowski, S. T. Gallagher, K. Bishop, K. R. Regner, Y. Liu and M. Liang, *J. Am. Soc. Nephrol.*, 2017, **28**, 2985–2992.
129. J. C. He, P. Y. Chuang, A. Ma'ayan and R. Iyengar, *Kidney Int.*, 2012, **81**, 22–39.

CHAPTER 6

Involvement of MicroRNAs in Autoimmune Diseases

RADHIKA RAHEJA, MAI FUJIWARA, LUCIEN P. GARO AND GOPAL MURUGAIYAN*

Ann Romney Center for Neurologic Diseases, Brigham and Women's Hospital and Harvard Medical School, 60 Fenwood Road, Hale Building for Transformative Medicine, Boston, MA 02115, USA
*E-mail: mgopal@rics.bwh.harvard.edu

6.1 Introduction

6.1.1 miRNA Biogenesis

MicroRNAs (miRNAs) are small, non-coding RNAs that are endogenous gene regulatory molecules. Their biogenesis has been extensively elucidated. The key molecules required for miRNA processing include (1) the RNAse II enzyme, Drosha (DGCR8/Pasha), that cleaves the pri-miRNA transcript in the nucleus into a double-stranded hairpin structure called pre-miRNA; (2) exportin-5 (XPO5), that is required for the export of the pre-miRNA into the cytoplasm in a manner dependent on the Ran-guanosine triphosphate (GTP) pathway; and finally, (3) the RNase III enzyme, Dicer, that cleaves the hairpin strand into two short independent miRNAs, one of which is the active strand, and the other of which is the inactive strand and subject to degradation. The active RNA strand performs its gene regulatory functions by forming a part

Drug Discovery Series No. 69
MicroRNAs in Diseases and Disorders: Emerging Therapeutic Targets
Edited by Philip V. Peplow, Bridget Martinez, George A. Calin and Aurora Esquela-Kerscher

Published by the Royal Society of Chemistry, www.rsc.org

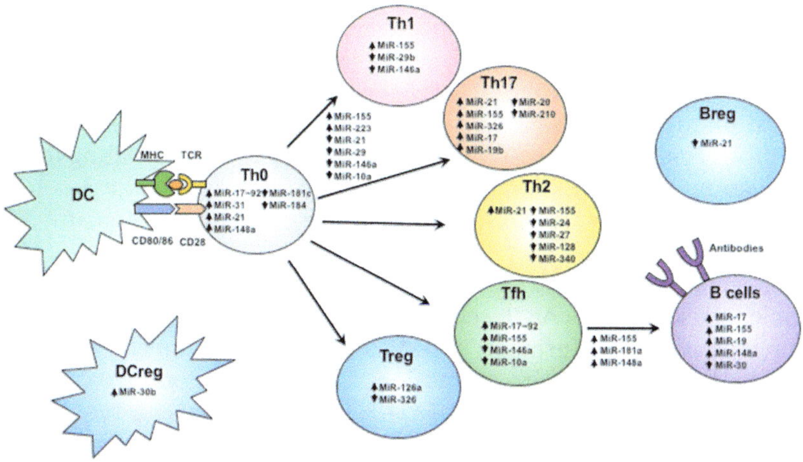

Figure 6.1 MicroRNA-mediated control of pathogenic and regulatory cells in the immune system. Cell-intrinsic miRNAs (indicated inside the cell) can either positively (red) or negatively (blue) regulate the function and/or development of specific immune cell subsets. miRNAs indicated outside the cell in the environment can also regulate immune cells in a cell-extrinsic manner.

of the RNA-induced silencing complex, which includes the Argonaute (AGO) protein.[1] miRNAs generally impart their regulatory function by binding to the 3′ untranslated region (UTR) of target messenger RNAs (mRNAs) and inhibiting their translation. In addition, miRNAs have also been shown to bind 5′ UTR as well as open reading frames of target mRNAs to affect translation.[2,3] miRNAs target specific mRNAs based on complementary strand binding. Moreover, this regulation by miRNAs is dependent on types of cells and tissues.[4] miRNAs control a wide range of biological processes, such as the development of immune cell subsets and responses (Figure 6.1). Most importantly, dysregulation of miRNA expression is observed in a variety of immune-mediated diseases, including autoimmunity, highlighting the involvement of miRNAs in disease pathogeneses.

6.1.2 Cells Regulated by miRNAs During Autoimmunity

6.1.2.1 T Cells

T cell development in the thymus progresses through various stages. In an early differentiation step, highly proliferative double-negative cells lack the expression of the glycoprotein immune cell surface markers, CD4 and CD8. This is followed by the double-positive stage where cells express both CD4 and CD8. At this stage, cells undergo rearrangement of the T cell receptor (TCR) and are positively selected to commit to either $CD4^+$ helper T cells (Th) or $CD8^+$ cytotoxic T cells.[5] $CD4^+$ T cells are implicated in most autoimmune diseases. The activation of $CD4^+$ T cells is brought about by a cascade of events involving an antigen (Ag)-specific TCR signaling response in synergy with the co-stimulatory molecules: inducible co-stimulator (ICOS), CD28, and cytotoxic T-lymphocyte antigen 4 (CTLA4), which regulate T cell proliferation and expansion into specific effector subsets. miRNAs, such as the miR-17~92 cluster, regulate various aspects of TCR signaling (*e.g.*, T cell activation, proliferation, and survival).[6,7] While miR-101 targets ICOS and destabilizes its expression,[8,9] miR-155 targets CTLA4 to ensure TCR signaling.[10,11]

miRNAs also regulate the expression of the cytokine interleukin (IL)-2, a key molecule secreted upon T cell activation critical for subsequent effector differentiation. For instance, miR-181c is known to directly target the 3′ UTR of IL-2, inhibiting its expression.[12] Alternatively, miR-184 inhibits nuclear factor of activated T cells (NFAT), a transcription factor required for IL-2 expression.[13] Conversely, overexpression of miR-31 in human T cells enhances NFAT, and subsequently IL-2 expression.[14]

Upon Ag stimulation, naive $CD4^+$ T cells differentiate into specific T-effector or Th subsets to generate appropriate immune responses.[15–17] Each of these subsets has a unique transcriptional program resulting in the secretion of effector cytokines. miRNAs can regulate the differentiation of naive T cells into specific effector subsets, which are essential for the protection of the host against a variety of pathogens and tumors. Therefore, dysregulated miRNA-mediated Th differentiation can subsequently lead to tissue inflammation and autoimmunity.[15–17] These subsets are discussed in detail later.

6.1.2.1.1 Th1 Cells. The signature cytokine, interferon-γ (IFN-γ), secreted by Th1 cells, plays a vital role in generating immune responses against viral, intracellular bacterial pathogens (*e.g.*, *Mycobacterium tuberculosis* and *Listeria monocytogenes*), tumor Ags, and self-Ags.[18] Interestingly, dysregulated Th1 responses are associated with autoimmune diseases such as rheumatoid arthritis, multiple sclerosis, psoriasis, and type 1 diabetes, as well as various cancers.[18] The exposure to IL-12 secreted from dendritic cells (DCs) and macrophages induces the expression of signal transducer and activator of transcription 4 (STAT4), which in turn stabilizes Th1 cells by upregulation of the characteristic transcription factor, T-box expressed in T cells (T-bet).[19,20] T-bet then promotes the secretion of IFN-γ. miRNAs can affect Th1 cells in a cell-intrinsic manner by directly modulating this transcriptional program, or they can have T cell-extrinsic effects by affecting cytokine secretion from DCs.

Dicer-deficient mice, as well as a $CD4^+$ T cell specific deletion of Drosha, result in aberrant Th1 differentiation, indicating the critical role of miRNAs in governing this process.[21] Of note, miR-29a and miR-29b inhibit Th1 differentiation by directly targeting T-bet and IFN-γ in a T cell-intrinsic manner.[22,23] miR-21, on the other hand, targets IL-12 within DCs to subsequently limit Th1 differentiation.[24] miR-146a also inhibits Th1 differentiation by indirectly suppressing the expression of STAT4.[25]

6.1.2.1.2 Th2 Cells. Th2-mediated responses protect the body against extracellular pathogens (*e.g.*, Helminth) and mediate chronic allergic inflammatory diseases, such as asthma.[26] IL-4 is the key cytokine that perpetuates Th2 cell differentiation in a positive feedback loop.[26] In response to IL-4, STAT6 is phosphorylated and induces the expression of the Th2 specific transcription factor, GATA binding protein 3 (GATA-3), which in turn induces avian musculoaponeurotic fibrosarcoma oncogene homolog (c-Maf), a transactivator of the IL-4 promoter.[27–30] IL-4-independent signaling involving thymic stromal lymphopoietin, IL-25, and IL-33 also play a role in Th2 differentiation. Together with IL-4, Th2 cells can secrete other cytokines such as IL-5 and IL-13.

$CD4^+$ T cells lacking Drosha and AGO exhibit dysregulated Th2 differentiation, implicating the crucial role of miRNAs in Th2 cells.[31] miR-24 and miR-27 have been found to inhibit Th2 differentiation, wherein miR-27 directly targets GATA-3.[32] The loss of miR-155 leads to excess Th2 responses due to the upregulation of the transcription factor, c-Maf.[33] miR-340, miR-27, and miR-128 also target B cell-specific Moloney murine leukemia virus integration site 1 (BMI1), which reduces GATA-3 expression and subsequently inhibits Th2 polarization.[34] miR-21 suppresses Sprouty 1 (SPRY1), an inhibitor of mitogen-activated protein kinase (MAPK), enabling GATA-3 expression and Th2 differentiation.[35,36]

6.1.2.1.3 Th17 Cells. The Th17 subset of helper cells eliminate extracellular bacteria and fungi.[37] Autoimmune diseases, such as inflammatory bowel diseases (IBD), rheumatoid arthritis, psoriasis, and multiple sclerosis, are associated with strong inflammatory responses mediated by uncontrolled Th17 responses.[37] Th17 cells secrete a wide range of inflammatory cytokines, namely, IL-17A, IL-17F, IL-21, IL-22, and granulocyte-macrophage colony-stimulating factor. In response to transforming growth factor-β (TGF-β) and the inflammatory cytokine, IL-6, the transcription factor STAT3 is activated, which then promotes the expression of the Th17-specific transcription factor, retinoic acid-related orphan receptor-γt (ROR-γt). ROR-γt, in turn, induces the expression of the signature cytokine, IL-17A.[38–40] There are several other transcription factors involved in Th17 differentiation, including retinoic acid-related orphan receptor α (RORα), interferon regulatory factor (IRF4), basic leucine zipper transcriptional factor ATF-like (BATF), and hypoxia inducible factor-1 alpha (HIF1α).[41–44] The cytokines IL-23 and IL-1β are critical for maintaining the pathogenic functions of polarized Th17 cells during autoimmunity.[45] A number of miRNAs are implicated in Th17 differentiation, including miR-326, miR-155, miR-21, miR-210, and miR-20b. These miRNAs are discussed in detail in Section 6.2.

6.1.2.1.4 Tfh Cells. A specialized subset of $CD4^+$ T cells, T follicular helper (Tfh) cells, assists B cells by promoting the survival and differentiation of germinal center (GC) B cells *via* cytokine secretion and interactions with ICOS–ICOS ligand (ICOSL).[46] In response to such interactions, B cells become antibody-secreting plasma cells. Tfh cells are therefore essential for normal B cell antibody responses. Consequently, increased Tfh cells are associated with systemic lupus erythematosus (SLE) and rheumatoid arthritis.[47] Tfh differentiation is mediated by DC–T cell and ICOS–ICOSL interactions, which induce the key transcription factor, B cell lymphoma 6 protein (BCL6), and subsequent induction of Tfh genes (*e.g.*, BATF, IRF4, STAT3, achaete-scute homolog 2 (ASCL2), and IL-21).[48–54] The miR-17~92 cluster and miR-155 are implicated in Tfh differentiation by targeting phosphatase and tensin homolog (PTEN) and PH domain leucine-rich repeat protein phosphatase 2 (PHLPP2).[55,56] miR-146a and miR-10a limit Tfh cells by targeting interleukin-1 receptor-associated kinase 1 (IRAK-1) and tumor necrosis factor receptor-associated factor 6 (TRAF-6), upstream of nuclear factor kappa-light-chain-enhancer of activated B cells (NF-κB), and BCL6, respectively.[57–60]

6.1.2.2 B Cells

B cells play a crucial role in immune responses through antibody and cytokine secretion, as well as Ag presentation. They are key for acute and long-term protection against bacterial and viral infections.[61] However, self-reactive B cells that escape tolerance mechanisms can be pathogenic, as seen in patients with rheumatoid arthritis and SLE.[61] B cells transit through pre-B cell, pro-B cell, and naive B cell stages during development. When naive B cells encounter their cognate Ag in a lymph node and are stimulated through the B cell receptor (BCR), they undergo clonal expansion, somatic hypermutation, and class switch recombination (CSR) in a lymphoid compartment known as a germinal center.[62]

miRNAs are critical regulators of B cell development and function. For instance, deficiency in Dicer or miR-17 in B cell progenitors blocks the pro-B to pre-B cell transition.[63,64] Importantly, dysregulation of miRNAs in B cells leads to the development of autoimmunity in mice. Specifically, mice lacking Dicer in mature B cells show a self-reactive BCR repertoire, elevated autoantibodies, and lupus-like kidney pathology.[65] miR-19a and 19b-1 have been shown to inhibit B cell tolerance by targeting PTEN, a key regulator of B cell tolerance.[63]

6.1.2.3 Dendritic Cells and Macrophages

6.1.2.3.1 Dendritic Cells. Dendritic cells are initiators of adaptive immune responses and tolerance. DCs can be classified into conventional DCs (cDCs), monocyte-derived DCs (moDCs), or plasmacytoid DCs (pDCs).[66] cDCs are classified as professional antigen presenting cells of the innate immune system.[67] They constitutively uptake Ags in their environment, process them into peptides, and present them on major histocompatibility complex (MHC) molecules to stimulate T cells.[66] Upon detection of pathogen- or danger-associated molecular patterns (PAMPs or DAMPs, respectively) through pattern recognition receptors such as toll-like receptors (TLRs), cDCs upregulate MHC and co-stimulatory molecules, and migrate to lymph nodes to prime T cells and promote Th differentiation.[67] Differentiation towards specific T cell subsets is influenced by the cytokines secreted by cDCs, as well as other factors in the microenvironment.[15–17] moDCs arise from infiltrating monocytes in response to inflammation or infection.[66] Although moDCs and cDCs show similar expression patterns of MHC class II, CD11b (a monocytic lineage marker), and CD11c (a DC lineage marker), moDCs also express CD64, the Fcγ receptor 1 (FcγR1), suggesting their monocytic origin.[66] pDCs are a minor subset of DCs known for their secretion of type I interferons (IFNs) in response to a viral infection.[68] pDCs express high levels of TLR7 and TLR9, which allow them to recognize virus or self-derived single-stranded RNAs and unmethylated CpG motifs of DNA, respectively.[68]

miRNAs control the functional responses of DCs that subsequently determine the nature of T cell responses. For example, miR-21 directly targets IL-12 in DCs, and suppresses Th1 differentiation.[24] TLR signaling rapidly induces miR-155 in DCs[69] and miR-155 promotes Th1/Th17-polarizing cytokine expression by inhibiting the suppressor of cytokine signaling 1 (SOCS1) and src homology 2 domain-containing inositol-5-phosphase 1 (SHIP-1).[70,71] Accordingly, miR-155-overexpressing DCs drive Th1 and Th17 cell-mediated autoimmunity in mice.[72]

6.1.2.3.2 Macrophages. Macrophages are phagocytic innate immune cells that maintain tissue homeostasis and respond to pathogens and tissue injury.[73] They primarily polarize towards M1 or M2 phenotypes depending on their environmental cues.[73] Exposure to PAMPs, such as lipopolysaccharide, as well as Th1/Th17-polarizing cytokines, favors the differentiation of M1 macrophages that further amplify inflammation.[74] Alternatively activated M2 macrophages develop in the presence of TGF-β and Th2-polarizing cytokines and promote resolution of inflammation.[74] Interestingly, miR-21 inhibits the formation of M2 macrophages by directly targeting prostaglandin E2 (PGE2), a key factor that promotes M2 polarization.[75]

6.1.2.4 Regulatory Cells

6.1.2.4.1 Regulatory T Cells. A subset of $CD4^+$ T cells, called regulatory T cells (Tregs), maintain immune homeostasis and exert immunosuppressive functions. These cells are characterized by the expression of the surface marker CD25 and the transcription factor Forkhead box P3 (FOXP3). Tregs are primarily induced by TGF-β, while IL-2 signaling mediated by the molecule STAT5, plays a role in maintaining Treg stability.[76,77] Conditional deletion of Drosha and Dicer within $FOXP3^+$ Tregs result in autoimmunity, implicating miRNAs in the regulation of Treg development and function.[78] In addition, certain miRNAs regulate the suppressive function of Tregs required to restrain expansion of pathogenic cell populations. For example, miR-326 targets the transcription factor protein avian erythroblastosis oncogene homolog (ETS1), which stabilizes FOXP3 expression in Tregs,[79] and possibly compromises the suppressive function of Tregs. These miRNAs are discussed in detail in Section 6.2.

6.1.2.4.2 Regulatory Dendritic Cells. Regulatory DCs are a subset of DCs characterized by expression of the regulatory cytokines TGF-β and IL-10, as well as the immunosuppressive molecules such as PGE2, nitric oxide, indoleamine 2,3-dioxygenase, and programmed death ligand-1 (PD-L1), all of which act to suppress immune responses.[80] In addition, miRNAs regulate the development and function of regulatory DCs. For instance, miR-30b is induced by TGF-β signaling mediated by the molecule mothers against decapentaplegic homolog 3 (SMAD3) in regulatory DCs, and promotes secretion of IL-10 and nitric oxide by inhibiting neurogenic locus notch homolog protein 1 (NOTCH1).[81]

6.1.2.4.3 Regulatory B Cells. Regulatory B cells (Bregs) exert their immunosuppressive properties through secretion of IL-10, IL-35, and TGF-β, and contribute to the prevention of autoimmunity. Although studies describing the roles of miRNAs in regulating Breg functions are limited, a recent study has shown that B cell-intrinsic miR-21 directly targets IL-10, thereby diminishing Breg suppressive function in mice.[82]

6.2 The Role of miRNAs in Autoimmune Diseases

Dysregulated miRNAs within immune and non-immune cells are associated with various forms of tissue inflammation and damage. In this section, we discuss a few major autoimmune diseases where miRNAs play key pathophysiological roles.

6.2.1 Inflammatory Bowel Disease

Inflammatory bowel disease is divided into two types of chronic inflammatory disorders of the intestines: ulcerative colitis and Crohn's disease. Emerging evidence supports the notion that IBD is caused by dysbiosis between the commensal microbiota in the gastrointestinal tract and the mucosal immune system, resulting in subsequent loss of intestinal immune homeostasis.[83] In Crohn's disease, multiple genetic variants have been identified in IBD patients that regulate immune functions. These include *NOD2*, encoding nucleotide oligomerization domain 2 (NOD2), which acts as an intracellular sensor for muramyl dipeptide (MDP), a component of the bacterial cell wall; *IL23R*, encoding IL-23 receptor which promotes Th17 responses; and *ATG16L1*, encoding autophagy-related protein 16-1 (ATG16L1), which is involved in autophagy, an intracellular system that maintains cellular

Table 6.1 Differentially expressed microRNAs in inflammatory bowel disease patients.[a]

miRNA	Change in expression compared to control	Tissue/cell type	Potential target(s)	Effect of change in expression	References
miR-21	Up	PBMCs	SMAD7	Enhanced Th17 cell differentiation	85, 86
miR-126	Up	Colonic tissues	IκBα	Activation of NF-κB-driven inflammation	87
miR-214	Up	Colonic tissues	PDLM2, PTEN	Activation of NF-κB-driven inflammation	88
miR-106b	Up	Colonic tissues	ATG16L1	Impaired autophagy of IECs	84, 96
miR-10a	Down	Colonic tissues	NOD2, IL-12/IL-23p40	Enhanced Th1 and Th17 differentiation	94, 95

[a]IECs: intestinal epithelial cells; NF-κB: nuclear factor-κB; PBMCs: peripheral blood mononuclear cells; Th: T-helper.

homeostasis.[83] The hallmark of IBD is excessive infiltration of the intestinal lamina propria by DCs, macrophages, neutrophils, T cells, and B cells, as well as increased colonic levels of tumor necrosis factor-α (TNF-α), and Th1-(IL-12)/Th17-(IL-1β, IL-6, and IL-23) polarizing cytokines.[83]

miRNAs have been implicated in IBDs, as expression profiles differ between ulcerative colitis patients and healthy controls, as well as Crohn's disease patients and healthy individuals. Partial overlap between differential miRNAs in ulcerative colitis and Crohn's disease patients suggests dysregulation of common pathways by these miRNAs in IBD pathogenesis.[84] Some specific miRNAs have been thoroughly studied in mice and in *in vitro* cellular systems to elucidate their role in mediating intestinal inflammation (Table 6.1).

Interestingly, miR-21 expression is elevated in patients with IBD, as well as in a mouse model of dextran sodium sulfate (DSS)-induced colitis.[85] miR-21-deficient mice develop attenuated disease, signifying its functional importance.[85] In general, miR-21 is associated with pro-inflammatory conditions. For example, in $CD4^+$ T cells, miR-21 promotes differentiation toward Th17 cells, which play a pathogenic role in IBD.[83] Specifically, miR-21 enhances Th17 differentiation by directly targeting the molecule SMAD7, which is a negative regulator of Th17 polarization.[86] miR-21-deficient mice show accumulation of M2 macrophages,[75] which may also partly explain their protection from DSS-induced colitis.

miR-126 is another miRNA upregulated in the colonic tissues of ulcerative colitis patients.[87] miR-126 promotes NF-κB-driven intestinal inflammation by directly targeting nuclear factor of kappa light polypeptide gene enhancer in B cell inhibitor, alpha (IκBα), a critical inhibitor of the NF-κB pathway.[87] Importantly, miR-126 expression levels are negatively correlated with IκBα levels in ulcerative colitis patients with active disease, and miR-126 overexpression results in reduced expression of IκBα in HT29 cells, a human colorectal adenocarcinoma cell line.[87] Together, these results suggest that silencing miR-126 may be therapeutically beneficial in both ulcerative colitis and Crohn's disease.

Regulation of miR-214 is also of interest, as this miRNA is consistently increased in the colonic tissues of ulcerative colitis patients.[88] IL-6 is one of the prevalent pro-inflammatory cytokines involved in IBD pathogenesis.[89] IL-6-induced STAT3 signaling promotes expression of miR-214, and in turn, miR-214 suppresses PDZ and LIM domain protein 2 (PDLIM2), and PTEN to promote NF-κB-mediated inflammation.[88] Notably, miR-214 inhibitor treatment results in amelioration of disease in a mouse model of DSS-induced colitis.[88]

The IL-23–Th17 axis implicated in IBD is regulated by multiple miRNAs, including miR-29a, miR-29b, and miR-29c.[90] NOD2 stimulation induces miR-29 in human DCs, and miR-29 in turn inhibits IL-23 expression by targeting IL-12/IL-23p40 directly and IL-23p19 indirectly through activating transcription factor 2 (ATF2).[90] miR-29-deficient mice develop exacerbated DSS-induced colitis associated with enhanced colonic levels of IL-23 and other Th17-polarizing cytokines.[90] Interestingly, DCs from Crohn's disease patients expressing *NOD2* polymorphisms incapable of inducing miR-29 upon stimulation secrete higher levels of IL-23, suggesting that defective miR-29 control of IL-23-driven Th17 responses may partly contribute to the pathogenesis of IBD in subsets of Crohn's disease patients.[90] These data also suggest that the sensing of commensal microbiota *via* NOD2 at steady-state may be critical for miR-29-driven maintenance of immunological tolerance against the commensal microbiota and prevention of IBD.

Another miRNA involved in regulation of IL-23–Th17 axis is let-7f, which directly inhibits IL-23R expression in $CD4^+$ T cells.[91] let-7f overexpression leads to reduced IL-17 expression in human $CD4^+$ T cells.[91] Interestingly, a Crohn's disease-associated variant of the *IL23R* gene, rs10889677, contains a 3′ UTR that results in impaired binding of let-7f and increased IL-23R expression.[92] This study suggests that let-7f control of the IL-23–Th17 axis may be defective in a subset of patients with this particular *IL23R* variant.

A hypoxia-induced miRNA, miR-210, can also inhibit Th17 differentiation by targeting HIF1α.[93] Accordingly, miR-210 deficiency in mice results in elevated Th17 responses and exacerbated disease in a model of T cell-mediated colitis.[93] These data suggest that exogenous administration of miR-210 may prove to be efficacious in ameliorating IBD.

In addition to these miRNAs, miR-10a has been identified as an important negative regulator of innate and adaptive immune responses in IBD.[94] The intestinal mucosa of IBD patients shows lower miR-10a expression compared to those of healthy controls, and miR-10a expression in these patients is negatively correlated with NOD2 and IL-12/IL-23p40 expression.[94,95] Commensal bacteria and pro-inflammatory cytokines, such as TNF-α and IFN-γ, suppress miR-10a expression in human DCs and release the miR-10a-mediated inhibition of NOD2 and IL-12/IL-23p40 expression.[94,95] Accordingly, the overexpression of miR-10a inhibits IL-12/IL-23p40 and NOD2 in DCs, leading to the dampening of Th1 and Th17 cell responses in IBD patients *ex vivo*.[94] These data suggest that miR-10a may be protective in the pathogenesis and progression of IBD.

Beyond their direct role in immune cells, miRNAs also contribute to disease by regulating non-immune cells in the gut. Levels of miR-106b and miR-130a are elevated in the mucosal tissues of Crohn's disease patients.[84] Both of these miRNAs directly target ATG16L1, a molecule involved in autophagy.[96,97] Autophagy is a key process that maintains cellular homeostasis through turnover of damaged organelles, which is essential for clearance of intracellular pathogens from intestinal epithelial cells (IECs).[98] In fact, miR-106b and miR-130a levels negatively correlate with ATG16L1 in the colonic tissues of Crohn's disease patients.[96,97] Therefore, elevated expression of miR-106b and miR-130a is likely to result in lower expression of ATG16L1, thus impairing autophagy processes in IECs, and potentially contributing to the pathogenesis of Crohn's disease.

Interestingly, miRNAs that are present in feces can regulate the commensal microbiota in the gut and influence IBD pathogenesis. IEC-specific ablation of Dicer results in more severe disease in a model of colitis, accompanied by uncontrolled bacterial growth.[99] This severe colitis phenotype in IEC-specific Dicer-deficient mice can be reversed with the administration of miRNAs isolated from feces of wild-type mice.[99]

Although only a few studies so far demonstrate the efficacy of miRNA-based therapeutics in mouse models of IBD, there are many miRNAs known to regulate IBD pathogenesis (including those not covered here). Mechanistic insights from each miRNA highlight the significant potential of miRNA-based therapeutics for patients with IBD.

6.2.2 Multiple Sclerosis

Multiple sclerosis is a chronic autoimmune disease of the central nervous system (CNS) in which a cascade of immune responses initiated in the periphery leads to infiltration of activated T cells into the CNS. Inflammation in the CNS is further perpetuated by uncontrolled immune responses mediated by infiltrating peripheral T cells, as well as resident immune cells in the brain, leading to extensive neuronal damage and demyelination.[100] The immune response in multiple sclerosis is mediated by upregulation of proinflammatory Th1 and Th17 cells, downregulation of Tregs, and activation of myeloid cells in the periphery (macrophages and DCs) and in the CNS (microglia). The functions of all these immune cells are regulated by miRNAs. As a consequence, several studies have focused on identifying miRNA expression within multiple sclerosis patients to study the molecular pathogenesis of the disease.

miRNA expression has been profiled in whole-blood components, cerebrospinal fluid (CSF) and lesions from multiple sclerosis patients and compared to healthy individuals (Table 6.2) (reviewed in ref. 101–103). Some notable miRNAs modulated in whole blood from relapsing remitting multiple sclerosis, primary progressive multiple sclerosis, and secondary progressive multiple sclerosis patients include members of the miR-17~92 cluster. In this cluster, miR-20a and miR-20b are downregulated,[104] while miR-17 is upregulated in $CD4^+$ T cells,[104–107] and miR-19b is upregulated in B cells and $CD4^+CD25^+$ regulatory T cells.[108,109] Interestingly, these miRNAs are also dysregulated in a mouse model of multiple sclerosis, experimental autoimmune encephalomyelitis (EAE), indicating their role in disease pathogenesis.

Further studies on functional consequences of the dysregulation of the above miRNAs have revealed specific pathways and molecular targets that might contribute to disease pathogenesis. For instance, miR-17 and miR-19b are known to induce the polarization of inflammatory Th17 cells by targeting IKAROS family zinc finger 4 (IKZF4) and PTEN, respectively.[110] PTEN is a known inhibitor of the phosphoinositide 3-kinase (PI3K)-AKT pathway whose activation induces STAT3 phosphorylation and subsequent expression of the Th17-specific transcription factor, RORγt.[110] PTEN is also implicated in the Th1 pathway; therefore, its downregulation by miR-19b might also confer Th1 responses in mice. The specific mechanism of IKZF4 action is not entirely known.[110] In addition to Th17 cells, miR-17 overexpression is associated with reduced Treg suppressive function by downregulating FOXP3 co-regulatory molecules.[111] In B cells, miR-19 also suppresses B cell central tolerance mediated *via* receptor editing by directly targeting PTEN in mice.[63] miR-20, which as mentioned is downregulated in multiple sclerosis and EAE, targets RORγt and STAT3, both of which are critical for Th17 polarization.[112]

Within mononuclear cells, $CD14^+$ monocytes, and lesions from multiple sclerosis patients, the expression of miR-155 is upregulated.[113–115] To correlate disease-relevant functions with miRNA expression, studies in mice have shown that miR-155-deficient mice are resistant to EAE with reduced amounts of IFN-γ-producing Th1 cells and IL-17-producing Th17 cells.[70] Interestingly, miR-155 functions in a T cell-intrinsic as well as a T cell-extrinsic manner in modulating EAE. For instance, within $CD4^+$ T cells, miR-155 promotes Th17 differentiation by targeting jumonji AT-rich interactive domain 2 (JARID2)[116] and ETS1.[117] JARID2 is a DNA-binding protein involved in chromatin modification,[116] and ETS1 is a transcription factor known to inhibit Th17 differentiation.[79] In addition, miR-155 targets the expression of SOCS1 and SHIP-1 within DCs and macrophages, which in turn leads to enhanced secretion of IL-12, IL-1β, IL-6, and IL-23 secretion, all cytokines required for Th1 and Th17 differentiation.[71] miR-155 also enhances the expression of the co-stimulatory molecules CD80, CD86, and C–C-chemokine receptor type 7 (CCR7) on DCs and macrophages by targeting the transcription factor, Finkel–Biskis–Jinkins murine osteogenic sarcoma (c-Fos).[114] miR-155-mediated suppression of c-Fos expression is critical for DC maturation in humans and

Table 6.2 Differentially expressed microRNAs in multiple sclerosis patients.

miRNA	Change in expression compared to control	Tissue/cell type	Potential target(s)	Effect of change in expression	References
miR-17	Up	$CD4^+$ T cells	IKZF4, PTEN	Polarization of Th17 cells	103–107, 110, 111
miR-19b	Up	B cells, Tregs	PTEN	Polarization of Th1 and Th17 cells, Suppression of B cell tolerance	63, 104, 106, 108–110
miR-155	Up	$CD14^+$ monocytes, CNS lesions	SOCS1, SHIP-1, c-Fos, CD47, JARID2, ETS1	Activation of Th17 differentiation	71, 113–118
miR-146a	Up	PBMCs	TRAF-6, IRAK-1	Activation of NF-κB-driven inflammation	59, 113, 115, 119, 120
miR-326a	Up	PBMCs, $CD4^+$ T cells	ETS1	Activation of Th17 differentiation	113, 115, 123, 124
miR-21	Up	PBMCs, CNS lesions	SMAD7, IL-10	Activation of Th17 differentiation and	82, 86, 126
miR-223	Up	Myeloid cells		Secretion of Th1 and Th17 polarizing cytokines by mDCs	133–135
miR-125a-3p	Up	CSF	SMAD4, p38, RHOA, NRG1, FYN	Impaired oligodendrocyte maturation	138
miR-219	Down	CNS lesions	HES5	Reduced oligodendrocyte maturation and aberrant myelination	113, 136, 137
miR-338	Down	CNS lesions	SOX6	Reduced oligodendrocyte maturation and aberrant myelination	113, 136, 137

[a]CNS: central nervous system; CSF: cerebrospinal fluid; mDCs: myeloid DCs; NF-κB: nuclear factor-κB; PBMCs: peripheral blood mononuclear cells; Th: T-helper; Tregs: T-regulatory cells.

mice.[118] Furthermore, miR-155 also appears to mediate its pathogenic role within the CNS by targeting CD47. This is a ubiquitously expressed surface molecule that protects cells from macrophage phagocytic activity. By downregulating CD47 expression, miR-155 makes cells of the CNS, including myelinated cells, susceptible to macrophage-mediated breakdown.[113]

Another miRNA strongly implicated in multiple sclerosis is miR-146a, whose expression is elevated in mononuclear cells and lesions from multiple sclerosis patients.[113,115,119] miR-146a is rapidly upregulated in response to TLR- and cytokine-induced NF-κB activation, and in turn, suppresses further NF-κB activation in a negative feedback loop.[59] Studies in mice have shown that miR-146a in both myeloid cells and $CD4^+$ T cells inhibits NF-κB signaling molecules TNF receptor associated factor 6 (TRAF-6) and interleukin-1 receptor associated kinase 1 (IRAK-1), which subsequently impairs pathogenic Th17 cell differentiation in both $CD4^+$ T cell-extrinsic and -intrinsic manners.[59,120] Therefore, mice deficient in miR-146a develop severe EAE due to uncontrolled Th17-mediated immune responses.[120] In other studies, miR-146a has been shown to promote Treg development and suppressive function by blocking STAT1, which limits subsequent Th1 responses. This type of regulation is critical for maintaining immunological tolerance.[121] However, in human astroglial cells studies have shown that the downregulation of IRAK-1 triggers IRAK-2-mediated NF-κB activation and the maintenance of immune responses.[122] This might suggest an alternative pathogenic role for miR-146a in multiple sclerosis lesions. Although due to its predominantly anti-inflammatory role, these contradictory functions of miR-146a warrant further investigation in the context of multiple sclerosis.

One of the strongly upregulated miRNAs in active lesions in multiple sclerosis patients, miR-326a, is also upregulated in peripheral blood lymphocytes, mononuclear cells, and $CD4^+$ T cells.[113,115,123–125] miR-326 is a pro-inflammatory miRNA, as mice deficient in miR-326 are resistant to EAE and produce fewer Th17 cells. miR-326 targets the transcription factor ETS1, a known inhibitor of Th17 differentiation.[123]

Another pro-inflammatory miRNA that is upregulated in PBMCs and brain lesions of multiple sclerosis patients is miR-21.[126] Studies in mice have shown that miR-21 can promote pathogenic Th17 differentiation by downregulating SMAD7, as described previously.[86] Consistent with this role, silencing miR-21 ameliorates EAE in mice.[86] In B cells, miR-21 can also inhibit IL-10 expression, thereby dampening the inflammatory suppression mediated by Bregs in EAE.[82]

miRNA regulation of Tregs, which play a key role in suppressing pathogenic immune responses that are dysregulated in multiple sclerosis, is another promising area of study.[127] miR-125a enhances Treg suppressive ability by negatively regulating STAT3, IFN-γ, IL-13, and IL-6 receptor (IL-6R), which are required for pathogenic Th1 and Th17 immune responses.[128,129] In fact, miR-125a deficiency leads to severe EAE in mice.[129]

In addition to miRNAs that primarily regulate T cell pathology, studies have identified miRNAs that are downregulated in B cells from multiple sclerosis patients.[94] These include miR-25, miR-106b, miR-93, and miR-181a, all of which play a role in B cell development.[130] For example, miR-181a-1/b-1 deletion in mice can enhance the reactivity of naive T cells to self-antigens. Specifically, miR-181a-1/b-1-deficient mice have more myelin oligodendrocyte glycoprotein-reactive T cells compared to wild-type mice.[131] While the role of miR-181a in B cells in multiple sclerosis/EAE needs to be investigated, studies in B cell lymphoma cell lines have revealed that miR-181a represses NF-κB signaling, resulting in enhanced B cell activation.[132] Overall, a more thorough investigation on the functional relevance of miRNA dysregulation in B cells in multiple sclerosis patients is needed.

Cells of the innate immune system, such as DCs, play a critical role in maintaining the pro-inflammatory response in multiple sclerosis.[100] miR-223 is highly expressed in myeloid cells of multiple sclerosis patients compared to healthy controls.[133] Consistent with this observation, miR-223-deficient mice present with ameliorated EAE, characterized by reduced myeloid DCs (mDCs) and Th17 cells in the CNS.[133] miR-223-deficient mDCs express lower MHC-II and CD86, increase PD-L1, a co-inhibitory ligand, and reduced IL-12, IL-1β, IL-6, and IL-23 expression, all of which may contribute to impaired Th1 and Th17 differentiation.[134,135] This makes miR-223 a strong promoter of mDC-driven pathogenic Th1 and Th17 responses in CNS autoimmunity.[134,135]

During multiple sclerosis, uncontrolled inflammation within the CNS leads to destruction of myelin. miRNAs play a vital role in remyelination by regulating oligodendrocyte development, maturation, and maintenance. Among the miRNAs, miR-219 and miR-338 are required for differentiation of oligodendrocyte precursors into mature myelinating oligodendrocytes by targeting the transcription factors, hes family bHLH transcription factor (HES5) and sex determining region Y-box 6 (SOX6), both of which inhibit differentiation.[136,137] Interestingly, expression of both these miRNAs is downregulated in multiple sclerosis tissue lesions.[113] Another notable miRNA, miR-125a-3p, is upregulated in CSF from multiple sclerosis patients.[138] *In vitro* studies with oligodendroglial precursor cells indicate that miR-125a impairs oligodendrocyte maturation by targeting multiple transcription factors.[138] However, more studies are required to understand how miRNAs regulate remyelination in multiple sclerosis.

Multiple sclerosis is one of the autoimmune diseases in which miRNA expression and function are well characterized in both patient populations and animal models, providing insight not only into multiple sclerosis pathophysiology, but other autoimmune conditions.

6.2.3 Rheumatoid Arthritis

Rheumatoid arthritis is a chronic autoimmune disease characterized by inflammation of the synovial tissue, which leads to damaged bone and cartilage, and eventually disability. The synovium in patients is a highly inflammatory environment consisting of cytokines, chemokines, and transcription factors secreted by T and B cells, and macrophage-like synoviocytes.[139]

Resident rheumatoid arthritis synovium fibroblasts (RASF) and fibroblast-like synoviocytes exhibit an aggressive phenotype with enhanced secretion of matrix metalloproteinases (MMPs) and adhesion molecules that contribute to bone destruction.

More recently, miRNAs have been elucidated for use as potential diagnostic biomarkers, as well as novel therapeutic targets, in rheumatoid arthritis. This understanding has been revealed by extensive profiling of blood, joints, and synovium from rheumatoid arthritis patients (Table 6.3).[140] For example, miR-146a is consistently upregulated in various patient samples, including synovial tissue, $CD68^+$ macrophages, $CD3^+$ and $CD4^+$ T cell subsets, $CD79^+$ B cells, and peripheral blood mononuclear cells (PBMCs) from rheumatoid arthritis patients.[141–145] Increased levels of miR-146a positively correlate with the pro-inflammatory cytokine, TNF-α. *In vitro* studies using Jurkat T cells indicate that miR-146a targets Fas -associated factor 1 (FAF1) and blocks T cell apoptosis, an essential process in maintaining tolerance, as $CD4^+$ T cells

Table 6.3 Differentially expressed microRNAs in rheumatoid arthritis patients.[a]

miRNA	Change in expression compared to control	Tissue/cell type	Potential target(s)	Effect of change in expression	References
miR-146a	Up	Synovial tissue, macrophages, T cells, B cells, PBMCs	TRAF-6, IRAK-1, FAF1	Reduced T cell apoptosis, increased joint destruction	80, 142–146
miR-155	Up	Synovial fluid monocytes, $CD14^+$ cells, monocytes	SHIP-1, SOCS1	Activation of Th17 differentiation	147–151
miR-223	Up	$CD4^+$ T cells, synovium	IGF-1R, ROQUIN, NF1-A	Increased Th17 differentiation and osteoclastogenesis	135, 152, 153
miR-203	Up	Synovial fibroblasts		Enhanced NF-κB signaling	154
miR-124	Down	Synoviocytes	CDK2, MCP-1	Aberrant cell proliferation and inflammatory cell recruitment	139
miR-23b	Down	Synoviocytes	TAB2, TAB3, IKK-α	Activation of NF-κB signaling	155
miR-34a*	Down	Synoviocytes	XIAP	Reduced cellular apoptosis	156

[a]NF-κB: nuclear factor-κB; PBMCs: peripheral blood mononuclear cells; Th: T-helper.

fail to undergo apoptosis in a murine model of proteoglycan-induced arthritis.[141,146] Further studies in synovium from rheumatoid arthritis patients have revealed that miR-146a is expressed in IL-17A-expressing cells.[144] In addition, miR-146a has been shown to suppress osteoclastogenesis, and *in vivo* administration of miR-146a reduces joint destruction in collagen-induced arthritis (CIA) in mice without ameliorating joint inflammation, suggesting an immune-independent function for this miRNA. Although controversial, it indicates immense therapeutic potential for miR-146a in rheumatoid arthritis.[143]

Another extensively studied miRNA, miR-155, is also enriched in RASF, rheumatoid arthritis synovial fluid monocytes, $CD14^+$ cells, and PBMCs.[147–150] miR-155 levels correlate positively with TNF-α and IL-1β. miR-155 is known to target SHIP-1, an anti-inflammatory molecule, thereby leading to activation of mDCs and pro-inflammatory cytokine secretion.[147] Consistent with this pathogenic role of miR-155 in rheumatoid arthritis, studies in a mouse model of CIA show that miR-155-deficient mice fail to develop disease, and have fewer autoreactive T and B cells with decreased levels of the pro-inflammatory cytokines, IL-6 and IL-17. This is accompanied by decreased joint destruction with reduced osteoclast differentiation.[151] However, in RASF, the enforced expression of miR-155 reduces the expression of MMP-3 and MMP-1, molecules that contribute to joint destruction. Further validation of this mechanism of function in CIA mice is required.[150]

The expression of miR-223 is also upregulated in $CD4^+$ T cells and synovium in rheumatoid arthritis patients.[152,153] One possible mechanism by which miR-223 might contribute to the pathogenesis of rheumatoid arthritis is by dampening the expression of insulin-like growth factor-1 receptor (IGF-1R), which regulates the secretion of anti-inflammatory IL-10 from activated T cells.[153] In mDCs that are known to play a crucial role in rheumatoid arthritis, miR-223 has been shown to downregulate the expression of Roquin, a repressor of Th17-polarizing cytokines (*e.g.*, IL-6, IL-1β, and IL-23). As a result, miR-223 promotes Th17 differentiation.[135] In the CIA mouse model of rheumatoid arthritis, lentivirus-mediated silencing of miR-223 reduces arthritis score and joint destruction. miR-223 is also highly expressed in bone marrow-derived macrophages and is required for osteoclast differentiation by targeting nuclear factor 1-A (NF1-A), a molecule that upregulates macrophage-colony stimulator factor receptor (M-CSFR). Therefore, silencing miR-223 can reduce joint destruction by inhibiting osteoclastogenesis.[152]

In addition, miR-203 is upregulated in synovial fibroblasts from rheumatoid arthritis patients.[154] miR-203 is believed to regulate rheumatoid arthritis pathogenesis by modulating chronic inflammation and joint destruction. RASFs overexpressing miR-203 exhibit elevated secretion of IL-6 due to enhancement of the NF-κB signaling pathway, as miR-203 potentially targets negative regulators of the pathway. In addition, these RASFs also secrete large amounts of MMP-1, which contributes to joint destruction.[150]

In contrast to miRNAs that are consistently upregulated in rheumatoid arthritis patients, other miRNAs such as miR-124a, miR-34a*, and miR-23b are downregulated in synoviocytes from rheumatoid arthritis patients compared to controls.[140] miR-124a is a known inhibitor of the cell cycle regulator, cyclin dependent kinase 2 (CDK2), and the chemokine monocyte chemoattractant protein 1 (MCP-1), which attracts T cells to the synovium and contributes to inflammation. Aberrant upregulation of these two molecules during rheumatoid arthritis results in uncontrolled cellular proliferation and enhanced chemoattractants, which recruit inflammatory T cells to the synovium.[139] The expression of miR-23b is also downregulated in the joints of CIA mice. Further investigation has revealed that miR-23b targets the expression of TGF-β-activated kinase 1/MAP3K7 binding protein 2, 3 (TAB2, TAB3), and an inhibitor of NF-κB kinase subunit α (IKK-α). Because these miR-23b targets are downstream of IL-1, TNF-α, and IL-17, they are required for NF-κB activation in synoviocytes.[155] Lastly, miR-34a* regulates apoptosis in RASF by targeting X-linked inhibitor of apoptosis (XIAP). Consequently, the downregulation of miR-34a* in patients is associated with higher levels of XIAP in synovial tissue from patients and reduced apoptosis.[156]

As in other autoimmune conditions, miRNAs appear to be dysregulated in various tissues from rheumatoid arthritis patients, indicating their role not only in mediating inflammatory responses, but also regulating osteoclast differentiation.

6.2.4 Systemic Lupus Erythematosus

Systemic lupus erythematosus (SLE) is an autoimmune disease that affects multiple organs, including the skin, joints, kidneys, and CNS.[157] The hallmarks of SLE are elevated levels of serum autonuclear autoantibodies and immune complex depositions in affected tissues. T cells, B cells, and DCs, especially pDCs, are involved in the pathogenesis of SLE. miRNA profiling studies in PBMCs and resident cells of affected tissues in SLE patients so far have identified 16 upregulated and six downregulated miRNAs.[158,159] Many of these miRNAs regulate DC, T cell, and B cell responses, highlighted herein (Table 6.4).

Type I IFN signatures correlate with disease severity in SLE patients.[160] Therefore, pDCs are significantly implicated in the pathogenesis of SLE.[161] miR-155 is upregulated in PBMCs of SLE patients, and it acts to promote pDC activation.[162] TLR7 stimulation results in rapid upregulation of miR-155-3p and miR-155-5p in human pDCs, after which miR-155-3p promotes IFN-α/β expression by suppressing IRAK-M, an inhibitor of TLR signaling. In addition, miR-155-5p inhibits IFN-α/β expression by targeting TAB2, a molecule implicated in NF-κB activation.[162] In addition to pDCs, miR-155 also controls cDC functions. miR-155, as described previously, promotes secretion of Th1- and Th17-promoting cytokines from cDCs.[70,71] Consistent with this role, treatment with miR-155 inhibitor reduces pro-inflammatory cytokines and disease pathology in a pristane-induced C57BL/6 model of SLE.[163]

miR-155 also regulates B cell responses by directly affecting the expression of activation induced cytidine deaminase (AID) and PU.1, both of which are involved in B cell maturation and antibody secretion. Disruption

Table 6.4 Differentially expressed microRNAs in systemic lupus erythematosus patients.[a]

miRNA	Change in expression compared to control	Tissue/cell type	Potential target(s)	Effect of change in expression	References
miR-155	Up	PBMCs	IRAK-M, TAB2, AID, PU.1, PELI1	Increased Tfh, Th1, Th17 differentiation, and B cell responses	70, 71, 162–167, 170
miR-21	Up	PBMCs, T cells	PDCD4	Activation of T cell and B cells	171–173
miR-326	Up	Tregs	ETS1	Decreased Tregs and increased autoreactive B cells	174–176
miR-148a	Up	PBMCs	PTEN, BIM, GADD45α, DMNT1	Activation of B cells	129, 177
miR-130b-3p	Up	Serum	ERBB2IP	Promotes EMT	178
miR-146a	Down	PBMCs	IRF5, STAT1, ICOS-ICOSL	Enhanced type 1 IFN signaling and autoantibodies secretion	58, 159, 179, 182

[a]EMT: epithelial–mesenchymal transition; IFN: interferon; PBMCs: peripheral blood mononuclear cells; Tfh: T-follicular helper; Th: T-helper.

of a miR-155 binding site in AID mRNA 3′ UTR leads to dysregulated AID expression, resulting in aberrantly enhanced CSR and impaired affinity maturation.[164,165] Furthermore, miR-155 appears to promote germinal center responses, which are negatively regulated by its target, PU.1. In B cells, miR-155 expression is rapidly induced upon BCR stimulation.[166] miR-155-deficient mice on the *Fas^{lpr}* background develop an attenuated lupus phenotype, accompanied by reduced titers of autoantibodies and decreased kidney pathology.[167] This phenotype is probably caused by the impaired class-switched antibody responses and reduced germinal center responses observed in miR-155-deficient mice.[33,168,169] In addition to B cell-intrinsic modulation, miR-155 positively regulates Tfh responses. miR-155 targets E3 ubiquitin ligase pellino homolog 1 (PELI1) and relieves avian reticuloendotheliosis (c-Rel), a subunit of NF-κB, from ubiquitin-mediated degradation, promoting Tfh cell expansion and humoral responses.[170]

Another miRNA, miR-21, is upregulated in PBMCs and T cells from SLE patients, and its expression levels correlate positively with disease activity.[171,172] Silencing miR-21 in SLE T cells reverses the activated phenotype of these cells (*i.e.*, increased proliferation and CD40L expression that promotes B cell responses), while overexpression of miR-21 in healthy donor T cells induces an activated phenotype.[172] miR-21 targets programmed cell death 4 (PDCD4), a tumor suppressor and a protein translation inhibitor, to mediate this effect, as PDCD4 expression levels are decreased in SLE CD4$^+$ T cells and correlate negatively with miR-21 in these cells.[172] More interestingly, in a spontaneous genetic model of SLE using B6.Sle123 mice, administration of locked nucleic acid (LNA)-based miR-21 inhibitor results in diminished autoimmune splenomegaly.[173] However, other critical disease parameters, such as titers of antinuclear antibodies, have not been examined.

miR-326 expression is increased in Tregs of new-onset SLE patients and is associated with decreased levels of ETS1.[174] ETS1 is a transcription factor that stabilizes FOXP3 expression in Tregs,[79] suggesting that miR-326-mediated downregulation of ETS1 compromises the suppressive function of Tregs in SLE patients. In addition to suppressing Tregs, miR-326 may exert pathogenic functions by regulating B cells. miR-326 promotes B cell function by targeting ETS1 and exacerbates lupus-like pathology in a MRL/lpr mouse model of SLE.[175] Given that ETS1 is required for the maintenance of B cell tolerance mechanisms,[176] downregulation of ETS1 by miR-326 is likely to promote autoreactive B cell responses in SLE patients.

miR-148a levels are also increased in PBMCs from SLE patients, and miR-148a deficiency results in decreased survival rate and exacerbated disease in MRL/lpr mice.[177] Since miR-148a is involved in suppression of B cell tolerance by targeting PTEN, Bcl-2-like protein 11 (BIM), and growth arrest and DNA damage-inducible 45 (GADD45α), the elevated levels of miR-148a observed in SLE patients probably contribute to their aberrant B cell responses. Furthermore, miR-148a directly targets DNA (cytosine-5)-methyltransferase 1 (DMNT1), a DNA methyltransferase in CD4$^+$ T cells, and leads to overexpression of autoimmunity-associated, methylation-sensitive genes, such as CD70 and lymphocyte function-associated-1 (LFA-1).[129]

In addition to immune cells, tissue-resident cells, such as renal tubular cells in kidneys, are involved in the pathogenesis of SLE, particularly in lupus nephritis.[157,158] Of note, miR-130b-3p, which is upregulated in the sera of SLE patients, promotes epithelial–mesenchymal transition (EMT) by targeting Erbb2 interacting protein (ERBB2IP), an inhibitor of TGF-β-driven EMT.[178]

In contrast to these miRNAs, miR-146a expression levels are decreased in PBMCs from SLE patients and correlate negatively with disease activity;[159] they are associated with kidney involvement, called lupus nephritis.[179] Further highlighting the involvement of miR-146a in SLE, two single nucleotide polymorphisms in a gene encoding *miR-146a* confer increased susceptibility to SLE.[180] miR-146a is induced upon TLR7/9 signaling in human pDCs, and its forced expression impairs TLR–NF–κB signaling, pro-inflammatory cytokine expression, and MHC-II and co-stimulatory molecule expression in pDCs.[181] Importantly, miR-146 directly suppresses type I IFN pathways by targeting IRF5 and STAT1. Therefore, reduced expression of miR-146a in SLE patient PBMCs may contribute to dysregulated type I IFN signatures in SLE.[159] In addition, miR-146a has a T cell-intrinsic role wherein it negatively regulates Tfh differentiation by inhibiting ICOS-ICOSL pathway.[58] Tfh cells promote pathogenic B cell responses in SLE by driving T cell-dependent autoreactive antibody production. Furthermore, the delivery of exogenous anti-inflammatory miR-146a through bacteriophage-based virus-like particles to BXSB mice results in reduced titers of autoantibodies and pro-inflammatory cytokines.[182]

The efficacy of miRNA-based therapy has been demonstrated in multiple mouse models of SLE.[158] Although there is no ongoing clinical trial to test miRNA-based therapeutics in patients with SLE, these findings in mouse models are promising and may pave the way for future development of such strategies.

6.2.5 Other Autoimmune Diseases

In addition to the diseases discussed, other autoimmune diseases have also been found to be regulated by miRNAs. Type 1 diabetes (T1D) is an autoimmune condition characterized by the destruction of β-pancreatic cells *via* T cells and macrophages. A comprehensive analysis of dysregulated miRNAs in serum/plasma and PBMCs from T1D patients has revealed the following six miRNAs are consistently upregulated: miR-21-5p, miR-24-3p, miR-148a-3p, miR-181a-5p, miR-210-5p, and miR-375; while five miRNAs are consistently downregulated: miR-100-5p, miR-146a-5p, miR-150-5p, miR-342-3p, and miR-1275. All these miRNAs play a critical role in modulating the TCR signaling pathway, including PI3K/serine/threonine kinase (Akt), Janus kinase (JAK)/STAT, and NF-κB signaling.[183]

Psoriasis is a chronic autoimmune disorder affecting the skin. The pathogenesis of psoriasis is complex, as it involves interactions between the commensal microbiota residing on the skin, DCs, Th cell subsets, keratinocytes, and genetic and environmental factors influencing these populations.[184] miRNA profiling studies in the skin and PBMCs of psoriasis patients have identified 12 upregulated miRNAs: miR-21, miR-31, miR-135b, miR-136, miR-138, miR-146a, miR-155, miR-184, miR-203, miR-210, miR-221/222, and miR-214; and two downregulated miRNAs: miR-99a and miR-125b.[184] Most of these miRNAs regulate either keratinocyte differentiation and proliferation and/or T cell-mediated inflammation. Of note is miR-21, which is increased in the lesions of psoriasis patients.[185] In a patient-derived skin xenograft model of psoriasis in severe combined immunodeficient (SCID) mice, administration of LNA-modified miR-21 inhibitor has been found to ameliorate disease pathology in the transplanted tissues.[185]

Asthma is a chronic inflammatory disease that involves the infiltration of inflammatory cells, such as Th2 cells, into the lungs.[26] miRNA profiling studies in patients have revealed that miR-223-3p, miR-142-3p, and miR-629-3p are increased in sputum of patients with severe asthma.[186] miR-126 and miR-21 are upregulated in patient bronchial epithelial cells.[187] Studies in mice have demonstrated that miR-21 is also upregulated in murine models of asthma and human eosinophilic esophagitis.[35] Interestingly, it has been shown that the loss of miR-155 leads to excessive Th2 responses, with mice developing Th2-mediated spontaneous lung inflammation.[33] These miRNAs appear to play a critical role in modulating Th2 differentiation and inflammatory responses in the lungs.

6.3 miRNA Therapeutics in Preclinical Models of Autoimmune Diseases

miRNAs regulate gene expression in both immune and non-immune cells. Depending on the tissue and autoimmune context, the overexpression of immunoregulatory miRNAs might combat disease, and/or the inhibition of pathogenic miRNAs could be beneficial. Analogous to proteins and genes, miRNAs are emerging as potential therapeutic targets. As of 2017, approximately 20 miRNA-based clinical trials against various clinical indications, both immune- and non-immune-mediated, have been initiated.[188] Interestingly, the first human clinical trial involving anti-miR-122 has been successful against hepatitis C virus infection without any adverse events, underscoring the utility of miRNA as therapeutic molecules.[189] In contrast, the administration of liposomal miR-34a mimic (MRX34 clinical trial) in patients with a variety of solid tumors or hematologic malignancies led to serious immune-related adverse events in a small subset of patients. Such effects could be attributed to several factors that remain to be investigated, including the complex biology of miR-34a, the liposomal carrier used for delivery, and clinical trial design (inclusion and exclusion criteria of patients).[190]

miRNAs have been proposed as diagnostic markers for differentiating between patients and healthy controls. In addition, miRNAs have demonstrated promising therapeutic efficacy upon overexpression or inhibition in mouse models of autoimmunity (Table 6.5). For instance, silencing miR-326 reduces the severity of a mouse model of multiple sclerosis by inhibiting Th17 cells.[123] Due to the inflammatory role of miR-155 in promoting Th1 and Th17 cells, the use of anti-miR-155 also ameliorates EAE in mice.[70] In addition, silencing miR-155 in a pristane-induced model of SLE significantly reduces inflammation and diffuse alveolar hemorrhage that is commonly associated with SLE.[163] Silencing of miR-21, a pathogenic miRNA that promotes expression of pro-inflammatory cytokines, has also shown therapeutic efficacy in EAE.[86] Furthermore, miR-21 antagonists have been effective in ameliorating pathology in a mouse model of psoriasis with patient tissue xenotransplantation.[185] Silencing miR-21 also ameliorates splenomegaly in a mouse model of lupus,[173] indicating its diverse utility as a therapeutic.

Other miRNAs whose inhibition confers reduction in disease severity include miR-214 and miR-223. The use of a chemical inhibitor targeting miR-214 is effective in suppressing inflammation in ulcerative colitis patient-derived colonic tissue, as well as DSS-induced colitis in mice.[88] Lentivirus-mediated inhibition of miR-223, on the other hand, reduces disease severity of CIA in

Table 6.5 MicroRNA-based therapeutics in preclinical models of autoimmune diseases.[a]

miRNA	Agent	Disease model; indication	Potential target(s)	References
miR-155	Synthetic antagomir	Pristane C57BL/6; SLE	BCL6, SOCS1, PPARα,	158
	LNA-based inhibitor	EAE; multiple sclerosis	SOCS1, SHIP-1	69–71
miR-146a	Virus-like particle containing mimic	BXSB; SLE	IRF-5, STAT-1	182
	Double-stranded RNA mimic	CIA; rheumatoid arthritis		143
miR-21	LNA-based inhibitor	EAE; multiple sclerosis	SMAD7	86
	LNA-based inhibitor	B6.Sle123; SLE	PDCD4	162, 173
	LNA-based inhibitor	Patient-derived |skin xenotransplants; psoriasis	TIMP-1	185
miR-214	LNA-based inhibitor	DSS-induced colitis/IBD	PDLIM2, PTEN	88
miR-223	Lentivirus vector-based inhibitor	CIA; rheumatoid arthritis	NF1-A	152
miR-326	Lentivirus vector-based inhibitor	EAE; multiple sclerosis	ETS1	123

[a]CIA: collagen-induced arthritis; DSS: dextran sodium sulfate; EAE: experimental autoimmune encephalomyelitis; IBD: inflammatory bowel diseases; LNA: locked nucleic acid; SLE: systemic lupus erythematosus.

mice.[152] As mentioned previously, miRNAs can either be inhibited to block pro-inflammatory signals, or overexpressed to promote anti-inflammatory molecules. One such example of a miRNA whose overexpression is protective in autoimmunity is miR-146a. The administration of double-stranded miR-146a in CIA mice dramatically reduces bone destruction.[143]

A major remaining challenge in the field is optimizing delivery of miRNA therapeutics to specific target tissue. As a result, several delivery techniques are being tested, including nanoparticle, lentivirus-mediated, and adenoviral-mediated approaches, each with their own limitations and benefits.[191] Studies establishing modes of delivery, stability, reduced toxicity, and enhanced efficacy in target tissues are ongoing in cancer and other diseases, such as cardiovascular disease.[192] As a result, miRNA therapeutics, currently in the clinical pipeline, employ a variety of delivery strategies.[188,192] Based on findings from preclinical autoimmune models and research optimizing miRNA delivery *in vivo*, the possibility of utilizing effective miRNA-based therapeutics for autoimmune diseases remains possible in the near future.

6.4 Concluding Remarks

Findings on miRNAs in autoimmunity underscore their importance as valuable molecules governing disease processes. In many autoimmune conditions, including, but not limited to those covered in this chapter, dysregulation of miRNA expression is apparent. While the number of studies using patient samples to identify specific dysregulated miRNAs in particular diseases continues to grow, the functional outcomes for each miRNA are still not completely understood. In addition, although the utility of miRNAs as therapeutics has been well established in preclinical mouse models, significant work remains to be done in translating these findings to the clinic. Therefore, further investigation into precise mechanisms of miRNA dysregulation and elucidation of pathological consequences is paramount for a better understanding of the pathogenesis of human autoimmunity, and prevention and treatment of disease.

Acknowledgements

This work was supported by grants from the National Multiple Sclerosis Society (RG 1507-05164) and the National Institutes of Health (R01AI127853).

References

1. L. He and G. J. Hannon, *Nat. Rev. Genet.*, 2004, **5**, 522–531.
2. J. R. Lytle, T. A. Yario and J. A. Steitz, *Proc. Natl. Acad. Sci. U. S. A.*, 2007, **104**, 9667–9672.
3. I. Rigoutsos, *Cancer Res.*, 2009, **69**, 3245–3248.
4. S. Jonas and E. Izaurralde, *Nat. Rev. Genet.*, 2015, **16**, 421–433.
5. J. C. Zuniga-Pflucker, *Nat. Rev. Immunol.*, 2004, **4**, 67–72.
6. S. Jiang, C. Li, V. Olive, E. Lykken, F. Feng, J. Sevilla, Y. Wan, L. He and Q. J. Li, *Blood*, 2011, **118**, 5487–5497.
7. C. Xiao, L. Srinivasan, D. P. Calado, H. C. Patterson, B. Zhang, J. Wang, J. M. Henderson, J. L. Kutok and K. Rajewsky, *Nat. Immunol.*, 2008, **9**, 405–414.
8. E. Glasmacher, K. P. Hoefig, K. U. Vogel, N. Rath, L. Du, C. Wolf, E. Kremmer, X. Wang and V. Heissmeyer, *Nat. Immunol.*, 2010, **11**, 725–733.
9. D. Yu, A. H. Tan, X. Hu, V. Athanasopoulos, N. Simpson, D. G. Silva, A. Hutloff, K. M. Giles, P. J. Leedman, K. P. Lam, C. C. Goodnow and C. G. Vinuesa, *Nature*, 2007, **450**, 299–303.
10. C. E. Rudd, A. Taylor and H. Schneider, *Immunol. Rev.*, 2009, **229**, 12–26.
11. E. Sonkoly, P. Janson, M. L. Majuri, T. Savinko, N. Fyhrquist, L. Eidsmo, N. Xu, F. Meisgen, T. Wei, M. Bradley, J. Stenvang, S. Kauppinen, H. Alenius, A. Lauerma, B. Homey, O. Winqvist, M. Stahle and A. Pivarcsi, *J. Allergy Clin. Immunol.*, 2010, **126**, 581–589.e20.

12. Q. Xue, Z. Y. Guo, W. Li, W. H. Wen, Y. L. Meng, L. T. Jia, J. Wang, L. B. Yao, B. Q. Jin, T. Wang and A. G. Yang, *Mol. Immunol.*, 2011, **48**, 592–599.
13. R. P. Weitzel, M. L. Lesniewski, P. Haviernik, S. Kadereit, P. Leahy, N. J. Greco and M. J. Laughlin, *Blood*, 2009, **113**, 6648–6657.
14. W. Fan, D. Liang, Y. Tang, B. Qu, H. Cui, X. Luo, X. Huang, S. Chen, B. W. Higgs, B. Jallal, Y. Yao, J. B. Harley and N. Shen, *Arthritis Rheum.*, 2012, **64**, 3715–3725.
15. L. P. Garo and G. Murugaiyan, *Cell. Mol. Life Sci.*, 2016, **73**, 2041–2051.
16. M. Pagani, G. Rossetti, I. Panzeri, P. de Candia, R. J. Bonnal, R. L. Rossi, J. Geginat and S. Abrignani, *Immunol. Rev.*, 2013, **253**, 82–96.
17. D. Baumjohann and K. M. Ansel, *Nat. Rev. Immunol.*, 2013, **13**, 666–678.
18. R. B. Moss, T. Moll, M. El-Kalay, C. Kohne, W. Soo Hoo, J. Encinas and D. J. Carlo, *Expert Opin. Biol. Ther.*, 2004, **4**, 1887–1896.
19. A. C. Mullen, F. A. High, A. S. Hutchins, H. W. Lee, A. V. Villarino, D. M. Livingston, A. L. Kung, N. Cereb, T. P. Yao, S. Y. Yang and S. L. Reiner, *Science*, 2001, **292**, 1907–1910.
20. V. T. Thieu, Q. Yu, H. C. Chang, N. Yeh, E. T. Nguyen, S. Sehra and M. H. Kaplan, *Immunity*, 2008, **29**, 679–690.
21. S. A. Muljo, K. M. Ansel, C. Kanellopoulou, D. M. Livingston, A. Rao and K. Rajewsky, *J. Exp. Med.*, 2005, **202**, 261–269.
22. K. M. Smith, M. Guerau-de-Arellano, S. Costinean, J. L. Williams, A. Bottoni, G. Mavrikis Cox, A. R. Satoskar, C. M. Croce, M. K. Racke, A. E. Lovett-Racke and C. C. Whitacre, *J. Immunol.*, 2012, **189**, 1567–1576.
23. D. F. Steiner, M. F. Thomas, J. K. Hu, Z. Yang, J. E. Babiarz, C. D. Allen, M. Matloubian, R. Blelloch and K. M. Ansel, *Immunity*, 2011, **35**, 169–181.
24. T. X. Lu, J. Hartner, E. J. Lim, V. Fabry, M. K. Mingler, E. T. Cole, S. H. Orkin, B. J. Aronow and M. E. Rothenberg, *J. Immunol.*, 2011, **187**, 3362–3373.
25. P. Mohnle, S. V. Schutz, V. van der Heide, M. Hubner, B. Luchting, J. Sedlbauer, E. Limbeck, L. C. Hinske, J. Briegel and S. Kreth, *Eur. J. Immunol.*, 2015, **45**, 260–272.
26. T. Nakayama, K. Hirahara, A. Onodera, Y. Endo, H. Hosokawa, K. Shinoda, D. J. Tumes and Y. Okamoto, *Annu. Rev. Immunol.*, 2017, **35**, 53–84.
27. I. C. Ho, D. Lo and L. H. Glimcher, *J. Exp. Med.*, 1998, **188**, 1859–1866.
28. M. Kurowska-Stolarska, P. Kewin, G. Murphy, R. C. Russo, B. Stolarski, C. C. Garcia, M. Komai-Koma, N. Pitman, Y. Li, W. Niedbala, A. N. McKenzie, M. M. Teixeira, F. Y. Liew and D. Xu, *J. Immunol.*, 2008, **181**, 4780–4790.
29. Y. H. Wang, P. Angkasekwinai, N. Lu, K. S. Voo, K. Arima, S. Hanabuchi, A. Hippe, C. J. Corrigan, C. Dong, B. Homey, Z. Yao, S. Ying, D. P. Huston and Y. J. Liu, *J. Exp. Med.*, 2007, **204**, 1837–1847.
30. W. Zheng and R. A. Flavell, *Cell*, 1997, **89**, 587–596.
31. Y. Bronevetsky, A. V. Villarino, C. J. Eisley, R. Barbeau, A. J. Barczak, G. A. Heinz, E. Kremmer, V. Heissmeyer, M. T. McManus, D. J. Erle, A. Rao and K. M. Ansel, *J. Exp. Med.*, 2013, **210**, 417–432.

32. H. H. Pua, D. F. Steiner, S. Patel, J. R. Gonzalez, J. F. Ortiz-Carpena, R. Kageyama, N. T. Chiou, A. Gallman, D. de Kouchkovsky, L. T. Jeker, M. T. McManus, D. J. Erle and K. M. Ansel, *Immunity*, 2016, **44**, 821–832.
33. A. Rodriguez, E. Vigorito, S. Clare, M. V. Warren, P. Couttet, D. R. Soond, S. van Dongen, R. J. Grocock, P. P. Das, E. A. Miska, D. Vetrie, K. Okkenhaug, A. J. Enright, G. Dougan, M. Turner and A. Bradley, *Science*, 2007, **316**, 608–611.
34. K. Podshivalova and D. R. Salomon, *Crit. Rev. Immunol.*, 2013, **33**, 435–476.
35. R. F. Chen, H. C. Huang, C. Y. Ou, T. Y. Hsu, H. Chuang, J. C. Chang, L. Wang, H. C. Kuo and K. D. Yang, *Clin. Exp. Allergy*, 2010, **40**, 1482–1490.
36. D. V. Sawant, H. Wu, M. H. Kaplan and A. L. Dent, *Mol. Immunol.*, 2013, **54**, 435–442.
37. K. Ghoreschi, A. Laurence, X. P. Yang, K. Hirahara and J. J. O'Shea, *Trends Immunol.*, 2011, **32**, 395–401.
38. I. I. Ivanov, B. S. McKenzie, L. Zhou, C. E. Tadokoro, A. Lepelley, J. J. Lafaille, D. J. Cua and D. R. Littman, *Cell*, 2006, **126**, 1121–1133.
39. P. R. Mangan, L. E. Harrington, D. B. O'Quinn, W. S. Helms, D. C. Bullard, C. O. Elson, R. D. Hatton, S. M. Wahl, T. R. Schoeb and C. T. Weaver, *Nature*, 2006, **441**, 231–234.
40. X. O. Yang, A. D. Panopoulos, R. Nurieva, S. H. Chang, D. Wang, S. S. Watowich and C. Dong, *J. Biol. Chem.*, 2007, **282**, 9358–9363.
41. E. V. Dang, J. Barbi, H. Y. Yang, D. Jinasena, H. Yu, Y. Zheng, Z. Bordman, J. Fu, Y. Kim, H. R. Yen, W. Luo, K. Zeller, L. Shimoda, S. L. Topalian, G. L. Semenza, C. V. Dang, D. M. Pardoll and F. Pan, *Cell*, 2011, **146**, 772–784.
42. M. Huber, A. Brustle, K. Reinhard, A. Guralnik, G. Walter, A. Mahiny, E. von Low and M. Lohoff, *Proc. Natl. Acad. Sci. U. S. A.*, 2008, **105**, 20846–20851.
43. B. U. Schraml, K. Hildner, W. Ise, W. L. Lee, W. A. Smith, B. Solomon, G. Sahota, J. Sim, R. Mukasa, S. Cemerski, R. D. Hatton, G. D. Stormo, C. T. Weaver, J. H. Russell, T. L. Murphy and K. M. Murphy, *Nature*, 2009, **460**, 405–409.
44. X. O. Yang, B. P. Pappu, R. Nurieva, A. Akimzhanov, H. S. Kang, Y. Chung, L. Ma, B. Shah, A. D. Panopoulos, K. S. Schluns, S. S. Watowich, Q. Tian, A. M. Jetten and C. Dong, *Immunity*, 2008, **28**, 29–39.
45. K. Ghoreschi, A. Laurence, X. P. Yang, C. M. Tato, M. J. McGeachy, J. E. Konkel, H. L. Ramos, L. Wei, T. S. Davidson, N. Bouladoux, J. R. Grainger, Q. Chen, Y. Kanno, W. T. Watford, H. W. Sun, G. Eberl, E. M. Shevach, Y. Belkaid, D. J. Cua, W. Chen and J. J. O'Shea, *Nature*, 2010, **467**, 967–971.
46. S. Crotty, *Immunity*, 2014, **41**, 529–542.
47. J. E. Craft, *Nat. Rev. Rheumatol.*, 2012, **8**, 337–347.
48. A. T. Bauquet, H. Jin, A. M. Paterson, M. Mitsdoerffer, I. C. Ho, A. H. Sharpe and V. K. Kuchroo, *Nat. Immunol.*, 2009, **10**, 167–175.

49. N. Bollig, A. Brustle, K. Kellner, W. Ackermann, E. Abass, H. Raifer, B. Camara, C. Brendel, G. Giel, E. Bothur, M. Huber, C. Paul, A. Elli, R. A. Kroczek, R. Nurieva, C. Dong, R. Jacob, T. W. Mak and M. Lohoff, *Proc. Natl. Acad. Sci. U. S. A.*, 2012, **109**, 8664–8669.
50. W. Ise, M. Kohyama, B. U. Schraml, T. Zhang, B. Schwer, U. Basu, F. W. Alt, J. Tang, E. M. Oltz, T. L. Murphy and K. M. Murphy, *Nat. Immunol.*, 2011, **12**, 536–543.
51. R. J. Johnston, Y. S. Choi, J. A. Diamond, J. A. Yang and S. Crotty, *J. Exp. Med.*, 2012, **209**, 243–250.
52. R. J. Johnston, A. C. Poholek, D. DiToro, I. Yusuf, D. Eto, B. Barnett, A. L. Dent, J. Craft and S. Crotty, *Science*, 2009, **325**, 1006–1010.
53. C. S. Ma, D. T. Avery, A. Chan, M. Batten, J. Bustamante, S. Boisson-Dupuis, P. D. Arkwright, A. Y. Kreins, D. Averbuch, D. Engelhard, K. Magdorf, S. S. Kilic, Y. Minegishi, S. Nonoyama, M. A. French, S. Choo, J. M. Smart, J. Peake, M. Wong, P. Gray, M. C. Cook, D. A. Fulcher, J. L. Casanova, E. K. Deenick and S. G. Tangye, *Blood*, 2012, **119**, 3997–4008.
54. R. I. Nurieva, Y. Chung, G. J. Martinez, X. O. Yang, S. Tanaka, T. D. Matskevitch, Y. H. Wang and C. Dong, *Science*, 2009, **325**, 1001–1005.
55. R. Hu, D. A. Kagele, T. B. Huffaker, M. C. Runtsch, M. Alexander, J. Liu, E. Bake, W. Su, M. A. Williams, D. S. Rao, T. Moller, G. A. Garden, J. L. Round and R. M. O'Connell, *Immunity*, 2014, **41**, 605–619.
56. S. G. Kang, W. H. Liu, P. Lu, H. Y. Jin, H. W. Lim, J. Shepherd, D. Fremgen, E. Verdin, M. B. Oldstone, H. Qi, J. R. Teijaro and C. Xiao, *Nat. Immunol.*, 2013, **14**, 849–857.
57. G. Curtale, F. Citarella, C. Carissimi, M. Goldoni, N. Carucci, V. Fulci, D. Franceschini, F. Meloni, V. Barnaba and G. Macino, *Blood*, 2010, **115**, 265–273.
58. A. Pratama, M. Srivastava, N. J. Williams, I. Papa, S. K. Lee, X. T. Dinh, A. Hutloff, M. A. Jordan, J. L. Zhao, R. Casellas, V. Athanasopoulos and C. G. Vinuesa, *Nat. Commun.*, 2015, **6**, 6436.
59. K. D. Taganov, M. P. Boldin, K. J. Chang and D. Baltimore, *Proc. Natl. Acad. Sci. U. S. A.*, 2006, **103**, 12481–12486.
60. H. Takahashi, T. Kanno, S. Nakayamada, K. Hirahara, G. Sciume, S. A. Muljo, S. Kuchen, R. Casellas, L. Wei, Y. Kanno and J. J. O'Shea, *Nat. Immunol.*, 2012, **13**, 587–595.
61. D. Allman and S. Pillai, *Curr. Opin. Immunol.*, 2008, **20**, 149–157.
62. K. Pieper, B. Grimbacher and H. Eibel, *J. Allergy Clin. Immunol.*, 2013, **131**, 959–971.
63. M. Lai, A. Gonzalez-Martin, A. B. Cooper, H. Oda, H. Y. Jin, J. Shepherd, L. He, J. Zhu, D. Nemazee and C. Xiao, *Nat. Commun.*, 2016, **7**, 12207.
64. S. B. Koralov, S. A. Muljo, G. R. Galler, A. Krek, T. Chakraborty, C. Kanellopoulou, K. Jensen, B. S. Cobb, M. Merkenschlager, N. Rajewsky and K. Rajewsky, *Cell*, 2008, **132**, 860–874.
65. L. Belver, V. G. de Yebenes and A. R. Ramiro, *Immunity*, 2010, **33**, 713–722.
66. A. Mildner and S. Jung, *Immunity*, 2014, **40**, 642–656.

67. D. Sichien, B. N. Lambrecht, M. Guilliams and C. L. Scott, *Mucosal Immunol.*, 2017, **10**, 831–844.
68. M. Swiecki and M. Colonna, *Nat. Rev. Immunol.*, 2015, **15**, 471–485.
69. M. Ceppi, P. M. Pereira, I. Dunand-Sauthier, E. Barras, W. Reith, M. A. Santos and P. Pierre, *Proc. Natl. Acad. Sci. U. S. A.*, 2009, **106**, 2735–2740.
70. G. Murugaiyan, V. Beynon, A. Mittal, N. Joller and H. L. Weiner, *J. Immunol.*, 2011, **187**, 2213–2221.
71. R. M. O'Connell, D. Kahn, W. S. Gibson, J. L. Round, R. L. Scholz, A. A. Chaudhuri, M. E. Kahn, D. S. Rao and D. Baltimore, *Immunity*, 2010, **33**, 607–619.
72. E. F. Lind, D. G. Millar, D. Dissanayake, J. C. Savage, N. K. Grimshaw, W. G. Kerr and P. S. Ohashi, *J. Immunol.*, 2015, **195**, 4632–4640.
73. C. Varol, A. Mildner and S. Jung, *Annu. Rev. Immunol.*, 2015, **33**, 643–675.
74. J. B. Self-Fordham, A. R. Naqvi, J. R. Uttamani, V. Kulkarni and S. Nares, *Front. Immunol.*, 2017, **8**, 1062.
75. Z. Wang, S. Brandt, A. Medeiros, S. Wang, H. Wu, A. Dent and C. H. Serezani, *PLoS One*, 2015, **10**, e0115855.
76. Q. Chen, Y. C. Kim, A. Laurence, G. A. Punkosdy and E. M. Shevach, *J. Immunol.*, 2011, **186**, 6329–6337.
77. T. S. Davidson, R. J. DiPaolo, J. Andersson and E. M. Shevach, *J. Immunol.*, 2007, **178**, 4022–4026.
78. A. Liston, L. F. Lu, D. O'Carroll, A. Tarakhovsky and A. Y. Rudensky, *J. Exp. Med.*, 2008, **205**, 1993–2004.
79. E. Mouly, K. Chemin, H. V. Nguyen, M. Chopin, L. Mesnard, M. Leite-de-Moraes, O. Burlen-defranoux, A. Bandeira and J. C. Bories, *J. Exp. Med.*, 2010, **207**, 2113–2125.
80. J. Liu and X. Cao, *J. Autoimmun.*, 2015, **63**, 1–12.
81. X. Su, C. Qian, Q. Zhang, J. Hou, Y. Gu, Y. Han, Y. Chen, M. Jiang and X. Cao, *Nat. Commun.*, 2013, **4**, 2903.
82. H. Wang, W. Xu, Q. Shao and Q. Ding, *Oncotarget*, 2017, **8**, 94069–94079.
83. C. Abraham and J. H. Cho, *N. Engl. J. Med.*, 2009, **361**, 2066–2078.
84. X. M. Xu and H. J. Zhang, *World J. Gastroenterol.*, 2016, **22**, 2206–2218.
85. D. G. W. Johnston, M. A. Williams, C. A. Thaiss, R. Cabrera-Rubio, M. Raverdeau, C. McEntee, P. D. Cotter, E. Elinav, L. A. J. O'Neill and S. C. Corr, *J. Crohn's Colitis*, 2018, **12**, 835–848.
86. G. Murugaiyan, A. P. da Cunha, A. K. Ajay, N. Joller, L. P. Garo, S. Kumaradevan, N. Yosef, V. S. Vaidya and H. L. Weiner, *J. Clin. Invest.*, 2015, **125**, 1069–1080.
87. X. Feng, H. Wang, S. Ye, J. Guan, W. Tan, S. Cheng, G. Wei, W. Wu, F. Wu and Y. Zhou, *PLoS One*, 2012, **7**, e52782.
88. C. Polytarchou, D. W. Hommes, T. Palumbo, M. Hatziapostolou, M. Koutsioumpa, G. Koukos, A. E. van der Meulen-de Jong, A. Oikonomopoulos, W. K. van Deen, C. Vorvis, O. B. Serebrennikova, E. Birli, J. Choi, L. Chang, P. A. Anton, P. N. Tsichlis, C. Pothoulakis, H. W. Verspaget and D. Iliopoulos, *Gastroenterology*, 2015, **149**, 981–992. e11.

89. M. Allocca, M. Jovani, G. Fiorino, S. Schreiber and S. Danese, *Curr. Drug Targets*, 2013, **14**, 1508–1521.
90. O. Brain, B. M. Owens, T. Pichulik, P. Allan, E. Khatamzas, A. Leslie, T. Steevels, S. Sharma, A. Mayer, A. M. Catuneanu, V. Morton, M. Y. Sun, D. Jewell, M. Coccia, O. Harrison, K. Maloy, S. Schonefeldt, S. Bornschein, A. Liston and A. Simmons, *Immunity*, 2013, **39**, 521–536.
91. Z. Li, F. Wu, S. R. Brant and J. H. Kwon, *J. Immunol.*, 2011, **186**, 6182–6190.
92. A. Zwiers, L. Kraal, T. C. van de Pouw Kraan, T. Wurdinger, G. Bouma and G. Kraal, *J. Immunol.*, 2012, **188**, 1573–1577.
93. H. Wang, H. Flach, M. Onizawa, L. Wei, M. T. McManus and A. Weiss, *Nat. Immunol.*, 2014, **15**, 393–401.
94. W. Wu, C. He, C. Liu, A. T. Cao, X. Xue, H. L. Evans-Marin, M. Sun, L. Fang, S. Yao, I. V. Pinchuk, D. W. Powell, Z. Liu and Y. Cong, *Gut*, 2015, **64**, 1755–1764.
95. X. Xue, T. Feng, S. Yao, K. J. Wolf, C. G. Liu, X. Liu, C. O. Elson and Y. Cong, *J. Immunol.*, 2011, **187**, 5879–5886.
96. C. Lu, J. Chen, H. G. Xu, X. Zhou, Q. He, Y. L. Li, G. Jiang, Y. Shan, B. Xue, R. X. Zhao, Y. Wang, K. D. Werle, R. Cui, J. Liang and Z. X. Xu, *Gastroenterology*, 2014, **146**, 188–199.
97. H. T. Nguyen, G. Dalmasso, S. Muller, J. Carriere, F. Seibold and A. Darfeuille-Michaud, *Gastroenterology*, 2014, **146**, 508–519.
98. K. K. Patel and T. S. Stappenbeck, *Annu. Rev. Physiol.*, 2013, **75**, 241–262.
99. S. Liu, A. P. da Cunha, R. M. Rezende, R. Cialic, Z. Wei, L. Bry, L. E. Comstock, R. Gandhi and H. L. Weiner, *Cell Host Microbe*, 2016, **19**, 32–43.
100. C. A. Dendrou, L. Fugger and M. A. Friese, *Nat. Rev. Immunol.*, 2015, **15**, 545–558.
101. N. M. Baulina, O. G. Kulakova and O. O. Favorova, *Acta Nat.*, 2016, **8**, 21–33.
102. O. de Faria Jr, C. S. Moore, T. E. Kennedy, J. P. Antel, A. Bar-Or and A. S. Dhaunchak, *Front. Genet.*, 2012, **3**, 311.
103. K. U. Tufekci, M. G. Oner, S. Genc and K. Genc, *Autoimmune Dis.*, 2010, **2011**, 807426.
104. A. Keller, P. Leidinger, J. Lange, A. Borries, H. Schroers, M. Scheffler, H. P. Lenhof, K. Ruprecht and E. Meese, *PLoS One*, 2009, **4**, e7440.
105. M. B. Cox, M. J. Cairns, K. S. Gandhi, A. P. Carroll, S. Moscovis, G. J. Stewart, S. Broadley, R. J. Scott, D. R. Booth, J. Lechner-Scott and ANZgene Multiple Sclerosis Genetics Consortium, *PLoS One*, 2010, **5**, e12132.
106. A. Keller, P. Leidinger, F. Steinmeyer, C. Stahler, A. Franke, G. Hemmrich-Stanisak, A. Kappel, I. Wright, J. Dorr, F. Paul, R. Diem, B. Tocariu-Krick, B. Meder, C. Backes, E. Meese and K. Ruprecht, *Mult. Scler.*, 2014, **20**, 295–303.
107. D. Otaegui, S. E. Baranzini, R. Armananzas, B. Calvo, M. Munoz-Culla, P. Khankhanian, I. Inza, J. A. Lozano, T. Castillo-Trivino, A. Asensio, J. Olaskoaga and A. Lopez de Munain, *PLoS One*, 2009, **4**, e6309.

108. G. De Santis, M. Ferracin, A. Biondani, L. Caniatti, M. Rosaria Tola, M. Castellazzi, B. Zagatti, L. Battistini, G. Borsellino, E. Fainardi, R. Gavioli, M. Negrini, R. Furlan and E. Granieri, *J. Neuroimmunol.*, 2010, **226**, 165–171.
109. C. Sievers, M. Meira, F. Hoffmann, P. Fontoura, L. Kappos and R. L. Lindberg, *Clin. Immunol.*, 2012, **144**, 70–79.
110. S. Q. Liu, S. Jiang, C. Li, B. Zhang and Q. J. Li, *J. Biol. Chem.*, 2014, **289**, 12446–12456.
111. H. Y. Yang, J. Barbi, C. Y. Wu, Y. Zheng, P. D. Vignali, X. Wu, J. H. Tao, B. V. Park, S. Bandara, L. Novack, X. Ni, X. Yang, K. Y. Chang, R. C. Wu, J. Zhang, C. W. Yang, D. M. Pardoll, H. Li and F. Pan, *Immunity*, 2016, **45**, 83–93.
112. E. Zhu, X. Wang, B. Zheng, Q. Wang, J. Hao, S. Chen, Q. Zhao, L. Zhao, Z. Wu and Z. Yin, *J. Immunol.*, 2014, **192**, 5599–5609.
113. A. Junker, M. Krumbholz, S. Eisele, H. Mohan, F. Augstein, R. Bittner, H. Lassmann, H. Wekerle, R. Hohlfeld and E. Meinl, *Brain*, 2009, **132**, 3342–3352.
114. C. S. Moore, V. T. Rao, B. A. Durafourt, B. J. Bedell, S. K. Ludwin, A. Bar-Or and J. P. Antel, *Ann. Neurol.*, 2013, **74**, 709–720.
115. A. Waschbisch, M. Atiya, R. A. Linker, S. Potapov, S. Schwab and T. Derfuss, *PLoS One*, 2011, **6**, e24604.
116. T. M. Escobar, C. Kanellopoulou, D. G. Kugler, G. Kilaru, C. K. Nguyen, V. Nagarajan, R. K. Bhairavabhotla, D. Northrup, R. Zahr, P. Burr, X. Liu, K. Zhao, A. Sher, D. Jankovic, J. Zhu and S. A. Muljo, *Immunity*, 2014, **40**, 865–879.
117. S. Y. Na, M. J. Park, S. Park and E. S. Lee, *Clin. Exp. Rheumatol.*, 2016, **34**, S56–S63.
118. I. Dunand-Sauthier, M. L. Santiago-Raber, L. Capponi, C. E. Vejnar, O. Schaad, M. Irla, Q. Seguin-Estevez, P. Descombes, E. M. Zdobnov, H. Acha-Orbea and W. Reith, *Blood*, 2011, **117**, 4490–4500.
119. C. Fenoglio, C. Cantoni, M. De Riz, E. Ridolfi, F. Cortini, M. Serpente, C. Villa, C. Comi, F. Monaco, L. Mellesi, S. Valzelli, N. Bresolin, D. Galimberti and E. Scarpini, *Neurosci. Lett.*, 2011, **504**, 9–12.
120. B. Li, X. Wang, I. Y. Choi, Y. C. Wang, S. Liu, A. T. Pham, H. Moon, D. J. Smith, D. S. Rao, M. P. Boldin and L. Yang, *J. Clin. Invest.*, 2017, **127**, 3702–3716.
121. L. F. Lu, M. P. Boldin, A. Chaudhry, L. L. Lin, K. D. Taganov, T. Hanada, A. Yoshimura, D. Baltimore and A. Y. Rudensky, *Cell*, 2010, **142**, 914–929.
122. J. G. Cui, Y. Y. Li, Y. Zhao, S. Bhattacharjee and W. J. Lukiw, *J. Biol. Chem.*, 2010, **285**, 38951–38960.
123. C. Du, C. Liu, J. Kang, G. Zhao, Z. Ye, S. Huang, Z. Li, Z. Wu and G. Pei, *Nat. Immunol.*, 2009, **10**, 1252–1259.
124. M. A. Honardoost, A. Kiani-Esfahani, K. Ghaedi, M. Etemadifar and M. Salehi, *Gene*, 2014, **544**, 128–133.
125. R. L. Lindberg, F. Hoffmann, M. Mehling, J. Kuhle and L. Kappos, *Eur. J. Immunol.*, 2010, **40**, 888–898.

126. X. Ma, J. Zhou, Y. Zhong, L. Jiang, P. Mu, Y. Li, N. Singh, M. Nagarkatti and P. Nagarkatti, *Int. J. Med. Sci.*, 2014, **11**, 810–818.
127. K. Venken, N. Hellings, M. Thewissen, V. Somers, K. Hensen, J. L. Rummens, R. Medaer, R. Hupperts and P. Stinissen, *Immunology*, 2008, **123**, 79–89.
128. D. Li, C. Kong, A. Tsun, C. Chen, H. Song, G. Shi, W. Pan, D. Dai, N. Shen and B. Li, *Sci. Rep.*, 2015, **5**, 14615.
129. W. Pan, S. Zhu, D. Dai, Z. Liu, D. Li, B. Li, N. Gagliani, Y. Zheng, Y. Tang, M. T. Weirauch, X. Chen, W. Zhu, Y. Wang, B. Chen, Y. Qian, Y. Chen, J. Fang, R. Herbst, L. Richman, B. Jallal, J. B. Harley, R. A. Flavell, Y. Yao and N. Shen, *Nat. Commun.*, 2015, **6**, 7096.
130. T. J. Liang and C. Y. Qin, *APMIS*, 2009, **117**, 635–643.
131. S. A. Schaffert, C. Loh, S. Wang, C. P. Arnold, R. C. Axtell, E. W. Newell, G. Nolan, K. M. Ansel, M. M. Davis, L. Steinman and C. Z. Chen, *J. Immunol.*, 2015, **195**, 1470–1479.
132. G. A. Kozloski, X. Jiang, S. Bhatt, J. Ruiz, F. Vega, R. Shaknovich, A. Melnick and I. S. Lossos, *Blood*, 2016, **127**, 2856–2866.
133. C. Cantoni, F. Cignarella, L. Ghezzi, B. Mikesell, B. Bollman, M. M. Berrien-Elliott, A. R. Ireland, T. A. Fehniger, G. F. Wu and L. Piccio, *Acta Neuropathol.*, 2017, **133**, 61–77.
134. T. Satoorian, B. Li, X. Tang, J. Xiao, W. Xing, W. Shi, K. H. Lau, D. J. Baylink and X. Qin, *Immunology*, 2016, **148**, 326–338.
135. I. Ifergan, S. Chen, B. Zhang and S. D. Miller, *J. Immunol.*, 2016, **196**, 1455–1459.
136. J. C. Dugas, T. L. Cuellar, A. Scholze, B. Ason, A. Ibrahim, B. Emery, J. L. Zamanian, L. C. Foo, M. T. McManus and B. A. Barres, *Neuron*, 2010, **65**, 597–611.
137. X. Zhao, X. He, X. Han, Y. Yu, F. Ye, Y. Chen, T. Hoang, X. Xu, Q. S. Mi, M. Xin, F. Wang, B. Appel and Q. R. Lu, *Neuron*, 2010, **65**, 612–626.
138. D. Lecca, D. Marangon, G. T. Coppolino, A. M. Mendez, A. Finardi, G. D. Costa, V. Martinelli, R. Furlan and M. P. Abbracchio, *Sci. Rep.*, 2016, **6**, 34503.
139. Y. Nakamachi, S. Kawano, M. Takenokuchi, K. Nishimura, Y. Sakai, T. Chin, R. Saura, M. Kurosaka and S. Kumagai, *Arthritis Rheum.*, 2009, **60**, 1294–1304.
140. E. Salehi, R. Eftekhari, M. Oraei, A. Gharib and K. Bidad, *Clin. Rheumatol.*, 2015, **34**, 615–628.
141. J. Li, Y. Wan, Q. Guo, L. Zou, J. Zhang, Y. Fang, J. Zhang, J. Zhang, X. Fu, H. Liu, L. Lu and Y. Wu, *Arthritis Res. Ther.*, 2010, **12**, R81.
142. T. Nakasa, S. Miyaki, A. Okubo, M. Hashimoto, K. Nishida, M. Ochi and H. Asahara, *Arthritis Rheum.*, 2008, **58**, 1284–1292.
143. T. Nakasa, H. Shibuya, Y. Nagata, T. Niimoto and M. Ochi, *Arthritis Rheum.*, 2011, **63**, 1582–1590.
144. T. Niimoto, T. Nakasa, M. Ishikawa, A. Okuhara, B. Izumi, M. Deie, O. Suzuki, N. Adachi and M. Ochi, *BMC Musculoskeletal Disord.*, 2010, **11**, 209.

145. K. M. Pauley, M. Satoh, A. L. Chan, M. R. Bubb, W. H. Reeves and E. K. Chan, *Arthritis Res. Ther.*, 2008, **10**, R101.
146. J. Zhang, T. Bardos, K. Mikecz, A. Finnegan and T. T. Glant, *J. Immunol.*, 2001, **166**, 4981–4986.
147. M. Kurowska-Stolarska, S. Alivernini, L. E. Ballantine, D. L. Asquith, N. L. Millar, D. S. Gilchrist, J. Reilly, M. Ierna, A. R. Fraser, B. Stolarski, C. McSharry, A. J. Hueber, D. Baxter, J. Hunter, S. Gay, F. Y. Liew and I. B. McInnes, *Proc. Natl. Acad. Sci. U. S. A.*, 2011, **108**, 11193–11198.
148. L. Long, P. Yu, Y. Liu, S. Wang, R. Li, J. Shi, X. Zhang, Y. Li, X. Sun, B. Zhou, L. Cui and Z. Li, *Clin. Dev. Immunol.*, 2013, **2013**, 296139.
149. K. Smigielska-Czepiel, A. van den Berg, P. Jellema, R. J. van der Lei, J. Bijzet, J. Kluiver, A. M. Boots, E. Brouwer and B. J. Kroesen, *Genes Immun.*, 2014, **15**, 115–125.
150. J. Stanczyk, D. M. Pedrioli, F. Brentano, O. Sanchez-Pernaute, C. Kolling, R. E. Gay, M. Detmar, S. Gay and D. Kyburz, *Arthritis Rheum.*, 2008, **58**, 1001–1009.
151. S. Bluml, M. Bonelli, B. Niederreiter, A. Puchner, G. Mayr, S. Hayer, M. I. Koenders, W. B. van den Berg, J. Smolen and K. Redlich, *Arthritis Rheum.*, 2011, **63**, 1281–1288.
152. Y. T. Li, S. Y. Chen, C. R. Wang, M. F. Liu, C. C. Lin, I. M. Jou, A. L. Shiau and C. L. Wu, *Arthritis Rheum.*, 2012, **64**, 3240–3245.
153. M. C. Lu, C. L. Yu, H. C. Chen, H. C. Yu, H. B. Huang and N. S. Lai, *Clin. Exp. Immunol.*, 2014, **177**, 641–651.
154. J. Stanczyk, C. Ospelt, E. Karouzakis, A. Filer, K. Raza, C. Kolling, R. Gay, C. D. Buckley, P. P. Tak, S. Gay and D. Kyburz, *Arthritis Rheum.*, 2011, **63**, 373–381.
155. S. Zhu, W. Pan, X. Song, Y. Liu, X. Shao, Y. Tang, D. Liang, D. He, H. Wang, W. Liu, Y. Shi, J. B. Harley, N. Shen and Y. Qian, *Nat. Med.*, 2012, **18**, 1077–1086.
156. F. Niederer, M. Trenkmann, C. Ospelt, E. Karouzakis, M. Neidhart, J. Stanczyk, C. Kolling, R. E. Gay, M. Detmar, S. Gay, A. Jungel and D. Kyburz, *Arthritis Rheum.*, 2012, **64**, 1771–1779.
157. A. Kaul, C. Gordon, M. K. Crow, Z. Touma, M. B. Urowitz, R. van Vollenhoven, G. Ruiz-Irastorza and G. Hughes, *Nat. Rev. Dis. Primers*, 2016, **2**, 16039.
158. X. Le, X. Yu and N. Shen, *Curr. Opin. Rheumatol.*, 2017, **29**, 450–457.
159. Y. Tang, X. Luo, H. Cui, X. Ni, M. Yuan, Y. Guo, X. Huang, H. Zhou, N. de Vries, P. P. Tak, S. Chen and N. Shen, *Arthritis Rheum.*, 2009, **60**, 1065–1075.
160. K. B. Elkon and A. Wiedeman, *Curr. Opin. Rheumatol.*, 2012, **24**, 499–505.
161. V. Sisirak, D. Ganguly, K. L. Lewis, C. Couillault, L. Tanaka, S. Bolland, V. D'Agati, K. B. Elkon and B. Reizis, *J. Exp. Med.*, 2014, **211**, 1969–1976.
162. H. Zhou, X. Huang, H. Cui, X. Luo, Y. Tang, S. Chen, L. Wu and N. Shen, *Blood*, 2010, **116**, 5885–5894.

163. S. Zhou, Y. Wang, Y. Meng, C. Xiao, Z. Liu, P. Brohawn, B. W. Higgs, B. Jallal, Q. Jia, B. Qu, X. Huang, Y. Tang, Y. Yao, J. B. Harley and N. Shen, *Arthritis Rheumatol.*, 2016, **68**, 953–964.
164. Y. Dorsett, K. M. McBride, M. Jankovic, A. Gazumyan, T. H. Thai, D. F. Robbiani, M. Di Virgilio, B. Reina San-Martin, G. Heidkamp, T. A. Schwickert, T. Eisenreich, K. Rajewsky and M. C. Nussenzweig, *Immunity*, 2008, **28**, 630–638.
165. G. Teng, P. Hakimpour, P. Landgraf, A. Rice, T. Tuschl, R. Casellas and F. N. Papavasiliou, *Immunity*, 2008, **28**, 621–629.
166. G. A. Calin and C. M. Croce, *Nat. Rev. Cancer*, 2006, **6**, 857–866.
167. T. H. Thai, H. C. Patterson, D. H. Pham, K. Kis-Toth, D. A. Kaminski and G. C. Tsokos, *Proc. Natl. Acad. Sci. U. S. A.*, 2013, **110**, 20194–20199.
168. M. Turner and E. Vigorito, *Biochem. Soc. Trans.*, 2008, **36**, 531–533.
169. E. Vigorito, K. L. Perks, C. Abreu-Goodger, S. Bunting, Z. Xiang, S. Kohlhaas, P. P. Das, E. A. Miska, A. Rodriguez, A. Bradley, K. G. Smith, C. Rada, A. J. Enright, K. M. Toellner, I. C. Maclennan and M. Turner, *Immunity*, 2007, **27**, 847–859.
170. W. H. Liu, S. G. Kang, Z. Huang, C. J. Wu, H. Y. Jin, C. J. Maine, Y. Liu, J. Shepherd, M. Sabouri-Ghomi, A. Gonzalez-Martin, S. Xu, A. Hoffmann, Y. Zheng, L. F. Lu, N. Xiao, G. Fu and C. Xiao, *J. Exp. Med.*, 2016, **213**, 1901–1919.
171. W. Pan, S. Zhu, M. Yuan, H. Cui, L. Wang, X. Luo, J. Li, H. Zhou, Y. Tang and N. Shen, *J. Immunol.*, 2010, **184**, 6773–6781.
172. E. Stagakis, G. Bertsias, P. Verginis, M. Nakou, M. Hatziapostolou, H. Kritikos, D. Iliopoulos and D. T. Boumpas, *Ann. Rheum. Dis.*, 2011, **70**, 1496–1506.
173. B. G. Garchow, O. Bartulos Encinas, Y. T. Leung, P. Y. Tsao, R. A. Eisenberg, R. Caricchio, S. Obad, A. Petri, S. Kauppinen and M. Kiriakidou, *EMBO Mol. Med.*, 2011, **3**, 605–615.
174. X. G. Sun, J. H. Tao, N. Xiang, X. M. Li, G. S. Wang, X. Fang, C. Dai, M. Zhang, Z. Chen and X. P. Li, *Inflammation*, 2016, **39**, 822–829.
175. Y. Xia, J. H. Tao, X. Fang, N. Xiang, X. J. Dai, L. Jin, X. M. Li, Y. P. Wang and X. P. Li, *Mol. Ther.--Nucleic Acids*, 2018, **11**, 284–291.
176. L. Russell, S. John, J. Cullen, W. Luo, M. J. Shlomchik and L. A. Garrett-Sinha, *J. Immunol.*, 2015, **195**, 3574–3583.
177. A. Gonzalez-Martin, B. D. Adams, M. Lai, J. Shepherd, M. Salvador-Bernaldez, J. M. Salvador, J. Lu, D. Nemazee and C. Xiao, *Nat. Immunol.*, 2016, **17**, 433–440.
178. W. Wang, S. Mou, L. Wang, M. Zhang, X. Shao, W. Fang, R. Lu, C. Qi, Z. Fan, Q. Cao, Q. Wang, Y. Fang and Z. Ni, *Sci. Rep.*, 2015, **5**, 12644.
179. Y. J. Su, I. C. Lin, L. Wang, C. H. Lu, Y. L. Huang and H. C. Kuo, *Oncotarget*, 2018, **9**, 27911–27919.
180. R. Park, W. J. Lee and J. D. Ji, *Autoimmunity*, 2016, **49**, 451–458.
181. J. J. Karrich, L. C. Jachimowski, M. Libouban, A. Iyer, K. Brandwijk, E. W. Taanman-Kueter, M. Nagasawa, E. C. de Jong, C. H. Uittenbogaart and B. Blom, *Blood*, 2013, **122**, 3001–3009.

182. Y. Pan, T. Jia, Y. Zhang, K. Zhang, R. Zhang, J. Li and L. Wang, *Int. J. Nanomed.*, 2012, **7**, 5957–5967.
183. T. S. Assmann, M. Recamonde-Mendoza, B. M. De Souza and D. Crispim, *Endocr. Connect.*, 2017, **6**, 773–790.
184. J. E. Hawkes, G. H. Nguyen, M. Fujita, S. R. Florell, K. Callis Duffin, G. G. Krueger and R. M. O'Connell, *J. Invest. Dermatol.*, 2016, **136**, 365–371.
185. J. Guinea-Viniegra, M. Jimenez, H. B. Schonthaler, R. Navarro, Y. Delgado, M. J. Concha-Garzon, E. Tschachler, S. Obad, E. Dauden and E. F. Wagner, *Sci. Transl. Med.*, 2014, **6**, 225re221.
186. T. Maes, F. A. Cobos, F. Schleich, V. Sorbello, M. Henket, K. De Preter, K. R. Bracke, G. Conickx, C. Mesnil, J. Vandesompele, L. Lahousse, F. Bureau, P. Mestdagh, G. F. Joos, F. L. Ricciardolo, G. G. Brusselle and R. Louis, *J. Allergy Clin. Immunol.*, 2016, **137**, 1433–1446.
187. X. B. Wu, M. Y. Wang, H. Y. Zhu, S. Q. Tang, Y. D. You and Y. Q. Xie, *Int. J. Clin. Exp. Med.*, 2014, **7**, 1307–1312.
188. C. Chakraborty, A. R. Sharma, G. Sharma, C. G. P. Doss and S. S. Lee, *Mol. Ther.--Nucleic Acids*, 2017, **8**, 132–143.
189. S. Thakral and K. Ghoshal, *Curr. Gene Ther.*, 2015, **15**, 142–150.
190. E. Slabakova, Z. Culig, J. Remsik and K. Soucek, *Cell Death Dis.*, 2017, **8**, e3100.
191. N. Yang, *Int. J. Pharm. Invest.*, 2015, **5**, 179–181.
192. R. Rupaimoole and F. J. Slack, *Nat. Rev. Drug Discovery*, 2017, **16**, 203–222.

CHAPTER 7

MicroRNAs in Ischemic Stroke, Hemorrhagic Stroke, and Traumatic Brain Injury

PHILIP V. PEPLOW*[a] AND BRIDGET MARTINEZ[b,c,d]

[a]Department of Anatomy, University of Otago, Dunedin, New Zealand; [b]Department of Molecular and Cellular Biology, University of California, Merced, Merced, California, USA; [c]Department of Medicine, St. Georges University School of Medicine, Grenada; [d]Department of Physics and Engineering, Los Alamos National Laboratory, Los Alamos, New Mexico, USA
*E-mail: phil.peplow@otago.ac.nz

7.1 Introduction

Stroke is one of the leading causes of death worldwide and a major cause of long-term disability. In the USA, nearly 75% of all strokes occur in people over the age of 65.[1] However, recent studies have indicated an increasing incidence of ischemic stroke in adults younger than 50–55 years.[2,3] Approximately 80% of all strokes are ischemic due to occlusion of a vessel, and comprise two main types, thombotic and embolic. A thombotic stroke or infarction occurs when a clot forms in an artery supplying the brain and accounts for approximately 50% of all strokes. An embolic stroke is caused by a clot formed elsewhere in the body and which is subsequently transported though the bloodstream to the brain.

Drug Discovery Series No. 69
MicroRNAs in Diseases and Disorders: Emerging Therapeutic Targets
Edited by Philip V. Peplow, Bridget Martinez, George A. Calin and Aurora Esquela-Kerscher

Published by the Royal Society of Chemistry, www.rsc.org

Hemorrhagic strokes account for approximately 10–20% of all strokes; 8–15% in Western countries like USA, UK and Australia; and 18–24% in Japan and Korea.[4] They are divided into categories depending on the site and cause of bleeding. In an intracerebral hemorrhage (ICH), bleeding occurs from a ruptured blood vessel within the brain; hypertension, current smoking, excessive alcohol intake, hypocholesterolemia, drugs, and advanced age are all important risk factors.[4] ICH is more common in Asian patients, advanced age, males, and low- and middle-income countries and the case fatality rate of ICH is high: 40% at 1 month and 54% at 1 year, and only 12–39% of survivors can achieve long-term functional independence.[4] Ischemic strokes can convert to an ICH shortly afterwards in the form of hemorrhagic transformation,[5] and this progression is especially common for embolic strokes that are related to infective endocarditis.[6] A subarachnoid hemorrhage (SAH) involves bleeding from a damaged blood vessel causing blood to accumulate at the surface of the brain. As blood flows into the cerebral arteries, the bleeding increases pressure on the brain. In the days immediately following the bleeding, irritation from the blood clot around the brain can cause arteries supplying the brain and close to this area to go into spasm, which can damage the brain tissue. Most often, a SAH occurs because of a leaking saccular aneurysm. Hemorrhagic stroke is life theatening, with up to 50% of all people with ICH dying, many within the first 2 days. For those who survive a brain hemorrhage, recovery is slow.[7]

Traumatic brain injury (TBI) due to a violent blow to the head is the leading cause of death and disability in adults under the age of 45 years in Western countries,[8–10] with a very high number of brain injuries resulting from motor vehicle accidents and sporting activities. In addition, there has been a marked increase in the incidence of fall-related TBI in persons aged >75 years.[11] The severity of TBI is determined by the nature, speed, and location of the impact, and by complications such as hypotension, intracranial hemorrhage, or increased intracranial pressure. TBI is characterized by primary damage to the brain from the external mechanical force and by subsequent secondary injury due to various molecular and pathophysiological responses that eventually lead to neuronal cell death. These responses include brain edema, intracranial hypertension, and subsequent neurological dysfunction.[12] Secondary brain injury events may occur minutes, hours, or even days after the trauma.[10]

7.2 Diagnosis and Treatment of Stroke

The only United States Federal Drug Administration-approved intervention for acute ischemic stroke is intravenous administration of tissue plasminogen activator (tPA), which acts by dissolving the blood clot (recanalization), thereby improving blood flow to the part of the brain being deprived of blood. Thombolysis and/or mechanical thombectomy may be considered in patients with severe neurological symptoms and early presentation within 4.5 h of symptom onset. However, many stroke patients are not treated by intravenous thombolysis because of the narrow time window for treatment. In patients with minor stroke, while the immediate brain damage and associated symptoms are limited, there is a significant risk of a major stroke within 90 days,[13] even when taking aspirin antiplatelet therapy.[14] Early intervention in stroke is desirable, to limit the serious consequence of a major stroke and to lower the risk of a minor stroke progressing to a seriously disabling stroke. Surgical removal of the hematoma as an early-stage treatment for ICH may improve long-term prognosis, but no effective targeted therapy exists for hemorrhagic stroke.[15]

Proper medical treatment of a stroke patient relies on accurate and rapid differentiation between ischemic and hemorrhagic stroke. Ischemic and hemorrhagic stroke have completely divergent therapeutic options, and tPA treatment can convert ischemic stroke to hemorrhagic stroke. Clinically, it is therefore crucial to monitor and distinguish between ischemic and hemorrhage stroke within the first week of symptom onset to prevent adverse outcome. In current practice, diagnosis of ischemic *versus* hemorrhage stroke is performed using computed tomography (CT) or magnetic resonance imaging (MRI) scans. While imaging technology is sensitive, it is time consuming and difficult to implement in an extremely sick and immobile patient; it also requires the availability of expensive equipment, highly skilled radiological expertise to interpret the results, and repeat radiation exposure. There is a need for a reliable, relatively inexpensive method for differentiating between ischemic and hemorrhagic stroke in patients—potentially a point-of-care assay that can be performed daily within the first week of stroke onset.

7.3 Diagnosis and Treatment of TBI

The initial severity and prognosis of TBI in humans is classified on the Glasgow Coma Scale (GCS) which is based on thee tests: ocular, verbal, and motor responses. The sum of the thee values is the GCS score. The injury is classified as severe (GCS 3–8), moderate (GCS 9–12), or mild (GCS 13–15).[16] The GCS covers a broad severity of symptoms from subtle to more serious conditions in some cases accompanied with temporary loss of consciousness.[17] Sports-related and other minor traumatic brain injuries with GCS scores in the 13–15 range occur in 1.6–3.8 million young healthy individuals annually.[18] Mild TBI is characterized by loss of consciousness for <30 min and post-traumatic amnesia for <24 h and accounts for ~75–90% of all TBI cases.[19] CT and MRI have limited sensitivity to detect mild TBI.[20,21] Mild TBI is often unnoticed or misdiagnosed due to the absence of specific diagnostic markers. Moreover, treatment options for secondary brain damage following TBI have failed to show efficacy in clinical trials,[22] thereby emphasizing a need for new therapeutic targets.

It is helpful to distinguish between traumatic injuries involving penetration of the brain ("penetrating" injuries) and those not penetrating the brain (often referred to as "closed" head injuries). The profile of injury involving penetration of the brain depends on the location and trajectory of the object involved. In these injuries damage typically results from displacement or destruction of brain tissue by the object, fragmentation and deposition of bone or an object within brain tissue, or introduction of potential infectious material on the object.[23]

7.4 Changes in Cellular Structure and Function in Ischemic Stroke

Many changes take place at the cellular level when a stroke occurs. The obstruction of a major cerebral vessel (usually middle cerebral artery obstruction, MCAO), if not resolved within a short period of time, will lead to a core of severely ischemic brain tissue that may not be salvageable. The ultimate size of the brain infarct depends on the penumbra, a zone of tissue around the core of the infarct where neuronal electrophysiology is not compromised and blood flow is still maintained above a neuronal disabling level (*i.e.*, the critical 20–25% of normal blood flow). If blood flow in the penumbral zone decreases below this critical level and/or energy requirements are exceeded, the infarct zone will expand. Decreased blood flow leads to a reduction in phosphocreatinine and adenosine triphosphate (ATP) and if ischemia is prolonged, the energy depletion will be sufficient to cause severe impairment of cellular function by disruption of ATP-dependent processes. The disruption of ionic gradients across excitable (neuronal) and nonexcitable (glial) membranes owing to loss of ATP is characterized by efflux of K^+ from the cells, cellular depolarization, and influx of Na^+, Cl^-, and Ca^{2+} into the cells.[24,25] In the phase of increased extracellular K^+ together with a decrease in pH, ATP stores are rapidly depleted and ultimately result in marked changes in ion conductance. The increase in K^+ can reach levels sufficient to release neurotransmitters such as glutamate, which in turn will stimulate Na^+/Ca^{2+} channels coupled to the *N*-methyl-D-aspartate receptor; these will further lead to Na^+, Cl^-, and H_2O accumulation, cell swelling, and cytotoxic edema. Transient extracellular K^+-induced depolarization can also contribute to the expansion of the neuronal infarct. Peri-infarct depolarization produces disruption of ionic gradients and transmitter release, with the associated accumulation of Ca^{2+} in the cells leading to rapid and extensive breakdown of phospholipids, proteins, and nucleic acids by activation of calcium-dependent phospholipases, proteases, and endonucleases. Depolarization causes an increase in intracellular Ca^{2+} and an increase in extracellular glutamate. Glutamate, an excitatory neurotransmitter implicated in ischemic neuronal damage, results in excitotoxicity in which excessive extracellular glutamate kills neurons though an increase in intracellular Ca^{2+}.[26]

Accumulation of products such as free fatty acids that are metabolized to toxic lipid peroxides further contribute to structural and functional alterations of the membrane and cell function. Free radicals, which are a group of highly reactive oxygen species, are formed during ischemia and cause considerable damage to lipids, DNA, and proteins, and contribute to the process of neuronal death. In addition, free radicals contribute to the breakdown of the blood–brain barrier and brain edema. Levels of free-radical scavenging enzymes (*e.g.*, superoxide dismutase) decrease during ischemia and nitric oxide levels are elevated. Nitric oxide produced primarily by neuronal and inducible nitric oxide synthases promotes neuronal damage after ischemia.[26]

Evidence indicates that inflammation and immune response play an important role in the outcome of ischemic stroke.[27] Inflammation after stroke involves leukocyte infiltration in brain parenchyma that contributes to cerebral damage. Peripherally derived mononuclear phagocytes, T-lymphocytes, natural killer cells, and polymorphonuclear leukocytes, which produce and secrete cytokines, can all contribute to central nervous system (CNS) inflammation and gliosis.[28] Blood-derived leukocytes and resident microglia are the more activated inflammatory cells, accumulating in the brain tissue after cerebral ischemia, leading to inflammatory injury.[29] Microglia, the major source of cytokines and other immune molecules of the CNS, are the first non-neuronal cells that respond to CNS injury, becoming phagocytic when fully activated by neuronal death. It has been shown that neuroinflammation (*e.g.*, induced by systemic injection of lipopolysaccharide) and cerebral ischemia (*e.g.*, MCAO) in mice induced two different types of astrocytes, termed "A1" and "A2", respectively.[30] Ischemia-induced A2 reactive astrocytes upregulate many neurotrophic factors that promote survival and growth of neurons.[31]

7.5 Pathophysiology of TBI

Cerebral ischemia is an important contributor to increased morbidity and mortality in patients with head injuries. Similar mechanisms in cerebral ischemia and brain trauma contribute to loss of cellular integrity and tissue destruction. Mechanisms of cellular damage include free radical production, oxidative stress, inflammation, excitotoxicity, and apoptosis. Blunt head trauma produces shear forces that result in primary membrane damage to neuronal cell bodies, white matter structures and vascular beds as well as secondary injury mechanisms. Severe cerebral ischemia leads to metabolic stress, ionic perturbations, and a complex cascade of biochemical and molecular events, ultimately causing neuronal death.[32]

Mitochondrial dysfunction may be central to the pathophysiology of TBI and can be recognized by changes in mitochondrial ultrastructure. Changes in mitochondrial morphology were found in surgically resected human TBI tissues from thee contiguous cortical injury zones. The zones were designated as injury center, periphery, and penumbra. The distribution of mitochondrial morphological phenotypes varied significantly between the thee injury zones and when compared with control cortical tissue. Within the early injury interval (<1 day post-injury), the central zone of injury exhibits predominantly reactive/degenerative and end-stage degenerative changes in mitochondrial morphology. In the central zone there is moderate to severe swelling of parenchymal cells and processes. When the GCS is low (GSC 3–5), total disruption of tissue architecture and presence of primarily end-stage degenerating mitochondria are characteristic for all injury zones. Within the intermediate post-injury interval (1–3 days post-injury), cellular edematous changes are maximal, with a peak at ~36–42 h post-injury. Within the late post-injury phase (≥3 days post-injury), normal-appearing mitochondria are infrequent, even within the penumbra zone. The central zone contains primarily end-stage degenerating mitochondria, and reactive/degenerating mitochondria are abundant in both the penumbra and periphery zones.[33]

7.6 miRNAs in Ischemic Stroke

MicroRNAs (miRNAs) are abundant, endogenous, short, noncoding RNAs that act as important post-transcriptional regulators of gene expression by binding to the 3′-untranslated region of their target messenger RNAs (mRNAs), thereby interfering with translation or causing destabilization or preferential cleavage of target mRNAs.[34,35] During the past decade, substantial knowledge has accumulated regarding the biogenesis of miRNAs, their molecular mechanisms, and functional roles in a variety of cellular contexts. Altered expression of certain miRNA molecules suggests that they could have a crucial regulatory role in various diseases and disorders. In animals, miRNAs are synthesized from primary miRNAs (pri-miRNA) in two stages by the action of two RNase III-type enzymes: Drosha in the nucleus and Dicer in the cytoplasm. The pri-miRNA is processed within the nucleus by Drosha to generate a precursor microRNA (pre-miRNA). The transport of pre-miRNAs to the cytoplasm is mediated by exportin-5 in the cytoplasm. In the cytoplasm, they are further processed by Dicer to generate mature miRNAs of ~22 nucleotides in length. The mature miRNAs are then loaded onto an Argonate protein to produce the effector RNA-induced silencing complex.[36] miRNA therapeutics have been investigated in recent preclinical and clinical trials, *e.g.*, for cancer, hepatitis C, and melanoma.[36]

As indicated above, miRNAs are a family of small, genome-encoded endogenous RNAs that are transcribed, but not translated into proteins. They serve essential roles in virtually every aspect of brain function, including neurogenesis, neural development, and cellular responses leading to changes in synaptic plasticity. They are also implicated in neurodegeneration and neurological disorders, in responses to hypoxia and ischemia, and in ischemic tolerance induced by ischemic preconditioning. Complex interplay among multiple pathways including excitotoxicity, mitochondrial dysfunction, ionic imbalance, oxidative stress, and inflammation are involved in the mechanism of cerebral ischemia/reperfusion injury. These processes lead to both necrotic death in the anoxic core and delayed apoptosis-mediated cell death in the surrounding penumbra. While the fate of neurons in the anoxic core is probably fixed relatively soon after the initial insult, the cells in the peri-ischemic penumbra represent targets for rescue from delayed cell death. In more recent developments, miRNA expression profiling has been examined in stroke, and these studies indicate that miRNAs have emerged as key mediators in ischemic stroke biology.[37]

Studies targeting miRNAs for therapies in cancer have established that circulating miRNAs are largely stable in blood,[38] indicating that they likely represent a sampling of cellular activity, a route of intercellular communication, and as diagnostic and prognostic biomarkers for cerebral injury.[39] It has been shown that serum or plasma miRNAs are indicative of the disease state.[40] Much interest has developed in the use of circulating cell-free miRNAs as novel markers in the clinical diagnosis of disease, especially in cancer.[41] Reviews of miRNA profiles in ischemic stroke have been presented by Martinez and Peplow[37] and Chandran *et al.*[42]

miRNAs that have been identified in studies of acute and chonic phase of ischemic stroke are presented in Table 7.1, which summarizes the subjects and age, samples and time of collection, and alterations in levels of miRNAs. The studies have been conducted with human patients and experimental animal models of cerebral ischemia.

7.6.1 Human Studies for miRNAs and Ischemic Stroke

Four large-scale human studies have been described to determine a role for miRNAs during ischemic stroke. Yang *et al.*[43] reported that the circulating levels of miR-107, miR-128b, and miR-153 were significantly elevated in ischemic stroke patients (age 61 ± 11 years; 68% male) in comparison to healthy volunteers (age 56 ± 4 years; 60% male), with the peripheral blood samples collected from patients within 24 h of hospital admission. The circulating miRNAs were only measured in the acute phase, and although the impact of age, smoking, and hyperlipidemia on these was ruled out, it did not rule out the interference of other factors such as hypertension and diabetes mellitus. Additionally, while these circulating miRNAs might be used as potential non-invasive biomarkers for the diagnosis of ischemic stroke, the prognostic values of the circulating miRNAs for ischemic stroke could not be evaluated without the data at the chronic phase of stroke and the functional outcomes. Jia *et al.*[44] examined nine previously reported stroke-associated miRNAs (miR-21, miR-23a, miR-29b, miR-124, miR-145, miR-210, miR-221, miR-223, and miR-483-5p) in 146 acute ischemic stroke patients and 96 control participants without prior history of stroke, myocardial infarction, or peripheral vascular disease. Blood was collected from each patient within 24 h of stroke onset or control group. Serum miR-145 was significantly increased and serum miR-23a and miR-221 were decreased in patients. Serum miR-145 was strong positively correlated with plasma high-sensitivity C-reactive protein and moderate positively correlated with serum interleukin (IL)-6. Plasma high-sensitivity C-reactive protein and IL-6 are pro-inflammation markers in brain injury. Another study by

Table 7.1 MicroRNA (miRNA) alterations in ischemic stroke of human patients and animal models.[a]

Reference	Subjects and age	Sample and time	miRNAs increased	miRNAs decreased
43	Human, 114 patients with IS, age 61 ± 11 years; 68% M	Blood collected within 24 h of hospital admission	miR-107, -128b, -153 (acute phase)	
50	Human, 5 patients with IS, age 50–83 years, 40% M	Blood, time of collection not reported	miR-145	miR-122
47	Human, 30 patients with IS acute phase age 63 ± 12 years, 53% M; 30 patients with IS subacute phase, age 60 ± 10 years, 50% M	Blood collected at 1–5 days and 6–30 days from stroke	miR-146a (subacute phase)	miR-146a (acute phase), mIR-185 (acute and sub-acute phases)
44	Human, 146 patients with IS, age 67 ± 14 years, 77% M	Blood collected within 24 h of stroke onset	miR-145 (acute phase)	miR-23a, miR-221
45	Human, 169 patients with IS, mean age 54 years	Blood collected within 1–7 days after stroke	miR-125b-2*, -27a*, -422a, -488, -627 (acute phase)	
49	Human, 24 patients with IS, age 62 ± 8 years, 50% M	Blood collected 28 h from stroke onset	miR-363, miR-487b (acute phase)	miR-122, -148a, -*let-7i*, -19a, -320d, -4429 (acute phase)
51	Human, patients with IS, age 18–49 years	Blood collected between 2 and 24 months from stroke onset		miR-25*, -34b, -483-5p, -498
48	Human, 32 patients with IS, age 18–49 years	Blood, time of collection not reported	miR-145	
46	Human, 112 patients with IS, mean age 68 years, 70% M	Blood collected within 3, 7 and 14 days after stroke		miR-210

52	Human, stroke patients, age 18–49 years	Blood collected within 6–18 months from index stroke	miR-130b, -29b, -301a, -339-5p, -532-5p, -634, -886-5p in small artery stroke, miR-25*	miR-25*
100	Rat, Sprague Dawley strain, M, subjected to MCAO	Blood collected after MCAO		miR-122
45	Rat, Wistar strain, adult M, subjected to MCAO	Blood and brain samples collected at 0–72 h after MCAO	miR-125b-2*, -27a*, -422a, -488, -627 (acute phase)	
55	Mouse, C57BL/6 strain, M, MCAO for 15 min	Brain	miR-200, miR-182	
53	Rat, Sprague Dawley strain, adult M, subjected to MCAO	Blood, brain samples collected at 24 and 48 h of reperfusion	Blood: miR-19b, -290, -292-5p (at 24 h reperfusion) Brain: miR-138, -145 (at 24 and 48 h), *let-7* family, miR-150, -125a (at 48 h)	Blood: miR-103, -107, -191 (at 24 and 48 h)
54	Rat SH, adult M, subjected to MCAO	Brain samples collected, at 3–72 h reperfusion	miR-206, -145, -290	miR-218, -27a, -137

[a]IS: ischemic stroke; M: male; MCAO: middle cerebral artery occlusion; SH: spontaneously hypertensive.

Sepramaniam *et al.*[45] has shown that the highest expression of circulating miR-125b-2*, miR-27a*, miR-422a, miR-488, and miR-627 occurred within the acute phase of ischemic stroke irrespective of age or severity or confounding metabolic complications, with blood collected from patients (age 60 ± 1 year) within 1–7 days of stroke. Patients from outpatient clinics in recovery phases 2–24 months from stroke onset were also included in this study. Cluster analysis showed segregation among patients within 6 and 24 months of recovery and these clusters were distinct from acute patients (within 1–7 days) indicating temporal regulation of miRNAs during the progression of stroke pathogenesis and with the "recovery" patients clustered closer with the healthy controls, suggesting a tendency for the miRNA expression to return to normalcy. A study by Zeng *et al.*[46] showed that miR-210 measured in blood from ischemic stroke patients within 3, 7, and 14 days after stroke was significantly decreased compared to controls, especially at 7 and 14 days after stroke onset. There were no changes in miR-210 for different stroke subtypes such as large artery atherosclerotic stroke, cardioembolic stroke, and small artery lacunar stroke. The miR-210 level in stroke patients with good outcome (modified Rankin Scale, mRS ≤2) was significantly higher than patients with poor outcome (mRS >2). A positive correlation was found between blood and brain miR-210 in MCAO stroke mice. This study suggests that miR-210 is a sensitive biomarker for diagnosis and prognosis in acute cerebral ischemia.

Several smaller-scale studies have also been reported. Li *et al.*[47] examined the levels of thee miRNAs, miR-185, miR-146a, and miR-145, in the blood of ischemic stroke patients aged ≥50 years in the acute (1–5 days) or subacute (6–30 days) phase. More of these patients had smoking and hypertension as risk factors when compared to healthy controls. The miR-145 levels in stroke patients were similar to controls, whereas miR-146a and miR-185 levels were downregulated compared with controls. The miR-146a levels were downregulated in the acute phase but upregulated in the subacute phase. For miR-185, the levels were downregulated in both the acute and subacute phases. Among various miRNAs, miR-146a and miR-185 are inflammation-related miRNAs. However, in an earlier study by Gan *et al.*,[48] circulatory miR-145 expression was increased in ischemic stroke patients aged 18–49 years compared to healthy control subjects. It is not clear at what time the blood samples were collected after stroke onset and the difference between this study and that of Li *et al.*[47] may be due to different methodologies or existing risk factors. Jickling *et al.*[49] found that eight miRNAs were differentially expressed in blood collected 28 h from stroke onset between acute stroke patients and vascular risk factor controls matched for age and gender (age 63 years; 50% female). Six miRNAs had decreased expression (miR-122, miR-148a, *let-7i*, miR-19a, miR-320d, and miR-4429) and two had increased expression (miR-363 and miR-487b). These miRNAs are either known to regulate or are predicted with high probability to regulate the expression of several genes in pathways associated with ischemic stroke including nuclear factor-κB (NF-κB) signaling, toll-like receptor signaling, leukocyte extravasation, interleukin signaling, transforming growth factor-β signaling, and the prothombin activation pathway.

Among other studies, He *et al.*[50] identified 16 miRNAs as significantly differentially expressed in the peripheral blood of ischemic stroke patients (2 males and 3 females, aged 50–83 years), including 13 upregulated and three downregulated miRNAs. Among these miRNAs, the most significantly upregulated miRNA was miR-145 and the most significantly downregulated miRNA was miR-122. However, it was not indicated at what time the blood samples were collected from stroke onset, and the comparison used five other individuals as normal samples and were diagnosed as having other cerebral diseases that resulted from injury or anxiety.

In a study by Tan *et al.*[51] of ischemic stroke patients aged 18–49 years without any or minimal risk factor, and blood collected between 2 and 24 months from stroke onset, four miRNAs (miR-25*, miR-34b, miR-483-5p, and miR-498) were downregulated. These four miRNAs could prove to be specific for stroke pathogenesis in low-risk stroke patients. In a previous study by Tan *et al.*[52] on young stroke patients with existing risk factors, only miR-25* was found to be expressed, but it was upregulated. Additionally, in the earlier study by Tan *et al.*[52] of chonic stroke patients 18–49 years and for which blood samples were collected within 6–18 months from the index stroke, it was shown that among the 157 miRNAs identified for stroke samples, 138 were highly expressed (upregulated, fold change >1.0) and 19 miRNAs were poorly expressed (downregulated, fold change <1.0). Among the highly expressed miRNAs observed for stroke, 17 miRNAs could be identified as highly expressed in the subtypes large-vessel atherosclerosis, small-vessel disease, and cardioembolism. The miRNA profile of small artery stroke samples showed a distinctly different pattern from that of the large artery stroke samples. 79 miRNAs could be identified as differentially regulated among these two subtypes. Of these, in the small artery stroke samples, 71 constituted the highly expressed miRNAs (fold change >1), seven were found to be expressed at the basal level (fold change 1.0–1.04), and one to be poorly expressed (fold change 0.92). An opposite trend was observed in the large artery stroke samples. Of the 79 miRNAs, 77 were downregulated, one was undetectable and one was upregulated. Of the highly upregulated miRNAs (fold change >1.5) in small artery stroke samples, seven miRNAs (miR-130b, miR-29b, miR-301a, miR-339-5p, miR-532-5p, miR-634, and miR-886-5p) showed more than two-fold change. More miRNAs were downregulated in all good outcome (mRS <2) stroke samples compared to normal controls, irrespective of stroke subtype.[52]

7.6.2 Animal Studies for miRNAs and Ischemic Stroke

Sepramaniam *et al.*[45] used a rat middle cerebral artery stroke model, with animals sacrificed over a period of 0–72 h following MCAO and brain samples collected. The highest expression of miR-125b-2*, miR-27a*, miR-422a, miR-488, and miR-627 occurred in the acute phase (0–24 h) in the brain and in the corresponding blood, confirming that upregulation of expression of these miRNAs is a consequence of acute cerebral ischemia. Jeyaseelan *et al.*[53] observed that transient focal ischemia in male rats by MCAO within 24 h of reperfusion resulted in miR-19b, miR-290, and miR-292-5p to be highly expressed, whereas miR-103 and miR-107 were poorly expressed, in peripheral blood. An entirely new group of miRNAs was observed in the 48-h reperfusion blood samples. Among the 14 miRNAs that appeared at both time points, miR-150, miR-195, and miR-320 showed an opposite trend in expression at 24 and 48 h, whereas miR-103, miR-107, and miR-191 showed almost the same level of downregulation at both time points. 56 common miRNAs were present in brain samples of the rats subjected to MCAO, 32 miRNAs were found to be present in the 24-h samples only and eight miRNAs were present exclusively in the 48-h samples. Upregulation was observed for miR-138 and miR-145 at 24 and 48 h of reperfusion, respectively. The *let-7* family was upregulated in the 48-h-reperfused MCAO brain samples together with miR-150 and miR-125a. A study by Dharap *et al.*[54] profiled miRNAs in adult brain of adult male spontaneously hypertensive rats as a function of reperfusion time after transient MCAO. Of the 238 miRNAs evaluated, eight showed increased expression and 12 showed decreased expression at a minimum four out of five reperfusion time points studied between 3 and 72 h compared with sham. MiR-206, miR-145, and miR-290 showed increased expression while miR-218, miR-27a, and miR-137 exhibited decreased expression, and miR-*let-7a* did not show any change in expression at 24-h reperfusion compared to sham. Antagomir-mediated prevention of miR-145 expression led to an increased protein expression of its downstream target superoxide dismutase-2 in the post-ischemic brain of male MCAO rats.[54]

miRNAs may have an important role in neuroprotection brought on by ischemic preconditioning. Lee *et al.*[55] showed that ischemic preconditioning was induced in mice by MCAO for 15 min, and two miR families, miR-200 and miR-182, were selectively upregulated at 3 h after ischemic preconditioning. The miR-200 family was neuroprotective mainly by downregulating prolyl hydroxylase 2 levels and increasing hypoxia-inducible factor-1α.

7.7 miRNAs in Hemorrhagic Stroke

Martinez and Peplow[56] have recently reviewed miRNA profiles in hemorrhagic stroke. Leung *et al.*[57] compared miR-124-3p and miR-16 plasma levels in 93 stroke patients, median age 72 years, 51% male. Ischemic stroke was diagnosed in 74 patients and 19 patients were diagnosed with hemorrhagic stroke, with blood samples collected within 24 h of symptom onset. 23 age- and sex-matched healthy individuals were recruited as controls. Hypertension was a major stroke risk factor in the patients. Hemorrhagic stroke patients had higher median plasma miR-124-3p levels than ischemic stroke patients and healthy controls (1.94 and 2.55 fold change, respectively). Ischemic stroke patients had higher median plasma level of miR-16 compared with hemorrhagic stroke patients and healthy controls (1.24 and 1.35 fold change, respectively). The number of hemorrhagic stroke patients diagnosed with ICH or SAH was not reported, and therefore it is not known whether plasma levels of miR-124-3p and miR-16 can differentiate between these two categories.

miRNAs that have been identified in studies of intracerebral hemorrhagic and subarachnoid hemorrhagic stroke are summarized in Table 7.2. The studies were conducted with human patients and experimental animal models of hemorrhagic stroke.

7.7.1 Human Studies for miRNAs and Hemorrhagic Stroke

7.7.1.1 *Intracerebral Hemorrhage*

Thee large-scale human studies have been described. Wang *et al.*[58] reported that miR-130a was upregulated in serum of ICH patients (70% male, mean age 60 years) collected at 0–1, 2–3, 5–8, and 9–14 days after symptom onset. A large proportion (54%) of the patients were being treated with antihypertensive medications. Zhu *et al.*[59] found that miR-126, miR-146a, *let-7a*, and miR-26a were downregulated in blood of ICH patients (39% male, median age 57 years) with the blood collected on admission. Zheng *et al.*[60] observed an upregulation of miR-29c and miR-122 in plasma of ICH patients (30 patients with hematoma enlargement, 87% male, mean age 58 years; 49 patients without hematoma enlargement, 65% male, mean age 53 years) with blood samples taken from patients immediately after admission. Hypertension and smoking were confounding factors. In a smaller-scale study, Guo *et al.*[61] found an upregulation of miR-27a, miR-365, miR-150*, and miR-34c-3p in plasma of ICH patients (53% male, age 53 ± 2 years).

7.7.1.2 *Subarachnoid Hemorrhage*

Four small-scale studies have been described. Stylli *et al.*[62] reported an upregulation of miR-451a in cerebrospinal fluid (CSF) of patients with aneurysmal SAH (56% male, no cerebral vasospasm, median age 51 years; 30% male, with vasospasm, median age 56 years) with CSF collected from 1 to 18 days and centrifuged. Powers *et al.*[63] observed an upregulation of miR-491 and a downregulation of miR-92a and *let-7b* in CSF of patients with aneurysmal SAH (100% female, 30–74 years) with CSF collected from 3 to 12 days post-hemorrhage and not centrifuged. Su *et al.*[64] found an upregulation of miR-132-3p and miR-324 in blood of SAH patients (55% male, with delayed cerebral infarction (DCI), mean age 59 years; 35% male, without DCI, mean age 59 years) with blood collected after 7 days following onset of SAH. Hypertension was a confounding factor. Liu *et al.*[65] found altered expression of miRNAs in aneurysmal tissue of patients with ruptured intracranial aneurysms (17% male, age range 45–50 and 60–65 years). The changed miRNAs included endothelium-enriched miRNAs such as members of *let-7* family, miR-17, miR-23b, miR-126, miR-24-1, and miR-222, and vascular smooth muscle-enriched miRNAs such as miR-143 and miR-145. Also, levels of miR-1, miR-10a, miR-125b, and miR-26a were altered.

Table 7.2 MicroRNA alterations in hemorrhagic stroke of human patients and animal models.[a]

Reference	Subjects and age	Sample and time	miRNAs increased	miRNAs decreased
Intracerebral hemorrhage				
58	Human, 50 patients, mean age 60 years, 70% M; 32 had deep ICH and 18 had lobar ICH, 54% treated with antihypertensives	Serum collected at 0–1, 2–3, 5–8, and 9–14 days after symptom onset	miR-130a	
59	Human, 33 patients, median age 57 years, 39% M; presenting within 24 h of onset of a primary basal ganglia hemorrhage and absence of coma, hypertension	Blood collected from patients on admission		miR-126, -146a, -*let-7a*, -26a
61	Human, 15 patients; age 53 ± 2 years, 53% M; hypertension and smoking	Plasma	miR-27a, -365, -150*, -34c-3p	
60	Human, 79 patients admitted within 6 h of symptom onset 30 with hematoma enlargement within 24 h after hospitalization, mean age 58 years, 87% M 49 without hematoma enlargement, mean age 53 years, 65% M; hypertension and smoking	Plasma analyzed from blood samples taken from patients immediately after admission	miR-29c, -122	
58	Rat, adult Sprague Dawley strain, M, with ICH	Serum, perihematomal tissue at 1–7 days after ICH	miR-130a	
66	Rat, Sprague Dawley strain, sex not reported, with ICH	Primary microglia		miR-367

67	Rat, Sprague Dawley strain, M, with brain hemorrhage	Blood, brain tissue at 24 h after i.c.v. injection of lysed blood	In blood: miR-96, -152, -298, -333, -505 In brain: miR-542-3p	In blood: miR-125a-5p, -130b, -142-3p, -330, 342-5p, -685, -347 In brain: miR-155, -362-3p, 122, -450a-5p
Subararachnoid hemorrhage				
62	Human, 9 patients with aneurysmal SAH and no cerebral vasospasm, median age 51 years, 56% M; 10 patients with an aneurysmal SAH and vasospasm, median age 56 years, 30% M	CSF collected from 1 to 18 days and centrifuged	miR-451a	
63	Human, 8 patients with aneurysmal SAH, ages 30–74 years, 0% M	CSF collected from 3–12 days post-hemorrhage and not centrifuged	miR-491	miR-92a. *let-7b*
64	Human, 20 SAH patients with DCI, mean age 59 years, 55% M; 20 SAH patients without DCI, mean age 59 years, 35% M; hypertension	Blood collected after 7 days following onset of SAH	miR-132-3p, -324	
65	Human, 6 patients with ruptured intracranial aneurysms, ages in the ranges 45–50 and 60–65 years, 17% M; hypertension and smoking,	Aneurysmal tissue	The changed miRNAs included endothelium-enriched miRNAs such as members of *let-7* family, miR-17, -23b, -126, -24-1, -222, and vascular smooth muscle-enriched miRNAs such as miR-143 and -145. Also, miR-1, -10a, -125b, -26a	
68	Rat, Sprague Dawley strain, adult M, with SAH	MCAs and basilar artery	miR-30a.-143	

[a]CSF: cerebrospinal fluid; DCI: delayed cerebral infarction; HS, hemorrhagic stroke; ICH: intracerebral hemorrhage; M: male; i.c.v.: intracerebroventricular; MCAs: middle cerebral arteries; SAH: subarachnoid hemorrhage.

7.7.2 Animal Studies for miRNAs and Hemorrhagic Stroke

7.7.2.1 *Intracerebral Hemorrhage*

Thee studies with Sprague Dawley rats have been described. Wang *et al.*[58] observed an upregulation of miR-130a in serum and perihematomal tissue at 1–7 days after ICH in adult animals. Yuan *et al.*[66] found miR-367 to be downregulated in primary microglia from animals with ICH. Liu *et al.*[67] reported an upregulation of miR-96, miR-152, miR-298, miR-333, and miR-505 in blood and miR-542-3p in brain tissue, and a downregulation of miR-125a-5p, miR-130b, miR-142-3p, miR-330, miR-342-5p, miR-685, and miR-347 in blood and miR-155, miR-362-3p, miR-122, and miR-450a-5p in brain tissue of animals with brain hemorrhage.

7.7.2.2 *Subarachnoid Hemorrhage*

One study with adult Sprague Dawley rats has been described. Muller *et al.*[68] reported an upregulation of miR-30a and miR-143 in the middle cerebral arteries and basilar artery of animals with SAH.

7.8 miRNAs in Traumatic Brain Injury

miRNA profiles in traumatic brain injury have recently been reviewed by Martinez and Peplow.[69] miRNAs that have been identified in studies of traumatic brain injury are summarized in Table 7.3. The studies were conducted with human patients and experimental animal models of traumatic brain injury.

In experimental animal models, the extent of brain injury is classified according to the parameters used, together with behavioral testing and histological examination of brain sections. In the controlled cortical impact (CCI) procedure, which uses a pneumatic controlled cortical impact device to deliver mechanical energy to the intact dura, the amount of brain deformation is a function of the impactor tip size, contact velocity and duration time.[70–72] By varying these parameters, different severity grades of TBI injury have been produced in mice and rats. In the weight-drop injury method, a metal rod is dropped from a known height above the exposed skull of the animal to create a brain injury.[73] Different severity grades of TBI injury have been produced in mice by varying the weight, diameter of tip, and fall height of the impactor rod. The weight drop method was found to recreate the cognitive and behavioral deficits seen in human patients with mild TBI.[74,75] The fluid percussion injury method involves craniotomy to expose the dura and placement of a female Luer lock or plastic cap that is cemented in place over the craniotomy site; this is followed by transient compression and/or deformation of the underlying brain using a fluid-mediated pressure pulse, with a pulse duration time of ~20 ms.[76,77]

Table 7.3 MicroRNA alterations in traumatic brain injury of human patients and animal models.[a]

Reference	Subjects and age	Sample and time	miRNAs increased	miRNAs decreased
78	Human, 27 M/8 F mild TBI with EC injury, mean age 40 years; 24 M/11 F, severe TBI with EC injury, mean age 41 years	Serum collected 0–72 h and 15 days post-injury	miR-21 and miR-335 at 4–12 h, 48–72 h and day 15 in severe TBI	miR-425p and miR-502 at 0–1 h and 4–12 h in mild TBI
81	Human, 14 M/12 F severe (coma) TBI, mean age 49 years	CSF	miR-141, -257, -181a*, -27b*, -483-5p, -30b, -1289, -431*, -193b*, -499-3p, of which miR-141 and miR-257 had the greatest change in severe TBI (coma)	mIR-1297, -33b, -933, -449b, of which miR-1297 had the greatest change in severe TBI (coma)
79	Human, 59 M/17 F, mean age 49 years: 25 mild TBI, 26 moderate TBI, 25 severe TBI	Serum collected within 24 h and then daily for up to 21 days after injury	miR-93, -191, and -499 in TBI over 24 h to day 14 with the levels highest in severe TBI	
80	Human, 28 M/20 F, age 18–50 years: 16 mild TBI, 16 moderate TBI, 16 severe TBI with injury within preceding 24 h	Plasma		miR-23b was decreased in TBI with the levels lowest in severe TBI
82	Human, 7 M/1 F, 75% mild TBI and 25% moderate TBI, mean age 36 years; 8 M severe TBI, mean age 39 years	Serum samples of mild/moderate TBI were collected within 24 h from injury. For severe TBI, serum samples were collected within 48 h (mean 26 h) from injury and CSF samples collected within 48 h of injury	miR-151-5p, -195, -20a, -328, -362-3p, -30d, -451, -486, -505*, and -92a in serum; miR-328, -362-3p, -451, and -486 in CSF	

(*continued*)

Table 7.3 (*continued*)

Reference	Subjects and age	Sample and time	miRNAs increased	miRNAs decreased
83	Human, 11 mild TBI, age >14 years 10 severe TBI, age 14–65 years	Plasma was collected from mild TBI patients within 10 h of injury. For severe TBI patients, plasma was obtained at the earliest possible time after admission and patient stabilization	miR-16 and miR-92a at 0–10 h in mild TBI; miR-765 at 68 h in severe TBI	miR-16 and miR-92a at 68 h in severe TBI
91	Mouse, C57Bl6 strain, M, age 12 weeks, controlled cortical impact to produce moderate TBI	Cerebral cortex at 1–72 h post-injury	miR-711 in moderate TBI as early as 1 h post-injury, peaking at 6 h, and persisting though to 72 h	
86	Mouse, C57Bl6 strain, M, 9–12 weeks, controlled cortical impact to produce mild–moderate TBI	Cerebral cortex at 1–12 h post-injury	miR-2137, -451, -21*, -144, and -184 in mild–moderate TBI	miR-190, -137, -541, and -107 in mild–moderate TBI
89	Mouse, C57Bl6 strain, M, age 17–19 weeks housed in cages equipped with RW, controlled cortical impact to produce moderate TBI	Cerebral cortex	miR-21, -92a, and -874 in moderate TBI-non-runners	miR-138, -124, and *let-7c* in moderate TBI-non-runners
90	Mouse, C57Bl6 strain, M, age 17–19 weeks housed in cages equipped with RW, controlled cortical impact to produce moderate TBI	Hippocampus	miR-21 in moderate TBI-non-runners	

85	Mouse, C57Bl6 strain, M, age 20–24 weeks and 22–24 months, controlled cortical impact to produce mild-moderate TBI	Cerebral cortex at 24 h post-injury	miR-21 in mild–moderate TBI with maximum increase at 24 h post-injury (adult)	miR-21 in mild–moderate TBI (aged)
87	Mouse, C57Bl6 strain, M, age 12 weeks, controlled cortical impact to produce moderate TBI	Cerebral cortex at 1–24 h post-injury		miR-23a-3p and miR-27a-3p in moderate TBI at 1 h, 6 h, and 24 h post-injury
88	Rat, Sprague Dawley strain, M, 9–11 week, controlled cortical impact to produce moderate TBI	Hippocampus at 12 h post-injury	Mitochondrial-associated miR-155 and miR-223 in moderate TBI at 12 h post-injury; miR-155, -223, -146a, -142-3p, -142-5p, and -150 expression highest in enriched microglia and astrocytes, miR-124a expression highest in cortical neurons	
102	Rat, Sprague Dawley strain, M/F, age 7–8 weeks, controlled cortical impact to produce moderate TBI	Hippocampus at 1–6 days post-injury	miR-142-3p in moderate TBI at day 5 post-injury; miR-221 in moderate TBI at days 1, 3, and 5 post-injury	miR-142-3p in moderate TBI at days 1 and 3 post-injury
84	Rat, Sprague Dawley strain, M/F, age 7–8 weeks, controlled cortical impact to produce moderate TBI	Hippocampus at 1–7 days post-injury	miR-142-3p, -144, -340-5p, -674-5p, -153, -186, -190, -132*, -138-1* in moderate TBI at 1 h and days 1, 3, 5, and 7 post-injury	*let-7b* in moderate TBI at 1 h and days 1, 3, 5, and 7 post-injury
103	Rat, Sprague Dawley strain, M, age 8–9 weeks, controlled cortical impact to produce moderate TBI	Dorsal hippocampus at 1 and 7 days post-injury	miR-153 in moderate TBI at 24 h post-injury	miR-222 in moderate TBI at 24 h post-injury; miR-135a and miR-135b in moderate TBI at day 7 post-injury

(*continued*)

Table 7.3 (*continued*)

Reference	Subjects and age	Sample and time	miRNAs increased	miRNAs decreased
92	Mouse, C57Bl/6J strain, M, age 10–12 weeks, weight-drop injury to produce increasing grade of mild TBI	Brain tissue biopsy at 24 h and 7 days post-injury	miR-296-5p, -154*,and -466 in mild TBI in mild TBI at 24 h and day 7 post-injury	
75	Mouse, C57Bl/6J strain, M, age 11–12 weeks, weight-drop injury to produce increasing grade of mild TBI	Serum at 3 h post-injury	Number of microRNAs increased with increasing grade of injury, except for IS4 injury group in mild TBI at 3 h post-injury	Number of microRNAs increased with increasing grade of injury, except for IS4 injury group in mild TBI at 3 h post-injury
80	Rat, Sprague Dawley strain, M, age 12 weeks, weight-drop injury to produce moderate TBI	Plasma, cerebral cortex, hippocampus at 1, 3, 7, and 14 days post-injury		miR-23b in moderate TBI at days 1, 3, 7, and 14 post-injury
94	Rat, Sprague Dawley strain, M, age 12 weeks, fluid percussion injury to produce moderate TBI	Cerebral cortex/hippocampus at 6 h to 14 days post-injury	miR-21 in moderate TBI at 6 h and day 3 post-injury	
95	Rat, Wistar strain, sex not reported, age 8–9 weeks, fluid percussion injury to produce moderate TBI	Brain tissue at 6–72 h post-injury	miR-21 in moderate TBI at 6 h and days 1, 2 and 3 post-injury	

[a]CSF: cerebrospinal fluid; EC: extracranial; F: female; M: male; RW: running wheel; TBI: traumatic brain injury.

7.8.1 Human Studies for miRNAs and Traumatic Brain Injury

Three large-scale studies have been described. Di Pietro *et al.*[78] reported an upregulation of miR-21 and miR-335 at 4–12 h, 48–72 h and day 15 in the serum of patients with severe TBI (69% male, mean age 41 years) and a downregulation of miR-425p and miR-502 at 0–1 h and 4–12 h in patients with mild TBI (77% male, mean age 40 years). Serum was collected at 0–72 h and 15 days post-injury. Yang *et al.*[79] observed an upregulation of miR-93, miR-191, and miR-499 in TBI over 24 h to day 14 in patients with mild TBI and severe TBI, with the levels being highest in severe TBI. Sun *et al.*[80] found a downregulation of miR-23b in the plasma of patients with mild, moderate, and severe TBI, with the lowest levels in severe TBI. Injury occurred within the preceding 24 h.

There were also three smaller-scale studies. You *et al.*[81] reported an upregulation of miR-141, miR-257, miR-181a*, miR-27b*, miR-483-5p, miR-30b, miR-1289, miR-431*, miR-193b*, and miR-499-3p, of which miR-141 and miR-257 had the greatest change in CSF of patients with severe TBI (coma) (54% male, mean age 49 years). Bhomia *et al.*[82] found an upregulation of miR-151-5p, miR-195, miR-20a, miR-328, miR-362-3p, miR-30d, miR-451, miR-486, miR-505*, and miR-92a in serum and miR-328, miR-362-3p, miR-451, and miR-486 in CSF of patients with mild, moderate, and severe TBI. Serum samples of mild/moderate TBI were collected within 24 h of injury. For severe TBI, serum samples were collected within 48 h (mean 26 h) from the time of injury and CSF samples collected within 48 h of injury. Redell *et al.*[83] observed an upregulation of miR-16 and miR-92a at 0–10 h in mild TBI and miR-765 at 68 h in severe TBI in plasma of TBI patients. There was a downregulation of miR-16 and miR-92a at 68 h in plasma of patients with severe TBI. Plasma was collected from mild TBI patients within 10 h of injury. For severe TBI patients, plasma was obtained at the earliest possible time after admission and patient stabilization. microRNAs with upregulated and downregulated expression in patients with mild and severe TBI are shown in Figure 7.1.

7.8.2 Animal Studies for miRNAs and Traumatic Brain Injury

15 studies were found, which had mostly used male mice and rats. 10 of these studies had used the CCI method, three the weight-drop injury method, and two the fluid percussion injury method to create the brain injury.

7.8.2.1 Controlled Cortical Impact Studies

The CCI parameters ranged from 3 to 5 mm for diameter of impactor tip, from 0.8 to 6 m s^{-1} for impact velocity, from 85 to 200 ms for duration, and 1 to 2 mm for deformation depth of cortex. By reference to the parameters

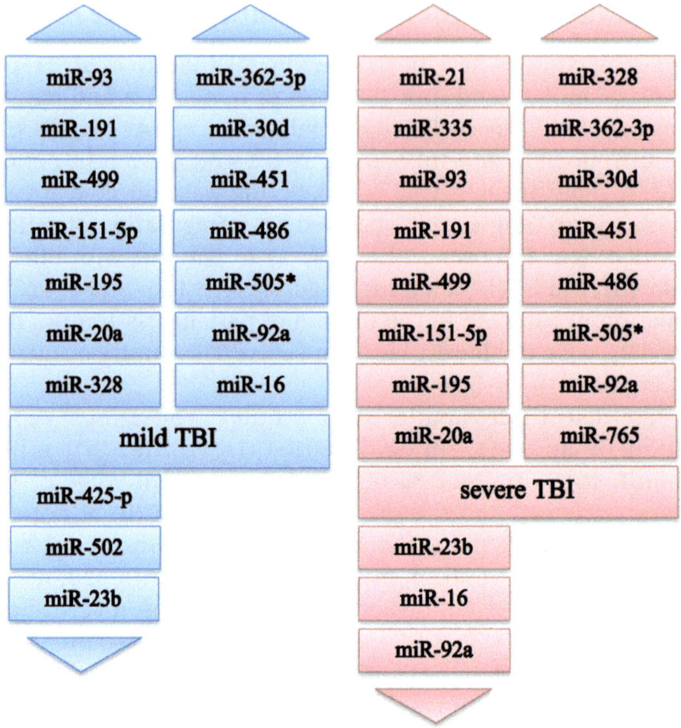

Figure 7.1 Altered expression of circulating microRNAs in human patients with mild and severe traumatic brain injury (TBI) (based on the studies reviewed). For mild TBI: increased levels of miR-93, -191, and -499 at 0–24 h and days 2–7 post-injury; miR-151-5p, -195, -20a, -328, -362-3p, -30d, -451, -486, -505*, and -92a at 0–24 h post-injury; and miR-16 at 0–10 h post-injury; decreased levels of miR-425-p and -502 at 0–1 h and 4–12 h post-injury and miR-23b at 0–24 h post-injury. For severe TBI: increased levels of miR-21 at 4–12 h and 48–72 h post-injury; miR-335 at 0–1 h, 4–12 h, and 48–72 h post-injury; miR-93, -191, and -499 at 0–24 h and days 2–7 post-injury; miR-151-5p, -195, -20a, -328, -362-3p, -30d, -451, -486, and -505* at 0–48 h post-injury; miR-92a at 0–48 h (mean 26 h) post-injury; and miR-765 mean 68 h post-injury; decreased levels of miR-23b at 0–24 h post-injury; and miR-16 and -92a at mean 68 h post-injury. Adapted from ref. 69 with permission.

used by Liu *et al.*[84] to induce a moderate TBI by CCI (equivalent to a GCS score of 9–12 for humans), most of the studies had used parameters to create a moderate TBI. In two of the studies by Sandhir *et al.*[85] and Meissner *et al.*,[86] the parameters chosen would more likely be associated with a mild–moderate brain injury. Meissner *et al.*[86] reported upregulated expression in the cortex of miR-2137, miR-451, miR-21*, miR-144, and miR-184, and downregulated expression of miR-190, miR-137, miR-541, and miR-107 at 1, 6, and

12 h after TBI in mice compared to control. Similarly, miR-23a and miR-27a were downregulated in cortex of TBI mice as early as 1 h and lasting up to 24 h post-injury compared to control.[87] Liu *et al.*[84] found that miR-142-3p, miR-144, miR-340-5p, miR-674-5p, miR-153, miR-186, miR-190, miR-132*, and miR-138-1* were upregulated, whereas *let-7b* was downregulated in the hippocampus of TBI rats compared to controls. Wang *et al.*[88] reported that miR-155 and miR-223 levels were significantly increased in the mitochondrial fraction of hippocampus of TBI rats compared to controls. Two studies examined miRNA profiles in mice given free access to a running wheel to exercise. Miao *et al.*[89] observed the cortical levels of miR-21, miR-92a, and miR-874 decreased, while those of miR-138, miR-124, and *let-7c* increased compared with the TBI non-runner group. Hu *et al.*[90] found that the level of miR-21 was significantly elevated in the hippocampus of TBI non-runners compared with that of sham animals and the increase in miR-21 expression in hippocampus was reversed in TBI runners. Sandhir *et al.*[85] showed that basal expression of miR-21 was higher in the brain of aged mice compared to adult brains. miR-21 levels increased following TBI in adult mice, with maximum increase at 24 h post-injury, whereas in the aged mice miR-21 expression in the brain was decreased after TBI. Strategies employing miRNA mimics or agomirs to increase the levels of downregulated miRNAs or miRNA antagomirs to decrease the levels of upregulated miRNAs were also examined as possible ways to improve outcomes in TBI.[87,91]

7.8.2.2 Weight-drop Injury Studies

The parameters used ranged from 246 g/333 g (mice) to 450 g (rats) for weight of rod 1 mm diameter, and from 2 cm/3 cm (mice) to 150 cm (rats) for fall height. Lesions were not observed in stained sections of injured mouse brains, consistent with creating a mild TBI in these animals.[75,92] A modified neurological severity score (mNSS) (see Zhang *et al.*[93]) of 10 in injured rats indicated a moderate TBI.[80] Both Sharma *et al.*[75] and Chandran *et al.*[92] reported that varying intensities of mild TBI induced a differential miRNA expression profile in mouse brain post-injury. Sharma *et al.*[75] observed that a set of 13 miRNAs identified mild TBI regardless of its severity: mmu-miR-376a, hsa-miR-214, mmu-miR-214, mmu-miR-337-5p, mmu-miR-574-3p, mmu-miR-434-3p, mmu-miR-671-3p, mmu-miR-218, and mmu-miR-199a-3p were all upregulated, whereas hsa-miR-106b*, mmu-miR-106b, mmu-miR-31, and mmu-miR-196c were all downregulated. In the rat study by Sun *et al.*[80] levels of miR-23b were decreased in the plasma, hippocampus, and cortex of TBI animals post-injury compared to sham. Pretreatment of TBI animals with lentivirus-miR-23b significantly decreased the lesion volume of the contused hemisphere and brain edema compared with TBI group. In addition, mNSS score at 3, 7, and 14 days post-injury was lowered in TBI animals pretreated with lentivirus-miR-23b.

7.8.2.3 Fluid Percussion Injury Studies

In one of the rat studies, the pressure applied was 1.8–2.0 atm, while in the other it was 1.8–2.2 atm. The duration of pulse was not reported in either study. Based on an mNSS score of 8 on day 1 post-injury[94] and histological sections of cerebral cortex after injury,[95] both studies had utilized animals with moderate TBI. Expression of miR-21 was increased in the impacted brain area[95] and was significantly increased or decreased in the agomir- or antagomir-injected animals post-injury compared to controls, respectively.[94]

No studies were found on miRNA expression levels in animals subjected to severe TBI, presumably because of the likely high mortality rate and ethical concerns.

7.9 Targeting miRNAs as a Therapeutic Approach

7.9.1 Therapeutics for Ischemic Stroke

For prevention and treatment of stroke, increased or decreased miRNA levels may be needed. Novel approaches are being developed to transport miRNA-related therapeutics into the brain across an intact blood–brain barrier, including chemical modification and use of targeting molecules, and methods to disrupt the blood–brain barrier. The potential for antagomir therapy has been shown in the context of acute ischemic stroke in mice, in which regulation of miR-181 affects the extent of brain injury. Knockdown of miR-181 by intracerebroventricular infusion of its antagomir effectively decreased the infarction size and protected the penumbra.[96] Yin *et al.*[97] showed that intracerebroventricular administration of miR-497 antagomir resulted in smaller infarct area following MCAO in mice. Antagomir-based therapy has demonstrated gender-specific miRNA interactions whereby intracerebro-ventricular infusion of *let7f* antagomir promoted neuroprotection in MCAO female rats but not MCAO males.[98] Intravenous injection of a specific miR-155 inhibitor at 48 h after MCAO in male mice was found to support brain microvasculature, reduce brain tissue damage, and improve functional recovery.[99] Liu *et al.*[100] showed that intravenous but not intracerebroventricular injection of miR-122 mimic in MCAO rat models decreased neurological deficits and brain infarction, and maintained vessel integrity after MCAO. The miR-122 mimic also downregulated direct and indirect target genes in blood after MCAO that are predicted to affect cell adhesion, leukocyte extravasation, eicosanoid and atherosclerosis signaling (Figure 7.2).

7.9.2 Therapeutics for Hemorrhagic Stroke

Using an experimental model of ICH, injection of miR-130a inhibitor into the right lateral ventricle before ICH induction in male rats significantly reduced endogenous expression of miR-130a, decreased brain edema, and alleviated

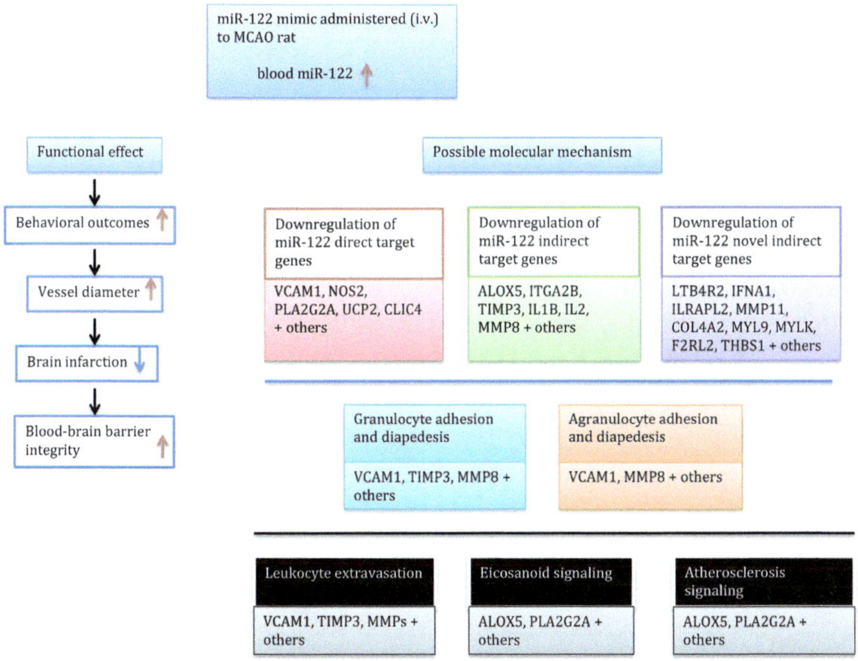

Figure 7.2 Possible molecular mechanism of miR-122 mimic-mediated support of post-stroke recovery. Upregulation of miR-122 results in stabilization of blood–brain barrier integrity, maintenance of vessel diameter, decreased brain infarction, and improved behavioral outcomes in treated animals. Possible molecular mechanisms mediating these processes could be associated with downregulation of miR-122 direct and indirect target genes by miR-122 mimic (upper panel) that are predicted to affect cell adhesion, diapedesis (middle panel), leukocyte extravasation, eicosanoid and atherosclerosis signaling (lower panel). Adapted from ref. 37 with permission.

blood–brain barrier disruption at 1 day after ICH. Neurological function was significantly improved.[58] In a mouse model of ICH, intracerebroventricular injection of miR-367 mimic significantly increased the miR-367 level *in vivo*, and significantly inhibited interleukin-1 receptor-associated kinase 4, NF-κB, p65, IL-6, IL-1β and tumor necrosis factor-alpha (TNF-α) expression in brain tissues after ICH, indicating that miR-367 could inhibit inflammation response *in vivo*. A miR-367 mimic significantly decreased brain edema and neurological injury (Figure 7.3).[66] Overexpression of miR-223 following intracerebroventricular injection of miR-223 mimic in ICH mice resulted in reduced brain edema and improved neurological functions. miR-223 significantly inhibited caspase-1 p20, NLRP3, TNF-α, IL-1β, and IL-6 expression in brain tissues after ICH, showing that miR-223 could inhibit inflammation response *in vivo*.[101]

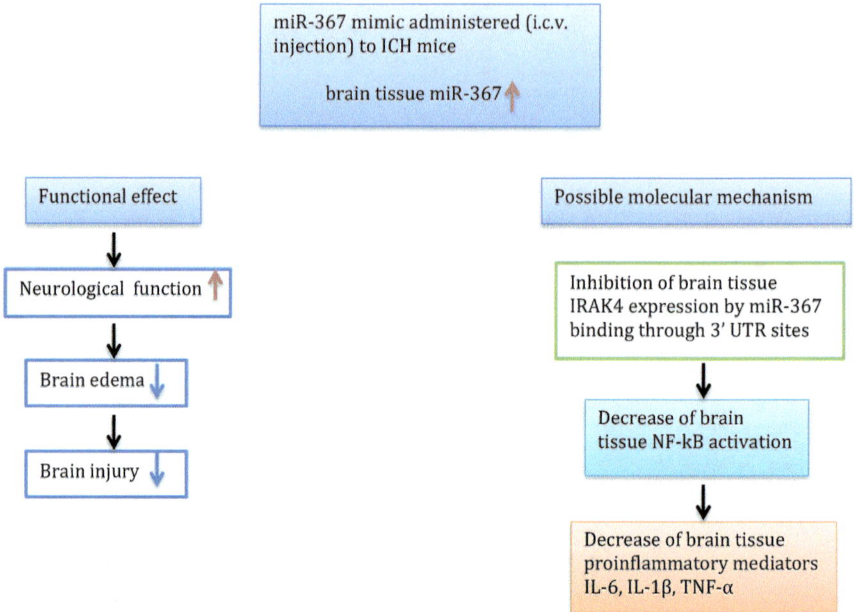

Figure 7.3 Possible molecular mechanism of miR-367 mimic-mediated support of post-stroke recovery in intracerebral hemorrhage (ICH) mice. Upregulation of brain tissue miR-367 results in decreased brain edema and brain injury, and improved neurological function in treated animals. Possible molecular mechanisms mediating these processes could be associated with inhibition of brain tissue interleukin-1 receptor-associated kinase 4 (IRAK4) expression and decreased brain tissue nuclear factor κB (NF-κB) activation by miR-367 mimic, leading to a decrease in the pro-inflammatory mediators interleukin-6 (IL-6), IL-1β, and tumor necrosis factor-α (TNF-α). i.c.v: intracerebroventricular; UTR: untranslated region. Adapted from ref. 56 with permission.

7.9.3 Therapeutics for Traumatic Brain Injury

The animal studies have examined specific miRNAs as biomarkers and therapeutic targets for moderate and mild TBI. Upregulation and downregulation of specific miRNAs in brains of TBI animals occurred at different time points post-injury.[95] miR-21 and miR-23b could be potential therapeutic targets for intervention after TBI. Several of the studies had examined the effect of voluntary exercise on miRNA profiling and shown it to be altered by exercise. In addition, differences in basal miRNA expression in the brain of adult and aged animals and alterations in response to TBI (*e.g.* miR-21) have been reported. The latter is very relevant as persons aged ≥75 years have an increasing incidence of fall-related TBI.

7.10 Conclusion

The data presented in the human studies indicate that individual or combinations of circulating miRNAs can serve as important diagnostic and prognostic biomarkers for stroke and TBI. They can distinguish between the subacute and acute phase of ischemic stroke, and also between mild and severe TBI. It also seems likely that they can differentiate between ischemic and hemorrhagic stroke, which is important regarding the most appropriate clinical treatment for the patient.

The animal studies have indicated with the various models used that miRNAs are biomarkers for ischemic stroke, hemorrhagic stroke, and TBI. Importantly, they have shown that new miRNAs are present in 48 h reperfusion blood samples compared to 24 h reperfusion blood samples in MCAO rats. In addition, it is possible to alter the levels of downstream targets of miRNAs by agomir or antagomir administration to increase or decrease, respectively, the level of a specific miRNA. In the TBI studies, exercise using a running wheel and aging were two important factors that influenced the levels of specific miRNAs in the adult brain.

References

1. www.strokecenter.org/patients/about-stroke/stroke-statistics/, accessed 8 December 2017.
2. Y. Bejot, B. Delpont and M. Giroud, Rising stroke incidence in young adults: More epidemiological evidence, more questions to be answered, *J. Am. Heart Assoc.*, 2016, **5**, e003661.
3. M. Tibæk, C. Dehlendorff, H. S. Jørgensen, H. B. Forchhammer, S. P. Johnsen and L. P. Kammersgaard, Increasing evidence for hospitalization for stroke and transient ischemic attack in young adults: A registry-based study, *J. Am. Heart Assoc.*, 2016, **5**, e003158.
4. S. J. An, T. J. Kim and B. W. Yoon, Epidemiology, risk factors and clinical features of intracerebral hemorrhage: An update, *J. Stroke*, 2017, **19**, 3–10.
5. E. B. Marsh, R. H. Llinas, A. E. Hillis and R. F. Gottesman, Hemorrhagic transformation in patients with acute ischaemic stroke and an indication for anticoagulation, *Eur. J. Neurol.*, 2013, **20**, 962–967.
6. I. J. Cho, J. S. Kim, H. J. Chang, Y. J. Kim, S. C. Lee, J. H. Choi, S. Shin, C. Y. Shim, G. R. Hong, J. W. Ha and N. Chung, Prediction of hemorrhagic transformation following embolic stroke in patients with prosthetic valve endocarditis, *J. Cardiovasc. Ultrasound*, 2013, **21**, 123–129.
7. Hemorrhagic stroke, https://www.drugs.com/health-guide/hemorrhagic-stroke.html, accessed 18 April 2018.
8. B. Jennett, Epidemiology of head injury, *Arch. Dis. Child.*, 1998, **78**, 403–406.

9. A. I. Maas, N. Stocchetti and R. Bullock, Moderate and severe traumatic brain injury in adults, *Lancet Neurol.*, 2008, **7**, 728–741.
10. G. M. Ribbers, Brain Injury: Long term outcome after traumatic brain injury, in *International Encyclopedia of Rehabilitation*, ed. J. H. Stone and M. Blouin, 2010, http://cirrie.buffalo.edu/encyclopedia/en/article/338, accessed 14 July 2017.
11. N. Adekoya, D. J. Thurman, D. D. White and K. W. Webb, Surveillance for traumatic brain injury deaths–United States, 1989–1998, *MMWR Surveill. Summ.*, 2002, **51**, 1–14.
12. K. Zweckberger, C. Erös, R. Zimmermann, S. W. Kim, D. Engel and N. Plesnila, Effect of early and delayed decompressive craniotomy on secondary brain damage after controlled cortical impact in mice, *J. Neurotrauma*, 2006, **23**, 1083–1093.
13. J. Emberson, K. R. Lees, P. Lyden, L. Blackwell, G. Albers, E. Bluhmki, T. Brott, G. Cohen, S. Davis, G. Donnan, J. Grotta, G. Howard, M. Kaste, M. Koga, R. von Kummer, M. Lansberg, R. I. Lindley, G. Murray, J. M. Olivot, M. Parsons, B. Tilley, D. Toni, K. Toyoda, N. Wahlgren, J. Wardlaw, W. Whiteley, G. J. del Zoppo, C. Baigent, P. Sandercock and W. Hacke, Effect of treatment delay, age and stroke severity on the effects of intravenous thombolysis with alteplase for acute ischemic stroke: a meta-analysis of individual patient data from randomised trials, *Lancet*, 2014, **384**, 1929–1935.
14. Y. Wang, Y. Wang, X. Zhao, L. Liu, D. Wang, C. Wang, C. Wang, H. Li, X. Meng, L. Cui, J. Jia, Q. Dong, A. Xu, J. Zeng, Y. Li, Z. Wang, H. Xia and S. C. Johnston, Clopidogrel with aspirin in acute minor stroke or transient ischemic attack, *N. Engl. J. Med.*, 2013, **369**, 11–19.
15. D. S. Liebeskind, Hemorhagic stroke treatment & management, http://emedicine.medscape.com/article/1916662-treatment, accessed 18 December 2017.
16. S. Bilgin, A. Guclu-Gunduz, H. Oruckaptan, N. Kose and B. Celik, Gait and Glasgow Coma Scale scores can predict functional recovery in patients with traumatic brain injury, *Neural Regener. Res.*, 2012, **7**, 1978–1984.
17. T. A. Gennarelli, Future directions in brain injury research, *Prog. Neurol. Surg.*, 2014, **28**, 243–250.
18. P. Leo and M. McCrea, Epidemiology, in *Translational Research in Traumatic Brain Injury*, ed. D. Laskowitz and G. Grant, CRC Press/Taylor and Francis Group, Boca Raton, 2016, ch. 1.
19. H. Fischer, U.S. Military Casualty Statistics: Operation New Dawn, Operation Iraqi Freedom, and Operation Enduring Freedom, CRS Report for Congress. Prepared for Members and Committees of Congress: Congressional Research Service, 2013, pp. 1–12.
20. S. R. Kesler, H. F. Adams and E. D. Bigler, SPECT, MR and quantitative MR imaging: correlates with neuropsychological and psychological outcome in traumatic brain injury, *Brain Inj.*, 2000, **14**, 851–857.

21. E. L. Yuh, S. R. Cooper, A. R. Ferguson and G. T. Manley, Quantitative CT improves outcome prediction in acute traumatic brain injury, *J. Neurotrauma*, 2012, **29**, 735–746.
22. A. I. Maas, E. W. Steyerberg, G. D. Murray, R. Bullock, A. Baethmann, L. F. Marshall and G. M. Teasdale, Why have recent trials of neuroprotective agents in head injury failed to show convincing efficacy? A pragmatic analysis and theoretical considerations, *Neurosurgery*, 1999, **44**, 1286–1298.
23. T. W. McAllister, Neurobiological consequences of traumatic brain injury, *Dialogues Clin. Neurosci.*, 2011, **13**, 287–300.
24. B. K. Siesjo, Pathophysiology and treatment of focal cerebral ischemia Part I: Pathophysiology, *J. Neurosurg.*, 1992, **77**, 169–184.
25. B. K. Siesjo, Pathophysiology and treatment of focal cerebral ischemia Part II: Mechanisms of damage and treatment, *J. Neurosurg.*, 1992, 77, 337–354.
26. F. C. Barone and G. Z. Feuerstein, Inflammatory mediators and stroke: new opportunities for novel therapeutics, *J. Cereb. Blood Flow Metab.*, 1999, **19**, 819–834.
27. B. M. Famakin, The immune response to acute focal cerebral ischemia and associated post-stroke immunodepression: A focused review, *Aging Dis.*, 2014, **5**, 307–326.
28. D. Brea, T. Sobrino, P. Ramos-Cabrer and J. Castillo, Inflammatory and neuroimmunomodulatory changes in acute cerebral ischemia, *Cerebrovasc. Dis.*, 2009, **27**(Suppl 1), 48–64.
29. S. E. Akopov, N. A. Simonian and G. S. Grigorian, Dynamics of polymorphonuclear leukocyte accumulation in acute cerebral infarction and their correlation with brain tissue damage, *Stroke*, 1996, **27**, 1739–1743.
30. J. L. Zamanian, L. Xu, L. C. Foo, N. Nouri, L. Zhou, R. G. Giffard and B. A. Barres, Genomic analysis of reactive astrogliosis, *J. Neurosci.*, 2012, **32**, 6391–6410.
31. S. A. Liddelow and B. A. Barres, Reactive astrocytes: Production, function, and therapeutic potential, *Immunity*, 2017, **46**, 957–967.
32. H. M. Bramlett and W. D. Dietrich, Pathophysiology of cerebral ischemia and brain trauma: similarities and differences, *J. Cereb. Blood Flow Metab.*, 2004, **24**, 133–150.
33. I. S. Balan, A. J. Saladino, B. Aarabi, R. J. Castellani, C. Wade, D. M. Stein, H. M. Eisenberg, H. H. Chen and G. Fiskum, Cellular alterations in human traumatic brain injury: changes in mitochondria morphology reflect regional levels of injury severity, *J. Neurotrauma*, 2013, **30**, 367–381.
34. D. Baek, J. Villen, C. Shin, F. D. Camargo, S. P. Gygi and D. P. Bartel, The impact of microRNAs on protein output, *Nature*, 2008, **455**, 64–71.
35. M. Ha and V. N. Kim, Regulation of microRNA biogenesis, *Nat. Rev. Mol. Cell Biol.*, 2014, **15**, 509–524.

36. F. Wahid, A. Shehzad, T. Khan and Y. Y. Kim, MicroRNAs: synthesis, mechanism, function, and recent clinical trials, *Biochim. Biophys. Acta*, 2010, **1803**, 1231–1243.
37. B. Martinez and P. V. Peplow, Blood microRNAs as potential diagnostic and prognostic markers in cerebral ischemic injury, *Neural Regener. Res.*, 2016, **11**, 1375–1378.
38. P. S. Mitchell, R. K. Parkin, E. M. Kroh, B. R. Fritz, S. K. Wyman, E. L. Pogosova-Agadjanyan, A. Peterson, J. Noteboom, K. C. O'Briant, A. Allen, D. W. Lin, N. Urban, C. W. Drescher, B. S. Knudsen, D. L. Stirewalt, R. Gentleman, R. L. Vessella, P. S. Nelson, D. B. Martin and M. Tewari, Circulating microRNAs as stable blood-based markers for cancer detection, *Proc. Natl. Acad. Sci. U. S. A.*, 2008, **105**, 10513–10518.
39. C. M. Stary, J. D. Bell, J. E. Cho and R. G. Gifford, Identification of microRNAs as targets for treatment of ischemic stroke, in *Acute Brain Impairment: Scientific Discoveries and Translational Research*, ed. P. V. Peplow, S. A. Dambinova, T. A. Gennarelli and B. Martinez, Royal Society of Chemistry Publishing, London, 2018, pp.105–127.
40. X. Chen, Y. Ba, L. Ma, X. Cai, Y. Yin, K. Wang, J. Guo, Y. Zhang, J. Chen, X. Guo, Q. Li, X. Li, W. Wang, Y. Zhang, J. Wang, X. Jiang, Y. Xiang, C. Xu, P. Zheng and J. Zhang, *et al.*, Characterization of microRNAs in serum: a novel class of biomarkers for diagnosis of cancer and other diseases, *Cell Res.*, 2008, **18**, 997–1006.
41. A. S. Ho, X. Huang, H. Cao, C. Chistman-Skieller, K. Bennewith, Q. T. Le and A. C. Koong, Circulating miR-210 as a novel hypoxia marker in pancreatic cancer, *Transl. Oncol.*, 2010, **3**, 109–113.
42. R. Chandran, S. L. Mehta and R. Vemuganti, Non-coding RNAs and neuroprotection after acute CNS injuries, *Neurochem. Int.*, 2017, **111**, 12–22.
43. Z. B. Yang, T. B. Li, Z. Zhang, K. D. Ren, Z. F. Zheng, J. Peng and X. J. Luo, The diagnostic value of circulating brain-specific microRNAs for ischemic stroke, *Intern. Med.*, 2016, **55**, 1279–1286.
44. L. Jia, F. Hao, W. Wang and Y. Qu, Circulating miR-145 is associated with plasma high-sensitivity C-reactive protein in acute ischemic stroke patients, *Cell Biochem. Funct.*, 2015, **33**, 314–319.
45. S. Sepramaniam, J. R. Tan, K. S. Tan, D. A. DeSilva, S. Tavintharan, F. P. Woon, C. W. Wang, F. L. Yong, D. S. Karolina, P. Kaur, F. J. Liu, K. Y. Lim, A. Armugam and K. Jeyaseelan, Circulating microRNAs as biomarkers of acute stroke, *Int. J. Mol. Sci.*, 2014, **15**, 1418–1432.
46. L. Zeng, J. Liu, Y. Wang, L. Wang, S. Weng, Y. Tang, C. Zheng, Q. Cheng, S. Chen and G. Y. Yang, MicroRNA-210 as a novel blood biomarker in acute cerebral ischemia, *Front. Biosci.*, 2011, **E3**, 1265–1272.
47. S. H. Li, S. Y. Su and J. L. Liu, Differential regulation of microRNAs in patients with ischemic stroke, *Curr. Neurovasc. Res.*, 2015, **11**, 214–221.
48. C. S. Gan, C. W. Wang and K. S. Tan, Circulatory microRNA-145 expression is increased in cerebral ischemia, *Genet. Mol. Res.*, 2012, **11**, 147–152.

49. G. C. Jickling, B. P. Ander, X. Zhan, D. Noblett, B. Stamova and D. Liu, microRNA expression in peripheral blood cells following acute ischemic stroke and their predicted gene targets, *PLoS One*, 2014, **9**, e99283.
50. W. He, S. Chen, X. Chen, S. Li and W. Chen, Bioinformatic analysis of potential microRNAs in ischemic stroke, *J. Stroke Cerebrovasc. Dis.*, 2016, **25**, 1753–1759.
51. J. R. Tan, K. S. Tan, Y. X. Koo, F. L. Yong, C. W. Wang, A. Armugam and K. Jeyaseelan, Blood microRNAs in low or no risk ischemic stroke patients, *Int. J. Mol. Sci.*, 2013, **14**, 2072–2084.
52. K. S. Tan, A. Armugam, S. Sepramaniam, K. Y. Lim, K. D. Setyowati, C. W. Wang and K. Jeyaseelan, Expression profile of microRNAs in young stroke patients, *PLoS One*, 2009, **4**, e7689.
53. K. Jeyaseelan, K. Y. Lim and A. Armugam, MicroRNA expression in the blood and brain of rats subjected to transient focal ischemia by middle cerebral artery occlusion, *Stroke*, 2008, **39**, 959–966.
54. A. Dharap, K. Bowen, R. Place, L. C. Li and R. Vemuganti, Transient focal ischemia induces extensive temporal changes in rat cerebral MicroRNAome, *J. Cereb. Blood Flow Metab.*, 2009, **29**, 675–687.
55. S. T. Lee, K. Chu, K. H. Jung, H. J. Yoon, D. Jeon, K. M. Kang, K. H. Park, E. K. Bae, M. Kim, S. K. Lee and J. K. Roh, MicroRNAs induced during ischemic preconditioning, *Stroke*, 2010, **41**, 1646–1651.
56. B. Martinez and P. V. Peplow, Blood microRNAs as potential diagnostic markers for hemorrhagic stroke, *Neural Regener. Res.*, 2017, **12**, 13–18.
57. L. Y. Leung, C. P. Chan, Y. K. Leung, H. L. Jiang, J. M. Abrigo, D. F. Wang, J. S. Chung, T. H. Rainer and C. A. Graham, Comparison of miR-1214-3p and miR-16 for early diagnosis of hemorrhagic and ischemic stroke, *Clin. Chim. Acta*, 2014, **433**, 139–144.
58. M. D. Wang, Y. Wang, Y. P. Xia, J. W. Dai, L. Gao, S. Q. Wang, H. J. Wang, L. Mao, M. Li, S. M. Yu, Y. Tu, Q. W. He, G. P. Zhang, L. Wang, G. Z. Xu, H. B. Xu, L. Q. Zhu and B. Hu, High serum miR-130a levels are associated with severe perihematomal edema and predict adverse outcome in acute ICH, *Mol. Neurobiol.*, 2016, **53**, 1310–1321.
59. Y. Zhu, J. L. Wang, Z. Y. He, F. Jin and L. Tang, Association of altered serum microRNAs with perihematomal edema after acute intracerebral hemorrhage, *PLoS One*, 2015, **10**, e0133783.
60. H. W. Zheng, Y. L. Wang, J. X. Lin, N. Li, X. Q. Zhao, G. F. Liu, L. P. Liu, Y. Jiao, W. K. Gu, D. Z. Wang and Y. J. Wang, Circulating microRNAs as potential risk biomarkers for hematoma enlargement after intracerebral hemorrhage, *CNS Neurosci. Ther.*, 2012, **18**, 1003–1010.
61. D. Guo, J. Liu, W. Wang, F. Hao, X. Sun, X. Wu, P. Bu, Y. Zhang, Y. Liu, F. Liu, Q. Zhang and F. Jiang, Alteration in abundance and compartmentalization of inflammation-related miRNAs in plasma after intracerebral hemorrhage, *Stroke*, 2013, **44**, 1739–1742.
62. S. S. Stylli, A. A. Adamides, R. M. Koldej, R. B. Luwor, D. S. Ritchie, J. Ziogas and A. H. Kaye, miRNA expression profiling of cerebrospinal fluid

in patients with aneurysmal subarachnoid hemorrhage, *J. Neurosurg.*, 2016, **126**, 1131–1139.
63. C. J. Powers, R. Dickerson, S. W. Zhang, C. Rink, S. Roy and C. K. Sen, Human cerebrospinal fluid microRNA: temporal changes following subarachnoid hemorrhage, *Physiol. Genomics*, 2016, **48**, 361–366.
64. X. W. Su, A. H. Chan, G. Lu, M. Lin, J. Sze, J. Y. Zhou, W. S. Poon, Q. Liu, V. Z. Zheng and G. K. Wong, Circulating microRNA 132-3p and 324-3p profiles in patients after acute aneurysmal subarachnoid hemorrhage, *PLoS One*, 2015, **10**, e0144724.
65. D. Liu, L. Han, X. Wu, X. Yang, Q. Zhang and F. Jiang, Genome-wide microRNA changes in human intracranial aneurysms, *BMC Neurol.*, 2014, **14**, 188.
66. B. Yuan, H. Shen, L. Lin, T. Su, L. Zhong and Z. Yang, MicroRNA367 negatively regulates the inflammatory response of microglia by targeting IRAK4 in intracerebral hemorrhage, *J. Neuroinflammation*, 2015, **12**, 206.
67. D. Z. Liu, Y. Tian, B. P. Ander, H. Xu, B. S. Stamova, X. Zhan, R. J. Turner, G. Jickling and F. R. Sharp, Brain and blood microRNA expression profiling of ischemic stroke, intracerebral hemorrhage, and kainate seizures, *J. Cereb. Blood Flow Metab.*, 2010, **30**, 92–101.
68. A. H. Muller, G. K. Povlsen, C. H. Bang-Berthelsen, L. S. Kruse, J. Nielsen, K. Warfvinge and L. Edvinsson, Regulation of microRNAs miR-30a and miR-143 in cerebral vasculature after experimental subarachnoid hemorrhage in rats, *BMC Genomics*, 2015, **16**, 119.
69. B. Martinez and P. V. Peplow, MicroRNAs as diagnostic markers and therapeutic targets for traumatic brain injury, *Neural Regener. Res.*, 2017, **12**, 1749–1761.
70. J. W. Lighthall, Controlled cortical impact: a new experimental brain injury model, *J. Neurotrauma*, 1988, **5**, 1–15.
71. L. M. Davis, J. R. Pauly, R. D. Readnower, J. M. Rho and P. G. Sullivan, Fasting is neuroprotective following traumatic brain injury, *J. Neurosci. Res.*, 2008, **86**, 1812–1822.
72. L. K. Gilmer, K. N. Roberts, K. Joy, P. G. Sullivan and S. W. Scheff, Early mitochondrial dysfunction after cortical contusion injury, *J. Neurotrauma*, 2009, **26**, 1271–1280.
73. A. Marmarou, M. A. Foda, W. van den Brink, J. Campbell, H. Kita and K. Demetriadou, A new model of diffuse brain injury in rats. Part 1: Pathophysiology and biomechanics, *J. Neurosurg.*, 1994, **80**, 291–300.
74. S. H. Yang, J. Gustafson, M. Gangidine, D. Stepien, R. Schuster, T. A. Pritts, M. D. Goodman, D. G. Remick and A. B. Lentsch, A murine model of mild traumatic brain injury exhibiting cognitive and motor deficits, *J. Surg. Res.*, 2013, **184**, 981–988.
75. A. Sharma, R. Chandran, E. S. Barry, M. Bhomia, M. A. Hutchison, N. S. Balakathiresan, N. E. Grunberg and R. K. Maheshwari, Identification of serum microRNA signatures for diagnosis of mild traumatic brain injury in a closed head injury model, *PLoS One*, 2014, **9**, e112019.

76. S. V. Kabadi, G. D. Hilton, B. A. Stoica, D. N. Zapple and A. I. Faden, Fluid-percussion-induced traumatic brain injury model in rats, *Nat. Protoc.*, 2010, **5**, 1552–1563.
77. J. Alder, W. Fujioka, J. Lifshitz, D. P. Crockett and S. Thakker-Varia, Lateral fluid percussion: model of traumatic brain injury in mice, *J. Visualized Exp.*, 2011, **54**, 3063.
78. V. Di Pietro, M. Ragusa, D. Davies, Z. Su, J. Hazeldine, G. Lazzarino, L. J. Hill, N. Crombie, M. Foster, M. Purrello, A. Logan and A. Belli, MicroRNAs as novel biomarkers for the diagnosis and prognosis of mild and severe traumatic brain injury, *J. Neurotrauma*, 2017, **34**, 1948–1956.
79. T. Yang, J. Song, X. Bu, C. Wang, J. Wu, J. Cai, S. Wan, C. Fan, C. Zhang and J. Wang, Elevated serum miR-93, miR-191, and miR-499 are noninvasive biomarkers for the presence and progression of traumatic brain injury, *J. Neurochem.*, 2016, **137**, 122–129.
80. L. Sun, A. Liu, J. Zhang, W. Ji, Y. Li, X. Yang, Z. Wu and J. Guo, miR-23b improves cognitive impairments in traumatic brain injury by targeting ATG12-mediated neuronal autophagy, *Behav. Brain Res.*, 2016, **340**, 126–136.
81. W. D. You, Q. L. Tang, L. Wang, J. Lei, J. F. Feng, Q. Mao, G. Y. Gao and J. Y. Jiang, Alteration of microRNA expression in cerebrospinal fluid of unconscious patients after traumatic brain injury and a bioinformatic analysis of related single nucleotide polymorphisms, *Chin. J. Traumatol.*, 2016, **19**, 11–15.
82. M. Bhomia, N. S. Balakarthiresan, K. K. Wang, L. Papas and B. R. Maheshwari, A panel of serum miRNA biomarkers for the diagnosis of severe to mild traumatic brain injury in humans, *Sci. Rep.*, 2016, **6**, 28148.
83. J. B. Redell, A. N. Moore, N. H. Ward, G. W. Hergenroader and R. K. Dash, Human traumatic brain injury alters plasma microRNA levels, *J. Neurotrauma*, 2010, **27**, 2147–2156.
84. L. Liu, T. Sun, Z. Liu, X. Chen, L. Zhao, G. Qu and Q. Li, Traumatic brain injury dysregulates microRNAs to modulate cell signaling in rat hippocampus, *PLoS One*, 2014, **9**, e103948.
85. R. Sandhir, E. Gregory and N. E. Berman, Differential response of miRNA-21 and its targets after traumatic brain injury in aging mice, *Neurochem. Int.*, 2014, **78**, 117–121.
86. L. Meissner, M. Gallozzi, M. Balbi, S. Schwarzmaier, S. Tiedt, N. A. Terpolilli and N. Plesnila, Temporal profile of microRNA expression in contused cortex after traumatic brain injury in mice, *J. Neurotrauma*, 2016, **33**, 713–720.
87. B. Sabirzhanov, Z. Zhao, B. A. Stoica, D. J. Loane, J. Wu, C. Borroto, S. G. Dorsey and A. I. Faden, Downregulation of miR-23a and miR-27a following experimental traumatic brain injury induces neuronal cell death though activation of proapoptotic Bcl-2 proteins, *J. Neurosci.*, 2014, **34**, 10055–10071.

88. W. X. Wang, N. P. Visavadiya, J. D. Pandya, P. T. Nelson, P. G. Sullivan and J. E. Springer, Mitochondria-associated microRNAs in rat hippocampus following traumatic brain injury, *Exp. Neurol.*, 2015, **265**, 84–93.
89. W. Miao, T. H. Bao, J. H. Han, M. Yin, Y. Yan, W. W. Wang and Y. H. Zhu, Voluntary exercise prior to traumatic brain injury alters miRNA expression in the injured mouse cerebral cortex, *Braz. J. Med. Biol. Res.*, 2015, **48**, 433–439.
90. T. Hu, F. J. Zhou, Y. F. Chang, Y. S. Li, G. C. Liu, Y. Hong, H. L. Chen, Y. B. Xiyang and T. H. Bao, miR21 is associated with the cognitive improvement following voluntary running wheel exercise in TBI mice, *J. Mol. Neurosci.*, 2015, **57**, 114–122.
91. B. Sabirzhanov, B. A. Stoica, Z. Zhao, D. J. Loane, J. Wu, S. G. Dorsey and A. I. Faden, miR-711 upregulation induces neuronal cell death after traumatic brain injury, *Cell Death Differ.*, 2016, **23**, 654–658.
92. R. Chandran, A. Sharma, M. Bhomia, N. S. Balakathiresan, B. E. Knollmann-Ritschel and R. K. Maheshwari, Differential expression of microRNAs in the brains of mice subjected to increasing grade of mild traumatic brain injury, *Brain Inj.*, 2017, **31**, 106–119.
93. R. Zhang, Y. Liu, K. Yan, L. Chen, X. R. Chen, P. Li, F. F. Chen and X. D. Jiang, Anti-inflammatory and immunomodulatory mechanisms of mesenchymal stem cell transplantation in experimental traumatic brain injury, *J. Neuroinflammation*, 2013, **10**, 106.
94. X. T. Ge, P. Lei, H. C. Wang, A. L. Zhang, Z. L. Han, X. Chen, S. H. Li, R. C. Jiang, C. S. Kang and J. N. Zhang, miR-21 improves the neurological outcome after traumatic brain injury in rats, *Sci. Rep.*, 2014, **4**, 6718.
95. P. Lei, Y. Li, X. Chen, S. Yang and J. Zhang, Microarray based analysis of microRNA expression in rat cerebral cortex after traumatic brain injury, *Brain Res.*, 2009, **12841**, 191–201.
96. Y. B. Ouyang, Y. Lu, S. Yue, L. J. Xu, X. X. Xiong, R. E. White, X. Sun and R. G. Giffard, miR-181 regulates GRP78 and influences outcome from cerebral ischemia *in vitro* and *in vivo*, *Neurobiol. Dis.*, 2012, **45**, 555–563.
97. K. J. Yin, Z. Deng, H. R. Huang, M. Hamblin, C. Q. Xie, J. F. Zhang and Y. E. Chen, miR-497 regulates neuronal death in mouse brain after transient focal cerebral ischemia, *Neurobiol. Dis.*, 2010, **38**, 17–26.
98. A. Selvamani, P. Sathyan, R. C. Miranda and F. Sohabji, An antagomir to microRNA let7f promotes neuroprotection in an ischemic stroke model, *PLoS One*, 2012, **7**, e32662.
99. E. Caballero-Garrido, J. C. Pena-Philippides, T. Lordkipanidze, D. Bragin, Y. Yang, E. B. Erhardt and T. Roitbak, *In vivo* inhibition of miR-155 promotes recovery after experimental mouse stroke, *J. Neurosci.*, 2015, **9**, 12446–12464.
100. D. Z. Liu, G. C. Jickling, B. P. Ander, H. Hull, X. Zhan, C. Cox, N. Shoff, C. Dykstra-Aiello, B. Stamova and F. R. Sharp, Elevating microRNA-122 in blood improves outcomes after temporary middle cerebral artery occlusion in rats, *J. Cereb. Blood Flow Metab.*, 2016, **36**, 1374–1383.

101. Z. Yang, L. Zhong, R. Xian and B. Yuan, MicroRNA-223 regulates inflammation and brain injury *via* feedback to NLRP3 inflammasome after intracerebral hemorrhage, *Mol. Immunol.*, 2015, **65**, 267–276.
102. T. Y. Sun, X. R. Chen, Z. L. Liu, L. L. Zhao, Y. X. Jiang, G. Q. Qu, R. S. Wang, S. Z. Huang and L. Liu, Expression profiling of microRNAs in hippocampus of rats following traumatic brain injury, *J. Huazhong Univ. Sci. Technol., Med. Sci.*, 2014, **34**, 548–553.
103. Z. Hu, D. Yu, C. Almeida-Suhett, K. Tu, A. M. Marini, L. Eiden, M. F. Braga, J. Zhu and Z. Li, Expression of miRNAs and their cooperative regulation of the pathophysiology in traumatic brain injury, *PLoS One*, 2012, **7**, e39357.

CHAPTER 8

MicroRNAs as Biomarkers for Neurological Disorders

CHI-KWAN LEUNG[a,c], GANG LU[a,c], GEORGE KWOK-CHU WONG[b] AND WAI-YEE CHAN*[a,c]

[a]CUHK-SDU Joint Laboratory on Reproductive Genetics, School of Biomedical Sciences, The Chinese University of Hong Kong, Hong Kong, China; [b]Division of Neurosurgery, Department of Surgery, The Chinese University of Hong Kong, Hong Kong, China; [c]National Research Center for Assisted Reproductive Technology and Reproductive Genetics, Jinan, Shandong, 250001, China
*E-mail: chanwy@cuhk.edu.hk

8.1 Introduction

According to the World Health Organization, neurological disorders are defined as diseases of the central and peripheral nervous systems including the brain, spinal cord, cranial and peripheral nerves, neuromuscular junction, and muscle.[1] A total of 445 major and rare neurological disorders have been documented by the US National Institutes of Health National Institute of Neurological Disorders and Stroke (NIH-NINDS). These include dementia (*e.g.*, Alzheimer's disease, Creutzfeldt–Jakob disease, and Huntington's disease), epilepsy, headache disorders (*e.g.*, migraine, tension-type, and cluster headaches), multiple sclerosis, viral-/bacterial-/parasitic-associated neuroinfections (*e.g.*, HIV-associated neuropathy, poliomyelitis, encephalitis, leprosy,

Drug Discovery Series No. 69
MicroRNAs in Diseases and Disorders: Emerging Therapeutic Targets
Edited by Philip V. Peplow, Bridget Martinez, George A. Calin and Aurora Esquela-Kerscher

Published by the Royal Society of Chemistry, www.rsc.org

meningitis, neurocysticercosis, and neurosyphilis), neurological disorders associated with malnutrition (*e.g.*, ataxic polyneuropathy, toxic optic neuropathy, and spastic paraparesis), alcohol (*e.g.*, fetal alcohol syndrome, alcohol-related polyneuropathy, and Wernicke's encephalopathy), pain (*e.g.*, complex regional pain syndrome and neuropathic pain), Parkinson's disease, stroke (*e.g.*, ischemic stroke, primary intracerebral hemorrhage, and subarachnoid hemorrhage), brain and nervous system cancers (*e.g.*, astrocytic tumor, oligodendroglial tumor, and craniopharyngiomas), and traumatic brain injuries, *etc.*[2]

A recent systematic analysis for the Global Burden of Disease Study 2015 (known as GBD 2015),[3] an international collaboration study coordinated by the Institute for Health Metrics and Evaluation at the University of Washington in Seattle (WA, USA), comprehensively compared the burden of 315 diseases and injuries across 195 countries and territories from 1995 to 2015.[4,5] Neurological disorders represented the leading cause of disability-adjusted life-years (DALYs, an internationally standardized measure of the future stream of healthy life lost due to the incidence of specific diseases and injuries)[6] out of 250.7 million cases, constituting 10.2% of global DALYs. This class of diseases is the second leading cause of mortality, contributing to 16.8% of global deaths (*i.e.*, 94 million). The total number of DALYs from all neurological disorders exceeded those from other diseases and injuries, such as cardiovascular diseases (228.8 million) and cancers (209.4 million). Stroke constitutes the largest proportion of total DALYs (47.3%) and mortality (67.3%) among other neurological disorders. Migraine (13.1%), meningitis (10.1%), and Alzheimer's disease and other forms of dementia (9.5%) ranked as the second, third, and fourth contributors to overall DALYs. Alzheimer's disease and other forms of dementia (20.3%), meningitis (4%), and brain cancers (2.4%) are the second, third, and fourth contributors to mortality. Tension-type headache is the most prevalent neurological disorder (1508.9 million), followed by migraine (958.8 million), medication-use headache (58.5 million), and Alzheimer's disease and other forms of dementia (46 million). With rapid population increase and aging worldwide, it is almost certain that the global burden due to neurological diseases is going to grow.[7] It is of utmost importance to identify disease-specific diagnostic, prognostic, predictive, and therapeutic agents to improve public health and neurological care.

MicroRNAs (miRNAs) are an actively studied class of small noncoding RNA molecules (usually 19–23 nucleotides) that essentially govern all biological processes *via* post-transcriptional mRNA destabilization or translational repression.[8,9] The biogenesis of miRNAs has been excellently reviewed elsewhere and is not discussed in this chapter.[10–12] Since the 1993 discovery of the first miRNA, *lin-4*, which controls temporal larval development in *Caenorhabditis elegans*, the inventory of miRNA across phyla has been expanding rapidly.[13] According to the miRBase 22, a total of 38 589 hairpin precursor miRNAs producing 48 885 mature miRNAs from >250 species have been identified.[14] Mature miRNA is derived from the 5′ arm or the 3′ arm of the same precursor miRNA and is denoted with a -5p or -3p suffix, respectively.[15] Recent studies indicate that both mature miR#-5p and miR#-3p arms confer tissue or cell-specific functions in numerous physiological processes.[16–18] The human genome (GRCh38) has 2693 mature miRNAs, and 724 of them are classified as high-confidence miRNAs.[19,20] Many miRNAs function as promising diagnostic, prognostic, predictive, and therapeutic biomarkers with strong clinical evidence for myriad diseases, such as autoimmune disorders, viral infectious diseases, cardiovascular diseases, diabetes, and certain forms of cancer and neurodegenerative diseases.[21–23] Currently, 137 active miRNA-related interventional and observational studies are registered in the ClinicalTrials.gov results database maintained by the National Institutes of Health National Library of Medicine.[24] The first miRNA-targeted therapeutic is miravirsen (study SPC3649-207), developed by Santaris Pharma in Denmark. It targets miR-122 for the treatment of genotype 1 chronic hepatitis C that is refractory to PEGylated-interferon alpha plus ribavirin, and a phase 2 clinical trial[25] has been completed. Several other miRNA-based therapeutics are in the preclinical phase and are almost ready to enter clinical trials.[26]

Despite the canonical role of miRNAs as genetic regulators to intracellularly control gene expression, previous studies have clearly demonstrated that cell-free circulating miRNAs are also stably present and differentially expressed in 12 biological fluids: plasma, tears, colostrum, saliva, urine, amniotic fluid, breast milk, cerebrospinal fluid (CSF), bronchial lavage, pleural fluid, seminal fluid, and peritoneal fluid.[27] Moreover, miR-335, miR-509-5p, and miR-515-3p are abundantly and ubiquitously expressed in 11 out of the tested 12 body fluids. In contrast, miR-224, miR-637, miR-193b, and miR-508-5p are exclusively detected in plasma, tears, breast milk, and seminal fluid, respectively. Circulating miRNAs have been found encapsulated in extracellular vesicles (*i.e.* exosomes, microvesicles, and apoptotic bodies) contributing to ~75% of the noncoding RNA fraction[28] and in vesicles constituted by high-density lipoproteins. The biogenesis, mechanism of miRNA release, size, and composition of surface markers of different extracellular vesicles are vastly different.[29] Formation and release of exosomes, a class of cell-derived secretory vesicles (30–150 nm) containing selected RNA and protein cargoes of endosomal origin *via* the endocytic recycling pathway is a highly ATP-dependent and selective process.[30] A distinct package of circulating miRNAs from parent cells are loaded in exosomes to confer functional specificity to regulate target gene expression in recipient cells.[31] The secretion rate of circulating miRNA-enriched-microvesicles, *via* the outward budding and fission of cell membrane of their parent cells, into biological fluids increases rapidly upon stimulation.[32] Numerous cell types have been shown to release functional circulating miRNA to engage intercellular communication such as cancer cells,[33] neuronal cells[34] and a panel of hematopoietic and vascular cells.[35] The release of apoptotic bodies, the largest extracellular vesicles with a size range of 1–5 μm in diameter, occurs during early apoptosis and under stress or stimulation.[36] The secretion of apoptotic bodies into biological fluids is mediated by Rho-associated coiled-coil kinase 1 (ROCK1) and caspase-3 (CASP3) to trigger membrane blebbing.[37,38] Additionally, circulating miRNAs are associated with RNA-binding proteins[39,40] and several protein members of the Argonaute (AGO) family[41] and released to extracellular milieu for cell–cell communication.

Circulating miRNAs are resistant to RNase degradation and have profound extracellular stability.[42] Circulating RNAs also remain stable under undesirable conditions, such as heat, extreme pH, prolonged storage, and multiple freeze–thaw cycles.[43] Once the miRNA-enriched extracellular vesicles are released from donor cells, the protected miRNA cargos undergo extravesicular transfer to recipient cells *via* the circulating system.[44–46] The still-active miRNAs then inhibit their *bona fide* target gene expression.[47,48] Circulating miRNAs indeed behave as hormones to nonautonomously regulate the physiology of distant recipient cells.[43]

Distinct expression profiles of circulating miRNAs collected from different biological fluids associated with certain pathophysiological conditions suggest that the circulating miRNAs are selectively released from specific cells/tissues/organs of a specific disease at distinct progression stages into targeted biological fluids.[49,50] Circulating organ-/tissue-/cell-specific/enriched miRNAs associated with specific diseases at different stages of progression allows precise diagnosis, prognosis, and personalized treatment.[51] For example, two plasma-specific and tissue-specific circulating miRNA-based signatures were identified for histological classification and tumor node metastasis staging in non-small cell lung cancer.[52]

Circulating miRNAs have been demonstrated to be robust diagnostic/prognostic biomarkers for numerous diseases, including gestational diabetes mellitus (*e.g.*, miR-330-3p);[53] tuberculosis (*e.g.*, miR-93 and miR-29a);[54] sepsis (*e.g.*, miR-150 and miR-146a);[55] hepatitis (*e.g.*, miR-122 and miR-99a-5p);[56] immunodeficiency virus (HIV)-associated neurological disorders (*e.g.*, miR-495-3p and *let-7b-5p*);[57] varicella (*e.g.*, miR-197 and miR-629);[58] influenza (*e.g.*, miR-1260 and miR-664);[59] hand, foot, and mouth disease (*e.g.*, miR-148a and miR-143);[60] and certain forms of cancer (*e.g.*, the miR-21/miR-375 ratio for esophageal cancer, miR-451 and miR-486 for gastric cancer, miR-378 for colorectal cancer, and miR-223 for pancreatic cancer).[48,61,62]

Additionally, circulating miRNAs have attracted increasing attention from neurologists for their promising roles in disease classification and monitoring disease progression. Sheinerman *et al.* reported a circulating miRNA-based classifier to detect and categorize different subtypes or the progression status of Alzheimer's disease (AD), frontotemporal dementia (FTD), Parkinson's disease (PD), and amyotrophic lateral sclerosis (ALS), highlighting the robustness and reproducibility of applying circulating miRNAs as biomarkers for diagnosis and prognosis.[63] Considering that the collection of circulating miRNAs is minimally invasive or noninvasive and that the quantitation of a specific or a panel of circulating miRNAs is fairly easy and inexpensive, these fluid-enriched miRNAs may be ideal candidates for promising noninvasive and reliable biomarkers for detecting and monitoring specific physiopathological conditions.[27]

Here, we review the current knowledge of the roles of circulating miRNAs in the detection, classification, and differentiation of neurological diseases.

8.2 Ischemic Stroke

According to the Centers for Disease Control and Prevention (CDC), ~6.5 million people died from stroke in 2013, one of the leading causes of mortality and long-term disability.[64] In the USA, about 795 000 stroke cases occur each year, with someone suffering a stroke every 40 s and dying every 4 min.[65] Ischemic stroke accounts for 87% of stroke, followed by hemorrhagic stroke (intracerebral hemorrhage and subarachnoid hemorrhage) and transient ischemic attack (aka mini-stroke).[66]

Previous studies have indicated that circulating miR-145 is consistently elevated in the whole-blood[67] and serum[68] samples of ischemic stroke patients. miR-210 was differentially regulated in the blood of acute cerebral ischemia patients.[69] The expression levels of miR-99a-5p in neutrophils and lymphocytes in the blood sample of acute ischemic stroke patients could be used to separate the stroke patients from healthy controls with accuracy levels of 82% and 85%, respectively.[70] Sepramaniam *et al.* reported a large-scale cohort study that profiled the expression of blood-derived miRNAs from 169 acute stroke patients by microarray. A total of five circulatory miRNAs (miR-125b-2, miR-27a, miR-422a, miR-488, and miR-627) were found to be elevated in stroke patients, with a profound discriminatory power from healthy controls and patients with metabolic complications.[71]

Recently, Vijayan *et al.* conducted small RNA sequencing and identified a total of 4656 differentially expressed miRNAs in the serum samples from patients with ischemic stroke. Among them, the expression of four miRNAs, *PC-3p-57664*, *PC-5p-12969*, miR-122-5p, and miR-211–5p, was confirmed in an independent validation cohort, with a strong discriminatory power of 76%, 80%, 72%, and 72%, respectively, highlighting their promising role in diagnosing ischemic stroke.[72]

A recent systematic review reported by Dewdney *et al.* indicated that a meta-analysis of eight clinical studies recruiting 572 ischemic stroke patients and 431 control subjects identified 22 circulatory miRNAs that were differentially expressed in body fluids. Despite the fact that miR-106b was the only one consistently identified in more than one study, the reported expression levels were either largely varied or the directionality of expression was opposite to that observed in other studies.[73] The marked discrepancy across studies hinders the clinical application of miRNA-based biomarkers and prompts the need to further research into these miRNAs-related pathophysiological mechanisms.

8.3 Aneurysmal Subarachnoid Hemorrhage

Aneurysmal subarachnoid hemorrhage (SAH) accounts for 3–5% of strokes. Approximately 44% of SAH patients develop delayed cerebral infarction (DCI), a well-established clinical marker for neurological consequence post-SAH, at day 4–7 following the initial hemorrhage.[74]

We and others have successfully demonstrated in SAH patients a dynamic shift of the circulatory miRNA profile, which could be a useful tool for clinicians to devise miRNA-based diagnostic, prognostic, and therapeutic tools for management of the disease.[75–78] Recently, we profiled the expressions of circulating miRNAs in SAH patients with or without DCI and determined that a four miRNA-based classifier (miR-4532, miR-4463, miR-1290, and miR-4793) collected in sera exhibited a powerful discriminatory power between SAH patients with and without DCI with an accuracy of 100% using the least absolute shrinkage and selection operator (LASSO) method of regression analysis. The miRNA panel also differentiated SAH patient with or without DCI from control groups with an accuracy of 99.3% and 82%, respectively, inferring its promising role in disease classification.[79]

Ye *et al.* profiled the plasma miRNAs in SAH patients after the onset (phase I), the peak period of cerebral vasospasm (phase II), and cerebral vasospasm dissipation stage (phase III). Elevated levels of miR-3195, miR-4788, and miR-1914 were shown to discriminate SAH patients at phase I and phase II with a predictive power of 70%, 81%, and 78%, respectively, and SAH patients at phase II and III with a predictive power of 70%, 89%, and 65%, respectively. The combination of the three miRNA signatures exhibited a higher predictive power to differentiate phase I *versus* II and II *versus* III of 98% and 90%, respectively.[80] Also, serum levels of miR-502-5p and miR-1297 were elevated in SAH patients on the day after onset and could discriminate SAH patients from controls with a predictive power of 96% and 95%, respectively.[75]

8.4 Alzheimer's Disease

Alzheimer's disease is the most common type of dementia (50–75%), characterized by progressive memory loss and cognitive impairment.[81] In 2017, 3.65 million AD cases with both amyloidosis and neurodegeneration were reported in the United States; 42% of them required special late-staged neurological care. Estimates suggest that the prevalence will surge to 9.3 million by 2060.[82]

Bekris *et al.* reported the identification of miR-15a from 26 plasma samples of AD patients, which correlates with AD-specific neuropathological hallmarks, such as the neuritic plaque score and the Braak stage, a six-stage clinical assessment and differentiation of AD-associated neurofibrillary pathology introduced by Braak *et al.* in 1991.[83] Serum miR-15a could act as a promising biomarker of neurological aberration in AD patients.[84] Bhatnagar *et al.* examined the expression levels of miR-34a and miR-34c in 110

AD plasma patients. miR-34a and miR-34c were found to be upregulated in AD samples, and they could distinguish the AD from control groups with an accuracy of 79% and 94%.[85] It was postulated that miR-34c regulates cell survival and oxidative stress response, the two mechanisms governing AD pathology, by mediating its target genes, such as B-cell lymphoma 2 (Bcl2), silent mating type information regulation 2 homolog 1 (Sirt1), presenilin-1 (Ps1), and one cut homeobox 2 (Onecut2).[85]

Liu *et al.* determined that miR-884 was downregulated in both CSF and serum samples from 77 AD patients with mild cognitive impairment and dementia of Alzheimer's type (DAT) and that it was negatively correlated with serum Aβ42. Most importantly, miR-884 targets amyloid precursor protein (APP) and the beta-site APP-cleaving enzyme (BACE-1) that is crucial for AD pathogenesis.[86] The same group also reported that both miR-135a and miR-200b in 38 serum and CSF samples from DAT patients were significantly downregulated compared to the control group. The authors found that the levels of miR-135a and miR-200b were decreased in hippocampi from APP/presenilin 1 (PS1) double-transgenic mice, targeting BACE-1 and APP, respectively.[87] Zhu *et al.* adopted a candidate gene approach to specifically evaluate the expression level of miR-210 in AD patients. They discovered that the level of miR-210 was decreased in 56 CSF and serum samples collected from mild cognitive impairment and AD patients. The degree of reduction was more severe in AD patients than in the mild cognitive impairment patients, suggesting that miR-210 is indicative of AD pathogenesis.[88] Gui *et al.* concluded that the miR-29c, miR-136-3p, miR-16-2, miR-331-5p, miR-132-5p, and miR-485-5p were differentially expressed in 28 CSF samples from AD patients.[89]

Cogswell *et al.* reported the profiling of circulating miRNAs in CSF of 10 AD patients (Braak 5) and 10 non-demented controls (Braak 1). A total of 60 out of 242 miRNAs tested were differentially expressed between the AD and control groups; 23 and 27 of them were downregulated and upregulated, respectively, compared to the control groups. Among them, miR-30c and miR-146b were also consistently shown to be significantly altered in the hippocampus and/or medial frontal gyrus, the regions mainly affected by AD pathology, in either early (Braak 3,4) or late (Braak 5,6) stages of the disease.[90] An aberrant downregulation of miR-146b expression was also implicated in the aggregation of Aβ42 in senile plaques and in the formation of neurofibrillary tangles by hyperphosphorylated tau, the histopathological hallmarks of AD.[91] The attenuation of miR-146b is expected to derepress the translational repression of IRAK1 and TRAF6, the downstream effectors of Toll-receptor (TLR) signaling, to promote TLR-mediated neurodegeneration.[92,93] Sala Frigerio *et al.* examined the expression levels of 728 CSF-miRNAs in two cohorts with a total of 35 AD patients with dementia and 37 control subjects. miR-27a-3p consistently proved to be significantly downregulated in both AD cohorts with elevated levels of Aβ42, total-tau, and phosphorylated-tau. An *in vitro* luciferase assay suggested that miR-27a-3p targets four AD-associated genes in AD pathogenesis: β-secretase 1 (Bace1), glycogen synthase kinase 3 beta

(Gsk3b), microtubule-associated protein tau (Mapt), and Ps1. The physiological relevance of miR-27a-3p-mediated gene regulation in AD progression remains elusive.[94] However, the correlation between the altered expression of miR-27a-3p and a higher level of tau and Aβ42 in CSF was not replicated in an independent study.[95]

A microarray-based profiling of circulating miRNAs in CSF in six AD tissues with a clinical dementia rating of 2 or 3 discerned that miR-9, miR-125b, miR-146a, and miR-155 were markedly upregulated in AD-associated CSF and brain tissue-derived extracellular fluid. The upregulation of these four miRNAs is attenuated by anti-nuclear factor kappa-light-chain-enhancer of activated B cells (NF-κB) agents, raising the therapeutic possibility that the expression of the AD-associated miRNAs could be manipulated using NF-κB inhibitors.[96] However, Kiko *et al.* revealed that both miR-145a and miR-34a were found to be downregulated in both plasma and the CSF of 10 AD patients.[97] Later, in 20 CSF samples derived from AD patients, Müller *et al.* reported a significant reduction in miR-146a levels, which can also be used as a classifier to distinguish AD [Mini Mental State Examination (MMSE) score of 20.2 ± 3.4] from control groups with an accuracy of 80%. MMSE is a 30-point questionnaire to measure cognitive impairment for dementia; any score <24 indicates cognitive impairment.[98] Additionally, in the same study, they demonstrated that the contamination of a minute amount of blood in a CSF sample with <100 erythrocytes can significantly alter the miRNA expression level. The effect of blood cells in CSF samples needs to be considered in data interpretation.[99] However, Denk *et al.* profiled the expression of 1881 miRNAs in CSF collected from 22 AD patients and identified six reliable (miR-4449, miR-106a, miR-1274a, miR-146a, miR-335, and miR-100) and nine informative (miR-103, miR-375, miR-505, miR-708, miR-4467, miR-219, miR-296, miR-766, and miR-3622b-3p) biomarkers for AD. Compared to control groups, the level of miR-146a was significantly elevated in the blood cell-free CSF samples of AD patients and to be negatively correlated with tau and Aβ1–42 abundance.[95]

Although Kiko *et al.* found a decrease in the level of miR-125b in AD-derived CSF samples,[97] an independent study indicated that miR-125b and miR-29a were indeed elevated in 18 cell-free CSF samples of AD patients, suggesting that they could function as AD classifiers with an accuracy of 69% and 79.5%, respectively.[100] In contrast, Galimberti *et al.* established that both miR-125b and miR-26b were repressed in 22 cell-free CSF samples of AD patients compared to 18 noninflammatory and inflammatory neurological disease samples in the control group. The serum level of miR-125b was also shown to differentiate AD from control groups with an accuracy of 82%.[101] Tan *et al.* conducted a larger cohort study to identify circulating miRNAs in the serum of 105 AD patients and 155 control subjects.[102] miR-125b was consistently found to be downregulated in AD patients.[102] Importantly, miR-125b expression was demonstrated to correlate negatively with MMSE scores (11.01–29.76) in recruited subjects and was suggested to distinguish AD patients from control subjects with an accuracy of 74.6%.[102] Furthermore, the level of miR-9 was elevated in AD serum samples, but this result was contradictory to an earlier small-scale cohort study by Geekiyanage *et al.*[103] They discovered that the levels of miR-9, as well as miR-137, miR-181, miR-29a, and miR-29b, were all significantly reduced in seven AD serum samples.

Burgos *et al.* conducted the first global profiling of circulating miRNA in CSF and serum in both AD and PD patients, using small-RNA next-generation sequencing.[104] A total of 41 miRNAs were downregulated in 69 CSF samples of AD patients. However, 17 miRNAs were upregulated and three were downregulated in the serum of 53 AD patients. A number of the circulating miRNAs have been consistently identified in other studies with the same direction of expression, including miR-181a-5p,[90] miR-708-5p,[101] miR-136-3p,[89] and miR-135a-5p.[104] The authors suggested that circulating RNAs are promising indicators of disease progression and can also be used to evaluate therapeutic efficacy.[104] Moreover, a total of five and 16 miRNAs in CSF and serum, respectively, can be used to effectively differentiate AD from PD (Table 8.1), highlighting the potential of these disease-specific circulating miRNAs in the clinical classification of various neurological disorders. Dong *et al.* performed another small-RNA sequencing profiling of 184 serum samples of AD patients followed by validation using quantitative reverse transcriptase polymerase chain reaction (qRT-PCR). Four miRNAs (miR-31, miR-93, miR-143, and miR-146a) were significantly downregulated in the AD serum samples.[105]

Single miRNA-based biomarkers for AD diagnosis or prognosis, such as miR-146a, miR-9, and miR-125b, could lead to conflicting results. A global unbiased survey of circulating miRNAs from AD bodily fluid was performed to identify a miRNA-based signature to the classifier to distinguish AD from controls or other neurological diseases. Kumar *et al.* conducted an miRNA expression array to identify the top 100 differentially expressed miRNAs in plasma samples from 57 AD patients. Seven miRNAs (*let-7d-5p*, *let-7g-5p*, miR-15b-5p, miR-142-3p, miR-191-5p, miR-301a-3p, and miR-545-3p) were significantly downregulated in AD patients compared to control groups. Authors also observed that these seven plasma-based miRNA signatures could robustly differentiate AD from controls groups in a validation study comprising 20 AD and 17 control subjects with an accuracy of >95%. This was of significant clinical interest for disease differentiation.[106] A plasma-based profiling of circulating miRNAs collected from 250 diseased subjects of AD, FTD, PD, and ALS identified miRNA-based classifiers out of 37 brain-enriched miRNAs to distinguish a specific neurological disease from the control groups, other diseases, and different subtypes of the same disease (Table 8.1).[63] For example, a four miRNA-based classifier (miR-99b, miR-874, miR-9-3p, and miR-155) can differentiate AD from controls with 89% accuracy, while a miR-31, miR-206, miR-125b, miR-335-5p, miR-107, miR-335-5p-based classifier can clearly separate AD from ALS with 93% accuracy. However, Denk *et al.* reported that a miRNA-based classifier (miR-100, miR-103, and miR-375) could robustly classify AD and control cases with accuracy levels of 99.5% and 96.4%, respectively.[95] Tan *et al.* reported the largest cohort study with 208 AD and 205 control subjects and identified miR-342-3p and a six miRNA-based classifier (miR-98-5p, miR-885-5p, miR-483-3p, miR-342-3p, miR-191-5p, and *let-7d-5p*) in serum to separate AD from controls with an accuracy >75% using next-generation sequencing.[107] Additionally, Cheng *et al.* performed miRNA sequencing followed by qRT-PCR validation discovered 16 differentially regulated serum miRNAs from 50 AD patients; three were downregulated (miR-1306-5p, miR-342-3p, and miR-18b-5p), and 13 (miR-20a, miR-30e, miR-582, miR-106a, miR-361, miR-143, miR-424, miR-93, miR-106b, miR-101, miR-15b, miR-335, and miR-15a) were downregulated. The authors discerned that a miRNA-based classifier comprising these 16 AD-associated serum miRNAs can differentiate AD from control groups with 82% accuracy. However, the authors mentioned that this miRNA signature failed to diagnose late AD with the highest (Pittsburg Compound B-standardized uptake value ratio) PiB-SUVR score of 3.03.[108] The PiB-SUVR is an *in vivo* quantitative measurement of amyloid-beta-protein deposits using PiB-based amyloid emission tomography for AD diagnosis;[109] any score >3 indicates dementia.[110] Nonetheless, the results hold promise for utilizing circulating miRNA-based signatures or classifiers to detect and differentiate different groups and subtypes of neurological diseases.

8.5 Parkinson's Disease

Parkinson's disease is characterized by cardinal motor symptoms, including impaired executive functioning, postural instability, resting tremor, bradykinesias, and memory loss. It is the second most common neurological disease worldwide, after AD. The prevalence of PD is ~0.1–0.2% of the world population at one time, representing up to 1% of the population aged >60.[111] The pathological hallmarks of PD include a loss of dopaminergic neurons and the assembly and progressive spreading of α-synuclein-containing Lewy bodies in the substantia nigra.[112] Lewy bodies are defined as the protein aggregates developed in nerve cells in the regions associated with cognitive functions and are considered a histological hallmark of neuronal degeneration.[113] Similar to other neurological diseases, when PD patients reach terminal stages with the exhibition of full-blown symptoms, the disease becomes irreversible and is not treatable. The rapid and accurate diagnosis of PD at an early stage could attenuate or even prevent worsening of the pathological process and result in a better quality of life.

Cardo *et al.* observed that miR-331-5p was significantly upregulated in the plasma of 31 PD patients and that it was predicted to target numerous regulators acting in the neurogenesis and neurodegeneration underlying the pathogenesis of PD.[114] miR-133b was found to be regulated in the serum of 46 PD patients and was correlated with the abundance of ceruloplasmin, a ferroxidase that modulates iron hemostasis, which is crucial for the development of Lewy bodies in PD patients. However, the serum level of miR-133b was not correlated with disease severity and motor impairment phenotypes.[115] Gui *et al.* discovered a total of 27 circulating miRNAs that were differentially expressed in CSF from 47 PD patients; 15

Table 8.1 Summary of the circulating miRNAs of neurological disorders.[a]

Neurological disease	Body fluid	Sub groups	# miRNAs tested	miRNA testing platform	Circulating miRNAs Control (accuracy %)
Alzheimer's disease	Plasma	N/A	37 (brain-enriched)	qRT-PCR	miR-99b/miR-874/miR-9-3p/miR-155 (89%)
			5		Up: miR-15a
			2		miR-34a (79%) miR-34c (94%)
				miRNA expression assay	*let-7d-5p*, *let-7g-5p*, miR-15b-5p, miR-142-3p, miR-191-5p, miR-301a-3p, miR-545-3p (95%)
			6	qRT-PCR	Up: miR-29a, miR-29b Down: miR-125b, miR-146a, miR-34a
	CSF	Braak 1 *versus* Braak 5	242		Up: miR-138-5p Down: miR-146b, miR-181a, miR-182-5p
		N/A	728		Down: miR-27a-3p
			N/A	Microarray	Up: miR-9, miR-125b, miR-146a, miR-155
			8	qRT-PCR	Down: miR-146a (80%)
			1178	OpenArray RT-qPCR	miR-100, miR-103 And miR-375 (96.4%)
			4	qRT-PCR	Up: miR-125b (69%), miR-29a (79.5%)
			84		Down: miR-125b (82%), miR-26b
			62		Down: miR-384
			N/A	Microarray	Down: miR-135a, miR-200b
			1	qRT-PCR	Down: miR-210
			746	qRT-PCR-based array	Down: miR-16-2, miR-331-5p, miR-136-3p, miR-29c Up: miR-132-5p, miR-485-5p, miR-151
			N/A	Small RNA sequencing	Down: miR-124-3p, miR-138-5p, miR-127-3p, miR-132-3p, miR-127-5p, miR-136-3p, miR-381, miR-101-5p, miR-199b-5p, miR-136-5p, miR-1842, miR-181a-5p, miR-598, miR-218-5p, miR-9-3p, miR-769-5p, miR-953, miR-760, miR-181a-3p, miR-181b-5p, miR-488-3p, miR-4953, miR-708-3p, miR-874, miR-873-5p, miR-129-5p, miR-181, miR-139-5p, miR-3200-3p, miR-431-3p, miR-9-5p, miR-326, miR-377-5p, miR-4331, miR-323a-3p, miR-134, miR-3293, miR-10a-5p, miR-33b-5p, miR-410, miR-708-5p

classifier from							
Other neurological diseases (accuracy %)	Other subtypes (accuracy %)	Diseased group n	Control group n	Age (years)	Sex (M/F)	Location of the study	Reference
FTD: miR-125b/miR-874/miR-107/miR-335-5p/miR-155/miR-29a (76%) PD: miR-31/miR-99b/miR-129-3p/miR-16/miR-204/miR-9-3p (89%) ALS: miR-31/miR-206/miR-125b/miR-335-5p/miR-107/miR-335-5p (93%)		50	50	53–87	24/26	USA	63
N/A		26	26	69–87	24/28	USA	84
		110	123	51–90	N/A	Canada	85
		57	35	61–91	35/57	USA	106
		10	10	67–86	7/13	Japan	97
		10	10	49–90	10/10	The Netherlands	90
		35	37	60–84	36/36	Sweden	94
		6	6	65–80	N/A	USA	96
		20	20	53.9–79.3	20/20	Nijmegen	99
		22	28	47.3–80.6	23/27	Germany	95
		18	20	47.3–80.6	16/22	Nijmegen	100
		22	18	68–72	21/19	Italy	101
		77	50	57–69	68/59	China	86
		5	5	70–78	8/2	China	87
		56	42	61–85	53/45	China	88
		28	27	42–79	24/31	China	89
PD: miR-19b-3p, miR-19a-3p, miR-127-5p, miR-873-5p, miR-32-5p		67	70	74–89	76/72	USA	104

(continued)

Table 8.1 (*continued*)

Neurological disease	Body fluid	Sub groups	# miRNAs tested	miRNA testing platform	Circulating miRNAs Control (accuracy %)
	Serum		62	qRT-PCR	Down: miR-384
			84		Down: miR-125b (82%), miR-26b
			N/A	Microarray	Down: miR-135a, miR-200b
			1	qRT-PCR	Down: miR-210
			N/A	Small-RNA sequencing	Up: miR-34b-3p, miR-219-2-3p, miR-22-5p, miR-125b-1-3p, miR-1307-5p, miR-34c-5p, miR-34b-5p, miR-887,miR-135a-5p, miR-184, miR-30c-2-3p, miR-873-3p, miR-125a-3p, miR-671-3p, miR-1285-3p, miR-3176, miR-127-3p Down: miR-182-5p, miR-21-5p, miR-375
			6	qRT-PCR	Up: miR-9 Down: miR-125b, miR-181c
			7		Down: miR-137, miR-181, miR-9, miR-29a, miR-29b
		AD *versus* MCI *versus* vascular dementia (VD)	N/A	Small-RNA sequencing/qRT-PCR	Down: miR-31, miR-93, miR-143, and miR-146a
		N/A	N/A		Up: [miR-20a, miR-30e, miR-582, miR-106a, miR-361, miR-143. miR-424, miR-93, miR-106b, miR-101, miR-15b, miR-335, miR-15a Down: miR-1306-5p, miR-342-3p, miR-18b-5p] (82%)
			N/A	Small-RNA sequencing/qRT-PCR	[Down: miR-98-5p, miR-885-5p, miR-483-3p, miR-342-3p, miR-191-5p and *let-7d-5p*] (74%) miR-342-3p (76%)
Parkinson's disease	Plasma	Total	37 (brain-enriched)	qRT-PCR	miR-9-3p/miR-129-3p/miR-127-3p/miR-411/miR-191-5p/miR-138 (91%)
		Dementia	37 (brain-enriched)	qRT-PCR	miR-206/miR-9 (70%)
		MCI	37 (brain-enriched)	qRT-PCR	miR-132 and miR-134
		N/A	N/A	Microarray	Up: miR-505, miR-1826
			4	qRT-PCR	miR-1826, miR-450b-3p, miR-505, and miR-626
			377	TaqMan low-density miRNA cards	Up: miR-331-5p
			N/A	Small-RNA sequencing	Up: miR-27 Down: *let-7a*, *let-7f*, miR-142-3p, and miR-222

classifier from							
Other neurological diseases (accuracy %)	Other subtypes (accuracy %)	Diseased group n	Control group n	Age (years)	Sex (M/F)	Location of the study	Refer-ence
N/A		77	50	57–69	68/59	China	86
		22	18	68–72	21/19	Italy	101
		69	30	69–83	36/63	China	87
		56	42	61–85	53/45	China	88
PD: miR-320a, miR-320b, miR-378d, miR-378f, miR-378a, miR-378b, miR-378c, miR-193a-5p, miR-320c, miR-1285-3p, miR-550a-3-5p, miR-874, miR-125a-5p, miR-21-5p, miR-671-3p, *let-7i-3p*	N/A	64	72	74–89	76/72	USA	104
N/A		105	155	66–76	132/128	China	102
N/A		7	7	N/A	6/8	USA	103
		187	123	73–88	184/126	China	105
		50	59	66–80	47/62	Australia	108
N/A		208	205	66–84	198/215	China	107
ALS: miR-9-3p/miR-206/miR-155/miR-206/miR-7/miR-206 (81%)		50	50	49–82	35/15	USA	63
N/A		15	50	61–82	11/4	USA	
		8	50	54–77	5/3	USA	
		104	62	55–80	74/92	UK	120
		30	7	60–90	19/18	Sweden	
		31	25	63–69	N/A	Asturias, Spain	114
N/A		25	25	56–72	16/9	China	117

(continued)

Table 8.1 (*continued*)

Neurological disease	Body fluid	Sub groups	# miRNAs tested	miRNA testing platform	Circulating miRNAs: Control (accuracy %)
	CSF		746	qRT-PCR-based array	Up: miR-103a, miR-30b, miR-16-2, miR-26a, miR-331-5p, miR-153, miR-132-5p, miR-485-5p, miR-127-3p, miR-409-3p, miR-433,miR-370, miR-7g-3p, miR-873-3p, miR-136-3p, miR-10a-5p Down: miR-1, miR-22, miR-29, miR-374, miR-119a, miR-126, miR-151,miR-28, miR-301a, miR-19b-3p, miR-29c
			N/A	Small-RNA sequencing	Down: miR-132-5p, miR-485-5p, miR-127-3p, miR-128, miR-409-3p, miR-4331, miR-370, miR-431-3p, miR-873-3p, miR-136-3p, miR-212-3p, miR-10a-5p, miR-1224-5p, miR-4448 Up: miR-19a-3p, miR-19b-3p, *let-7g-3p*
	Serum	N/A	N/A		Down: miR-16.2-3p, miR-1294 Up: miR-338-3p, miR-30e-3p, miR-30a-3p
			377	qRT-PCR	miR-29c, miR-29a, and miR-19a (75%)
			754	TaqMan low density array	Up: miR-24 Down: miR-34b, miR-148b
			1	qRT-PCR	Down: miR-133
ALS	Plasma	Total	37 (brain-enriched)		miR-206/miR-31/miR-335-5p/miR-338-3p (88%)
		With FTD	37 (brain-enriched)		N/A
		With MCI	37 (brain-enriched)		N/A
FTD		Total	37 (brain-enriched)		miR-9-3p/*let-7e*/miR-7/miR-451/miR-335-5p/*let-7e* (87%)
		Behavioral variant	37 (brain-enriched)		N/A
		PPA logopenic variant	37 (brain-enriched)		
		PPA/progressive nonfluent aphasia	37 (brain-enriched)		
		PPA/semantic dementia	37 (brain-enriched)		
		Progressive supranuclear palsy	37 (brain-enriched)		

classifier from								
Other neurological diseases (accuracy %)	Other subtypes (accuracy %)	Diseased group n	Control group n	Age (years)	Sex (M/F)	Location of the study	Reference
N/A		47	27	42–79	34/40	China	89
		65	70	75–85	84/71	USA	104
		60	72	75–85	84/71	USA	104
		197	122	41–79	146/173	Spain	118
MSA: miR-24, miR-34b, miR-148b		25	25	55–63	26/24	Italy	119
		46	46	57–62	53/39	China	115
		50	50	29–83	36/14	USA	63
	miR-218/miR-451/miR-29a/miR-9-3p/miR-491-5/*let-7e* (76%)	5	50	49–70	4/1	USA	
		3	50	54–63	2/1	USA	
		50	50	46–76	28/22	USA	
PD: miR-129-3p/miR-9-3p/miR-338-3p/miR-99b/miR-323-3p/miR-127-3p (87%)	miR-335-5p/miR-433 (66%)	23	50	46–74	15/8	USA	
ALS: miR-129-3p/miR-206/miR-338-3p/*let-7e*	miR-134/miR-29a/miR-181a/	1	50	70	1/0	USA	
N/A	miR-138/miR-338-3p/miR-29a (72%)	8	50	61–75	3/5	USA	
		8	50	49–76	4/4	USA	
	miR-155/miR-99b (74%)	10	50	54–75	5/5	USA	

(*continued*)

Table 8.1 (*continued*)

Neurological disease	Body fluid	Sub groups	# miRNAs tested	miRNA testing platform	Circulating miRNAs Control (accuracy %)
Multiple sclerosis	Serum	N/A	N/A	Small-RNA sequencing	miR-15b-5p, miR-23a-3p, miR-223-3p, miR-374a-5p, miR-30b-5p, miR-433-3p, miR-485-3p, miR-342-3p, miR-432-5p
Epilepsy		N/A	N/A		miR-106b-5p (81%)
SAH		SAH with or without DCI	28	qRT-PCR	DCI *versus* non-DCI (miR-4532, miR-4463, miR-1290, and miR-4793) 100%
		Day 3 after SAH	N/A	Microarray	miR-502-5p, miR-1297
	Plasma	SAH with phase 1 2, or 3	N/A		miR-3195, miR-4788, miR-1914
	CSF	SAH patients with or without vasospasm	800	nanoString nCounter expression assay	miR-204-5p, miR-223-3p, miR-337-5p, miR-451a, miR-489, miR-508-3p, miR-514-3p, miR-516-5p, miR-548 m, miR-599, miR-937, miR-1224-3p, and miR-1301
Stroke	Whole blood	N/A	1	qRT-PCR	Up: miR-145
		N/A	N/A	Microarray	Up: miR-125b-2, -27a, -422a, -488 and -627
		Acute stroke	1	qRT-PCR	Up: miR-99a-5p
		N/A	9		Down: miR-126, miR-21
	Serum		N/A	Small-RNA sequencing	Up: PC-3p-57664, PC-5p-12969, miR-122-5p,miR-211–5p
			9	qRT-PCR	Up: miR-145
			3100	Microarray	Up: miR-32-3p, miR-106-5p, miR-1246, miR-532-5p
			2	qRT-PCR	Up: *let-7e* (78%)
			1		miR-223
	Plasma	Ischemic stroke (24 h, 1 week, 4 weeks, 24 weeks)	3		Down: miR-30a, miR-126 and *let-7b*
		Hemorrhagic stroke *versus* ischemic stroke	2		miR-124-3p, miR-16
		N/A	2		Down: miR-21, miR-24
		N/A	1347	Microarray	Up: miR-106b-5p, miR-4306 Down: miR-320e

[a]AD: Alzheimer's disease; ALS: amyotrophic lateral sclerosis; CSF: cerebrospinal fluid; DCI: delayed cerebral infarction; F: female; FTD: frontotemporal dementia; M: male; MCI: mild cognitive impairment; MS: multiple sclerosis; MSA: multiple system atrophy; N/A: not available; PD: Parkinson's disease; PPA: primary progressive aphasia; qRT-PCR: quantitative reverse-transcriptase polymerase chain reaction; SAH: aneurysmal subarachnoid hemorrhage.

classifier from							
Other neurological diseases (accuracy %)	Other subtypes (accuracy %)	Diseased group n	Control group n	Age (years)	Sex (M/F)	Location of the study	Reference
N/A		25	11	19–67	12/24	Australia	132
		147	142	15–40	143/146	China	129
		20	20	33–71	9/21	Hong Kong	79
		60	10	N/A	28/32	China	75
		9	9	N/A	N/A	China	80
		20	4	43–71	9/15	Australia	78
		32	14	18–49	N/A	Malaysia	67
		169	24	31–47	N/A	Singapore	137
		21	N/A	N/A	N/A	China	137
		21	10	N/A	N/A	South Korea	138
		34	11	56–68	18/27	USA	72
		146	96	48–81	100/145	China	68
		117	82	66–69	60/139	China	139
		72	51	55–75	71/52	China	140
		50	33	55–73	52/31	China	141
		197	50	55–71	120/12	China	142
		93	23	65–72	59/60	Hong Kong	143
		68	21	N/A	N/A	China	144
		136	116	37–64	135/117	China	145

were upregulated and 12 were downregulated (Table 8.1). A pathway enrichment analysis of the deregulated miRNAs indicated that neurotrophin and mTOR signaling, ubiquitin-mediated proteasomal degradation, long-term potentiation, axon guidance, and cholinergic, dopaminergic and glutamatergic synapses are strongly associated with PD development.[89] Vallelunga *et al.* attempted to identify serum miRNAs to differentiate PD from control and multiple system atrophy (MSA), a commonly misdiagnosed neurological disorder due to a shared clinical manifestation with PD.[116] The study indicated that miR-24, miR-223, and miR-324-3p were upregulated in both PD and MSA patients, and miR-30c and miR-148b were specifically downregulated in PD, but not in MSA compared to control groups. Additionally, miR-24, miR-34b, and miR-148b were upregulated in MSA compared to PD patients.[116] However, the clinical relevance of the miRNA-based signature's ability to discriminate PD from the control, MSA from the control, and PD from MSA remains under-studied. Recently, Chen *et al.* performed miRNA profiling in plasma from 25 PD patients and noted that the discriminative powers of miR-27a, *let-7a*, *let-7f*, miR-142-3p, and miR-222 were >80% successful in differentiating PD from controls.[117]

Burgos *et al.* identified 14 miRNAs that were downregulated and three miRNAs that were upregulated in 65 CSF samples of PD patients.[104] The serum levels of three miRNAs were found to be elevated and those of two miRNAs were decreased in 60 samples of PD patients. Seven CSF miRNAs overlapped with the findings of Gui *et al.*;[89] however, with the opposite direction of expression. Burgos *et al.* found that the levels of miR-132-5p, miR-485-5p, miR-409-3p, miR-433, miR-873-3p, and miR-136-3p were decreased in the CSF samples of PD patients; they were also shown to be elevated in the study by Gui *et al.* In contrast, Burgos *et al.* indicated that the level of miR-19-3p was elevated in the CSF samples of AD patients, and it was indeed repressed in the study by Gui *et al.* The clinical relevance of these circulating miRNAs warrants further validation in a larger cohort study for disease classification and monitoring.

Botta-Orfila *et al.* conducted a qRT-PCR array to profile the expression of 377 miRNAs in blood serum; the levels of miR-29c, miR-29a, and miR-19a were consistently found to be decreased in both a pilot test and in two follow-up validation studies. miRNA-based classifiers miR-29c, miR-29a, and miR-19a were shown to separate PD from groups with an accuracy of 75.5%; the sex-specific discrimination analysis using the three miRNA-based classifiers can distinguish men and women PD patients from controls with an accuracy of 84% and 70%, respectively. This miRNA signature failed to classify AD from control groups, suggesting that this miRNA-based signature is specific to PD, not AD.[118] The same research group also determined that miR-19b can aid in the identification of patients prone to short-term conversion to PD and dementia with Lewy bodies.[119] Khoo *et al.* reported that a combination of the miR-1826, miR-450b-3p, miR-626, and miR-505 identified in plasma could separate PD from the control with an accuracy of 95.5% in the replication set. However, this miRNA-based combination exhibited an insignificant predictive power to discriminate PD from the control in an independent validation set.[120] Sheinerman *et al.* reported that a six miRNA-based miRNA combination can distinguish PD from controls with an accuracy of 91%; miR-9-3p, miR-206, miR-155, miR-206, miR-7, and miR-206 can separate PD from ALS with an accuracy of 81%; and a miRNA pair of miR-206 and miR-9 can differentiate PD with dementia from other subtypes, such as PD with mild cognitive impairment, with an accuracy of 70%.[63]

Petillo *et al.* reported a four miRNA-based classifier composed of miR-1826, miR-450b-3p, miR-505, and miR-626 identified in plasma that can separate PD from the control groups with an accuracy of 79%; this panel was tested in another validation set for PD classification.[121]

8.6 Frontotemporal Dementia

Frontotemporal dementia is a highly heterogeneous disorder with diversified clinical manifestations associated with numerous neuropathological entities, such as behavioral variant FTD, nonfluent/agrammatic variant primary progressive aphasia, semantic variant primary progressive aphasia, FTD with motor neuron disease, progressive supranuclear palsy syndrome, and corticobasal syndrome, resulting in myriad impairments in behavior, language, executive functioning and motor function.[122] It was estimated that the prevalence of FTD, progressive supranuclear palsy, and corticobasal syndrome was ~0.01%, which is equivalent to approximately 20 000–30 000 patients in the United States at one time.[123] Although FTD is less prevalent than AD, FTD is a common dementia in those aged <65 years.[124]

The same study by Sheinerman *et al.* revealed that a six miRNA-based classifier (miR-9-3p *let-7e*, miR-7, miR-451, miR-335-5p, and *let-7e*) can differentiate FTD from the control groups with an accuracy of 87%; a combination of miR-129-3p, miR-9-3p, miR-338-3p, miR-99b, miR-323-3p, miR-127-3a can separate FTD from PD with an accuracy of 87%; and an miRNA pair (miR-155 and miR-99b) can discriminate FTD with progressive supranuclear palsy from other subtypes, such as primary progressive aphasia, with an accuracy of 74%.[63]

8.7 Amyotrophic Lateral Sclerosis

According to NIH-NINDS, amyotrophic lateral sclerosis, one of the rare progressive motor neuron disorders, is caused by the progressive neurodegeneration of motor neurons, resulting in weakening muscles; fasciculations in the arm, leg, shoulder, or tongue; muscle cramps; spasticity; atrophy; slurred and nasal speech; difficulty in chewing or swallowing; and eventual complete abrogation of the control of voluntary movements.[2] The CDC estimated that 14 000–15 000 Americans suffered from ALS in 2016. Unfortunately, no effective cure is currently available to attenuate or reverse disease progression.[125–127]

It was demonstrated that an miR-206, miR-31, miR-335-5p, miR-338-3p-based classifier can differentiate ALS from controls with an accuracy of 88%, and a six miRNA-based classifier can distinguish ALS with FTD from ALS with mild cognitive impairment with an accuracy of 76%.[63] These miRNAs may hold promise for disease detection, diagnosis, prognosis, and treatment.

8.8 Epilepsy

Epilepsy, due to aberrant and synchronous neuronal firing, is a chronic neurological disorder resulting in recurring seizures.[128] This disorder was estimated to affect 4–10% (~50 million) of the population worldwide. Since the description of the clinical manifestation and the medical history are inadequate, it often results in delayed or inappropriate disease management.[129] miRNA-based biomarkers may provide an objective classification of the disease for better management and treatment.

Wang *et al.* identified six serum-derived miRNAs differentially regulated in 117 epilepsy patients compared to 112 control groups; four were upregulated (*let-7d-5p*, miR-106b-5p, miR-130a-3p, and miR-146a-5p), and two were downregulated (miR-15a-5p and miR-194-5p). Among them, miR-106b-5p was the best biomarker to distinguish the epilepsy group from controls in an independent validation cohort with an accuracy of 81%, highlighting the potential of the clinical application of miR-106b-5p for epilepsy classification.[129]

8.9 Multiple Sclerosis

Multiple sclerosis (MS), characterized by neuroinflammation, demyelination, and neuro-axonal injury, is one of the most common causes of neurological disability, resulting in progressive and long-term disability.[130] Relapsing-remitting MS is the most common subtype of MS, and a majority of relapsing-remitting MS patients will eventually transition to secondary progressive MS, manifested by progressive resistance to available clinical therapy.[131]

Ebrahimkhani *et al.* reported the identification of two four-miRNA classifier groups (miR-15b-5p, miR-451a, miR-30b-5p, and miR-342-3p and miR-127-3p, miR-370-3p, miR-409-3p, and miR-432-5p) that were aberrantly expressed in the sera of RRMS and progressive MS patients compared to the control subjects. Authors also identified a nine miRNA-based classifier (miR-15b-5p, miR-23a-3p, miR-223-3p, miR-374a-5p, miR-30b-5p, miR-433-3p, miR-485-3p, miR-342-3p, and miR-432-5p) to differentiate relapsing-remitting MS from progressive MS with a predictive power of 95%. Importantly, the functions of the predicted targets of the miRNA panel are enriched in a neurotrophin signaling and immunity pathway. The circulating miRNAs may facilitate the confirmation of the disease progression from relapsing-remitting MS and progressive MS.[132]

8.10 Conclusions and Perspectives

Although a substantial amount of effort has been dedicating to profiling differentially expressed circulating miRNAs of the same neurological disease, only a few of the miRNA signatures or classifiers are consistently identified across studies. Batistela *et al.* reported that 17 out of 120 miRNAs for AD and 13 out of 48 miRNAs for PD were identified in more than one report. Additionally, expression of 13 miRNAs in the same body fluid were found to be in the opposite directionality in AD.[133] The significant discrepancies across studies can be explained by small study sizes, poor homogeneity, and high interindividual variation of the recruited subjects in terms of body mass index, age, lifestyle, physical activity, gender, smoking habits, ethnicity, epigenetics, diet, environment factors, circadian rhythm, disease subtypes, disease stages, miRNA extraction methods, the choice of experimental platform (*e.g.*, small-RNA sequencing, microarray, or RT-qPCR), and the statistical methods used for normalization, data processing, and detection of differentially expressed miRNAs.[133,134] A strict clinical definition of patient inclusion and exclusion criteria is vital to minimize interindividual variation to improve data consistency and reproducibility across studies.[48] The standardization of sample handling, the method of expression profiling, statistical analysis, and data reporting (*e.g.*, area-under-the-curve values, sensitivity, specificity, and accuracy) is of paramount importance for communities to conduct meta-analysis across studies and to retrieve clinically relevant information to devise strategies for rapid and efficient disease classification across different disorders and the different subtypes of the same neurological disease for better disease management. Additionally, large-scale controlled validation studies are of urgent clinical need to confirm the clinical relevance of the miRNA-based classifiers for specific neurological diseases.

Circulating miRNAs have been shown to implicate other (patho)physiological states beyond neurological disorders, such as cancers and pregnancy,[135] and selectively profiling the expression of a panel of brain-enriched miRNAs could facilitate the identification of neurological disease-associated miRNA-based biomarkers for detection, prognosis, and therapeutics.[136]

Our list of miRNA-based biomarkers for neurological disorders is not exhaustive (Table 8.1), and numerous neurological diseases, such as stroke, have not been discussed thoroughly. However, this chapter provides an entry point for basic researchers and clinicians to probe into the highly promising approach to developing a minimally invasive method of utilizing circulatory miRNAs identified in different body fluids for disease differentiation, monitoring, and personalized treatment. The accumulating knowledge of neurological disease-associated circulatory miRNAs will advance our understanding of disease biology and promote the development of novel therapeutics.

Acknowledgements

This work was supported in part by funds from the One-off Funding for Joint Lab/Research Collaboration provided to CUHK-Shandong University Joint Laboratory on Reproductive Genetics (Project 3132970) and to the Hong Kong Branch of the Chinese National Center for Assisted Reproductive Technology and Reproductive Genetics (Project 3132973), and the Hospital for Reproductive Medicine, Shandong University provided to the CUHK-Shandong University Joint Laboratory on Reproductive Genetics (Project 7104693).

References

1. World Health Organization Neurological Disorders, Public Health Challenges. www.who.int/mental_health/neurology/neurological_disorders_report_web.pdf, 2016.
2. National Institute of Neurological Disorders and Stroke, All Disorders. www.ninds.nih.gov/Disorders/All-Disorders, 2018.
3. GBD Neurological Disorders Collaborative Group, *Lancet Neurol.*, 2017, **16**, 877–897.
4. GBD 2015 DALYs and HALE Collaborators, *Lancet*, 2016, **388**, 1603–1658.
5. GBD 2015 Disease and Injury Incidence and Prevalence Collaborators, *Lancet*, 2016, **388**, 1545–1602.
6. A. D. Lopez, C. D. Mathers, M. Ezzati, D. T. Jamison and C. J. L. Murray, in *Global Burden of Disease and Risk Factors*, ed. A. D. Lopez, C. D. Mathers, M. Ezzati, D. T. Jamison and C. J. L. Murray, Washington, DC, 2006.
7. J. J. Sejvar, *Lancet Neurol.*, 2017, **16**, 858–859.
8. D. Baek, J. Villen, C. Shin, F. D. Camargo, S. P. Gygi and D. P. Bartel, *Nature*, 2008, **455**, 64–71.
9. M. Selbach, B. Schwanhausser, N. Thierfelder, Z. Fang, R. Khanin and N. Rajewsky, *Nature*, 2008, **455**, 58–63.
10. N. P. Achkar, D. A. Cambiagno and P. A. Manavella, *Trends Plant Sci.*, 2016, **21**, 1034–1044.
11. M. Ha and V. N. Kim, *Nat. Rev. Mol. Cell Biol.*, 2014, **15**, 509–524.
12. E. F. Finnegan and A. E. Pasquinelli, *Crit. Rev. Biochem. Mol. Biol.*, 2013, **48**, 51–68.
13. R. C. Lee, R. L. Feinbaum and V. Ambros, *Cell*, 1993, **75**, 843–854.
14. A. Kozomara and S. Griffiths-Jones, *Nucleic Acids Res.*, 2011, **39**, D152–D157.
15. A. Kozomara and S. Griffiths-Jones, *Nucleic Acids Res.*, 2014, **42**, D68–D73.
16. R. Mitra, J. Sun and Z. Zhao, *Int. J. Cancer*, 2015, **137**, 1516–1518.
17. K. W. Tsai, C. M. Leung, Y. H. Lo, T. W. Chen, W. C. Chan, S. Y. Yu, Y. T. Tu, H. C. Lam, S. C. Li, L. P. Ger, W. S. Liu and H. T. Chang, *Sci. Rep.*, 2016, **6**, 28176.
18. R. Mitra, C. C. Lin, C. M. Eischen, S. Bandyopadhyay and Z. Zhao, *RNA*, 2015, **21**, 1055–1065.

19. G. Van Peer, S. Lefever, J. Anckaert, A. Beckers, A. Rihani, A. Van Goethem, P. J. Volders, F. Zeka, M. Ongenaert, P. Mestdagh and J. Vandesompele, *Database*, 2014, **2014**, 1–8.
20. A. Kozomara and S. Griffiths-Jones, *Nucleic Acids Res.*, 2014, **42**, D68–D73.
21. Y. Li and K. V. Kowdley, *Genomics, Proteomics Bioinf.*, 2012, **10**, 246–253.
22. P. Shah, S. K. Cho, P. W. Thulstrup, M. J. Bjerrum, P. H. Lee, J. H. Kang, Y. J. Bhang and S. W. Yang, *Mov. Disord.*, 2017, **10**, 18–28.
23. J. Wang, J. Chen and S. Sen, *J. Cell. Physiol.*, 2016, **231**, 25–30.
24. V. Anand, A. Cahan and S. Ghosh, *AMIA Jt. Summits Transl. Sci. Proc.*, 2017, **2017**, 37–47.
25. M. Lindow and S. Kauppinen, *J. Cell Biol.*, 2012, **199**, 407–412.
26. C. Chakraborty, A. R. Sharma, G. Sharma, C. G. P. Doss and S. S. Lee, *Mol. Ther.--Nucleic Acids*, 2017, **8**, 132–143.
27. J. A. Weber, D. H. Baxter, S. Zhang, D. Y. Huang, K. H. Huang, M. J. Lee, D. J. Galas and K. Wang, *Clin. Chem.*, 2010, **56**, 1733–1741.
28. L. Pigati, S. C. Yaddanapudi, R. Iyengar, D. J. Kim, S. A. Hearn, D. Danforth, M. L. Hastings and D. M. Duelli, *PLoS One*, 2010, **5**, e13515.
29. M. HasanSohel, *Achiev. Life Sci.*, 2016, **10**, 175–186.
30. G. Camussi, M. C. Deregibus, S. Bruno, V. Cantaluppi and L. Biancone, *Kidney Int.*, 2010, **78**, 838–848.
31. M. Yanez-Mo, P. R. Siljander, Z. Andreu, A. B. Zavec, F. E. Borras, E. I. Buzas, K. Buzas, E. Casal, F. Cappello, J. Carvalho, E. Colas, A. Cordeiro-da Silva, S. Fais, J. M. Falcon-Perez, I. M. Ghobrial, B. Giebel, M. Gimona, M. Graner, I. Gursel, M. Gursel, N. H. Heegaard, A. Hendrix, P. Kierulf, K. Kokubun, M. Kosanovic, V. Kralj-Iglic, E. M. Kramer-Albers, S. Laitinen, C. Lasser, T. Lener, E. Ligeti, A. Line, G. Lipps, A. Llorente, J. Lotvall, M. Mancek-Keber, A. Marcilla, M. Mittelbrunn, I. Nazarenko, E. N. Nolte-'t Hoen, T. A. Nyman, L. O'Driscoll, M. Olivan, C. Oliveira, E. Pallinger, H. A. Del Portillo, J. Reventos, M. Rigau, E. Rohde, M. Sammar, F. Sanchez-Madrid, N. Santarem, K. Schallmoser, M. S. Ostenfeld, W. Stoorvogel, R. Stukelj, S. G. Van der Grein, M. H. Vasconcelos, M. H. Wauben and O. De Wever, *J. Extracell. Vesicles*, 2015, **4**, 27066.
32. M. Ui, T. Katada, K. Nogimori, M. Tamura and M. Yajima, *Tanpakushitsu Kakusan Koso*, 1986, **31**, 300–323.
33. D. Castellana, C. Kunzelmann and J. M. Freyssinet, *Hamostaseologie*, 2009, **29**, 51–57.
34. A. M. Marzesco, P. Janich, M. Wilsch-Brauninger, V. Dubreuil, K. Langenfeld, D. Corbeil and W. B. Huttner, *J. Cell Sci.*, 2005, **118**, 2849–2858.
35. A. S. Shet, *Vasc. Health Risk Manage.*, 2008, **4**, 769–774.
36. L. Turiak, P. Misjak, T. G. Szabo, B. Aradi, K. Paloczi, O. Ozohanics, L. Drahos, A. Kittel, A. Falus, E. I. Buzas and K. Vekey, *J. Proteomics*, 2011, **74**, 2025–2033.
37. D. R. Croft, M. L. Coleman, S. Li, D. Robertson, T. Sullivan, C. L. Stewart and M. F. Olson, *J. Cell Biol.*, 2005, **168**, 245–255.
38. E. van der Pol, A. N. Boing, P. Harrison, A. Sturk and R. Nieuwland, *Pharmacol. Rev.*, 2012, **64**, 676–705.

39. K. C. Vickers, B. T. Palmisano, B. M. Shoucri, R. D. Shamburek and A. T. Remaley, *Nat. Cell Biol.*, 2011, **13**, 423–433.
40. K. Wang, S. Zhang, J. Weber, D. Baxter and D. J. Galas, *Nucleic Acids Res.*, 2010, **38**, 7248–7259.
41. J. D. Arroyo, J. R. Chevillet, E. M. Kroh, I. K. Ruf, C. C. Pritchard, D. F. Gibson, P. S. Mitchell, C. F. Bennett, E. L. Pogosova-Agadjanyan, D. L. Stirewalt, J. F. Tait and M. Tewari, *Proc. Natl. Acad. Sci. U. S. A.*, 2011, **108**, 5003–5008.
42. A. Turchinovich, L. Weiz and B. Burwinkel, *Trends Biochem. Sci.*, 2012, **37**, 460–465.
43. M. A. Cortez, C. Bueso-Ramos, J. Ferdin, G. Lopez-Berestein, A. K. Sood and G. A. Calin, *Nat. Rev. Clin. Oncol.*, 2011, **8**, 467–477.
44. M. L. Squadrito, C. Baer, F. Burdet, C. Maderna, G. D. Gilfillan, R. Lyle, M. Ibberson and M. De Palma, *Cell Rep.*, 2014, **8**, 1432–1446.
45. P. S. Mitchell, R. K. Parkin, E. M. Kroh, B. R. Fritz, S. K. Wyman, E. L. Pogosova-Agadjanyan, A. Peterson, J. Noteboom, K. C. O'Briant, A. Allen, D. W. Lin, N. Urban, C. W. Drescher, B. S. Knudsen, D. L. Stirewalt, R. Gentleman, R. L. Vessella, P. S. Nelson, D. B. Martin and M. Tewari, *Proc. Natl. Acad. Sci. U. S. A.*, 2008, **105**, 10513–10518.
46. E. Hergenreider, S. Heydt, K. Treguer, T. Boettger, A. J. Horrevoets, A. M. Zeiher, M. P. Scheffer, A. S. Frangakis, X. Yin, M. Mayr, T. Braun, C. Urbich, R. A. Boon and S. Dimmeler, *Nat. Cell Biol.*, 2012, **14**, 249–256.
47. M. Mittelbrunn, C. Gutierrez-Vazquez, C. Villarroya-Beltri, S. Gonzalez, F. Sanchez-Cabo, M. A. Gonzalez, A. Bernad and F. Sanchez-Madrid, *Nat. Commun.*, 2011, **2**, 282.
48. M. Hackl, U. Heilmeier, S. Weilner and J. Grillari, *Mol. Cell. Endocrinol.*, 2016, **432**, 83–95.
49. F. Mar-Aguilar, J. A. Mendoza-Ramirez, I. Malagon-Santiago, P. K. Espino-Silva, S. K. Santuario-Facio, P. Ruiz-Flores, C. Rodriguez-Padilla and D. Resendez-Perez, *Dis. Markers*, 2013, **34**, 163–169.
50. S. S. Noferesti, M. M. Sohel, M. Hoelker, D. Salilew-Wondim, E. Tholen, C. Looft, F. Rings, C. Neuhoff, K. Schellander and D. Tesfaye, *J. Ovarian Res.*, 2015, **8**, 81.
51. M. Zhou, H. Hara, Y. Dai, L. Mou, D. K. Cooper, C. Wu and Z. Cai, *Int. J. Mol. Sci.*, 2016, **17**, 1232.
52. Q. Pu, Y. Huang, Y. Lu, Y. Peng, J. Zhang, G. Feng, C. Wang, L. Liu and Y. Dai, *Thorac. Cancer*, 2016, **7**, 348–354.
53. G. Sebastiani, E. Guarino, G. E. Grieco, C. Formichi, C. Delli Poggi, E. Ceccarelli and F. Dotta, *Front. Endocrinol.*, 2017, **8**, 345.
54. H. Harapan, F. Fitra, I. Ichsan, M. Mulyadi, P. Miotto, N. A. Hasan, M. Calado and D. M. Cirillo, *Tuberculosis*, 2013, **93**, 596–605.
55. F. Benz, S. Roy, C. Trautwein, C. Roderburg and T. Luedde, *Int. J. Mol. Sci.*, 2016, **17**, 78.
56. L. S. Enache, E. L. Enache, C. Ramiere, O. Diaz, L. Bancu, A. Sin and P. Andre, *Int. J. Mol. Sci.*, 2014, **15**, 17644–17666.

57. F. Kadri, A. LaPlante, M. De Luca, L. Doyle, C. Velasco-Gonzalez, J. R. Patterson, P. E. Molina, S. Nelson, A. H. Zea, C. H. Parsons and F. Peruzzi, *J. Cell. Physiol.*, 2016, **231**, 829–836.
58. Y. Qi, Z. Zhu, Z. Shi, Y. Ge, K. Zhao, M. Zhou and L. Cui, *Viruses*, 2014, **6**, 1823–1836.
59. P. A. Tambyah, S. Sepramaniam, J. Mohamed Ali, S. C. Chai, P. Swaminathan, A. Armugam and K. Jeyaseelan, *PLoS One*, 2013, **8**, e76811.
60. L. Cui, Y. Qi, H. Li, Y. Ge, K. Zhao, X. Qi, X. Guo, Z. Shi, M. Zhou, B. Zhu, Y. Guo, J. Li, C. W. Stratton, Y. W. Tang and H. Wang, *PLoS One*, 2011, **6**, e27071.
61. C. N. Correia, N. C. Nalpas, K. E. McLoughlin, J. A. Browne, S. V. Gordon, D. E. MacHugh and R. G. Shaughnessy, *Front. Immunol.*, 2017, **8**, 118.
62. T. Kawaguchi, S. Komatsu, D. Ichikawa, M. Tsujiura, H. Takeshita, S. Hirajima, M. Miyamae, W. Okajima, T. Ohashi, T. Imamura, J. Kiuchi, H. Konishi, A. Shiozaki, K. Okamoto and E. Otsuji, *Int. J. Mol. Sci.*, 2016, **17**, 1459.
63. K. S. Sheinerman, J. B. Toledo, V. G. Tsivinsky, D. Irwin, M. Grossman, D. Weintraub, H. I. Hurtig, A. Chen-Plotkin, D. A. Wolk, L. F. McCluskey, L. B. Elman, J. Q. Trojanowski and S. R. Umansky, *Alzheimer's Res. Ther.*, 2017, **9**, 89.
64. A. G. Thrift, T. Thayabaranathan, G. Howard, V. J. Howard, P. M. Rothwell, V. L. Feigin, B. Norrving, G. A. Donnan and D. A. Cadilhac, *Int. J. Stroke*, 2017, **12**, 13–32.
65. Centers for Disease Control and Prevention, *Stroke Facts*, 2017, available at https://www.cdc.gov/stroke/facts.htm.
66. P. Amarenco, J. Bogousslavsky, L. R. Caplan, G. A. Donnan and M. G. Hennerici, *Cerebrovasc. Dis.*, 2009, **27**, 493–501.
67. C. S. Gan, C. W. Wang and K. S. Tan, *GMR, Genet. Mol. Res.*, 2012, **11**, 147–152.
68. L. Jia, F. Hao, W. Wang and Y. Qu, *Cell Biochem. Funct.*, 2015, **33**, 314–319.
69. L. Zeng, J. Liu, Y. Wang, L. Wang, S. Weng, Y. Tang, C. Zheng, Q. Cheng, S. Chen and G. Y. Yang, *Front. Biosci., Elite Ed.*, 2011, **3**, 1265–1272.
70. G. Li, H. Zhao, Q. Ma, Z. Tao, R. Wang, Z. Fan, Y. Feng, X. Ji and Y. Luo, *Brain Circ.*, 2017, **3**, 21–28.
71. S. Sepramaniam, J. R. Tan, K. S. Tan, D. A. DeSilva, S. Tavintharan, F. P. Woon, C. W. Wang, F. L. Yong, D. S. Karolina, P. Kaur, F. J. Liu, K. Y. Lim, A. Armugam and K. Jeyaseelan, *Int. J. Mol. Sci.*, 2014, **15**, 1418–1432.
72. M. Vijayan, S. Kumar, X. Yin, D. Zafer, V. Chanana, P. Cengiz and P. H. Reddy, *Hum. Mol. Genet.*, 2018, **27**, 2318–2329.
73. B. Dewdney, A. Trollope, J. Moxon, D. T. Manapurathe, E. Biros and J. Golledge, *J. Stroke Cerebrovasc. Dis.*, 2018, **27**, 522–530.
74. G. K. Wong, W. S. Poon, R. Boet, M. T. Chan, T. Gin, S. C. Ng and B. C. Zee, *Neurosurgery*, 2011, **68**, 1556–1561; discussion 1561.
75. N. S. Lai, J. Q. Zhang, F. Y. Qin, B. Sheng, X. G. Fang and Z. B. Li, *Biosci. Rep.*, 2017, **37**, BSR20160480.
76. S. Bache, R. Rasmussen, M. Rossing, F. P. Laigaard, F. C. Nielsen and K. Moller, *Stroke*, 2017, **48**, 2391–2398.

77. C. J. Powers, R. Dickerson, S. W. Zhang, C. Rink, S. Roy and C. K. Sen, *Physiol. Genomics*, 2016, **48**, 361–366.
78. S. S. Stylli, A. A. Adamides, R. M. Koldej, R. B. Luwor, D. S. Ritchie, J. Ziogas and A. H. Kaye, *J. Neurosurg.*, 2017, **126**, 1131–1139.
79. G. Lu, M. S. Wong, M. Z. Q. Xiong, C. K. Leung, X. W. Su, J. Y. Zhou, W. S. Poon, V. Z. Y. Zheng, W. Y. Chan and G. K. C. Wong, *J. Am. Heart Assoc.*, 2017, **6**, 1–9.
80. X. Zhou, F. Ye, J. Zhang, H. Gao, J. Jiang and N. Liu, *Int. J. Clin. Exp. Pathol.*, 2017, **10**, 7154–7160.
81. P. Scheltens, K. Blennow, M. M. Breteler, B. de Strooper, G. B. Frisoni, S. Salloway and W. M. Van der Flier, *Lancet*, 2016, **388**, 505–517.
82. R. Brookmeyer, N. Abdalla, C. H. Kawas and M. M. Corrada, *Alzheimer's Dementia*, 2018, **14**, 121–129.
83. H. Braak and E. Braak, *Acta Neuropathol.*, 1991, **82**, 239–259.
84. L. M. Bekris, F. Lutz, T. J. Montine, C. E. Yu, D. Tsuang, E. R. Peskind and J. B. Leverenz, *Biomarkers*, 2013, **18**, 455–466.
85. S. Bhatnagar, H. Chertkow, H. M. Schipper, Z. Yuan, V. Shetty, S. Jenkins, T. Jones and E. Wang, *Front. Mol. Neurosci.*, 2014, **7**, 2.
86. C. G. Liu, J. L. Wang, L. Li and P. C. Wang, *Int. J. Mol. Med.*, 2014, **34**, 160–166.
87. C. G. Liu, J. L. Wang, L. Li, L. X. Xue, Y. Q. Zhang and P. C. Wang, *Brain Res.*, 2014, **1583**, 55–64.
88. Y. Zhu, C. Li, A. Sun, Y. Wang and S. Zhou, *Exp. Ther. Med.*, 2015, **9**, 1013–1017.
89. Y. Gui, H. Liu, L. Zhang, W. Lv and X. Hu, *Oncotarget*, 2015, **6**, 37043–37053.
90. J. P. Cogswell, J. Ward, I. A. Taylor, M. Waters, Y. Shi, B. Cannon, K. Kelnar, J. Kemppainen, D. Brown, C. Chen, R. K. Prinjha, J. C. Richardson, A. M. Saunders, A. D. Roses and C. A. Richards, *J. Alzheimer's Dis.*, 2008, **14**, 27–41.
91. R. T. Mott and C. M. Hulette, *Neuroimaging Clin. N. Am.*, 2005, **15**, 755–765.
92. S. Lehnardt, L. Massillon, P. Follett, F. E. Jensen, R. Ratan, P. A. Rosenberg, J. J. Volpe and T. Vartanian, *Proc. Natl. Acad. Sci. U. S. A.*, 2003, **100**, 8514–8519.
93. K. D. Taganov, M. P. Boldin, K. J. Chang and D. Baltimore, *Proc. Natl. Acad. Sci. U. S. A.*, 2006, **103**, 12481–12486.
94. C. Sala Frigerio, P. Lau, E. Salta, J. Tournoy, K. Bossers, R. Vandenberghe, A. Wallin, M. Bjerke, H. Zetterberg, K. Blennow and B. De Strooper, *Neurology*, 2013, **81**, 2103–2106.
95. J. Denk, K. Boelmans, C. Siegismund, D. Lassner, S. Arlt and H. Jahn, *PLoS One*, 2015, **10**, e0126423.
96. P. N. Alexandrov, P. Dua, J. M. Hill, S. Bhattacharjee, Y. Zhao and W. J. Lukiw, *Int. J. Biochem. Mol. Biol.*, 2012, **3**, 365–373.
97. T. Kiko, K. Nakagawa, T. Tsuduki, K. Furukawa, H. Arai and T. Miyazawa, *J. Alzheimer's Dis.*, 2014, **39**, 253–259.
98. D. Mungas, *Geriatrics*, 1991, **46**, 54–58, 63, 66.

99. M. Muller, H. B. Kuiperij, J. A. Claassen, B. Kusters and M. M. Verbeek, *Neurobiol. Aging*, 2014, **35**, 152–158.
100. M. Muller, L. Jakel, I. B. Bruinsma, J. A. Claassen, H. B. Kuiperij and M. M. Verbeek, *Mol. Neurobiol.*, 2016, **53**, 2894–2899.
101. D. Galimberti, C. Villa, C. Fenoglio, M. Serpente, L. Ghezzi, S. M. Cioffi, A. Arighi, G. Fumagalli and E. Scarpini, *J. Alzheimer's Dis.*, 2014, **42**, 1261–1267.
102. L. Tan, J. T. Yu, Q. Y. Liu, M. S. Tan, W. Zhang, N. Hu, Y. L. Wang, L. Sun, T. Jiang and L. Tan, *J. Neurol. Sci.*, 2014, **336**, 52–56.
103. H. Geekiyanage, G. A. Jicha, P. T. Nelson and C. Chan, *Exp. Neurol.*, 2012, **235**, 491–496.
104. K. Burgos, I. Malenica, R. Metpally, A. Courtright, B. Rakela, T. Beach, H. Shill, C. Adler, M. Sabbagh, S. Villa, W. Tembe, D. Craig and K. Van Keuren-Jensen, *PLoS One*, 2014, **9**, e94839.
105. H. Dong, J. Li, L. Huang, X. Chen, D. Li, T. Wang, C. Hu, J. Xu, C. Zhang, K. Zen, S. Xiao, Q. Yan, C. Wang and C. Y. Zhang, *Dis. Markers*, 2015, **2015**, 625659.
106. P. Kumar, Z. Dezso, C. MacKenzie, J. Oestreicher, S. Agoulnik, M. Byrne, F. Bernier, M. Yanagimachi, K. Aoshima and Y. Oda, *PLoS One*, 2013, **8**, e69807.
107. L. Tan, J. T. Yu, M. S. Tan, Q. Y. Liu, H. F. Wang, W. Zhang, T. Jiang and L. Tan, *J. Alzheimer's Dis.*, 2014, **40**, 1017–1027.
108. L. Cheng, J. D. Doecke, R. A. Sharples, V. L. Villemagne, C. J. Fowler, A. Rembach, R. N. Martins, C. C. Rowe, S. L. Macaulay, C. L. Masters, A. F. Hill, B. Australian Imaging and G. Lifestyle Research, *Mol. Psychiatry*, 2015, **20**, 1188–1196.
109. R. L. McNamee, S. H. Yee, J. C. Price, W. E. Klunk, B. Rosario, L. Weissfeld, S. Ziolko, M. Berginc, B. Lopresti, S. Dekosky and C. A. Mathis, *J. Nucl. Med.*, 2009, **50**, 348–355.
110. P. J. Snyder, *Alzheimer's Dementia*, 2015, **1**, 1–4.
111. O. B. Tysnes and A. Storstein, *J. Neural Transm.*, 2017, **124**, 901–905.
112. H. Braak, U. Rub, W. P. Gai and K. Del Tredici, *J. Neural Transm.*, 2003, **110**, 517–536.
113. K. Wakabayashi, K. Tanji, F. Mori and H. Takahashi, *Neuropathology*, 2007, **27**, 494–506.
114. L. F. Cardo, E. Coto, L. de Mena, R. Ribacoba, G. Moris, M. Menendez and V. Alvarez, *J. Neurol.*, 2013, **260**, 1420–1422.
115. N. Zhao, L. Jin, G. Fei, Z. Zheng and C. Zhong, *Parkinsonism Relat. Disord.*, 2014, **20**, 1177–1180.
116. A. Vallelunga, M. Ragusa, S. Di Mauro, T. Iannitti, M. Pilleri, R. Biundo, L. Weis, C. Di Pietro, A. De Iuliis, A. Nicoletti, M. Zappia, M. Purrello and A. Antonini, *Front. Cell. Neurosci.*, 2014, **8**, 156.
117. L. Chen, J. Yang, J. Lu, S. Cao, Q. Zhao and Z. Yu, *Brain Behav.*, 2018, **8**, e00941.
118. T. Botta-Orfila, X. Morato, Y. Compta, J. J. Lozano, N. Falgas, F. Valldeoriola, C. Pont-Sunyer, D. Vilas, L. Mengual, M. Fernandez, J. L. Molinuevo, A. Antonell, M. J. Marti, R. Fernandez-Santiago and M. Ezquerra, *J. Neurosci. Res.*, 2014, **92**, 1071–1077.

119. R. Fernandez-Santiago, A. Iranzo, C. Gaig, M. Serradell, M. Fernandez, E. Tolosa, J. Santamaria and M. Ezquerra, *Ann. Neurol.*, 2015, **77**, 895–901.
120. S. K. Khoo, D. Petillo, U. J. Kang, J. H. Resau, B. Berryhill, J. Linder, L. Forsgren, L. A. Neuman and A. C. Tan, *J. Parkinson's Dis.*, 2012, **2**, 321–331.
121. D. Petillo, S. Orey, A. C. Tan, L. Forsgren and S. K. Khoo, *AIMS Med. Sci.*, 2015, **2**, 7–14.
122. N. T. Olney, S. Spina and B. L. Miller, *Neurol. Clin.*, 2017, **35**, 339–374.
123. I. T. Coyle-Gilchrist, K. M. Dick, K. Patterson, P. Vazquez Rodriquez, E. Wehmann, A. Wilcox, C. J. Lansdall, K. E. Dawson, J. Wiggins, S. Mead, C. Brayne and J. B. Rowe, *Neurology*, 2016, **86**, 1736–1743.
124. D. S. Knopman and R. O. Roberts, *J. Mol. Neurosci.*, 2011, **45**, 330–335.
125. J. Morris, *Neurodiagn. J.*, 2015, **55**, 180–194.
126. V. Sundaramoorthy, J. M. Sultana and J. D. Atkin, *Front. Neurosci.*, 2015, **9**, 400.
127. C. Z. de Paula, B. D. Goncalves and L. B. Vieira, *BioMed Res. Int.*, 2015, **2015**, 198612.
128. D. C. Henshall, *Curr. Opin. Neurol.*, 2014, **27**, 199–205.
129. J. Wang, J. T. Yu, L. Tan, Y. Tian, J. Ma, C. C. Tan, H. F. Wang, Y. Liu, M. S. Tan, T. Jiang and L. Tan, *Sci. Rep.*, 2015, **5**, 9522.
130. R. Nicholas and W. Rashid, *Am. Fam. Physician*, 2013, **87**, 712–714.
131. F. D. Lublin, S. C. Reingold, J. A. Cohen, G. R. Cutter, P. S. Sorensen, A. J. Thompson, J. S. Wolinsky, L. J. Balcer, B. Banwell, F. Barkhof, B. Bebo Jr, P. A. Calabresi, M. Clanet, G. Comi, R. J. Fox, M. S. Freedman, A. D. Goodman, M. Inglese, L. Kappos, B. C. Kieseier, J. A. Lincoln, C. Lubetzki, A. E. Miller, X. Montalban, P. W. O'Connor, J. Petkau, C. Pozzilli, R. A. Rudick, M. P. Sormani, O. Stuve, E. Waubant and C. H. Polman, *Neurology*, 2014, **83**, 278–286.
132. S. Ebrahimkhani, F. Vafaee, P. E. Young, S. S. J. Hur, S. Hawke, E. Devenney, H. Beadnall, M. H. Barnett, C. M. Suter and M. E. Buckland, *Sci. Rep.*, 2017, **7**, 14293.
133. M. S. Batistela, N. D. Josviak, C. D. Sulzbach and R. L. de Souza, *Int. J. Neurosci.*, 2017, **127**, 547–558.
134. J. Satoh, Y. Kino and S. Niida, *Biomarker Insights*, 2015, **10**, 21–31.
135. G. Reid, M. B. Kirschner and N. van Zandwijk, *Crit. Rev. Oncol. Hematol.*, 2011, **80**, 193–208.
136. K. S. Sheinerman and S. R. Umansky, *Front. Cell. Neurosci.*, 2013, **7**, 150.
137. B. Dewdney, A. Trollope, J. Moxon, D. Thomas Manapurathe, E. Biros and J. Golledge, *J. Stroke Cerebrovasc. Dis.*, 2018, **27**, 522–530.
138. K. Chu, J. S. Moon, K. H. Jung, S. T. Lee, J. J. Ban, J. Y. Huh, S. Y. Kim, J. M. Kim and J. K. Roh, *Mol. Ther.*, 2012, **20**, S197.
139. P. Li, F. Teng, F. Gao, M. Zhang, J. Wu and C. Zhang, *Cell. Mol. Neurobiol.*, 2015, **35**, 433–447.
140. G. P. Peng, Y. Yuan, S. S. Wu, F. P. He, Y. W. Hu and B. Y. Luo, *Transl. Stroke Res.*, 2015, **6**, 437–445.

141. Y. Chen, Y. Song, J. Huang, M. Qu, Y. Zhang, J. Geng, Z. Zhang, J. Liu and G. Y. Yang, *Front. Neurol.*, 2017, **8**, 57.
142. G. Long, F. Wang, H. Li, Z. Yin, C. Sandip, Y. Lou, Y. Wang, C. Chen and D. W. Wang, *BMC Neurol.*, 2013, **13**, 178.
143. L. Y. Leung, C. P. Chan, Y. K. Leung, H. L. Jiang, J. M. Abrigo, F. Wang de, J. S. Chung, T. H. Rainer and C. A. Graham, *Clin. Chim. Acta*, 2014, **433**, 139–144.
144. J. Zhou and J. Zhang, *Mol. Med. Rep.*, 2014, **10**, 971–976.
145. W. Wang, G. Sun, L. Zhang, L. Shi and Y. Zeng, *J. Stroke Cerebrovasc. Dis.*, 2014, **23**, 2607–2613.

CHAPTER 9

MicroRNAs in Endocrine Disorders

JEAN-FRANCOIS LANDRIER, ADEL DERGHAL, ESMA KARKENI AND LOURDES MOUNIEN*

Aix Marseille University, INSERM, INRA, C2VN, Marseille, France
*E-mail: lourdes.mounien@univ-amu.fr

9.1 Introduction

Gene expression can be controlled at a transcriptional level by the activity of DNA-binding transcription factors or controlled post-transcriptionally by changes in RNA stability or localization, protein translation, or biological half-life. Recently, a similar paradigm has emerged that operates at the RNA level, whereby small RNAs, microRNAs (miRNAs), play predominantly inhibitory regulatory roles by binding to *cis*-elements in the 3′ untranslated region (UTR) of message-encoding RNAs. The miRNAs are small noncoding RNA molecules of 21–23 nucleotides that post-transcriptionally repress the expression of genes by binding to 3′ UTRs of the target messenger RNAs (mRNAs; for review see ref. 1). They were first discovered in *Caenorhabditis elegans* in 1993 and, later on, in vertebrates and plants.[2,3] Today, thousands of miRNAs have been identified, making them one of the most abundant classes of gene-regulatory molecules in multicellular organisms. These noncoding RNAs behave as specific gene silencers by base pairing to the 3′

Drug Discovery Series No. 69
MicroRNAs in Diseases and Disorders: Emerging Therapeutic Targets
Edited by Philip V. Peplow, Bridget Martinez, George A. Calin and Aurora Esquela-Kerscher

Published by the Royal Society of Chemistry, www.rsc.org

UTR of target mRNAs, but have been also shown to bind anywhere along the length of the mRNA transcript to exert their effects. miRNAs exert their action by inhibiting translation and by affecting mRNA stability and degradation.[1,4] Most mammalian miRNAs are transcribed by RNA polymerase II as long precursor molecules containing stem–loop structures. This primary transcript is cleaved by the microprocessor complex containing the RNase III-type enzyme Drosha and the DGCR8/Pasha protein yielding a ≈70 nucleotide hairpin-structured precursor (pre-miRNA).[5] The pre-miRNA is transported into the cytoplasm and cleaved by another RNase III enzyme called Dicer to generate a ≈22 nucleotide mature miRNA (functional) (for review see ref. 6). Upon separation of the two strands, the guide strand binds to an Argonaute (Ago) protein and is incorporated into the RNA-induced silencing complex, allowing the identification of complementary sites within the 3′ UTRs of target mRNAs.[6] Based on computational algorithms, ~60% of human transcripts contain potential miRNA-binding sites within their 3′ UTRs.[7] A single miRNA can potentially bind to more than 100 target mRNAs, and multiple miRNAs can cooperate to finely tune the expression of the same transcript.[8–10] miRNAs play key roles in numerous physiological processes, including cell proliferation, apoptosis, neurodevelopment, and tissue differentiation, but also in pathological processes (for review see ref. 11).

Interestingly, miRNAs are now considered relevant actors in the modulation of endocrine aspects of normal and pathophysiological conditions. Related to endocrine disease, the expression of miRNAs is dysregulated in disorders affecting different neuroendocrine axes; such as the reproductive or thyroid axis.[12] (For review see ref. 13.) In addition, defects in miRNA biogenesis and function contribute in the development of metabolic disorders. For instance, miR-14, miR-278, and *let-7* are involved in the metabolism of lipid and glucose.[14] (For review see ref. 15.) Although the miRNAs are involved in several endocrine diseases, for the purpose of this chapter, we have focused our attention on the functions of miRNAs in metabolic syndromes (obesity and type 2 diabetes) and neuroendocrine diseases.

9.2 miRNAs and Metabolic Syndrome

9.2.1 Metabolic Syndrome

Obesity and overweight are two very common diseases in our modern societies. According to the World Health Organization (fact sheet n°311: https://www.who.int/news-room/fact-sheets/detail/obesity-and-overweight), worldwide obesity rates, measured by the body mass index (BMI), have nearly doubled since 1980. In 2008, 35% of adults were overweight and 11% were obese. More than 40 million children under the age of 5 years were overweight in 2011. While this phenomenon is well known in United States, which has an adult obesity rate of 35%, overweight and obesity are spreading massively in other industrialized or developing countries.[16] The incredible spread of obesity is also associated with a real public health problem, because of its costs and health effects. Indeed, excess body weight increases the likelihood of various metabolic diseases such as heart disease, type 2 diabetes, osteoarthritis, and certain types of cancer.[17] Obesity and overweight usually result from an imbalance between energy intake and energy output. It is caused by the interaction of multiple factors such as caloric and food intake, physical inactivity, genetic predisposition, and individual metabolism. Essential treatment of obesity consists of low-calorie low-fat diets, increased physical activity, and diverse strategies contributing to the modification of lifestyle. However, weight losses achieved with lifestyle intervention are modest and limited by high rates of recidivism and compensatory slowing metabolism. In this context, it is essential to understand the contribution of the genetic and epigenetic traits in the onset of obesity. In accordance with this fact, different studies have depicted that multiple loci on the human genome are associated with obesity and metabolic syndrome.[18–20] Interestingly, omics approaches suggested a correlation between the expression of several miRNAs in different tissues (*e.g.*, adipose tissue, liver, and pancreas) and obesity or metabolic syndrome.[19,21] (For review see ref. 22.) In addition, different studies reported that the expression of miRNAs directly correlated with diet and lifestyle (for review see ref. 23).[24]

The control of energy homeostasis is finely tuned by endocrine and neural mechanisms that cooperate to maintain the balance between caloric intake and energy expenditure (for review see ref. 25 and 26). In this respect, the central nervous system continuously monitors modifications in metabolic parameters (*i.e.*, blood glucose or free fatty acid levels) or hormones (insulin, leptin, and ghrelin) and elicits adaptive responses like food intake regulation or autonomic nervous system modulation.[25,26] Particularly, leptin is an adipose-derived hormone that is crucial to maintain both normal body weight and insulin sensitivity by acting in the different nuclei of the hypothalamus as arcuate or ventromedial nuclei.[27–29] In addition, obesity is associated with inflammation in adipose tissue. The production of inflammatory cytokines can interfere with insulin signalling and then can contribute to type 2 diabetes as well as many other obesity-related diseases (for reviews see ref. 30 and 31). In this context, the endocrine function of the adipose tissue is essential to maintain a normal weight and the regulation of energy homeostasis. Subsequently, it is not unexpected that miRNAs may be a new layer of regulation of the different functions of the adipose tissue in obesity (for review see ref. 32).

Altogether, these different results highlight the interest in the role of miRNAs in the field of obesity and related diseases. Particularly, the alteration in miRNA expression could induce changes in the pattern of genes controlling a range of biological processes, including adipogenesis as well as inflammation of the adipose tissue, lipid metabolism, and insulin resistance. In this part of the chapter, we develop aspects linked to the functions of miRNAs in the biology of the adipose tissue and the pancreas that are interdependent for lipid and glucose metabolism.

9.2.2 miRNAs and Adipogenesis

A better understanding of the regulation of adipogenesis is essential for the development of new therapies against the onset of obesity. In fact, mild obesity is associated with an increase of adipose cell size (hypertrophic obesity) while severe obesity or childhood obesity also involves an increase of the number of adipocytes (hyperplasia obesity). Adipogenesis appears as a complex process involving several hormones or transcription factors as peroxisome proliferator-activated receptor γ (PPARγ), CCAAT/enhancer-binding protein (C/EBP) family and Sterol regulatory element-binding protein 1 (SREBP1) (for review see ref. 33). Recently, numerous studies indicate that miRNAs play an important role in adipocyte differentiation and contribute to the development of obesity (Figure 9.1). The effect of miRNAs on adipogenesis seems to be complex, as there are multiple targets for miRNAs. Nevertheless, miRNAs regulate adipogenesis by modulating the expression of key factors of adipocyte differentiation (for review see ref. 34).[35]

The first evidence suggesting a role for miRNAs in adipogenesis was in *Drosophila* flies indicating that miR-14 and miR-278 regulate lipid metabolism.[36,37] Among the several miRNAs special attention has been paid to miR-143. Using antisense oligonucleotides transfected into human preadipocytes, miR-143 was identified as a regulator of adipocyte differentiation.[38] In accordance with this work, Takanabe *et al.* highlighted that miR-143 levels in adipose tissue of obese mice closely correlated with expression levels of adipocyte differentiation markers such as PPARγ.[39] Another example is miR-130a, which inhibits adipogenesis through the down-regulation of the expression of PPARγ.[40] In the context of

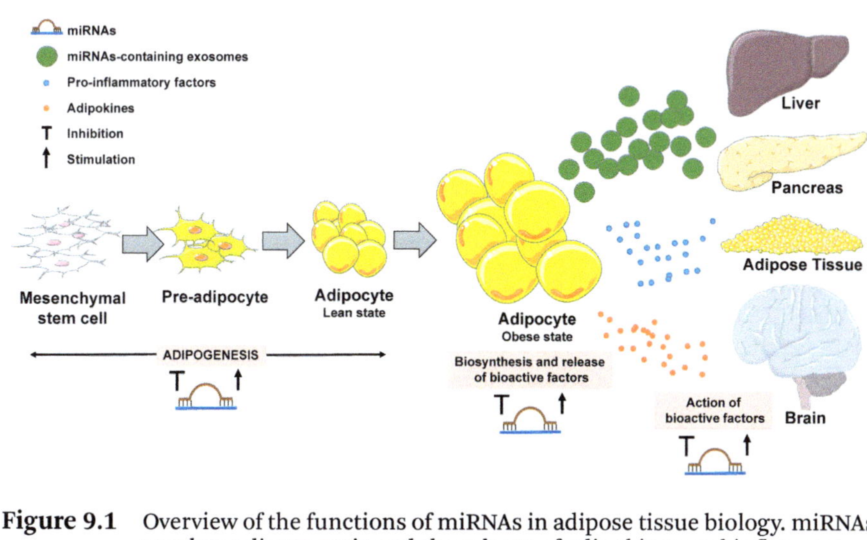

Figure 9.1 Overview of the functions of miRNAs in adipose tissue biology. miRNAs regulate adipogenesis and the release of adipokines and inflammatory factors. They also modulate the effects of bioactive signals released by the adipose tissue, directly or indirectly, *via* exosomes.

adipogenesis, Wang *et al.* demonstrated that the miRNA cluster miR-17-92, including miR-17, miR-18a, miR-19a, miR-19b-1, miR-20a, and miR-92a, is up-regulated during adipocyte clonal expansion and accelerated adipocyte differentiation by negatively regulating the key cell cycle regulator and tumor-suppressor gene Rb2/p130.[41] As indicated earlier, various miRNAs can affect adipogenesis by targeting C/EBP. For instance, miR-375 promotes 3T3-L1 adipocyte cell line differentiation by up-regulating the expression of C/EBP and PPARγ2.[42]

9.2.3 Dysregulation of miRNAs Expression in Obesity

Many miRNAs are up-regulated or down-regulated in the white adipose tissue of obese human subjects compared to non-obese individuals.[43,44] (For review see ref. 45 and 46.) Adipose tissue from the visceral area is more important for the metabolic aspects than the subcutaneous tissue (for review see ref. 47). In this context, profiling analysis showed that numerous miRNAs are differentially expressed in subcutaneous and visceral white adipose tissue.[48] In humans, several studies illustrated a correlation between the expression of miRNAs in adipose tissue and different metabolic parameters (BMI, adipogenesis, glycemia, and leptinemia).[43,49] In accordance with this, Heneghan *et al.* found that the expression of miR-17-5p and miR-132 differed significantly between obese and non-obese omental fat.[50] Interestingly, the expression of these two miRNAs in omental fat and blood from obese patients correlated significantly with glycosylated hemoglobin, leptin, BMI, and fasting blood glucose.[50] An increase in the expression of miR-21 was found in white adipose tissue of obese humans compared to lean controls and positively correlated with BMI.[51] In another set of experiments, a correlation between miR-519d and the protein levels of PPARα involved in fatty acid homeostasis was established.[44] miR-519d suppressed the translation of the PPARα protein, and increased lipid accumulation during pre-adipocyte differentiation.[44] Recently, the group of J.W. Helge evaluated the miRNA expression in subcutaneous adipose tissue from 19 individuals with severe obesity (10 women and nine men) before and after a 15-week weight-loss intervention.[52] This intervention led to up-regulation of miR-29a-3p and miR-29a-5p and down-regulation of miR-20b-5p.[52] In another study, it was shown that the expression levels of miR-221 is up-regulated in obese people and that this miRNA can modulate fat metabolism through leptin and tumor-necrosis factor-α (TNF-α).[53] Interestingly, additional groups also reported an increase in the expression of miR-221 in adipose tissue and liver from leptin-deficient ob/ob and diet-induced obesity (DIO) mouse models.[54,55] Other studies indicated that the expression of a large number of miRNAs is modified in mouse models of obesity (for review see ref. 56). For instance, Chartoumpekis *et al.* showed the up-regulation of miR-342-3p, miR-142-3p, miR-142-5p, miR-21, miR-146a, miR-146b, and miR-379 and the down-regulation of miR-122, miR-133b, miR-1, miR-30a, miR-192, and miR-203 during the development of obesity in mice.[57] The different mouse models for obesity (DIO or ob/ob) can be used as relevant models to test the involvement of one miRNA in obesity. As indicated above, miR-21 is up-regulated in obese humans.[51] In a novel approach, the group of S. Dimmeler highlighted that locked nucleic acid-miR-21 treatment led to significant weight loss and reduced adipocyte size, as well as repression of targets such as transforming growth factor-β-receptor 2 (TGFBR2) and phosphatase and tensin homolog (PTEN).[58] In the same vein, *let-7* knockout mice did not develop insulin resistance, despite diet-induced obesity.[14]

Endogenous miRNAs produced within adipocytes have been well characterized. However, an increasing number of exogenous miRNAs have been shown to exist in different bodily fluids, such as plasma, serum, urine, and saliva.[59,60] Many of these fluid-based miRNAs are present in small extracellular vesicles, called exosomes, that are secreted by various cell types including adipose cells. Adipose tissue can release exosomal miRNAs that can act as signalling molecules (for review see ref. 61) (Figure 9.1). In the context of obesity, DIO or ob/ob mice released more exosomes than control mice.[62] By this mechanism, adipose tissue can modulate the expression of genes located in other metabolic organs (Figure 9.1). For instance, the group of R. Kahn nicely demonstrated that exosomal miRNAs released from adipose tissue affected the hepatic expression of fibroblast growth factor 21 (FGF-21) involved in glucose homeostasis.[63]

Altogether, these different studies indicate that miRNAs can be used as biomarkers in the context of metabolic diseases, but they could also be targeted for pharmaceutical approaches.

9.2.4 miRNAs and Adipokines

Adipose tissue is an important endocrine organ known to secrete nearly 600 adipokines, which could be used as biomarkers in the clinical field (for review see ref. 64). The most important adipokines include leptin, adiponectin, resistin, and apelin (for review see ref. 65). These adipokines exert a large range of effects on diverse organs, suggesting that the impairment of adipose tissue function alters the normal physiology of these target organs.[65]

These adipokines modulate the expression of miRNAs involved in adipogenesis (for review see ref. 56). Xu *et al.* demonstrated that leptin increased the expression of miR-378 in human adipocytes.[66] Data from the literature clearly describes that miRNAs are important for the secretion or the effect of adipokines (leptin and adiponectin) (Figure 9.1). For example, miR-218 can target the 3′ UTR of the mRNA coding for the adiponectin receptor (AdipoR), which inhibits the effect of adiponectin on glucose uptake.[67] At the central level, several miRNAs are identified as targets of leptin in the context of feeding behavior.[68–71] Conversely, the group of M. Taouis showed that overexpression of miR-200a in the hypothalamus of leptin-deficient mice can down-regulate leptin receptor hypothalamic expression.[70]

9.2.5 miRNAs, Cytokines, and Inflammation of the Adipose Tissue

It is well established that chronic inflammation is a key feature of obesity, and this obesity-associated inflammation is involved in the onset of the metabolic syndrome (for review see ref. 72 and 73). A number of miRNAs act as regulators of the expression of inflammatory markers (for review see ref. 45, 74, and 75) (Figure 9.1). For instance, the over-expression of miR-132 in primary human adipose-derived stem cells leads to an increase in the production of interleukin-8 and monocyte chemoattractant protein-1 (MCP1).[76] In addition, it has been suggested that miR-126 and miR-193b regulate the expression and the release, respectively, of MCP1.[77] Another set of studies revealed that miRNAs can modulate the expression and/or the secretion of TNF-α. More precisely, the decrease in the expression of miR-221 is associated with a high level of TNF-α in the human adipose tissue-derived mesenchymal stem cells from obese females.[78] Lorente-Cebrián *et al.* observed that miR-145 stimulates the expression of TNF-α in adipocytes through the activation of the nuclear transcription factor kappa-B (NF-κB) pathway.[79] Interestingly, the group of J.F. Landrier demonstrated a pro-inflammatory connection between NF-κB and miR-155 which could participate in the amplification of inflammatory status in adipocytes.[80]

Several studies highlighted a dysregulation of miRNA expression during the inflammation of adipose tissue. The profile of expression of miRNAs is similar in the TNF-α-treated adipocyte 3T3-L1 cell line and in the adipocytes from the leptin-deficient ob/ob mice model.[81] In human pre-adipocytes, TNF-α or leptin induces a decrease of miR-221 and an increase of miR-335.[78,82] In the murine adipocyte model, TNF-α leads to an increase in the expression of miR-130, miR-146a, miR-146b, miR-150, miR-221, and miR-222 and a decrease of miR-103 and miR-143 levels.[81,83–85]

Altogether, these different studies show that the expression of miRNAs in adipose tissue can be modulated by inflammation and that miRNAs can be involved in the inflammation process induced by adipose tissue during the onset of obesity (Figure 9.1). These inflammatory processes are now well documented to be associated with the defect of glucose homeostasis that leads to type 2 diabetes. In particular, the pro-inflammatory factors can induce impairment of insulin secretion as well as sensitivity to insulin in peripheral organs. Recently, miRNAs have been considered as new actors in these pathophysiological aspects.

9.2.6 miRNAs, Insulin Synthesis, Secretion, and Sensitivity

Type 2 diabetes is characterized by a defect in insulin secretion, but also an impairment of insulin sensitivity. In this context, miRNAs play a role in insulin synthesis and secretion as well as in insulin sensitivity (for review see ref. 86). Different studies indicate that ablation of Dicer in β-cells leads to the onset of diabetes because of an impairment in glucose-stimulated insulin secretion and insulin biosynthesis.[87–89] Other studies have indicated that Ago is also crucial for the secretion of insulin.[90] Specific deletion of Ago in the β-cell line MIN6 enhances the biosynthesis of insulin.[90] Several miRNAs were originally identified to be expressed specifically in the pancreatic endocrine cell line.[91] Among them, the group of M. Stoffel described miR-375 as an important modulator of β-cell functions.[92] miR-375 overexpression in β-cells leads to a reduction of the number and viability of β-cells.[92,93] Interestingly, miR-375 regulates the release of insulin through direct targeting of genes such as apoptosis-inducing factor mitochondria-associated 1 (Aifm1), gephyrin (Gphn), tyrosine 3-monooxygenase/tryptophan 5-monooxygenase activation protein zeta (ywhaz), and myotrophin (MTPN), which are relevant for exocytosis.[90,91] It has also been reported that miR-9 exerted an inhibitory effect on insulin exocytosis by an action on the transcription factor one cut homeobox 2 (Onecut2) and granuphilin (Sytl4), a negative regulator of secretion.[94] Additionally, miR-124a and *let-7b*, both of which are abundantly expressed in pancreatic islet cells, are postulated to be important in the release of insulin through the modulation of MTPN expression.[95]

As indicated earlier, miRNAs are also important actors for the transcription of insulin. Using the MIN6 cell line, Tang *et al.* demonstrated that miR-30d is up-regulated in the presence of high glucose levels.[96] Interestingly, the overexpression of miR-30d increased insulin gene expression, while inhibition of miR-30d abolished glucose-stimulated insulin gene transcription.[96] The overexpression of miR-375 also suppressed glucose-stimulated insulin expression by targeting 3′-phosphoinositide-dependent protein kinase-1 (PDK1), indicating that miR-375 is one of the most relevant miRNAs in the pancreas.[91,93]

Skeletal muscle and the liver are the main users of glucose under the action of insulin (Figure 9.2). In these organs, several studies have described mechanisms of insulin resistance involving miRNAs. Recently, Zhou *et al.* demonstrated that miR-29a was involved in insulin resistance by targeting PPARδ, which led to a decrease in the levels of glucose transporter type 4 (GLUT4) in skeletal muscle.[97] Interestingly, several miRNAs are potentially involved in the expression of GLUT4 by targeting GLUT4 mRNA directly or indirectly *via* transcription factors such as peroxisome proliferator-activated receptor gamma coactivator 1-alpha (PGC1α), as well as Krüeppel-like factor 15 (KLF15).[98] The importance of miR-29a in insulin resistance was confirmed by a study indicating that miR-29a was up-regulated in the liver of db/db mice.[99] This increase in the expression of miR-29a was consequently accompanied by an attenuation of insulin inhibition of the expression of phosphoenol pyruvate carboxykinase (PEPCK), which is involved in the inhibition of gluconeogenesis.[99] In mice, a high fat diet increased the expression of miR-29a in myocytes, which led to an impairment in insulin signalling by decreasing insulin receptor substrate 1 (IRS1) expression.[100] In liver, insulin resistance was also associated with miR-33, which modulates the expression of IRS2.[101] Interestingly, overweight humans had decreased expression of

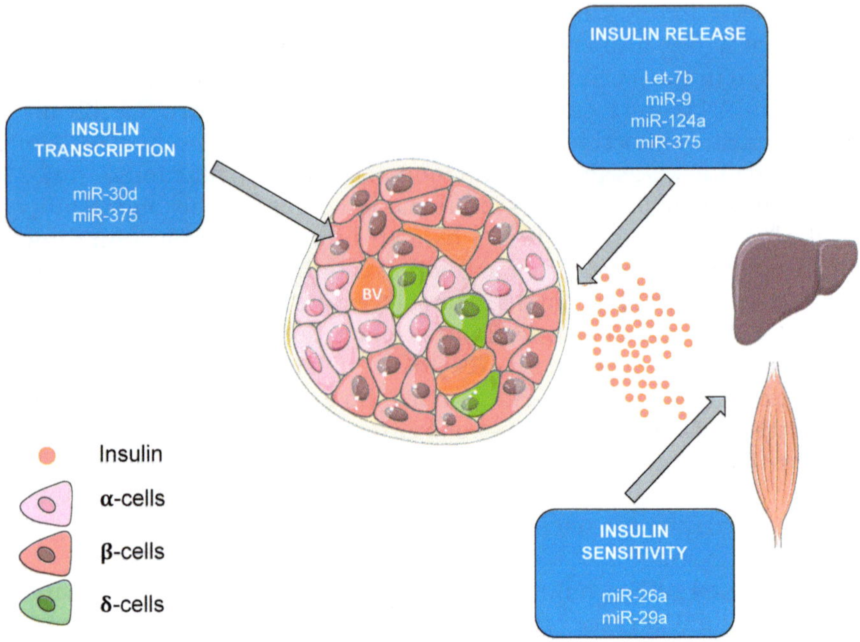

Figure 9.2 miRNAs involved in biosynthesis and effect of insulin. Several miRNAs positively or negatively regulated the transcription and the release of insulin, as well as sensitivity to insulin in the liver and skeletal muscle. BV: blood vessel.

miR-26a in the liver compared to lean individuals.[102] Moreover, liver-specific overexpression of miR-26a in mice fed a high-fat diet showed improved insulin sensitivity.[102]

9.3 miRNAS and Neuroendocrine Diseases

Neuroendocrine diseases are disorders that affect the interaction between the nervous system and the endocrine system. These disorders of neuroendocrine function include adrenal disorders, Kallmann syndrome, and neuroendocrine tumors, which are neoplasms that arise from cells of the endocrine and nervous systems. During recent decades, the functions of miRNAs in the field of neuroendocrinology have been investigated. In particular, it has been demonstrated that hormones can modulate the expression of miRNAs in different organs and in which some pathologies such as cancer can be induced.[103] (For review see ref. 104–106.) For instance, oestrogens or thyroid hormones can modulate the biogenesis of several miRNAs in different organs.[107–109] In this section, we mainly focus our attention on the functions of miRNAs in the impairment of the neuroendocrine axis (mainly the thyroid and reproductive axes), but also on the use of miRNAs as prognostic and diagnostic biomarkers in neuroendocrine pathologies.

9.3.1 miRNAs in the Thyroid Axis

Thyroid hormones (tri-iodothyronine (T3) and thyroxine (T4)) are important for brain development and the modulation of metabolism, and are associated with many aspects of normal adult physiology.[110] Thyroid cancer is the main form of endocrine cancer but, if diagnosed early, it is one of the most treatable forms of cancer. Four main types of thyroid cancer have been described: papillary thyroid cancer, follicular thyroid cancer, medullary cancer, and anaplastic thyroid cancer (for review see ref. 111). The most important tool in the diagnosis of thyroid cancer is fine-needle aspiration biopsy (FNAB), but this test is based on a description of the cells. FNAB cannot provide information on tissue architecture and vascular or neural invasion, and thus it is very limited in diagnosing malignancy.[112]

It is now recognized that miRNAs are important for thyroid carcinogenesis and, as indicated, they can be considered as potential diagnostic and prognostic biomarkers. In addition, as described later, they also appear as potential targets for treatment of these different forms of cancer.

In physiological conditions, the thyroid gland highly expresses *let-7* family miRNAs, the miR-15/miR-16 cluster, and other miRNAs such as miR-30a/miR-30c/miR-30d, miR-99b, miR-125b, miR-127, miR-200c, and miR-138.[113] Interestingly, this set of miRNAs is down-regulated in thyroid cancer.[13,114–117] These observations suggest that miRNAs are essential for the normal function of thyroid gland. The thyroid gland is important to produce T3 and T4 from iodine, which involves a set of genes important for iodine metabolism (for review see ref. 118). The impairment in the expression of genes involved in iodine metabolism leads to an imbalance in thyroid hormone synthesis and thyroid disease. The most important protein in this process is the natrium-iodide symporter (NIS), which allows the transport of ionic iodide from the blood to the cytoplasm of thyroid follicular cells.[119] Interestingly, the down-regulation of NIS expression is associated with papillary thyroid cancer. Thus, it has been clearly established that NIS is a target for some miRNAs. To date, the expression of NIS is linked to miR-146b and miR-339, which are over-expressed in thyroid cancer.[120–122] Interestingly, the miRNAs discussed here are also potential therapy tools for restoring thyroid function and radioiodine trapping during radiotherapy. In fact, down-regulation of miR-146b restores radioiodine sensitivity in thyroid cancer.[122] In the case of follicular thyroid cancer, miR-146b as well as miR-221 are up-regulated in comparison with normal tissue.[123]

Clearly, the dysregulation of miRNA expression is tightly associated with cancer initiation and progression of the thyroid gland. The mechanisms involve a network of miRNAs that target genes regulating the cell cycle. Boufraqech *et al.* reviewed the miRNAs that play a role in cancer cell proliferation, differentiation, invasion, migration, and cell death.[124] As in other cancers, miRNAs are promising diagnostic and prognostic markers in thyroid cancer.[124,125]

9.3.2 miRNAs in the Reproductive Axis

The gonadotropic axis plays an essential role in the regulation of reproductive functions (for review see ref. 126). The production of sex hormones is under the control of the hypothalamic gonadotropin-releasing hormone (GnRH) which acts on pituitary gonadotrophs to stimulate luteinizing hormone and follicle-stimulating hormone (FSH) synthesis and secretion.[126]

A large body of data has indicated that miRNAs modulate the gonadotropic axis at different levels. This regulation by miRNAs can be sustained by the sexual dimorphic expression of miRNAs observed during development.[127] For instance, *let-7a* and *let-7b* exhibited a sex-specific expression in the hypothalamus of rats.[128] It has been established that GnRH neurons originate in the olfactory placode and migrate to the hypothalamus during embryonic development. Defects in this process cause Kallmann syndrome characterized by delay or absence of sexual maturation.[129] Microarrays, antisense morpholinos, and knockdown approaches indicate that the miR-200 family and miR-9 contribute to the migration of GnRH neurons from the olfactory placode to the hypothalamus.[130,131]

It is interesting to note that miRNAs are also associated with the epigenetic mechanisms of infertility and human congenital idiopathic hypogonadotropic hypogonadism (IHH).[132] In fact, miRNAs are involved in a switch from repression to induction of GnRH expression during the onset of puberty.[133] More precisely, miR-200 and miR-155, respectively, regulated Zinc finger E-box-binding homeobox 1 (ZEB1), a repressor of GnRH transcriptional activators and GnRH itself, and C/EBP, a nitric oxide-mediated repressor of GnRH that acts both directly and through ZEB1 in GnRH neurons.[133]

Independently of GnRH, FSH and luteinizing hormone are also involved in the IHH. Specific deletion of Dicer in pituitary gonadotrophs leads to fertility defects by the loss of FSH and luteinizing hormone synthesis.[134,135] Recently, Ahmed *et al.* demonstrated that the invalidation of miR-7a2 in mice led to hypogonadism and infertility without affecting GnRH synthesis or release.[136] More precisely, miR-7a2 regulates sexual maturation and reproductive function through its effects on pituitary prostaglandin and bone morphogenetic protein 4 (BMP4) signaling.[136] Another study indicated that the simultaneous deletion of miR-200b and miR-429 induced a decrease in serum luteinizing hormone in female mice, but not in males.[137] Both miRNAs are strongly expressed in the pituitary gland, where they suppress expression of the transcriptional repressor ZEB1.[137] FSH expression was also found to be regulated by miRNAs. In fact, using an *in vitro* approach (mouse LβT2 gonadotrope cells), miR-132 and miR-212 regulation of FSH expression through the Forkhead box protein O1 (FOXO1) and Sirtuin 1 (SIRT1) has been described.[138]

Altogether, these studies suggest that miRNAs are essential for the accurate development of the reproductive axis in males as well as in females in different species. The field of miRNA research in reproductive axis is currently very active, with particular interest in the mechanisms involved in the onset of puberty and in gender-specific disparities.

9.3.3 miRNAs as Biomarkers in Neuroendocrine Tumours

Neuroendocrine tumors are most often found in the gastroenteropancreatic system and lungs (for review see ref. 139). However, they can occur in others organs, such as the adrenal gland (for review see ref. 140). While many therapeutic advances have been made in neuroendocrine tumors, ultimately, many patients with metastatic disease die due to progression of disease.[141,142] To date, it has been difficult to discern the benign or malignant character of neuroendocrine tumors. For this reason, it is important to have an accurate, affordable, and reliable biomarker to help inform prognosis and predict response to treatment. In this context, miRNAs have been investigated as potential biomarkers for neuroendocrine tumors.

The pattern of miRNA expression differs between the neuroendocrine tumors. For instance, the down-regulation of miR-133a in gastrointestinal neuroendocrine tumors was associated with metastatic disease.[143] In the case of small intestinal neuroendocrine tumors, Li *et al.* demonstrated that, out of nine miRNAs, five (miR-96, miR-182, miR-183, miR-196a, and miR-200a) were up-regulated during tumor progression, whereas four (miR-31, miR-129-5p, miR-133a, and miR-215) were down-regulated.[144] In general, an up-regulation of miR-10b, miR-133a, miR-145, miR-146, miR-222 and a down-regulation of miR-19a, miR-19b, miR-183, and miR-488 were observed in gastrointestinal neuroendocrine tumors.[145] (For review see ref. 146.) These studies described a correlation between the expression of a group of miRNAs and established prognostic factors, but they failed to predict progression-free survival or overall survival. However, one study involving 44 patients with small cell carcinoma of the cervix demonstrated an association of down-regulation of miR-100 and miR-125b with poorer overall survival.[147]

9.4 Conclusions

As illustrated in this chapter, miRNAs have emerged as key regulators of endocrine functions. Results from previous studies indicate that miRNA expression is associated with different aspects of endocrine diseases: development of glands, release of hormones, sensitivity to hormones, and tumorigenesis. We have also documented a potential role for miRNAs as biomarkers for the diagnosis or prognosis of different endocrine diseases. In our opinion, most relevant finding in the field of miRNAs research is the novel discovery that miRNAs packaged within cell-secreted exosomes could be important for the cross-talk between different organs. In particular, recent studies highlight the role of exosomes in the communication between adipose tissue and the liver. Additional studies are needed to entirely understand this novel mechanism of cell–cell communication.

Acknowledgements

This work was supported by funding obtained from Aix-Marseille University, INRA INSERM and Benjamin Delessert foundation. AD and EK were the recipient of a doctoral fellowship from the French Ministry of Education. The authors are grateful to S. Mounien for critical reading of the manuscript.

References

1. D. P. Bartel, *Cell*, 2004, **116**, 281–297.
2. R. C. Lee, R. L. Feinbaum and V. Ambros, *Cell*, 1993, **75**, 843–854.
3. B. Wightman, I. Ha and G. Ruvkun, *Cell*, 1993, **75**, 855–862.
4. H. Guo, N. T. Ingolia, J. S. Weissman and D. P. Bartel, *Nature*, 2010, **466**, 835–840.
5. J. Han, Y. Lee, K.-H. Yeom, Y.-K. Kim, H. Jin and V. N. Kim, *Genes Dev.*, 2004, **18**, 3016–3027.
6. H. Siomi and M. C. Siomi, *Mol. Cell*, 2010, **38**, 323–332.
7. R. C. Friedman, K. K.-H. Farh, C. B. Burge and D. P. Bartel, *Genome Res.*, 2009, **19**, 92–105.
8. M. Selbach, B. Schwanhäusser, N. Thierfelder, Z. Fang, R. Khanin and N. Rajewsky, *Nature*, 2008, **455**, 58–63.
9. A. Grimson, K. K.-H. Farh, W. K. Johnston, P. Garrett-Engele, L. P. Lim and D. P. Bartel, *Mol. Cell*, 2007, **27**, 91–105.
10. J. G. Doench and P. A. Sharp, *Genes Dev.*, 2004, **18**, 504–511.
11. N. Bushati and S. M. Cohen, *Annu. Rev. Cell Dev. Biol.*, 2007, **23**, 175–205.
12. M. M. Hossain, M. M. H. Sohel, K. Schellander and D. Tesfaye, *Cell Tissue Res.*, 2012, **349**, 679–690.
13. P. Pallante, S. Battista, G. M. Pierantoni and A. Fusco, *Nat. Rev. Endocrinol.*, 2014, **10**, 88–101.
14. R. J. A. Frost and E. N. Olson, *Proc. Natl. Acad. Sci. U. S. A.*, 2011, **108**, 21075–21080.
15. J. Krützfeldt and M. Stoffel, *Cell Metab.*, 2006, **4**, 9–12.
16. C. L. Ogden, M. D. Carroll, B. K. Kit and K. M. Flegal, *NCHS Data Brief.*, 2012, 1–8.
17. D. W. Haslam and W. P. T. James, *Lancet*, 2005, **366**, 1197–1209.
18. S. I. Berndt, S. Gustafsson, R. Mägi, A. Ganna, E. Wheeler, M. F. Feitosa, A. E. Justice, K. L. Monda, D. C. Croteau-Chonka, F. R. Day, T. Esko, T. Fall, T. Ferreira, D. Gentilini, A. U. Jackson, J. Luan, J. C. Randall, S. Vedantam, C. J. Willer, T. W. Winkler, A. R. Wood, T. Workalemahu, Y.-J. Hu, S. H. Lee, L. Liang, D.-Y. Lin, J. L. Min, B. M. Neale, G. Thorleifsson, J. Yang, E. Albrecht, N. Amin, J. L. Bragg-Gresham, G. Cadby, M. den Heijer, N. Eklund, K. Fischer, A. Goel, J.-J. Hottenga, J. E. Huffman, I. Jarick, Å. Johansson, T. Johnson, S. Kanoni, M. E. Kleber, I. R. König, K. Kristiansson, Z. Kutalik, C. Lamina, C. Lecoeur, G. Li, M. Mangino,

W. L. McArdle, C. Medina-Gomez, M. Müller-Nurasyid, J. S. Ngwa, I. M. Nolte, L. Paternoster, S. Pechlivanis, M. Perola, M. J. Peters, M. Preuss, L. M. Rose, J. Shi, D. Shungin, A. V. Smith, R. J. Strawbridge, I. Surakka, A. Teumer, M. D. Trip, J. Tyrer, J. V. Van Vliet-Ostaptchouk, L. Vandenput, L. L. Waite, J. H. Zhao, D. Absher, F. W. Asselbergs, M. Atalay, A. P. Attwood, A. J. Balmforth, H. Basart, J. Beilby, L. L. Bonnycastle, P. Brambilla, M. Bruinenberg, H. Campbell, D. I. Chasman, P. S. Chines, F. S. Collins, J. M. Connell, W. O. Cookson, U. de Faire, F. de Vegt, M. Dei, M. Dimitriou, S. Edkins, K. Estrada, D. M. Evans, M. Farrall, M. M. Ferrario, J. Ferrières, L. Franke, F. Frau, P. V. Gejman, H. Grallert, H. Grönberg, V. Gudnason, A. S. Hall, P. Hall, A.-L. Hartikainen, C. Hayward, N. L. Heard-Costa, A. C. Heath, J. Hebebrand, G. Homuth, F. B. Hu, S. E. Hunt, E. Hyppönen, C. Iribarren, K. B. Jacobs, J.-O. Jansson, A. Jula, M. Kähönen, S. Kathiresan, F. Kee, K.-T. Khaw, M. Kivimäki, W. Koenig, A. T. Kraja, M. Kumari, K. Kuulasmaa, J. Kuusisto, J. H. Laitinen, T. A. Lakka, C. Langenberg, L. J. Launer, L. Lind, J. Lindström, J. Liu, A. Liuzzi, M.-L. Lokki, M. Lorentzon, P. A. Madden, P. K. Magnusson, P. Manunta, D. Marek, W. März, I. Mateo Leach, B. McKnight, S. E. Medland, E. Mihailov, L. Milani, G. W. Montgomery, V. Mooser, T. W. Mühleisen, P. B. Munroe, A. W. Musk, N. Narisu, G. Navis, G. Nicholson, E. A. Nohr, K. K. Ong, B. A. Oostra, C. N. A. Palmer, A. Palotie, J. F. Peden, N. Pedersen, A. Peters, O. Polasek, A. Pouta, P. P. Pramstaller, I. Prokopenko, C. Pütter, A. Radhakrishnan, O. Raitakari, A. Rendon, F. Rivadeneira, I. Rudan, T. E. Saaristo, J. G. Sambrook, A. R. Sanders, S. Sanna, J. Saramies, S. Schipf, S. Schreiber, H. Schunkert, S.-Y. Shin, S. Signorini, J. Sinisalo, B. Skrobek, N. Soranzo, A. Stančáková, K. Stark, J. C. Stephens, K. Stirrups, R. P. Stolk, M. Stumvoll, A. J. Swift, E. V. Theodoraki, B. Thorand, D.-A. Tregouet, E. Tremoli, M. M. Van der Klauw, J. B. J. van Meurs, S. H. Vermeulen, J. Viikari, J. Virtamo, V. Vitart, G. Waeber, Z. Wang, E. Widén, S. H. Wild, G. Willemsen, B. R. Winkelmann, J. C. M. Witteman, B. H. R. Wolffenbuttel, A. Wong, A. F. Wright, M. C. Zillikens, P. Amouyel, B. O. Boehm, E. Boerwinkle, D. I. Boomsma, M. J. Caulfield, S. J. Chanock, L. A. Cupples, D. Cusi, G. V. Dedoussis, J. Erdmann, J. G. Eriksson, P. W. Franks, P. Froguel, C. Gieger, U. Gyllensten, A. Hamsten, T. B. Harris, C. Hengstenberg, A. A. Hicks, A. Hingorani, A. Hinney, A. Hofman, K. G. Hovingh, K. Hveem, T. Illig, M.-R. Jarvelin, K.-H. Jöckel, S. M. Keinanen-Kiukaanniemi, L. A. Kiemeney, D. Kuh, M. Laakso, T. Lehtimäki, D. F. Levinson, N. G. Martin, A. Metspalu, A. D. Morris, M. S. Nieminen, I. Njølstad, C. Ohlsson, A. J. Oldehinkel, W. H. Ouwehand, L. J. Palmer, B. Penninx, C. Power, M. A. Province, B. M. Psaty, L. Qi, R. Rauramaa, P. M. Ridker, S. Ripatti, V. Salomaa, N. J. Samani, H. Snieder, T. I. A. Sørensen, T. D. Spector, K. Stefansson, A. Tönjes, J. Tuomilehto, A. G. Uitterlinden, M. Uusitupa, P. van der Harst, P. Vollenweider, H. Wallaschofski, N. J. Wareham, H. Watkins, H.-E. Wichmann, J. F. Wilson, G. R. Abecasis, T. L. Assimes, I. Barroso, M. Boehnke, I. B. Borecki, P. Deloukas, C. S. Fox, T. Frayling, L. C. Groop, T. Haritunian, I. M. Heid, D. Hunter, R. C.

Kaplan, F. Karpe, M. F. Moffatt, K. L. Mohlke, J. R. O'Connell, Y. Pawitan, E. E. Schadt, D. Schlessinger, V. Steinthorsdottir, D. P. Strachan, U. Thorsteinsdottir, C. M. van Duijn, P. M. Visscher, A. M. Di Blasio, J. N. Hirschhorn, C. M. Lindgren, A. P. Morris, D. Meyre, A. Scherag, M. I. McCarthy, E. K. Speliotes, K. E. North, R. J. F. Loos and E. Ingelsson, *Nat. Genet.*, 2013, **45**, 501–512.
19. T. Kunej, D. Jevsinek Skok, M. Zorc, A. Ogrinc, J. J. Michal, M. Kovac and Z. Jiang, *J. Genomics*, 2013, **1**, 45–55.
20. R. J. F. Loos and G. S. H. Yeo, *Nat. Rev. Endocrinol.*, 2014, **10**, 51–61.
21. Q. Jiang, Y. Wang, Y. Hao, L. Juan, M. Teng, X. Zhang, M. Li, G. Wang and Y. Liu, *Nucleic Acids Res.*, 2009, **37**, D98–D104.
22. O. Dumortier, C. Hinault and E. Van Obberghen, *Cell Metab.*, 2013, **18**, 312–324.
23. J. D. Palmer, B. P. Soule, B. A. Simone, N. G. Zaorsky, L. Jin and N. L. Simone, *Ageing Res. Rev.*, 2014, **17**, 16–24.
24. M. L. Slattery, J. S. Herrick, L. E. Mullany, J. R. Stevens and R. K. Wolff, *Pharmacogenomics Pers. Med.*, 2017, **10**, 1–16.
25. A. Derghal, M. Djelloul, J. Trouslard and L. Mounien, *Front. Neurosci.*, 2017, **11**, 181.
26. G. J. Morton, D. E. Cummings, D. G. Baskin, G. S. Barsh and M. W. Schwartz, *Nature*, 2006, **443**, 289–295.
27. N. Balthasar, R. Coppari, J. McMinn, S. M. Liu, C. E. Lee, V. Tang, C. D. Kenny, R. A. McGovern, S. C. Chua Jr, J. K. Elmquist and B. B. Lowell, *Neuron*, 2004, **42**, 983–991.
28. R. Coppari, M. Ichinose, C. E. Lee, A. E. Pullen, C. D. Kenny, R. A. McGovern, V. Tang, S. M. Liu, T. Ludwig, S. C. Chua, B. B. Lowell and J. K. Elmquist, *Cell Metab.*, 2005, **1**, 63–72.
29. H. Dhillon, J. M. Zigman, C. Ye, C. E. Lee, R. A. McGovern, V. Tang, C. D. Kenny, L. M. Christiansen, R. D. White, E. A. Edelstein, R. Coppari, N. Balthasar, M. A. Cowley, S. Chua, J. K. Elmquist and B. B. Lowell, *Neuron*, 2006, **49**, 191–203.
30. J.-F. Landrier, E. Karkeni, J. Marcotorchino, L. Bonnet and F. Tourniaire, *Proc. Nutr. Soc.*, 2016, **75**, 38–46.
31. J.-F. Landrier, J. Marcotorchino and F. Tourniaire, *Nutrients*, 2012, **4**, 1622–1649.
32. S. Vienberg, J. Geiger, S. Madsen and L. T. Dalgaard, *Acta Physiol.*, 2017, **219**, 346–361.
33. B. M. Spiegelman and J. S. Flier, *Cell*, 1996, **87**, 377–389.
34. N. L. Price and C. Fernández-Hernando, *Biochim. Biophys. Acta*, 2016, **1861**, 2104–2110.
35. N. L. Price, B. Holtrup, S. L. Kwei, M. Wabitsch, M. Rodeheffer, L. Bianchini, Y. Suárez and C. Fernández-Hernando, *Mol. Cell. Biol.*, 2016, **36**, 1180–1193.
36. P. Xu, S. Y. Vernooy, M. Guo and B. A. Hay, *Curr. Biol.*, 2003, **13**, 790–795.
37. A. A. Teleman, S. Maitra and S. M. Cohen, *Genes Dev.*, 2006, **20**, 417–422.

38. C. Esau, X. Kang, E. Peralta, E. Hanson, E. G. Marcusson, L. V. Ravichandran, Y. Sun, S. Koo, R. J. Perera, R. Jain, N. M. Dean, S. M. Freier, C. F. Bennett, B. Lollo and R. Griffey, *J. Biol. Chem.*, 2004, **279**, 52361–52365.
39. R. Takanabe, K. Ono, Y. Abe, T. Takaya, T. Horie, H. Wada, T. Kita, N. Satoh, A. Shimatsu and K. Hasegawa, *Biochem. Biophys. Res. Commun.*, 2008, **376**, 728–732.
40. E. K. Lee, M. J. Lee, K. Abdelmohsen, W. Kim, M. M. Kim, S. Srikantan, J. L. Martindale, E. R. Hutchison, H. H. Kim, B. S. Marasa, R. Selimyan, J. M. Egan, S. R. Smith, S. K. Fried and M. Gorospe, *Mol. Cell. Biol.*, 2011, **31**, 626–638.
41. Q. Wang, Y. C. Li, J. Wang, J. Kong, Y. Qi, R. J. Quigg and X. Li, *Proc. Natl. Acad. Sci. U. S. A.*, 2008, **105**, 2889–2894.
42. H.-Y. Ling, G.-B. Wen, S.-D. Feng, Q.-H. Tuo, H.-S. Ou, C. H. Yao, B.-Y. Zhu, Z.-P. Gao, L. Zhang and D.-F. Liao, *Clin. Exp. Pharmacol. Physiol.*, 2011, **38**, 239–246.
43. F. J. Ortega, J. M. Moreno-Navarrete, G. Pardo, M. Sabater, M. Hummel, A. Ferrer, J. I. Rodriguez-Hermosa, B. Ruiz, W. Ricart, B. Peral and J. M. Fernández-Real, *PLoS One*, 2010, **5**, e9022.
44. R. Martinelli, C. Nardelli, V. Pilone, T. Buonomo, R. Liguori, I. Castanò, P. Buono, S. Masone, G. Persico, P. Forestieri, L. Pastore and L. Sacchetti, *Obesity*, 2010, **18**, 2170–2176.
45. P. Arner and A. Kulyté, *Nat. Rev. Endocrinol.*, 2015, **11**, 276–288.
46. H. M. Heneghan, N. Miller and M. J. Kerin, *Obes. Rev.*, 2010, **11**, 354–361.
47. G. Maurizi, L. Babini and L. Della Guardia, *J. Cell. Physiol.*, 2018, **233**(12), 9077–9086.
48. M.-C. Vohl, R. Sladek, J. Robitaille, S. Gurd, P. Marceau, D. Richard, T. J. Hudson and A. Tchernof, *Obes. Res.*, 2004, **12**, 1217–1222.
49. N. Klöting, S. Berthold, P. Kovacs, M. R. Schön, M. Fasshauer, K. Ruschke, M. Stumvoll and M. Blüher, *PLoS One*, 2009, **4**, e4699.
50. H. M. Heneghan, N. Miller, O. J. McAnena, T. O'Brien and M. J. Kerin, *J. Clin. Endocrinol. Metab.*, 2011, **96**, E846–E850.
51. P. Keller, V. Gburcik, N. Petrovic, I. J. Gallagher, J. Nedergaard, B. Cannon and J. A. Timmons, *BMC Endocr. Disord.*, 2011, **11**, 7.
52. M. M. Kristensen, P. K. Davidsen, A. Vigelsø, C. N. Hansen, L. J. Jensen, N. Jessen, J. M. Bruun, F. Dela and J. W. Helge, *Obesity*, 2017, **25**, 572–580.
53. A. Meerson, M. Traurig, V. Ossowski, J. M. Fleming, M. Mullins and L. J. Baier, *Diabetologia*, 2013, **56**, 1971–1979.
54. A. Meerson, M. Traurig, V. Ossowski, J. M. Fleming, M. Mullins and L. J. Baier, *Diabetologia*, 2013, **56**, 1971–1979.
55. S. Li, X. Chen, H. Zhang, X. Liang, Y. Xiang, C. Yu, K. Zen, Y. Li and C.-Y. Zhang, *J. Lipid Res.*, 2009, **50**, 1756–1765.
56. G. Iacomino and A. Siani, *Genes Nutr.*, 2017, **12**, 23.

57. D. V. Chartoumpekis, A. Zaravinos, P. G. Ziros, R. P. Iskrenova, A. I. Psyrogiannis, V. E. Kyriazopoulou and I. G. Habeos, *PLoS One*, 2012, **7**, e34872.
58. T. Seeger, A. Fischer, M. Muhly-Reinholz, A. M. Zeiher and S. Dimmeler, *Obesity*, 2014, **22**, 2352–2360.
59. A. Turchinovich, L. Weiz, A. Langheinz and B. Burwinkel, *Nucleic Acids Res.*, 2011, **39**, 7223–7233.
60. M. A. Cortez and G. A. Calin, *Expert Opin. Biol. Ther.*, 2009, **9**, 703–711.
61. C. Guay and R. Regazzi, *Diabetes, Obes. Metab.*, 2017, **19**(Suppl 1), 137–146.
62. Z. Deng, A. Poliakov, R. W. Hardy, R. Clements, C. Liu, Y. Liu, J. Wang, X. Xiang, S. Zhang, X. Zhuang, S. V. Shah, D. Sun, S. Michalek, W. E. Grizzle, T. Garvey, J. Mobley and H.-G. Zhang, *Diabetes*, 2009, **58**, 2498–2505.
63. T. Thomou, M. A. Mori, J. M. Dreyfuss, M. Konishi, M. Sakaguchi, C. Wolfrum, T. N. Rao, J. N. Winnay, R. Garcia-Martin, S. K. Grinspoon, P. Gorden and C. R. Kahn, *Nature*, 2017, **542**, 450–455.
64. G. Flehmig, M. Scholz, N. Klöting, M. Fasshauer, A. Tönjes, M. Stumvoll, B.-S. Youn and M. Blüher, *PLoS One*, 2014, **9**, e99785.
65. G. Maurizi, L. Della Guardia, A. Maurizi and A. Poloni, *J. Cell. Physiol.*, 2018, **233**, 88–97.
66. L. Xu, C. Shi, G. Xu, L. Chen, L. Zhu, L. Zhu, X. Guo, M. Xu and C. Ji, *Cell Biochem. Biophys.*, 2014, **70**, 771–776.
67. H. Du, Z. Fu, G. He, Y. Wang, G. Xia, M. Fang and T. Zhang, *Mol. Med. Rep.*, 2015, **11**, 4701–4705.
68. A. Derghal, M. Djelloul, C. Airault, C. Pierre, M. Dallaporta, J.-D. Troadec, V. Tillement, C. Tardivel, B. Bariohay, J. Trouslard and L. Mounien, *Front. Cell. Neurosci.*, 2015, **9**, 172.
69. C. Benoit, H. Ould-Hamouda, D. Crepin, A. Gertler, L. Amar and M. Taouis, *J. Endocrinol.*, 2013, **218**, 35–47.
70. D. Crépin, Y. Benomar, L. Riffault, H. Amine, A. Gertler and M. Taouis, *Mol. Cell. Endocrinol.*, 2014, **384**, 1–11.
71. I. A. Vinnikov, K. Hajdukiewicz, J. Reymann, J. Beneke, R. Czajkowski, L. C. Roth, M. Novak, A. Roller, N. Dörner, V. Starkuviene, F. J. Theis, H. Erfle, G. Schütz, V. Grinevich and W. Konopka, *J. Neurosci.*, 2014, **34**, 10659–10674.
72. Y.-H. Tseng, A. M. Cypess and C. R. Kahn, *Nat. Rev. Drug Discovery*, 2010, **9**, 465–482.
73. J. Marcotorchino, F. Tourniaire and J.-F. Landrier, *Horm. Mol. Biol. Clin. Invest.*, 2013, **15**, 123–128.
74. Q. Ge, S. Brichard, X. Yi and Q. Li, *J. Immunol. Res.*, 2014, **2014**, 987285.
75. M. Hulsmans, D. De Keyzer and P. Holvoet, *FASEB J.*, 2011, **25**, 2515–2527.
76. J. C. Strum, J. H. Johnson, J. Ward, H. Xie, J. Feild, A. Hester, A. Alford and K. M. Waters, *Mol. Endocrinol.*, 2009, **23**, 1876–1884.

77. E. Arner, N. Mejhert, A. Kulyté, P. J. Balwierz, M. Pachkov, M. Cormont, S. Lorente-Cebrián, A. Ehrlund, J. Laurencikiene, P. Hedén, K. Dahlman-Wright, J.-F. Tanti, Y. Hayashizaki, M. Rydén, I. Dahlman, E. van Nimwegen, C. O. Daub and P. Arner, *Diabetes*, 2012, **61**, 1986–1993.
78. W.-W. Chou, Y.-T. Wang, Y.-C. Liao, S.-C. Chuang, S.-N. Wang and S.-H. H. Juo, *Cell. Physiol. Biochem.*, 2013, **32**, 127–137.
79. S. Lorente-Cebrián, N. Mejhert, A. Kulyté, J. Laurencikiene, G. Åström, P. Hedén, M. Rydén and P. Arner, *PLoS One*, 2014, **9**, e86800.
80. E. Karkeni, J. Astier, F. Tourniaire, M. El Abed, B. Romier, E. Gouranton, L. Wan, P. Borel, J. Salles, S. Walrand, J. Ye and J.-F. Landrier, *J. Clin. Endocrinol. Metab.*, 2016, **101**, 1615–1626.
81. H. Xie, B. Lim and H. F. Lodish, *Diabetes*, 2009, **58**, 1050–1057.
82. L. Zhu, L. Chen, C.-M. Shi, G.-F. Xu, L.-L. Xu, L.-L. Zhu, X.-R. Guo, Y. Ni, Y. Cui and C. Ji, *Cell Biochem. Biophys.*, 2014, **68**, 283–290.
83. C. Kim, H. Lee, Y. M. Cho, O.-J. Kwon, W. Kim and E. K. Lee, *FEBS Lett.*, 2013, **587**(23), 3853–3858.
84. C. Shi, L. Zhu, X. Chen, N. Gu, L. Chen, L. Zhu, L. Yang, L. Pang, X. Guo, C. Ji and C. Zhang, *J. Interferon Cytokine Res.*, 2014, **34**, 342–348.
85. E. Karkeni, L. Bonnet, J. Marcotorchino, F. Tourniaire, J. Astier, J. Ye and J.-F. Landrier, *Epigenetics*, 2018, **13**, 156–162.
86. X. Tang, G. Tang and S. Ozcan, *Biochim. Biophys. Acta*, 2008, **1779**, 697–701.
87. T. Melkman-Zehavi, R. Oren, S. Kredo-Russo, T. Shapira, A. D. Mandelbaum, N. Rivkin, T. Nir, K. A. Lennox, M. A. Behlke, Y. Dor and E. Hornstein, *EMBO J.*, 2011, **30**, 835–845.
88. A. D. Mandelbaum, T. Melkman-Zehavi, R. Oren, S. Kredo-Russo, T. Nir, Y. Dor and E. Hornstein, *Exp. Diabetes Res.*, 2012, **2012**, 470302.
89. M. Kalis, C. Bolmeson, J. L. S. Esguerra, S. Gupta, A. Edlund, N. Tormo-Badia, D. Speidel, D. Holmberg, S. Mayans, N. K. S. Khoo, A. Wendt, L. Eliasson and C. M. Cilio, *PLoS One*, 2011, **6**, e29166.
90. S. G. Tattikota, M. D. Sury, T. Rathjen, H.-H. Wessels, A. K. Pandey, X. You, C. Becker, W. Chen, M. Selbach and M. N. Poy, *Mol. Cell. Proteomics*, 2013, **12**, 1214–1225.
91. M. N. Poy, L. Eliasson, J. Krutzfeldt, S. Kuwajima, X. Ma, P. E. Macdonald, S. Pfeffer, T. Tuschl, N. Rajewsky, P. Rorsman and M. Stoffel, *Nature*, 2004, **432**, 226–230.
92. M. N. Poy, J. Hausser, M. Trajkovski, M. Braun, S. Collins, P. Rorsman, M. Zavolan and M. Stoffel, *Proc. Natl. Acad. Sci. U. S. A.*, 2009, **106**, 5813–5818.
93. A. El Ouaamari, N. Baroukh, G. A. Martens, P. Lebrun, D. Pipeleers and E. van Obberghen, *Diabetes*, 2008, **57**, 2708–2717.
94. V. Plaisance, A. Abderrahmani, V. Perret-Menoud, P. Jacquemin, F. Lemaigre and R. Regazzi, *J. Biol. Chem.*, 2006, **281**, 26932–26942.
95. A. Krek, D. Grün, M. N. Poy, R. Wolf, L. Rosenberg, E. J. Epstein, P. MacMenamin, I. da Piedade, K. C. Gunsalus, M. Stoffel and N. Rajewsky, *Nat. Genet.*, 2005, **37**, 495–500.

96. X. Tang, L. Muniappan, G. Tang and S. Ozcan, *RNA*, 2009, **15**, 287–293.
97. Y. Zhou, P. Gu, W. Shi, J. Li, Q. Hao, X. Cao, Q. Lu and Y. Zeng, *Int. J. Mol. Med.*, 2016, **37**, 931–938.
98. J. V. Esteves, F. J. Enguita and U. F. Machado, *J. Diabetes Res.*, 2017, **2017**, 7267910.
99. A. K. Pandey, G. Verma, S. Vig, S. Srivastava, A. K. Srivastava and M. Datta, *Mol. Cell. Endocrinol.*, 2011, **332**, 125–133.
100. W.-M. Yang, H.-J. Jeong, S.-Y. Park and W. Lee, *FEBS Lett.*, 2014, **588**, 2170–2176.
101. L. Goedeke, F. M. Vales-Lara, M. Fenstermaker, D. Cirera-Salinas, A. Chamorro-Jorganes, C. M. Ramírez, J. A. Mattison, R. de Cabo, Y. Suárez and C. Fernández-Hernando, *Mol. Cell. Biol.*, 2013, **33**, 2339–2352.
102. X. Fu, B. Dong, Y. Tian, P. Lefebvre, Z. Meng, X. Wang, F. Pattou, W. Han, X. Wang, F. Lou, R. Jove, B. Staels, D. D. Moore and W. Huang, *J. Clin. Invest.*, 2015, **125**, 2497–2509.
103. Z. Hu, W.-J. Shen, Y. Cortez, X. Tang, L.-F. Liu, F. B. Kraemer and S. Azhar, *PLoS One*, 2013, **8**, e78040.
104. A. Derghal, M. Djelloul, J. Trouslard and L. Mounien, *Front. Neurosci.*, 2016, **10**, 318.
105. B. E. Cameron, P. M. Craig and V. L. Trudeau, *Environ. Toxicol. Chem.*, 2016, **35**, 788–793.
106. C. M. Klinge, *Mol. Cell. Endocrinol.*, 2015, **418**(Pt 3), 273–297.
107. H. Dong, M. Paquette, A. Williams, R. T. Zoeller, M. Wade and C. Yauk, *PLoS One*, 2010, **5**, e12136.
108. A. Gupta, E. Caffrey, G. Callagy and S. Gupta, *Biochem. Soc. Trans.*, 2012, **40**, 752–758.
109. Y. S. Rao, N. N. Mott, Y. Wang, W. C. J. Chung and T. R. Pak, *Endocrinology*, 2013, **154**, 2795–2806.
110. M. Mallya and A. L. Ogilvy-Stuart, *Best Pract. Res., Clin. Endocrinol. Metab.*, 2018, **32**, 17–25.
111. T. Kondo, S. Ezzat and S. L. Asa, *Nat. Rev. Cancer*, 2006, **6**, 292–306.
112. H. Mazeh, *Oncologist*, 2012, **17**, 1032–1038.
113. F. Marini, E. Luzi and M. L. Brandi, *J. Thyroid Res.*, 2011, **2011**, 407123.
114. H. He, K. Jazdzewski, W. Li, S. Liyanarachchi, R. Nagy, S. Volinia, G. A. Calin, C.-G. Liu, K. Franssila, S. Suster, R. T. Kloos, C. M. Croce and A. de la Chapelle, *Proc. Natl. Acad. Sci. U. S. A.*, 2005, **102**, 19075–19080.
115. M. Swierniak, A. Wojcicka, M. Czetwertynska, E. Stachlewska, M. Maciag, W. Wiechno, B. Gornicka, M. Bogdanska, L. Koperski, A. de la Chapelle and K. Jazdzewski, *J. Clin. Endocrinol. Metab.*, 2013, **98**, E1401–E1409.
116. R. Visone, P. Pallante, A. Vecchione, R. Cirombella, M. Ferracin, A. Ferraro, S. Volinia, S. Coluzzi, V. Leone, E. Borbone, C.-G. Liu, F. Petrocca, G. Troncone, G. A. Calin, A. Scarpa, C. Colato, G. Tallini, M. Santoro, C. M. Croce and A. Fusco, *Oncogene*, 2007, **26**, 7590–7595.
117. J. Braun, C. Hoang-Vu, H. Dralle and S. Hüttelmaier, *Oncogene*, 2010, **29**, 4237–4244.

118. T. M. Ortiga-Carvalho, M. I. Chiamolera, C. C. Pazos-Moura and F. E. Wondisford, *Compr. Physiol.*, 2016, **6**, 1387–1428.
119. G. Dai, O. Levy and N. Carrasco, *Nature*, 1996, **379**, 458–460.
120. G. Riesco-Eizaguirre, L. Wert-Lamas, J. Perales-Patón, A. Sastre-Perona, L. P. Fernández and P. Santisteban, *Cancer Res.*, 2015, **75**, 4119–4130.
121. A. Lakshmanan, A. Wojcicka, M. Kotlarek, X. Zhang, K. Jazdzewski and S. M. Jhiang, *Endocr.-Relat. Cancer*, 2015, **22**, 11–21.
122. L. Li, B. Lv, B. Chen, M. Guan, Y. Sun, H. Li, B. Zhang, C. Ding, S. He and Q. Zeng, *Biochem. Biophys. Res. Commun.*, 2015, **462**, 314–321.
123. B. Wojtas, C. Ferraz, T. Stokowy, S. Hauptmann, D. Lange, H. Dralle, T. Musholt, B. Jarzab, R. Paschke and M. Eszlinger, *Mol. Cell. Endocrinol.*, 2014, **388**, 1–9.
124. M. Boufraqech, J. Klubo-Gwiezdzinska and E. Kebebew, *Best Pract. Res., Clin. Endocrinol. Metab.*, 2016, **30**, 603–619.
125. C. S. Fuziwara and E. T. Kimura, *Mol. Cell. Endocrinol.*, 2017, **456**, 44–50.
126. A. Kaprara and I. T. Huhtaniemi, *Metabolism*, 2018, **86**(3), 17.
127. V. Y. Topper, D. M. Walker and A. C. Gore, *Mol. Cell. Endocrinol.*, 2015, **414**, 42–52.
128. S. Sangiao-Alvarellos, M. Manfredi-Lozano, F. Ruiz-Pino, V. M. Navarro, M. A. Sánchez-Garrido, S. Leon, C. Dieguez, F. Cordido, V. Matagne, G. A. Dissen, S. R. Ojeda, L. Pinilla and M. Tena-Sempere, *Endocrinology*, 2013, **154**, 942–955.
129. L. Maione, A. A. Dwyer, B. Francou, A. Guiochon-Mantel, N. Binart, J. Bouligand and J. Young, *Eur. J. Endocrinol.*, 2018, **178**, R55–R80.
130. P. S. Choi, L. Zakhary, W.-Y. Choi, S. Caron, E. Alvarez-Saavedra, E. A. Miska, M. McManus, B. Harfe, A. J. Giraldez, H. R. Horvitz, A. F. Schier and C. Dulac, *Neuron*, 2008, **57**, 41–55.
131. G. Garaffo, D. Conte, P. Provero, D. Tomaiuolo, Z. Luo, P. Pinciroli, C. Peano, I. D'Atri, Y. Gitton, T. Etzion, Y. Gothilf, D. Gays, M. M. Santoro and G. R. Merlo, *Mol. Cell. Neurosci.*, 2015, **68**, 103–119.
132. T. Geach, *Nat. Rev. Endocrinol.*, 2016, **12**, 559.
133. A. Messina, F. Langlet, K. Chachlaki, J. Roa, S. Rasika, N. Jouy, S. Gallet, F. Gaytan, J. Parkash, M. Tena-Sempere, P. Giacobini and V. Prevot, *Nat. Neurosci.*, 2016, **19**, 835–844.
134. H. Wang, I. Graham, R. Hastings, S. Gunewardena, M. L. Brinkmeier, P. M. Conn, S. A. Camper and T. R. Kumar, *J. Biol. Chem.*, 2015, **290**, 2699–2714.
135. H. Wang, R. Hastings, W. L. Miller and T. R. Kumar, *Mol. Cell. Endocrinol.*, 2016, **419**, 124–138.
136. K. Ahmed, M. P. LaPierre, E. Gasser, R. Denzler, Y. Yang, T. Rülicke, J. Kero, M. Latreille and M. Stoffel, *J. Clin. Invest.*, 2017, **127**, 1061–1074.
137. H. Hasuwa, J. Ueda, M. Ikawa and M. Okabe, *Science*, 2013, **341**, 71–73.
138. J. Lannes, D. L'Hôte, G. Garrel, J.-N. Laverrière, J. Cohen-Tannoudji and B. Quérat, *Mol. Endocrinol.*, 2015, **29**, 364–372.

139. D. L. Chan, S. J. Clarke, C. I. Diakos, P. J. Roach, D. L. Bailey, S. Singh and N. Pavlakis, *Crit. Rev. Oncol. Hematol.*, 2017, **113**, 268–282.
140. N. Hassan, J. T. Zhao and S. B. Sidhu, *Mol. Cell. Endocrinol.*, 2017, **456**, 36–43.
141. J. Hallet, C. H. L. Law, M. Cukier, R. Saskin, N. Liu and S. Singh, *Cancer*, 2015, **121**, 589–597.
142. B. Lawrence, B. I. Gustafsson, A. Chan, B. Svejda, M. Kidd and I. M. Modlin, *Endocrinol. Metab. Clin. North Am.*, 2011, **40**, 1–18.
143. C. Vicentini, M. Fassan, E. D'Angelo, V. Corbo, N. Silvestris, G. J. Nuovo and A. Scarpa, *Molecules*, 2014, **19**, 2458–2468.
144. S.-C. Li, A. Essaghir, C. Martijn, R. V. Lloyd, J.-B. Demoulin, K. Oberg and V. Giandomenico, *Mod. Pathol.*, 2013, **26**, 685–696.
145. K. Ruebel, A. A. Leontovich, G. A. Stilling, S. Zhang, A. Righi, L. Jin and R. V. Lloyd, *Mod. Pathol.*, 2010, **23**, 367–375.
146. A. Karpathakis, H. Dibra and C. Thirlwell, *Endocr.-Relat. Cancer*, 2013, **20**, R65–R82.
147. L. Huang, J.-X. Lin, Y.-H. Yu, M.-Y. Zhang, H.-Y. Wang and M. Zheng, *PLoS One*, 2012, **7**, e33762.

CHAPTER 10

MicroRNAs in Skin Development, Function and Disorders

DURBA PAL[a] AND CHANDAN K. SEN*[b]

[a]Center for Biomedical Engineering, Indian Institute of Technology—Ropar, Punjab-140001, India; [b]Indiana Center for Regenerative Medicine & Engineering, Department of Surgery, Indiana University School of Medicine, Indianapolis, IN 46202, USA
*E-mail: cksen@iu.edu

10.1 Introduction

The skin is a sophisticated organ which has the primary function of providing a barrier between the inside of the body and its surrounding environment. Thus, skin may be viewed as being functional only when its barrier function is intact.[1,2] In addition to defending the body against external threats such as pathogens and allergens, the skin has essential thermoregulatory, sensory and metabolic functions. The skin undergoes continuous self-renewal to replace old cells and repair damaged tissue. During gastrulation, emergence of the neuroectoderm paves the way for the development of the neural system and the skin. Induction of the neural fate is reinforced by fibroblast growth factors.[3] The epidermal fate is regulated by the expression of bone morphogenetic proteins (BMPs), with additional Wnt signalling.[4] A single

Drug Discovery Series No. 69
MicroRNAs in Diseases and Disorders: Emerging Therapeutic Targets
Edited by Philip V. Peplow, Bridget Martinez, George A. Calin and Aurora Esquela-Kerscher

Published by the Royal Society of Chemistry, www.rsc.org

cell layer of ectoderm and neural crest are formed during gastrulation. Surface ectoderm initiates the formation of a single layer of basal keratinocytes that undergoes into terminal differentiation called stratification, which then goes on to generate the epidermis, a multilayered stratified epithelium that protects the body from injury, infection and dehydration.[5,6] The Notch 1 and 2 signaling pathway regulates the formation of hair follicles and epidermal interfollicular lineages.[7] In adults, skin homeostasis is maintained through self-renewal and stem cell differentiation.[8–11] The significance of microRNAs (miRNAs) in the regulation of skin development has recently started to unfold.[12–15] miRNAs mostly regulate gene expression by post-translational gene silencing.

10.2 miRNAs in Skin Development and Homeostasis

Skin development is an active regulatory process where epidermis is formed, including appendages such as hair follicles, nails and glands.[16] During fetal development, the epidermis, hair follicle and sebaceous gland arise from a single layer of epidermal progenitor cells.[8,9] Recent articles shed light on the novel miRNAs regulating gene expression in stem cell biology and developmental biology.[17,18] A list of specific miRNAs that regulate skin development[19,20] is presented in Table 10.1.

Several reports identified specific miRNAs contributing important roles in lineage commitment by regulating the gene expression in developing mouse skin and hair follicles.[21,22] Yi *et al.*[22] showed that a distinct group of miRNAs is involved in developing epidermis and hair follicles. Andl *et al.*[21] used microarray analyses of miRNA expression in skin of newborn control mice compared with that of newborn Wnt inhibitor Dkk1 transgenic mice (hairless), to show similar outcomes. These studies showed preferred expression of miRNAs in the follicular or epidermal compartment. The miR-199 family was shown to be enriched in the hair follicle compartment, as compared to epidermis. miR-203 was highly expressed in both epidermal and hair follicular fractions.[21,22] Interestingly, single-layered epithelium of mouse E13.5 skin had very low levels of miR-203 compared to E15.5, at onset of epidermal stratification. These observations suggest a differential expression of miR-203 during differentiation and stratification.[23] However, miR-203 has no role in terminal keratinocyte differentiation.[23,24] Senoo *et al.* identified p63 as a key target of miR-203, which has a role in epithelial stem cell self-renewability. Therefore, deletion of miR-203 in mice results in critical loss of stratified epithelia.[25] The unique spatio-temporal expression of miR-203 resulted in its being the first miRNA investigated in skin.[23,24,26,27] The abundance of miR-203 is most apparent in the layer between the dividing and differentiating zones of the epidermis.[28] The high miR-203 expression significantly inhibits the proliferative potential of skin stem/progenitor cells.[23,24] This was reversed using antagomir-203 treatment in differentiated cells, resulting in increased cell division in the epidermis.[23]

Table 10.1 Differentially expressed miRNAs with documented functions in skin development and homeostasis.[a]

miRNA	Site	Functions	Reference
miR-199 family	Hair follicles	Hair follicle development	21, 22
miR-203	Epidermal differentiated cells	Induces keratinocyte differentiation by repressing p63 expression and maintains epidermal homeostasis	23–27
miR-214	Epidermal cells and hair follicles	Controls skin and hair follicle development by modulating the activity of the Wnt pathway; maintains homeostasis	29
miR-31	Hair follicles	Controlling hair cycle-associated changes by targeting Wnt and BMP	28
miR-125b	Skin stem cells	Repressor of stem cell differentiation	31, 32
miR-34a/c	Epidermal differentiation		47

[a]BMP: bone morphogenetic protein.

Ahmed *et al.*[29] found a novel miRNA-214 that controls skin and hair follicle development in a spatial–temporal expression pattern. miR-214 overexpression causes restriction in keratinocyte proliferation and inhibits the formation of hair follicles. Mechanistically, miR-214 directly targets the β-catenin pathway in keratinocytes. Moreover, miR-214 regulates postnatal skin development by actively participating in multiple signaling pathways, including Wnt, Shh, Edar and BMP. Thus, miR-214 plays a significant role in the developing and postnatal skin, as well as hair follicles.

Hair follicles cycle through stages of growth (anagen), regression (catagen) and rest (telogen).[30] Mardaryev *et al.* identified miR-31 as a critical player in regulating the gene expression for hair follicle growth and hair cycle-associated changes. miR-31 expression increased during the growth phase and decreased during the regression and rest phases of the hair cycle.[28] Mechanistically, miR-31 targets several key signaling pathways, including the Wnt and BMP pathways.[28]

The search for miRNAs that regulate embryonic and adult skin stem cells identified miR-125b as highly expressed in skin stem cells,[31] but it acts as a repressor. The repressor function of miR-125b in skin stem cells resulted in hyper-thickened epidermis, enlarged sebaceous glands and caused failure to generate a hair coat.[32]

10.3 Skin Integrity and Function

Integrity of the skin is necessary for its barrier and other related functions. However, it is important to observe that integrity alone may not guarantee full functionality of the skin. For example, during wound repair, the repaired skin may cover the defect, but may not have a barrier function because of defects in its tight junctions.[1,2] Thus, while integrity is required, it may not be sufficient to ensure skin function. Re-establishment of functional skin is required to achieve wound healing.[1,2] The differential expression patterns of skin miRNAs have been widely studied recently. Some miRNAs have been causatively connected to skin morphogenesis and differentiation. It has become clear that biofilm infection-inducible miRNAs may render otherwise intact looking skin non-functional.[1,2] Next-generation deep sequencing with wild-type skin, Dicer and DiGeorge syndrome critical region 8 (DGCR8) knockout skin has provided insight into the role of miRNAs in skin development. It was found that ~70 miRNAs that are abundantly expressed in skin are strikingly dependent on DGCR8 and Dicer.[33]

The primary role of the cutaneous epidermis is to provide a barrier that prevents loss of body water. This barrier is augmented by E-cadherin, which maintains proper localization of key tight-junctional proteins. Compromised E-cadherin leads to permeable tight junctions with the loss of skin barrier function.[34] It is reported that both miRNA-200 and miRNA-205 are significantly expressed in the epidermis and maintain the epithelial stability. They target the transcriptional repressors of E-cadherin-Zinc finger E-box binding homeobox 1 (ZEB1) and ZEB2.[32,35] Ghatak *et al.* observed that miR-dependent regulation of p21wild-type p53-activated fragment 1 (Waf1)/CDK interacting protein 1 (Cip1) is key in restoring the barrier function of the skin.[36] p21Waf1/Cip1 is directly targeted by 28 miRNAs.[37] In addition, miRNAs are essential in regulating skin pigmentation *via* the production and release of melanin by melanocytes.[38] Among the differentially expressed miRNAs, miRNA-25[39] and miRNA-434-5p[40] influence skin pigmentation.

10.4 miRNA and Skin Aging

MicroRNAs regulate aging-related protein expression.[41] Mature miRNAs guide argonaute proteins to mRNA complexes. The corresponding coding genes are thus silenced, enabling tissue homeostasis or triggering pathophysiological processes, depending on the specific conditions. Although there appear to be links between aging and cellular senescence, the biological mechanisms are not well understood. A profiling study using fibroblasts isolated from young and old humans identified several miRNAs with positive correlations with donor age. miR-23a-3p is present in both aged as well as senescent fibroblasts. miR-23a-3p targets HAS2 mRNA and silences hyaluronan-synthase 2 expression.[42] Hyaluronan is a major component of the extracellular matrix that regulates fibroblast phenotype.[43] Thus, miR-23a-3p regulates both cellular senescence and skin aging. Lately, numerous miRNAs have been identified to have direct roles in the regulation of skin cell senescence. Mancini *et al.* identified that miR-152 and -181a remodel collagen, thus promoting cellular senescence in fibroblasts.[44]

di Val Cervo *et al.* identified high expression of miR-138, miR-181a, miR-181b and miR-130b in keratinocytes during replicative senescence.[45] Sirt-1mRNA is targeted by miR-138, miR-181a and miR-181b, leading to the hypothesis that low Sirt1 activity is implicated in the replicative senescence of keratinocytes. Conversely, miR-130 targets p63, the master gene regulating epithelial development. In another study, the upregulation of miR-29a and miR-30 induced fibroblast replicative senescence. In contrast, the miR-17-92 cluster and miR-106 were found to be downregulated during this process.[46a] Recent evidence reported that senescence was induced in replicating human dermal fibroblasts by miR-152 and miR-181a.[44] miR-191 was upregulated during keratinocyte senescence.[46b]

Skin aging is a dynamic biological process regulated by a combination of extrinsic (chronic light/UV exposure, pollution and chemicals) and intrinsic factors (cellular metabolism, hormones and metabolic processes). Structural and compositional remodelling in the dermis is key to skin aging and is regulated by microRNAs. Profiling of miRNA expression in the human young and aged dermis[47] led to the recognition of an miRNA–gene network that is directly implicated in skin aging (Figure 10.1). A pool of almost 40 miRNAs was found to be differentially expressed in the aged dermis compared to the young dermis. The miR-34 and miR-29 family were specifically identified as having differential expression between aged *versus* young dermis. Bioinformatics studies of the miRNA–gene network revealed that the miR-34, miR-29 and miR-424 families may have a critical role in skin aging by regulating genes involved in cell adhesion, collagen synthesis and several signalling pathways. It was found that miR-34 regulated the expression and function

Figure 10.1 miRNAs that are modulated during senescence in human (A) keratinocytes and (B) fibroblasts.

of matrix metalloproteinase 1, Collagen type 1 alpha chain 1 (COL1A1) and elastin. Furthermore, miR-126, a key angiomiR involved in blood vessel protection and neovascularization, was found to be significantly downregulated in aged human skin, lowering the number and functionality of cutaneous microvasculature with age.[48]

10.5 miRNAs in Skin Diseases and Disorders

10.5.1 Cutaneous Wound Healing

There are several recent reviews on the topic of the significance of miRNAs in cutaneous wound healing.[49–52] In this subsection, we present an overview with emphasis on current developments. Tissue regeneration and wound healing rely largely on cell proliferation. Dermal fibroblasts play a central role in cutaneous wound healing by not only enabling re-epithelialization but also by cross-talking with immune cells at the site of injury.[53] A multitude of growth factors act in a temporally resolved manner to enable tissue repair by stimulating tissue growth as well as helping mount and resolve the process of acute inflammation.[54] Gu and Iyer reported on the role of miRNAs in activated dermal fibroblasts.[55] These cells are known to be highly responsive to growth factors.[53,54] A cluster of 33 miRNAs has been recognized to be involved in transitioning the quiescent dermal fibroblast to its actively cycling form.[55]

Skin repair is a complex physiological process which may be broadly viewed as three overlapping phases: inflammation, proliferation (re-epithelialization, angiogenesis, granulation and tissue formation) and scar maturation.[56] The role of miRNAs at different steps of wound healing is depicted in Figure 10.2. The role of miRNAs in wound angiogenesis is reviewed in Shilo *et al.*[57] Altered miRNA expression in wound angiogenesis is critical to impairment of healing progression resulting in chronic wounds.[57] The hypoxamir miR-210 impairs keratinocyte proliferation and therefore stalls wound healing.[58] Wound healing is impaired in keratinocyte-specific and endothelial-specific Dicer-knockout mice. These findings lend credence to the notion that miRNAs are implicated in keratinocyte differentiation and wound angiogenesis.[36,59] miR-155 has an important role in the inflammatory phase of wound healing. Downregulation of miR-155 in mice accelerated the healing process, resulting in an increase in macrophages and type-1 collagen deposition at the wound site.[60] A critical role of miR-21 in determining the functional fate of wound macrophages has emerged.[61] miR-21 also regulates extracellular matrix homeostasis at the site of tissue injury.[62] Of outstanding interest is the observation that wound-site macrophages may convert to fibroblasts during the acute phase of wound inflammation.[63] miR-21 is critically important in enabling such cell conversions, which seem to play out as an integral component of the physiological tissue repair process. Keratinocyte miR-21 was packed into extracellular vesicles. Extracellular vesicles are packaged with proteins, nucleic acids and/or lipids that are known to play a

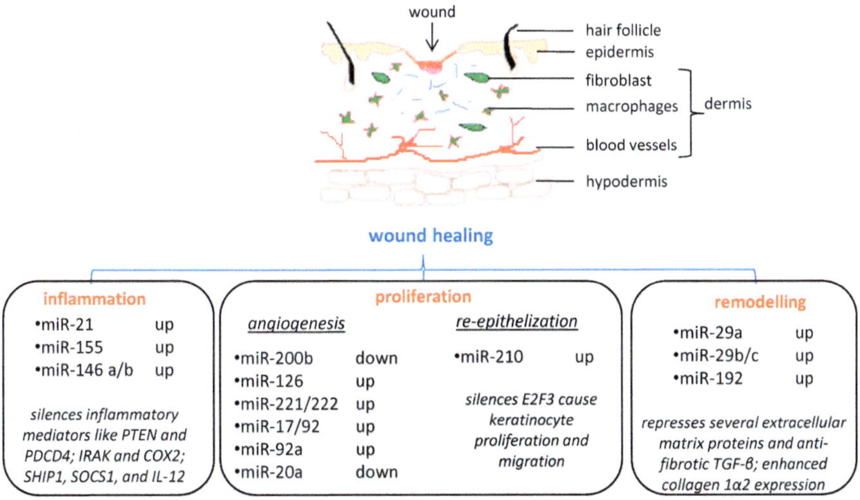

Figure 10.2 Schematic diagram of the structure of the skin showing miRNA regulations at three different stages of wound healing.

role in intercellular communication.[64] Extracellular vesicles carrying miR-21 may be delivered to wound-site macrophages in order to trigger macrophage to fibroblast conversion.[63] These converted fibroblasts reside in the repaired skin and are likely to contribute to the process of tissue remodelling long after the wound has healed. This observation also accounts for an interesting fate of macrophages at the site of wound inflammation.[63]

Chronic non-healing wounds increase with age. miRNA profiling study recognized miR-31 and miR-200 family as showing an age-dependent expression pattern during wound healing. It was also reported that miR-200c inhibits keratinocyte migration and thereby contributes to age-associated impairments in wound repair.[65]

10.5.2 Skin Carcinogenesis

Aberrant expressions of miRNAs cause tumorigenesis and cancer. miR-21, a potential oncogenic miRNA, is induced in a majority of cancers. Studies on miR-21 have led to the emergence of a parallel in molecular mechanisms between epithelial cell plasticity and skin carcinogenesis.[66,67] Analysis of mRNA and miRNA expression profiling in combination with target prediction algorithms has led to insights into miRNA–gene networks and their involvement in skin carcinogenesis.[68] Feramisco *et al.* showed unique miRNA expression profiles and DNA copy number modifications in human melanoma cell lines. miRNA copy number change may be linked to aberrant regulation of the miRNA gene in skin tumors.[69]

Malignant melanoma is a life-threatening skin cancer, and it is estimated to cause about 7230 (4740 men and 2490 women) deaths in the United States in 2019.[70] Profiling studies in malignant melanomas have identified numerous differentially regulated miRNAs associated with different phases of this cancer.[71–75] These lines of investigation have identified candidate miRNAs with potential to be diagnostic markers or targets for therapy.[76,77]

In contrast to malignant melanoma, the role of miRNAs in other skin cancers is not well studied. Compared to normal healthy skin, low miR-203 has been evident in cutaneous squamous cell carcinoma (SCC).[78] Conversely, miR-203 expression was significantly increased in normal human keratinocytes upon UVA irradiation.[78] Similar effects were observed in a head and neck SCC cell line after UVC treatment that results in cell death and cell cycle exit.[24] These findings support the proposed role of miR-203 as a tumor suppressor, but contrast with the elevated levels of miR-203 in other human epithelial tumors.[23] Pathogenic pathways induced by elevated miR-203 need to be better understood and warrant additional attention. miR-205 levels are often low in human epithelial tumors compared to normal skin;[79,80] however, in some human epithelial cancer studies, miR-205 were found to be upregulated along with the miR-200 family.[81–83] These apparently contrasting results support the notion that miRNAs are likely to have different functions in different cells. Members of the *let-7* family suppresses melanoma development,[71,84] reminiscent of their functions in other human cancers. Table 10.2 provides details of some important miRNAs involved in the development and progression of melanoma.[71–73,84–91]

10.5.3 Vitiligo

Over the past few years, microRNA profiling studies revealed the differential expression patterns of 20 microRNAs in vitiligo. Two of them, miR-146a and miR-191, were upregulated in the serum of vitiligo mice (Rag1KO) and humans. Rag1KO is a vitiligo mouse model with a deficiency in the generation of mature T-cells. *Via* the application of topical monobenzone cream, a skin depigmenting agent, the characteristics of vitiligo are induced.[92] Thus, microRNAs may serve as possible vitiligo biomarkers and also as potential therapeutic targets.[93] miR-146a regulates the immune cell differentiation that may be implicated in oxidative stress-dependent immune dysfunction in vitiligo. In vitiligo patients, miR-25 levels are increased in blood and skin lesions. Oxidative stress induces miR-25 expression, which downregulates stem cell factor synthesis and keratinocyte growth factor release. This negatively affects the proliferation and differentiation of melanocytes.[94,95] An independent study revealed that miR-224-3p and miR-4712-3p are elevated and miR-39-40-5p is lowered in blood of vitiligo patients. Such differentially expressed microRNAs have an adverse effect on immune cell function.[96] Patients with the TT and TC genotypes of the miR-196a-2 gene, in combination with elevated levels of tyrosinase, are more likely to develop vitiligo, compared to their counterparts with the CC genotype.[97] The CC genotype is associated with downregulation of tyrosinase-1 (TYR-1) and corresponding

Table 10.2 Differentially expressed miRNAs and their regulation in skin diseases.

Skin disease	miRNA	Regulation	Target	Reference
Melanoma	miR-221	Up	Targeting p27Kip1/CDKN1B	88
	miR-17	Up	Inhibiting ETV1 translation	87
	miR-18b	Down	MDM2-p53 pathway signalling	89
	miR-29c	Down	Hypermethylation of tumor-related genes	85
	Let-7a	Down	Increases the invasive behavior of melanocytes by integrin beta 3	84
	miR-196a	Down	Suppression increases cell migration by targeting HOX-B7	86
Squamous cell carcinoma	miR-125b	Down	Downregulates matrix metallopeptidase 13	90
	miR-203	Down	UV irradiation	23, 24, 78
	miR-193/365-1 cluster	Down	KRAS and MAX	91
Vitiligo	miR-25	Up	Degeneration of melanocytes	94, 95
	miR-146a	Up	Regulates the differentiation of immune cells	93
	miR-191	Up	Regulates the differentiation of immune cells	93
Scleroderma	miR-196a	Down	Upregulation of collagen type 1	99
	miR-92a	Up	Excessive collagen expression and accumulation	100
	Let-7a	Down	Excessive expression of type I collagen	102
Atopic dermatitis	miR-155	Up	Modulates T-cell proliferative responses by targeting CTLA-4.	104, 105
	miR-203	Up	Higher serum levels of IgE	109
	miR-146a	Up	Suppression of innate immune responses in keratinocytes	107
	miR-483-5p	Up		109
Psoriasis	miR-21	Up	Suppresses T-cell apoptosis	115–117
	miR-31	Up	NF-κB signalling and effects on inflammatory chemokines	113, 114
	miR-125b	Down	Regulates tumour necrosis factor-α	117, 120
	miR-210	Down	Alters transcription factor FOXP3 expression in $CD4^+$ T-cells	123
	miR-203	Up	Increases keratinocyte proliferation by targeting SOCS3	26, 121
	miR-221/222	Up	Reduction of TIMP3 levels	26, 121

lowered levels of reactive oxygen species (ROS), thus melanocytes are protected against ROS-mediated damage and the risk of developing vitiligo is reduced.[98]

10.5.4 Localized Scleroderma

Aberrant expressions of microRNA are also observed in scleroderma. Ihn and coworkers demonstrated that miR-196a levels were highly downregulated in the skin and serum of patients with localized scleroderma. In cultured dermal fibroblasts, miR-196a inhibition causes high expression of type 1 collagen, which then causes dermal fibrosis.[99] It was also revealed that scleroderma patients showed high miR-92a expression in the skin, resulting in excessive collagen expression by triggering the transforming growth factor-β pathway.[100] Conversely, miR-7 was decreased in patients with localized scleroderma. miR-7 inhibition in normal fibroblasts stimulated collagen formation and increased deposition of collagen type I.[101] In systemic and localized scleroderma, miR-let-7a is downregulated in mouse skin tissue. Delivery of miR-let-7a into mouse skin caused loss of dermal thickness and excessive collagen deposition.[102] microRNA studies of localized scleroderma have mostly focused on abnormalities of collagen synthesis. Currently, there are no microRNA profiling data on lesional skin in localized scleroderma; however, profiling studies and hierarchical cluster analysis in systemic scleroderma revealed that one microRNA cluster on chromosome 14q32.3 was primarily aberrant, with 18 out of 26 altered microRNAs identified belonging to that cluster.[103] Further widening of the scope of such research would lead to novel mechanistic insights.

10.5.5 Atopic Dermatitis

miR-155 is found to be highly expressed in atopic dermatitis, which causes persistent skin inflammation by directly targeting CTLA-4 protein, which negatively regulates T-cell activation.[104,105] miR-155 inhibition decreases the number of T-helper 17 (Th17) cells with interleukin (IL)-17 mRNA, whereas miR-155 over-expression elevates Th17 cell number and IL-17 expression in atopic dermatitis. Furthermore, in the atopic dermatitis skin, miR-155 expression negatively regulates suppressor of cytokine signalling-1 (SOCS-1) and its downstream targets IL-2, IL-3, IL-6 and interferon expression.[106] Thus, miR-155 plays a key role in atopic dermatitis by regulating differentiation and functioning of T-helper cells. Another important microRNA in atopic dermatitis is miR-146a, which targets the nuclear factor kappa B (NF-κB) signaling pathway and negatively regulates the expression of chemokines CCL5 and CCL8, thus regulating immune-mediated inflammation. miR-146-ablated mice suffer from skin inflammation characterized by elevated levels of interferon gamma, C–C motif chemokine ligand 5 (CCL5) and CCL8.[107] Keratinocyte proliferation and differentiation is imbalanced in atopic dermatitis skin. Suppression of miR-31 caused impaired keratinocyte differentiation by targeting the Notch signalling pathway.[108] Serum levels of microRNAs have been studied to obtain mechanistic understanding of the correlation between lesional skin and systemic response. High miR-203 serum levels were found in children with atopic dermatitis compared to adult patients, corresponding to higher serum levels of IgE. Furthermore, miR-483-5p levels are elevated in the serum of children with atopic dermatitis.[109]

10.5.6 Psoriasis

Psoriasis is an autoimmune skin disease that produces plaques of dry thickened scaling skin. It is estimated to be prevalent in 0.5–4.6% of the world's population, with variations depending on country and ethnicity. Different cell types of skin such as keratinocytes, fibroblasts, immune cells, T-cells and mast cells are pertinent in the formation of psoriatic skin. It is known that psoriasis is the result of derailed cross-talk between the immune cells and the other skin cells.[26,110,111] Comparison of miRNA expression profile of psoriasis-affected skin, atopic eczema-affected skin and healthy human skin provided early insights into the role of miRNAs in psoriasis.[26]

NF-κB regulates psoriatic keratinocyte proliferation.[112] Constitutive activation of NF-κB increases miR-31 levels in the skin of psoriatic patients. High miR-31 causes activation of endothelial cells and migration of leukocytes into the skin by inducing transforming growth factor-β.[113,114] Besides miR-31, miR-21 is also elevated in the psoriatic skin.[115] Inhibition of miR-21 causes apoptosis of T-cells, which could be targeted to manage unlimited T-cell activation in the psoriatic skin.[116] Cutaneous miR-21 and miR-125b levels are sensitive to UVB exposure.[117] In murine epidermal cells, UVB induces miR-21 expression, which downregulates tumor suppressor programmed cell death protein 4 (PCDP-4) by a pathway involving ROS and mitogen-activated protein kinase signaling.[118] miR-21 is multifunctional and must be understood depending on the context and tissue compartment. miR-21exhibits angiogenic function by increasing vascular endothelial growth factor levels in the psoriatic skin.[119] miR-125b is potently downregulated in the psoriatic skin. Low miR-125b upregulates the fibroblast growth factor receptor-2, thus suppressing keratinocyte proliferation and causing psoriatic keratinocyte differentiation.[120] Table 10.2 lists some of the currently known microRNAs that are directly implicated in psoriasis. Several microRNAs are directly linked to immunological abnormalities in the psoriatic skin.[26,121] For example, miR-138 influences Th1/Th2 imbalance in the psoriatic T-helper cells by regulating RUNX3, a key transcription factor for T-cell growth and maturation.[122] Lowered miR-210 modifies levels of the FOXP3 (transcription factor) expression in Th-cells of psoriatic patients, resulting in the impairment of the suppressive capacity of regulatory T-cells.[123] The expression of miR-193b and miR-223 are dysregulated in psoriatic dermal leukocytes and in circulating peripheral blood mononuclear cells.[124] Hence, the interaction of microRNAs with the target gene transcription factor may influence both activation and response of immune cells in psoriasis.

10.6 Conclusion

The significance of miRNAs in skin development, function and disorders is multifaceted and impacts regulatory steps that drive these processes. This chapter provides an overview of the current state of knowledge of miRNA-dependent mechanisms implicated in these processes. We highlight recent evidence implicating miRNA regulation of cutaneous wound healing, maintenance of skin barrier integrity and homeostasis. The role of miRNA–gene networks in promoting replicative senescence emphasizes an integral role for these regulatory RNAs in promoting aging. Finally, we cover pathological processes that are linked to derailment or alteration of miRNA function resulting in carcinogenesis and other acute and chronic skin diseases associated with aberrant immune reactions such as vitiligo, psoriasis, *etc.* The regulatory mechanisms of inflammation induced by miRNA are still under investigation. Understanding the basics of miRNA regulation of such pathological processes provides a good foundation for further mechanistic studies which may inform future therapeutic strategies. From a healthcare standpoint, it is important to generate more data from human tissue samples. The area of miRNAs as biomarkers is a fertile area that remains to be explored for skin-related disorders.

Acknowledgments

The authors acknowledge grant funding support from the National Institutes of Health R01NR015676, R01GM108014, R01GM095657.

References

1. K. G. Barki, A. Das, S. Dixith, P. D. Ghatak, S. Mathew-Steiner, E. Schwab, S. Khanna, D. J. Wozniak, S. Roy and C. K. Sen, *Ann. Surg.*, 2017, 1.
2. S. Roy, H. Elgharably, M. Sinha, K. Ganesh, S. Chaney, E. Mann, C. Miller, S. Khanna, V. K. Bergdall, H. M. Powell, C. H. Cook, G. M. Gordillo, D. J. Wozniak and C. K. Sen, *J. Pathol.*, 2014, **233**, 331–343.
3. N. Gaspard and P. Vanderhaeghen, *Curr. Opin. Neurobiol.*, 2010, **20**, 37–43.
4. P. A. Wilson and A. Hemmati-Brivanlou, *Nature*, 1995, **376**, 331–333.
5. M. I. Koster and D. R. Roop, *Annu. Rev. Cell Dev. Biol.*, 2007, **23**, 93–113.
6. S. E. Millar, K. Willert, P. C. Salinas, H. Roelink, R. Nusse, D. J. Sussman and G. S. Barsh, *Dev. Biol.*, 1999, **207**, 133–149.
7. N. Yamamoto, K. Tanigaki, H. Han, H. Hiai and T. Honjo, *Curr. Biol.*, 2003, **13**, 333–338.
8. R. Schmidt-Ullrich and R. Paus, *BioEssays*, 2005, **27**, 247–261.
9. C. Blanpain and E. Fuchs, *Annu. Rev. Cell Dev. Biol.*, 2006, **22**, 339–373.
10. E. Fuchs and V. Horsley, *Genes Dev.*, 2008, **22**, 976–985.

11. P. H. Jones, B. D. Simons and F. M. Watt, *Cell Stem Cell*, 2007, **1**, 371–381.
12. V. Ambros, *Nature*, 2004, **431**, 350–355.
13. D. P. Bartel, *Cell*, 2004, **116**, 281–297.
14. W. P. Kloosterman and R. H. Plasterk, *Dev. Cell*, 2006, **11**, 441–450.
15. N. Bushati and S. M. Cohen, *Annu. Rev. Cell Dev. Biol.*, 2007, **23**, 175–205.
16. C. Blanpain and E. Fuchs, *Nat. Rev. Mol. Cell Biol.*, 2009, **10**, 207–217.
17. V. K. Gangaraju and H. Lin, *Nat. Rev. Mol. Cell Biol.*, 2009, **10**, 116–125.
18. K. N. Ivey and D. Srivastava, *Cold Spring Harbor Perspect. Biol.*, 2015, **7**, a008144.
19. M. Selbach, B. Schwanhausser, N. Thierfelder, Z. Fang, R. Khanin and N. Rajewsky, *Nature*, 2008, **455**, 58–63.
20. R. Yi and E. Fuchs, *Cell Death Differ.*, 2010, **17**, 229–235.
21. T. Andl, E. P. Murchison, F. Liu, Y. Zhang, M. Yunta-Gonzalez, J. W. Tobias, C. D. Andl, J. T. Seykora, G. J. Hannon and S. E. Millar, *Curr. Biol.*, 2006, **16**, 1041–1049.
22. R. Yi, D. O'Carroll, H. A. Pasolli, Z. Zhang, F. S. Dietrich, A. Tarakhovsky and E. Fuchs, *Nat. Genet.*, 2006, **38**, 356–362.
23. R. Yi, M. N. Poy, M. Stoffel and E. Fuchs, *Nature*, 2008, **452**, 225–229.
24. A. M. Lena, R. Shalom-Feuerstein, P. Rivetti di Val Cervo, D. Aberdam, R. A. Knight, G. Melino and E. Candi, *Cell Death Differ.*, 2008, **15**, 1187–1195.
25. M. Senoo, F. Pinto, C. P. Crum and F. McKeon, *Cell*, 2007, **129**, 523–536.
26. E. Sonkoly, T. Wei, P. C. Janson, A. Saaf, L. Lundeberg, M. Tengvall-Linder, G. Norstedt, H. Alenius, B. Homey, A. Scheynius, M. Stahle and A. Pivarcsi, *PLoS One*, 2007, **2**, e610.
27. D. Aberdam, E. Candi, R. A. Knight and G. Melino, *Trends Biochem. Sci.*, 2008, **33**, 583–591.
28. A. N. Mardaryev, M. I. Ahmed, N. V. Vlahov, M. Y. Fessing, J. H. Gill, A. A. Sharov and N. V. Botchkareva, *FASEB J.*, 2010, **24**, 3869–3881.
29. M. I. Ahmed, M. Alam, V. U. Emelianov, K. Poterlowicz, A. Patel, A. A. Sharov, A. N. Mardaryev and N. V. Botchkareva, *J. Cell Biol.*, 2014, **207**, 549–567.
30. M. R. Schneider, R. Schmidt-Ullrich and R. Paus, *Curr. Biol.*, 2009, **19**, R132–R142.
31. L. Zhang, N. Stokes, L. Polak and E. Fuchs, *Cell Stem Cell*, 2011, **8**, 294–308.
32. P. A. Gregory, A. G. Bert, E. L. Paterson, S. C. Barry, A. Tsykin, G. Farshid, M. A. Vadas, Y. Khew-Goodall and G. J. Goodall, *Nat. Cell Biol.*, 2008, **10**, 593–601.
33. R. Yi, H. A. Pasolli, M. Landthaler, M. Hafner, T. Ojo, R. Sheridan, C. Sander, D. O'Carroll, M. Stoffel, T. Tuschl and E. Fuchs, *Proc. Natl. Acad. Sci. U. S. A.*, 2009, **106**, 498–502.
34. J. A. Tunggal, I. Helfrich, A. Schmitz, H. Schwarz, D. Gunzel, M. Fromm, R. Kemler, T. Krieg and C. M. Niessen, *EMBO J.*, 2005, **24**, 1146–1156.

35. S. M. Park, A. B. Gaur, E. Lengyel and M. E. Peter, *Genes Dev.*, 2008, **22**, 894–907.
36. S. Ghatak, Y. C. Chan, S. Khanna, J. Banerjee, J. Weist, S. Roy and C. K. Sen, *Mol. Ther.*, 2015, **23**, 1201–1210.
37. S. Wu, S. Huang, J. Ding, Y. Zhao, L. Liang, T. Liu, R. Zhan and X. He, *Oncogene*, 2010, **29**, 2302–2308.
38. M. G. Kosmadaki, A. Naif and P. Hee-Young, *G. Ital. Dermatol. Venereol.*, 2010, **145**, 47–55.
39. Z. Zhu, J. He, X. Jia, J. Jiang, R. Bai, X. Yu, L. Lv, R. Fan, X. He, J. Geng, R. You, Y. Dong, D. Qiao, K. B. Lee, G. W. Smith and C. Dong, *Domest. Anim. Endocrinol.*, 2010, **38**, 200–209.
40. D. Wu, J. S. Chen, D. C. Chang and S. L. Lin, *Clin., Cosmet. Invest. Dermatol.*, 2008, **1**, 19–35.
41. M. Mancini, A. M. Lena, G. Saintigny, C. Mahe, N. Di Daniele, G. Melino and E. Candi, *Ageing Res. Rev.*, 2014, **17**, 9–15.
42. K. Rock, J. Tigges, S. Sass, A. Schutze, A. M. Florea, A. C. Fender, F. J. Theis, J. Krutmann, F. Boege, E. Fritsche, G. Reifenberger and J. W. Fischer, *J. Invest. Dermatol.*, 2015, **135**, 369–377.
43. K. T. Dicker, L. A. Gurski, S. Pradhan-Bhatt, R. L. Witt, M. C. Farach-Carson and X. Jia, *Acta Biomater.*, 2014, **10**, 1558–1570.
44. M. Mancini, G. Saintigny, C. Mahe, M. Annicchiarico-Petruzzelli, G. Melino and E. Candi, *Aging*, 2012, **4**, 843–853.
45. P. Rivetti di Val Cervo, A. M. Lena, M. Nicoloso, S. Rossi, M. Mancini, H. Zhou, G. Saintigny, E. Dellambra, T. Odorisio, C. Mahe, G. A. Calin, E. Candi and G. Melino, *Proc. Natl. Acad. Sci. U. S. A.*, 2012, **109**, 1133–1138.
46. (a) M. Hackl, S. Brunner, K. Fortschegger, C. Schreiner, L. Micutkova, C. Muck, G. T. Laschober, G. Lepperdinger, N. Sampson, P. Berger, D. Herndler-Brandstetter, M. Wieser, H. Kuhnel, A. Strasser, M. Rinnerthaler, M. Breitenbach, M. Mildner, L. Eckhart, E. Tschachler, A. Trost, J. W. Bauer, C. Papak, Z. Trajanoski, M. Scheideler, R. Grillari-Voglauer, B. Grubeck-Loebenstein, P. Jansen-Durr and J. Grillari, *Aging Cell*, 2010, **9**, 291–296; (b) A. M. Lena, M. Mancini, P. Rivetti di Val Cervo, G. Saintigny, C. Mahé, G. Melino and E. Candi, *Biochem. Biophys. Res. Commun.*, 2012, **423**, 1763–1768.
47. T. Li, X. Yan, M. Jiang and L. Xiang, *J. Dermatol. Sci.*, 2016, **82**, 75–83.
48. J. Fiedler, E. Gronniger, A. Pfanne, S. Bronneke, K. Schmidt, C. S. Falk, H. Wenck, L. Terstegen, T. Thum and M. Winnefeld, *Exp. Dermatol.*, 2017, **26**, 284–286.
49. C. K. Sen and S. Roy, *MicroRNA in Cutaneous Wound Healing*, Springer Science+Business Media, New York, 2008.
50. J. Banerjee and C. K. Sen, *Adv. Exp. Med. Biol.*, 2015, **888**, 291–305.
51. J. Banerjee, Y. C. Chan and C. K. Sen, *Physiol. Genomics*, 2011, **43**, 543–556.
52. E. J. Mulholland, N. Dunne and H. O. McCarthy, *Mol. Ther.--Nucleic Acids*, 2017, **8**, 46–55.

53. M. Krasna, D. Domanovic, A. Tomsic, U. Svajger and M. Jeras, *Acta Dermatovenerol. Alp. Pannonica Adriat.*, 2007, **16**, 105–110.
54. P. Rozman and Z. Bolta, *Acta Dermatovenerol. Alp. Pannonica et Adriat.*, 2007, **16**, 156–165.
55. J. Gu and V. R. Iyer, *Genome Biol.*, 2006, **7**, R42.
56. C. K. Sen and S. Roy, *Wound Healing*, Elsevier, 3rd edn, 2012.
57. S. Shilo, S. Roy, S. Khanna and C. K. Sen, *DNA Cell Biol.*, 2007, **26**, 227–237.
58. S. Biswas, S. Roy, J. Banerjee, S. R. Hussain, S. Khanna, G. Meenakshisundaram, P. Kuppusamy, A. Friedman and C. K. Sen, *Proc. Natl. Acad. Sci. U. S. A.*, 2010, **107**, 6976–6981.
59. Y. Suarez, C. Fernandez-Hernando, J. Yu, S. A. Gerber, K. D. Harrison, J. S. Pober, M. L. Iruela-Arispe, M. Merkenschlager and W. C. Sessa, *Proc. Natl. Acad. Sci. U. S. A.*, 2008, **105**, 14082–14087.
60. C. van Solingen, E. Araldi, A. Chamorro-Jorganes, C. Fernandez-Hernando and Y. Suarez, *J. Cell. Mol. Med.*, 2014, **18**, 1104–1112.
61. A. Das, K. Ganesh, S. Khanna, C. K. Sen and S. Roy, *J. Immunol.*, 2014, **192**, 1120–1129.
62. S. Roy, S. Khanna, S. R. Hussain, S. Biswas, A. Azad, C. Rink, S. Gnyawali, S. Shilo, G. J. Nuovo and C. K. Sen, *Cardiovasc. Res.*, 2009, **82**, 21–29.
63. M. Sinha, C. K. Sen, K. Singh, A. Das, S. Ghatak, B. Rhea, B. Blackstone, H. M. Powell, S. Khanna and S. Roy, *Nat. Commun.*, 2018, **9**, 936.
64. G. van Niel, G. D'Angelo and G. Raposo, *Nat. Rev. Mol. Cell Biol.*, 2018, **19**, 213–228.
65. E. Aunin, D. Broadley, M. I. Ahmed, A. N. Mardaryev and N. V. Botchkareva, *Sci. Rep.*, 2017, **7**, 3257.
66. X. Ma, M. Kumar, S. N. Choudhury, L. E. Becker Buscaglia, J. R. Barker, K. Kanakamedala, M. F. Liu and Y. Li, *Proc. Natl. Acad. Sci. U. S. A.*, 2011, **108**, 10144–10149.
67. M. I. Ahmed, A. N. Mardaryev, C. J. Lewis, A. A. Sharov and N. V. Botchkareva, *J. Cell Sci.*, 2011, **124**, 3399–3404.
68. J. Zavadil, M. Narasimhan, M. Blumenberg and R. J. Schneider, *Cells Tissues Organs*, 2007, **185**, 157–161.
69. J. D. Feramisco, R. L. Cayce and H. Tsao, *Pediatr. Dermatol.*, 2008, **25**, 99–108.
70. R. L. Siegel, K. D. Miller and A. Jemal, *Ca-Cancer J. Clin.*, 2019, **69**, 7–34.
71. J. Schultz, P. Lorenz, G. Gross, S. Ibrahim and M. Kunz, *Cell Res.*, 2008, **18**, 549–557.
72. D. W. Mueller, M. Rehli and A. K. Bosserhoff, *J. Invest. Dermatol.*, 2009, **129**, 1740–1751.
73. M. F. Segura, D. Hanniford, S. Menendez, L. Reavie, X. Zou, S. Alvarez-Diaz, J. Zakrzewski, E. Blochin, A. Rose, D. Bogunovic, D. Polsky, J. Wei, P. Lee, I. Belitskaya-Levy, N. Bhardwaj, I. Osman and E. Hernando, *Proc. Natl. Acad. Sci. U. S. A.*, 2009, **106**, 1814–1819.
74. S. Caramuta, S. Egyhazi, M. Rodolfo, D. Witten, J. Hansson, C. Larsson and W. O. Lui, *J. Invest. Dermatol.*, 2010, **130**, 2062–2070.

75. J. Chen, H. E. Feilotter, G. C. Pare, X. Zhang, J. G. Pemberton, C. Garady, D. Lai, X. Yang and V. A. Tron, *Am. J. Pathol.*, 2010, **176**, 2520–2529.
76. D. W. Mueller and A. K. Bosserhoff, *Pigm. Cell Melanoma Res.*, 2010, **23**, 620–626.
77. D. W. Mueller and A. K. Bosserhoff, *Br. J. Cancer*, 2009, **101**, 551–556.
78. P. Dziunycz, G. Iotzova-Weiss, J. J. Eloranta, S. Lauchli, J. Hafner, L. E. French and G. F. Hofbauer, *J. Invest. Dermatol.*, 2010, **130**, 2686–2689.
79. A. Feber, L. Xi, J. D. Luketich, A. Pennathur, R. J. Landreneau, M. Wu, S. J. Swanson, T. E. Godfrey and V. R. Litle, *J. Thorac. Cardiovasc. Surg.*, 2008, **135**, 255–260; discussion 260.
80. G. Childs, M. Fazzari, G. Kung, N. Kawachi, M. Brandwein-Gensler, M. McLemore, Q. Chen, R. D. Burk, R. V. Smith, M. B. Prystowsky, T. J. Belbin and N. F. Schlecht, *Am. J. Pathol.*, 2009, **174**, 736–745.
81. M. V. Iorio, R. Visone, G. Di Leva, V. Donati, F. Petrocca, P. Casalini, C. Taccioli, S. Volinia, C. G. Liu, H. Alder, G. A. Calin, S. Menard and C. M. Croce, *Cancer Res.*, 2007, **67**, 8699–8707.
82. N. Tran, T. McLean, X. Zhang, C. J. Zhao, J. M. Thomson, C. O'Brien and B. Rose, *Biochem. Biophys. Res. Commun.*, 2007, **358**, 12–17.
83. J. Yu, D. G. Ryan, S. Getsios, M. Oliveira-Fernandes, A. Fatima and R. M. Lavker, *Proc. Natl. Acad. Sci. U. S. A.*, 2008, **105**, 19300–19305.
84. D. W. Muller and A. K. Bosserhoff, *Oncogene*, 2008, **27**, 6698–6706.
85. T. Nguyen, C. Kuo, M. B. Nicholl, M. S. Sim, R. R. Turner, D. L. Morton and D. S. Hoon, *Epigenetics*, 2011, **6**, 388–394.
86. S. Braig, D. W. Mueller, T. Rothhammer and A. K. Bosserhoff, *Cell. Mol. Life Sci.*, 2010, **67**, 3535–3548.
87. R. Cohen, E. Greenberg, Y. Nemlich, J. Schachter and G. Markel, *Oncotarget*, 2015, **6**, 19006–19016.
88. F. Felicetti, M. C. Errico, L. Bottero, P. Segnalini, A. Stoppacciaro, M. Biffoni, N. Felli, G. Mattia, M. Petrini, M. P. Colombo, C. Peschle and A. Care, *Cancer Res.*, 2008, **68**, 2745–2754.
89. A. A. Dar, S. Majid, C. Rittsteuer, D. de Semir, V. Bezrookove, S. Tong, M. Nosrati, R. Sagebiel, J. R. Miller 3rd and M. Kashani-Sabet, *J. Natl. Cancer Inst.*, 2013, **105**, 433–442.
90. N. Xu, L. Zhang, F. Meisgen, M. Harada, J. Heilborn, B. Homey, D. Grander, M. Stahle, E. Sonkoly and A. Pivarcsi, *J. Biol. Chem.*, 2012, **287**, 29899–29908.
91. C. Gastaldi, T. Bertero, N. Xu, I. Bourget-Ponzio, K. Lebrigand, S. Fourre, A. Popa, N. Cardot-Leccia, G. Meneguzzi, E. Sonkoly, A. Pivarcsi, B. Mari, P. Barbry, G. Ponzio and R. Rezzonico, *Carcinogenesis*, 2014, **35**, 1110–1120.
92. Y. Zhu, S. Wang and A. Xu, *Exp. Dermatol.*, 2013, **22**, 499–501.
93. Y. L. Shi, M. Weiland, H. W. Lim, Q. S. Mi and L. Zhou, *Exp. Dermatol.*, 2014, **23**, 140–142.
94. A. Y. Lee, N. H. Kim, W. I. Choi and Y. H. Youm, *J. Invest. Dermatol.*, 2005, **124**, 976–983.

95. Q. Shi, W. Zhang, S. Guo, Z. Jian, S. Li, K. Li, R. Ge, W. Dai, G. Wang, T. Gao and C. Li, *Cell Death Differ.*, 2016, **23**, 496–508.
96. Y. Wang, K. Wang, J. Liang, H. Yang, N. Dang, X. Yang and Y. Kong, *J. Dermatol.*, 2015, **42**, 193–197.
97. T. T. Cui, X. L. Yi, W. G. Zhang, C. Wei, F. B. Zhou, Z. Jian, G. Wang, T. W. Gao, C. Y. Li and K. Li, *Arch. Dermatol. Res.*, 2015, **307**, 683–692.
98. Y. Huang, X. Yi, Z. Jian, C. Wei, S. Li, C. Cai, P. Zhang, K. Li, S. Guo, L. Liu, Q. Shi, T. Gao and C. Li, *Pigm. Cell Melanoma Res.*, 2013, **26**, 338–347.
99. T. Makino, M. Jinnin, M. Etoh, K. Yamane, I. Kajihara, K. Makino, A. Ichihara, T. Igata, K. Sakai, S. Fukushima and H. Ihn, *Eur. J. Dermatol.*, 2014, **24**, 470–476.
100. T. Sing, M. Jinnin, K. Yamane, N. Honda, K. Makino, I. Kajihara, T. Makino, K. Sakai, S. Masuguchi, S. Fukushima and H. Ihn, *Rheumatology*, 2012, **51**, 1550–1556.
101. M. Etoh, M. Jinnin, K. Makino, K. Yamane, W. Nakayama, J. Aoi, N. Honda, I. Kajihara, T. Makino, S. Fukushima and H. Ihn, *Arch. Dermatol. Res.*, 2013, **305**, 9–15.
102. K. Makino, M. Jinnin, A. Hirano, K. Yamane, M. Eto, T. Kusano, N. Honda, I. Kajihara, T. Makino, K. Sakai, S. Masuguchi, S. Fukushima and H. Ihn, *J. Immunol.*, 2013, **190**, 3905–3915.
103. G. Salazar, S. Assassi, M. Wu, J. Hagan and M. D. Mayes, *Ann. Rheum. Dis.*, 2014, **73**, A55–A56.
104. E. Sonkoly, P. Janson, M. L. Majuri, T. Savinko, N. Fyhrquist, L. Eidsmo, N. Xu, F. Meisgen, T. Wei, M. Bradley, J. Stenvang, S. Kauppinen, H. Alenius, A. Lauerma, B. Homey, O. Winqvist, M. Stahle and A. Pivarcsi, *J. Allergy Clin. Immunol.*, 2010, **126**, 581–589. e1–20.
105. L. Ciszak, I. Frydecka, D. Wolowiec, A. Szteblich and A. Kosmaczewska, *Tumor Biol.*, 2016, **37**, 4143–4157.
106. P. J. Egan, K. E. Lawlor, W. S. Alexander and I. P. Wicks, *J. Clin. Invest.*, 2003, **111**, 915–924.
107. A. Rebane, T. Runnel, A. Aab, J. Maslovskaja, B. Ruckert, M. Zimmermann, M. Plaas, J. Karner, A. Treis, M. Pihlap, U. Haljasorg, H. Hermann, N. Nagy, L. Kemeny, T. Erm, K. Kingo, M. Li, M. P. Boldin and C. A. Akdis, *J. Allergy Clin. Immunol.*, 2014, **134**, 836–847. e11.
108. H. Peng, N. Kaplan, R. B. Hamanaka, J. Katsnelson, H. Blatt, W. Yang, L. Hao, P. J. Bryar, R. S. Johnson, S. Getsios, N. S. Chandel and R. M. Lavker, *Proc. Natl. Acad. Sci. U. S. A.*, 2012, **109**, 14030–14034.
109. Y. Lv, R. Qi, J. Xu, Z. Di, H. Zheng, W. Huo, L. Zhang, H. Chen and X. Gao, *PLoS One*, 2014, **9**, e115448.
110. M. Lebwohl, *Lancet*, 2003, **361**, 1197–1204.
111. M. A. Lowes, A. M. Bowcock and J. G. Krueger, *Nature*, 2007, **445**, 866–873.
112. H. Wang, T. Syrovets, D. Kess, B. Buchele, H. Hainzl, O. Lunov, J. M. Weiss, K. Scharffetter-Kochanek and T. Simmet, *J. Immunol.*, 2009, **183**, 4755–4763.

113. N. Xu, F. Meisgen, L. M. Butler, G. Han, X. J. Wang, C. Soderberg-Naucler, M. Stahle, A. Pivarcsi and E. Sonkoly, *J. Immunol.*, 2013, **190**, 678–688.
114. S. Yan, Z. Xu, F. Lou, L. Zhang, F. Ke, J. Bai, Z. Liu, J. Liu, H. Wang, H. Zhu, Y. Sun, W. Cai, Y. Gao, B. Su, Q. Li, X. Yang, J. Yu, Y. Lai, X. Z. Yu, Y. Zheng, N. Shen, Y. E. Chin and H. Wang, *Nat. Commun.*, 2015, **6**, 7652.
115. J. Xia and W. Zhang, *Physiol. Genomics*, 2014, **46**, 113–122.
116. F. Meisgen, N. Xu, T. Wei, P. C. Janson, S. Obad, O. Broom, N. Nagy, S. Kauppinen, L. Kemeny, M. Stahle, A. Pivarcsi and E. Sonkoly, *Exp. Dermatol.*, 2012, **21**, 312–314.
117. X. Gu, E. Nylander, P. J. Coates and K. Nylander, *Acta Derm.-Venereol.*, 2011, **91**, 392–397.
118. L. Hou, L. Bowman, T. G. Meighan, P. Pratheeshkumar, X. Shi and M. Ding, *Exp. Toxicol. Pathol.*, 2013, **65**, 1145–1148.
119. M. E. Killeen, L. Ferris, E. A. Kupetsky, L. Falo Jr and A. R. Mathers, *J. Immunol.*, 2013, **190**, 4324–4336.
120. N. Xu, P. Brodin, T. Wei, F. Meisgen, L. Eidsmo, N. Nagy, L. Kemeny, M. Stahle, E. Sonkoly and A. Pivarcsi, *J. Invest. Dermatol.*, 2011, **131**, 1521–1529.
121. J. R. Zibert, M. B. Lovendorf, T. Litman, J. Olsen, B. Kaczkowski and L. Skov, *J. Dermatol. Sci.*, 2010, **58**, 177–185.
122. D. Fu, W. Yu, M. Li, H. Wang, D. Liu, X. Song, Z. Li and Z. Tian, *Immunol. Lett.*, 2015, **166**, 55–62.
123. M. Zhao, L. T. Wang, G. P. Liang, P. Zhang, X. J. Deng, Q. Tang, H. Y. Zhai, C. C. Chang, Y. W. Su and Q. J. Lu, *Clin. Immunol.*, 2014, **150**, 22–30.
124. M. B. Lovendorf, H. Mitsui, J. R. Zibert, M. A. Ropke, M. Hafner, B. Dyring-Andersen, C. M. Bonefeld, J. G. Krueger and L. Skov, *Exp. Dermatol.*, 2015, **24**, 187–193.

CHAPTER 11

MicroRNAs in Haematological Diseases

A. C. P. MAFRA[a,b,c,d] AND G. A. CALIN*[a,b]

[a]Department of Experimental Therapeutics, The University of Texas MD Anderson Cancer Center, 1515 Holcombe Blvd. Unit 1950, Zip Code 77030, Houston, Texas, USA; [b]Center for RNA Inference and Non-coding RNAs, The University of Texas MD Anderson Cancer Center, 1515 Holcombe Blvd. Unit 1950, Zip Code 77030, Houston, Texas, USA; [c]Brazilian Biosciences National Laboratory (LNBio), Brazilian Center for Research in Energy and Materials (CNPEM), Rua Giuseppe Maximo Scolfaro 10000, Zip Code 13083-970, Campinas, Sao Paulo, Brazil; [d]Department of Genetics, Evolution and Bioagents, Institute of Biology, University of Campinas – UNICAMP, P.O. Box: 6109, Zip Code 13083-970, Campinas, Sao Paulo, Brazil
*E-mail: gcalin@mdanderson.org

11.1 Introduction

MicroRNAs (miRNAs) are non-protein-coding, small, 19–25-nucleotide transcripts involved in the regulation of a great number of protein-coding transcripts.[1] miRNAs are involved in multiple biological processes, from development, differentiation and cell cycle regulation, to senescence, metabolism and modulation of the immune system.[1,2] The first miRNAs described were lineage-defective 4 (*lin-4*) by Lee and colleagues in 1993,[3] and lethal 7 (*let-7*) by Reinhart and colleagues in 2000,[4] both essential for the normal

Drug Discovery Series No. 69
MicroRNAs in Diseases and Disorders: Emerging Therapeutic Targets
Edited by Philip V. Peplow, Bridget Martinez, George A. Calin and Aurora Esquela-Kerscher

Published by the Royal Society of Chemistry, www.rsc.org

development of *Caenorhabditis elegans*. *let-7* and *lin-4* were found to be responsible for regulating a series of biological processes during the embryonic development of worms, despite their lack of protein-coding regions. More recently, thousands of miRNAs were described, and their functions are still being studied.[5]

miRNAs not only regulate genes of several physiological functions; they are also involved in human diseases. Abnormal miRNA expression and function was found to be involved in cardiovascular, neurological, immune-related and genetic diseases, as well as cancer.[5,6] In fact, the first report on miRNA abnormality in human disease was in chronic lymphocytic leukaemia (CLL). Calin *et al.* showed that the *mir-15a/mir-16-1* cluster was significantly downregulated in a frequently deleted region of chromosome 13 (13q14), the most common cytogenetic aberration in CLL patients.[7] Since then, several other miRNAs have been found to play important roles in CLL, the most common type of leukaemia in the Western population, and also in other types of leukaemia, such as acute myeloid leukaemia (AML), which holds the poorest prognosis and death rates of the most common haematological diseases.

In this chapter, we focus on the biogenesis and regulatory function of miRNAs, their roles and the mechanisms of dysregulation in cancer, focusing on haematological neoplasms, specifically in CLL and AML.

11.2 miRNA Biogenesis and Function

miRNAs are usually transcribed by RNA polymerase II to produce large primary transcripts, with a 5′ cap structure and a 3′ poly(A) tail, termed pri-miRNAs.[8] Pri-miRNAs are then cleaved by a protein complex, composed of RNA-binding protein DCGR8 and a type III RNase Drosha, into an ~85-nucleotide stem–loop precursor structure called pre-miRNA. This hairpin structure is then ready to be exported from the nucleus to the cytoplasm, by the Ran/GTP/Exportin 5 complex. Pre-miRNAs are processed by another RNase III enzyme, Dicer, into a ~22-nucleotide molecule. The mature miRNA structure is formed when the duplex is unwound and incorporated into a protein complex called RNA-induced silencing complex (RISC). The miRNA/RISC complex mediates silencing through direct targeting of the mRNA sequence for cleavage and degradation, or by blocking the translation machinery to bind to the target mRNA. Usually, the mechanism by which the miRNA regulates mRNA expression depends on the complementarity between the miRNA and mRNA sequences.[9,10]

In most cases, miRNAs bind to the 3′ untranslated region of mRNAs, through a six-to-eight nucleotide long complementarity region in the 5′ end of miRNAs, called the seed sequence.[11] However, there have been reports on binding of miRNAs to their target mRNAs without the seed sequence.[12] Unconventional miRNA functions were also reported, but less investigated.[13] Regardless of the mechanism of binding, the complementarity between sequences can dictate the fate of the target mRNA. Generally, if there is complete alignment, mRNAs are cleaved and degraded by the RISC complex; if partial complementarity is achieved, the miRNA/RISC complex blocks translation of that mRNA. However, there are studies in zebrafish that report that both phenomena are responsible for mRNA regulation sequentially, with translational repression happening first, followed by mRNA degradation.[14,15]

11.3 miRNA Dysregulation in Cancer

Cancer is an extremely complex disease, and every type of cancer has its singularities, mainly due to different cell types and tissues that are affected. Not only that, but every type of cancer also has different subtypes. Patients present particularities based on their own genetic background, lifestyle and environment. Thus, it is important that scientists study and understand cancer considering its complex biology. Still, cancer cells share genetic and biological characteristics, as extensively described and reviewed by Hanahan and Weinberg in 'Hallmarks of cancer'.[16] The main processes that enable cancer cells to survive and disseminate are dysregulated. In most of these cellular processes, if not in all, involvement of miRNAs was described.[10]

Like oncogenes, which are frequently turned on, overly expressed and/or amplified, oncogenic miRNAs may also play a role in promoting tumour initiation, progression and dissemination. The *mir-17-92* gene cluster, for example, was found to be amplified in cancer cells compared to normal cells in several types of cancer.[10] The mir-17-92 cluster was overexpressed in B-cell lymphomas[17] and lung cancer[18] and there is a translocation of the gene region found in T-cell acute lymphoblastic leukaemia (ALL).[19] The mechanism with which mir-17-92 might promote cancer cell growth, survival and dissemination is still unknown, but there is evidence that two tumour suppressor genes (TSGs), Phosphatase and tensin homolog (*PTEN*) and Retinoblastoma-like protein 2 (*RB2*) are targets of this oncomiR.[18,20] There is also evidence that cancer-Myelocytomatosis (*c-Myc*), the most amplified gene in human cancer, induces transcription of mir-17-92,[21] enhancing the evidence that this miRNA cluster has an oncogenic role. mir-155 is overexpressed in cancer cells of almost all types of lymphomas, compared to expression levels in normal cells.[22,23] Its increased expression levels were associated with the oncogene B-cell integration cluster (*BIC*), also overly expressed in lymphomas. BIC is known for accelerating the pathogenesis of lymphomas and leukaemia. Through mechanisms that are still unresolved, mir-155 induces the expression of BIC, increasing gene levels and worsening disease.[22] mir-21 was found to be aberrantly expressed in human glioblastoma (GBM):[24] >100-fold higher in GBM cells compared to normal equivalent brain cells.[25] Knockdown of mir-21 in GBM cell lines induced caspase activation and apoptosis, indicating that this miRNA might be responsible for evading cell death in GBM cancer cells.[25]

As well as having an oncogenic role, miRNAs can also function as tumour suppressors.[10] The first cancer-related miRNAs described, the *mir-15a/16-1* gene cluster, are strongly downregulated in the majority of CLL cases.[7] Early studies found that mir-15a/16-1 in CLL mediated control of B-cell lymphoma 2 (BCL2) expression, which resulted in the inhibition of pro-apoptotic pathways and promotion of cancer cell survival.[26] Later, it was also found that mir-15a/16-1 control the expression of genes involved in cell-cycle entry in B-cell malignancies.[27] *let-7* is believed to play a crucial role in the pathogenesis of lung cancer.[28] Expression levels of *let-7* were found to be critically reduced in *in vitro* and *in vivo* studies, and downregulation of *let-7* is associated with decreased post-operative survival of all cancer types and stages of lung cancer.[28] It was later found that MYC and rat sarcoma (RAS), causally implicated in several types of cancer, including lung cancer, are direct targets of *let-7*.[29]

11.4 miRNAs in Haematopoiesis

Haematological malignancies comprise leukaemia, lymphomas and myelomas. Despite advances in treatment, leukaemia still has a considerable death rate: out of the 35 000 new cases diagnosed, approximately 15 000 will die.[30] Several miRNAs and their targets were described to be dysregulated in leukaemia. To better understand how miRNAs could be involved in the pathogenesis of leukaemia, it is imperative to know the mechanism by which they regulate haematopoiesis.

miRNAs have been found to play important roles in human haematopoiesis (Figure 11.1). mir-128a and mir-181a regulate the differentiation of the multipotent progenitor cell in both common myeloid (CMP) and common lymphoid (CLP) progenitors, whereas mir-146 blocks lymphoid differentiation and mir-155, mir-24a and mir-17 suppress myeloid differentiation at an early stage.[31]

During myeloid differentiation (Figure 11.1), mir-16, mir-103 and mir-107 were predicted to block early granulocyte and macrophage commitment,[31] while mir-223 positively regulates two genes responsible for human granulopoiesis, Nuclear factor 1-A (NFI-A) and CCAAT/enhancer-binding protein alpha (C/EBPalpha).[32] An increased differentiation of acute promyelocytic leukaemia (APL) cells is observed by ectopic expression of mir-223, while silencing of mir-223 inhibits the differentiation response to retinoic acid,[32] which is a current approved treatment for APL. mir-17-5p, mir-20a and mir-106a have also shown an inhibitory effect on monocytic differentiation and maturation, by targeting the transcription factor acute myeloid leukaemia 1 (AML1) and promoting downregulation of the macrophage colony-stimulating factor receptor (M-CSFR).[33] During the differentiation process of erythrocytes, downregulation of mir-150, mir-155, mir-221 and mir-222 occurs, while mir-451 and mir-16 levels increase at late stages of erythropoiesis.[34,35] A screening of miRNA expression performed in megakaryocytes showed that mir-10a, mir-130a, mir-126, mir-106, mir-10b, mir-17 and mir-20 are downregulated during cell differentiation.[36] In particular, mir-130a

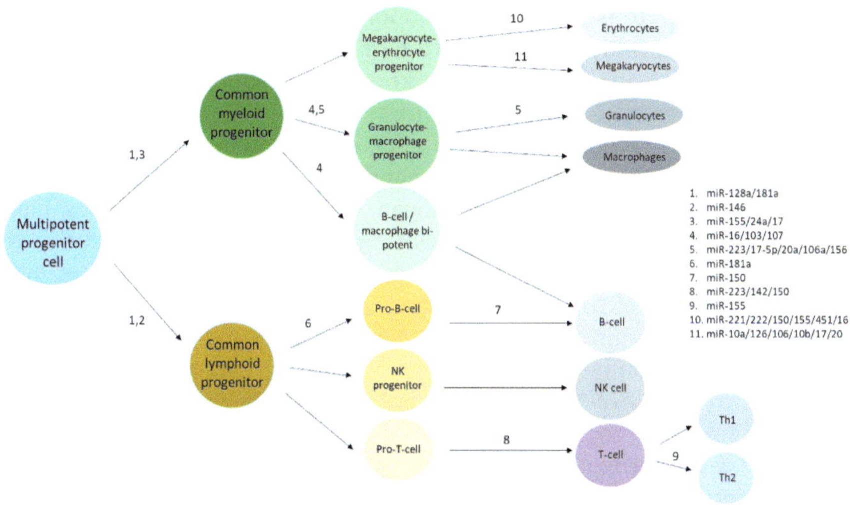

Figure 11.1 Briefly, a multipotent progenitor cell can give rise to a common myeloid or lymphoid progenitor. Common myeloid progenitors (CMPs) will develop megakaryocyte–erythrocyte progenitors, that will differentiate into either megakaryocytes or erythrocytes. CMPs will also develop into granulocyte–macrophage progenitors that will generate granulocytes and macrophages. Finally, CMPs can give rise to a B-cell/macrophage bi-potent cell, that can either generate macrophages or B-cells (of another lineage). Common lymphoid progenitors will generate pro-B-cells that will develop into B-cells; natural killer (NK) progenitors that will develop into NK cells; and pro-T-cells that will develop into T-cells. miRNAs involved in the process of lineage differentiation are marked by numbers, separated by each process in which they play a role. Th : T-helper cells.

targets the transcription factor V-maf musculoaponeurotic fibrosarcoma oncogene homolog B (MAFB), involved in the activation of glycoprotein IIb (*GPIIB*) promoter,[36] an important protein for platelet physiology.

In lymphoid differentiation (Figure 11.1), mir-150 was shown to play an important role in the maturation of B-cells and T-cells. Specifically, ectopic expression of mir-150 (which is selectively expressed in mature, resting B- and T-cells, but not in their progenitors)[37] in murine hematopoietic stem and progenitor cells significantly reduced mature B-cell levels in the circulation, spleen and lymph nodes, with no effect in the levels of T-cells and myeloid cells.[38] The observed effect on B-cell lymphopoiesis could be explained with a transition blockage from the pro-B to pre-B-cell stage. These results were confirmed by mir-150 transgenic and knockout mice models that supported the stage-specific expression of this miRNA during lymphocyte development. mir-150 directly targets the transcription factor cancer-myeloblastosis c-MYB, an important driver of hemopoiesis. Interestingly, it was shown that mir-150 controls c-MYB expression *in vivo* in a dose-dependent manner over a very narrow range, with a 30–35% reduction of c-MYB protein levels after overexpression of the miRNA.[39]

Another miRNA important in lymphopoiesis is mir-155, shown to reduce myeloid and erythroid colony formation *in vitro*.[31] These results were later confirmed *in vivo* with transgenic and knockout mice models of mir-155.[40] Pre-B-cell proliferation followed by progression to B-cell malignancy was described in the transgenic mouse model, characterized by B-cell-targeted overexpression of mir-155, indicating that deregulation of this miRNA could occur as an early event in leukaemogenesis. Two different groups have developed mir-155 knockout mice models and they both identified a critical role of this miRNA in the immune response.[41,42] One of them reported a biased polarization of T-helper (Th) cells toward Th2 differentiation,[41] largely responsible for allergic reactions. The other showed that mir-155-deficient mice had a compromised ability of dendritic cells to activate T-cells, most likely due to defective antigen presentation or inefficient co-stimulatory functions.[42]

Chen *et al.* identified three miRNAs that are specifically expressed in hematopoietic cells, mir-181a, mir-223 and mir-142, and showed that their expression was dynamically regulated during early hemopoiesis and lineage commitment.[43] Specifically, mir-181a was shown to play a role in lymphopoiesis. A scarcity of T-lymphocytes and increased B-cells was observed after ectopic expression of mir-181a. It seems that forced expression of mir-181a during thymocyte development could result in retention of positively selected thymocytes in the thymus and in an apparent decrease in peripheral T-cells.[44]

Ectopic expression of mir-223 and mir-142 induced a 30–50% increase in the proportion of the T-lineage cells, but not of the B-lineage or myeloid cells, despite the fact that these miRNAs are highly expressed in myeloid lineages (as well as mir-142 in B cells) (Figure 11.1).[43] Notably, a different pattern of expression of these miRNAs was observed in human haematopoiesis with respect to the mouse model.[45] Conversely to what has been described in human, in murine haematopoiesis, mir-150 is upregulated during B-cell maturation,[37] for example. Therefore, it is imperative to understand the role of certain miRNAs both in humans and mice, given the importance of mouse models for pre-clinical trials.

The increasing knowledge of the physiological implications of miRNAs in human haematopoiesis represents a mandatory background to properly design future studies aimed at identifying and correcting aberrancies of the microRNAome in haematological malignancies.[46]

11.5 Mechanisms of miRNA Dysregulation in Chronic Lymphocytic Leukaemia and Acute Myeloid Leukaemia

Leukaemias are malignancies of the bone marrow and blood that are classified into four main groups based on cell type and rate of growth: acute lymphocytic (ALL), acute myeloid (AML), chronic myeloid (CML), and chronic

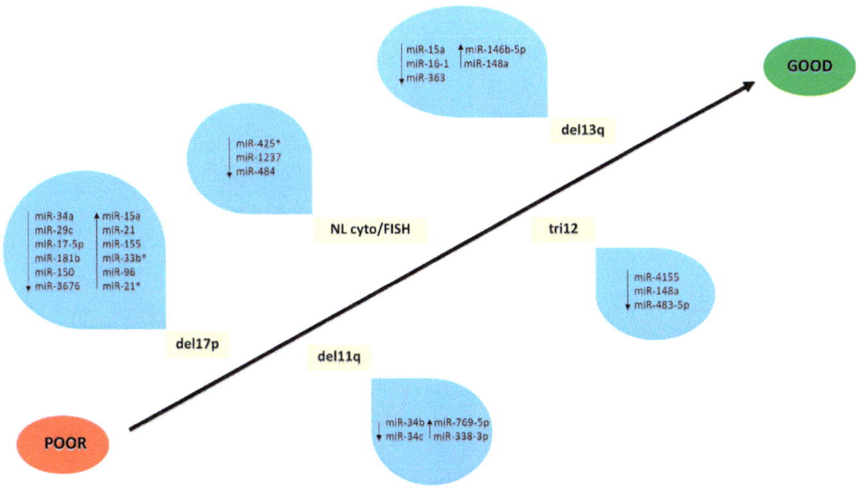

Figure 11.2 Chronic lymphocytic leukaemia (CLL) cytogenetic subtypes associated with their respective prognosis, beginning with the poorest, del17p, followed by del11q, NL/cyto FISH, tri12 and del 13q, the latter being the best. Associated with them, in blue balloons, the most common miRNAs accompanied by their expression variations, represented by arrows.

lymphocytic (CLL).[30] Lymphoid leukaemia originates from phenotypically differentiated but immune-incompetent B-cells,[47] while myelogenous leukaemia originates from different types of immature, myeloid progenitors.[48] In this section, we focus on two specific leukaemia types: CLL, the most common type of adult leukaemia[49] and AML, the haematological malignancy with the highest mortality rate to date.[30]

11.5.1 miRNA Implications in CLL

CLL has a tremendously variable clinical course, and complete stratification of the disease, along with accurate prognostic markers, has yet to be achieved. Usually the typical chromosomal aberrations found in CLL patients are used as prognostic markers, since the abnormalities are related to disease progression and aggressiveness.[50] This notwithstanding, non-coding transcripts, such as miRNAs, were also proven to play important roles in the disease course of CLL[51] (Figure 11.2 and Table 11.1).

The prognosis and outcome of CLL is highly variable. Some patients present the indolent form of the disease that usually does not require prompt treatment, while others develop the aggressive form of the disease, most often requiring immediate therapy.[50] To date, the prediction of the clinical course of CLL is largely dependent on the typical cytogenetic abnormalities found in these cases. Most patients can be categorized in five different

Table 11.1 Most common miRNAs involved with the pathogenesis of CLL, associated with CLL type, targets and mechanism of action and, if, available, clinical correlation.[a]

miRNA	Chromosomal location	Expression variation	CLL subtype	Targets	Clinical correlation	Reference
mir-15a/16-1	13q14	Downregulation	del13q	*BCL2/MCL1*	Associated with pathogenesis of CLL	26,57
mir-146-5p	5q33	Overexpression	del13q			58
mir-148a	7p15	Overexpression	del13q			58
mir-363	Xq26	Downregulation	del13q			58
mir-34b/c	11q23	Downregulation	del11q	*ZAP70*		60,64
mir-769-5p	19q13	Overexpression	del11q			58
mir-338-3p	17q25	Overexpression	del11q			58
mir-34a	1p36	Downregulation	del17p	Regulated by p53	Resistance to therapy and refractory disease	66,95,96
mir-29c	1q32	Downregulation	del17p		Poorer outcome	64,66, 67,81
mir-17-5p	13q31	Downregulation	del17p			64,66,67
mir-150	19q13	Downregulation	del17p		Associated with poor prognostic factors	58
mir-3676	17p	Deletion	del17p	Co-deleted with p53	Worse survival and involved in pathogenesis of CLL	70
mir-181 family		Downregulation			Resistance to therapy	68
mir-155	21q21	Upregulation			Poorer outcome	99

[a]CLL: chronic lymphoid leukaemia.

groups: deletion of the long arm of chromosome 13 (del13q); deletion of the long arm of chromosome 11 (del11q); deletion of the short arm of chromosome 17 (del17p); trisomy of chromosome 12 (tri12); and patients with normal cytogenetic and fluorescent *in situ* hybridization (FISH) status (NL cyto/FISH).[52]

The most common chromosomal aberration is del13q, found in ~55% of cases (Figure 11.2). The presence of this abnormality is considered a good prognostic factor, with patients presenting a median survival of 133 months.[52] This minimally deleted region was analysed for the identification of a TSG, but failed to identify any protein-coding genes.[53–56] In 2002, our group found that a cluster of miRNAs, *mir-15a/16-1*, located in the 13q14 deleted region, was deleted or downregulated in the majority of CLL patients with this chromosomal abnormality.[7] It was later found that this miRNA cluster targeted oncogenes involved in CLL initiation and progression, such as BCL2 and myeloid cell leukaemia 1 (MCL1).[26,57] Later, in 2010, it was found that the mir-15a/16-1 cluster mediated the regulation of cell-cycle entry genes.[27] This finding, of a non-protein-coding transcript that functioned as a tumour suppressor was revolutionary for the genetic cancer research.

A miRNA array analysis also found other miRNAs associated with del13q, such as overexpression of mir-146-5p and mir-148a, as well as downregulation of mir-16 and mir-363.[58] The specific roles of these miRNAs, their targets and their importance in the pathogenesis of this cytogenetic group are yet to be defined.

The second most common cytogenetic aberration found in CLL is del11q, present in ~18% of patients (Figure 11.2). This group is characterized by poor response to treatment and by a shorter progression-free survival of 79 months.[52] Ataxia–telangiectasia mutated (ATM), an important tumour suppressor, is located in this locus, and mutations in this gene correlate with poor prognosis in CLL.[59] The *mir-34b/c* cluster is also located in this deleted region and expression of this cluster is significantly lower in del11q cases.[60] This cluster is a direct target of p53, which is frequently mutated or deleted in CLL.[61] In addition, the *mir-34b/c* cluster is often downregulated through epigenetic mechanisms, such as DNA hypermethylation.[62,63] There is an inverse correlation between the presence of del11q and *mir-34b/c* hypermethylation, indicating that the silencing of this cluster through epigenetic mechanisms happens independently of the cytogenetic aberration.[62] One of the main targets of mir-34b/c is the 70 kDa zeta-associated protein (ZAP70), an oncogene that is also considered a prognostic factor of the molecular kind in CLL.[64] In the microRNAome study mentioned earlier, an overexpression of mir-769-5p and mir-338-3p in del11q cases compared to other cytogenetic groups was also found.[58]

Another cytogenetic group comprises the deletion of the short arm of chromosome 17, representing the group with the poorest prognosis, with a median survival of 32 months and a poor response to therapy[52] (Figure 11.2). *TP53* is located in this deleted region. Since *TP53* is the most deleted mutated TSG in cancer,[65] several studies were aimed at understanding the cellular consequences of the loss of function of this protein in this CLL group. For example, Mraz *et al.* analysed the expression of 35 miRNAs in patients with some type of abnormality in p53 *versus* patients with wild-type p53. They found that mir-34a was extremely downregulated in patients with mutated p53, which agrees with the fact that this family is regulated by p53 itself.[66] There was also significant downregulation of mir-29c and mir-17-5p.[64,66,67] In another study, p53 was found to be involved in a regulatory feedback loop with the mir-15a/16-1 cluster.[64] Rossi *et al.* found a series of miRNAs differentially expressed in del17p patients compared to other cytogenetic groups, such as mir-15a, mir-21, mir-34a, mir-155 and mir-181b.[68] Negrini *et al.*, in the microRNAome study, found significant downregulation of mir-34a and mir-150, along with upregulation of mir-33b, mir-96 and mir-21.[58] Balatti *et al.* showed that *mir-3673* is co-deleted with *TP53*, being downregulated in four cytogenetic groups compared to normal B-cells and mutated in 1% of CLL cases.[69] mir-3676 directly targets the T-cell leukaemia/lymphoma 1 (TCL1) oncogene, proven to be involved in survival and pathogenesis of leukaemia.[70]

The tri12 and NL cyto/FISH groups are intermediate prognostic factors and remain poorly described (Figure 11.2). Tri12 patients comprise ~16% of cases with a median survival of 114 months, while NL cyto/FISH represents 18% of patients with a median survival of 111 months.[52] It is believed that cancer initiation in tri12 cases is triggered by an oncogene in chromosome 12 that has an augmented function due to an extra copy, but that gene has yet to be identified.[71] Four genes differentially expressed were associated with tri12: Huntingtin-interacting protein 1-related protein (*HIP1R*), Myogenic factor 6 (*MYF6*), Purigenic receptor P2Y12 (*P2RY12*) and Cluster of differentiation 200 (*CD200*), but only the first two are located in chromosome 12.[72] The microRNAome study showed significant downregulation of mir-155, mir-148a and mir-483-5p in tri12; and of mir-425, mir-1237 and mir-484 in NL cyto/FISH, both compared to other cytogenetic groups.[58]

In an attempt to molecularly discriminate subtypes of CLL, Visone *et al.* identified 32 miRNAs differentially expressed between the groups del13q, del11q, del17p, tri12 and NL cyto/FISH.[73] However, the most interesting finding was the downregulation of mir-223, mir-29b/c and the mir-181 family associated with unmutated immunoglobulin heavy-chain variable region (IGHV) status and downregulation of ZAP70 in del17p, which all correlate with poor prognosis and disease progression.

Molecular-based prognostic factors, besides cytogenetic analysis, were developed to try to stratify better the course of CLL. The mutational status of IGHV and the expression levels of ZAP70, cluster of differentiation 38 (CD38) and β-2-microglobulin (B2M), independently, represent these prognostic factors. Unmutated IGHV, higher levels of ZAP70 (>20%), higher levels of CD38 (>30%) and increased gene expression of B2M (more than two-fold) is considered the poorest prognosis.[74] Usually, these markers are considered independently of cytogenetic status, although presence of del11q and del17p can worsen the prognosis. Nevertheless, establishing miRNA signatures that accompany these factors might aid in understanding the molecular mechanism behind the status of these factors.

With this in mind, our group identified a 13-miRNA signature that can differentiate patients between high ZAP70 *versus* low ZAP70 expression and unmutated IGHV *versus* mutated IGHV.[75] mir-15a, mir-195, mir-221, mir-23b, mir-155, mir-223, mir-29a-2, mir-24-1, mir-29b-2, mir-146, mir-16-1, mir-16-2 and mir-29c were differentially expressed between the two groups. The expression status of some of these miRNAs was validated and associated with the same prognostic factors in independent studies. For example, mir-15a[76,77] and mir-16[76] were upregulated and mir-29a/b[76,78,79] was downregulated in unmutated IGHV. In addition, downregulation of mir-29c[77,78,80,81] and mir-223[38,77,78,80–82] was associated with all poor molecular prognostic factors.

mir-150 downregulation is consistently associated with poor prognostic factors.[77,78,83] Interestingly, downregulation of mir-150 in B-cells in CLL, but higher concentrations of mir-150 in plasma of the same patients was correlated with poor prognosis, increase in tumour burden and disease aggressiveness.[80]

Besides associating specific miRNAs and building miRNA signatures that associate with the known prognostic factors of CLL, there has been some development in stipulating miRNA-based scoring systems for the stratification and for survival analysis of CLL. Stamatopoulos *et al.* developed a quantitative polymerase chain reaction (qPCR) scoring system based on the gene expression levels of *ZAP70*, lipoprotein lipase (*LPL*), mir-223 and mir-29c measured simultaneously. Briefly, the lowest scores indicate the poorest prognosis, achieved by high ZAP70 and LPL, and low mir-223 and mir-29c. This retrospective study predicted treatment-free survival and overall survival, and sheds light on the prognostic value of miRNAs. qPCRs are relatively simple and inexpensive methods that could be employed for disease course prediction, refining CLL stratification.[84]

Rossi *et al.* developed a score based on mir-21 expression and FISH or karyotype results. The combination of the results generated a score, the highest indicating a poor overall survival. High levels of mir-21, along with del17p identified by FISH or karyotyping generated a high score, therefore predicting a poor survival. Moreover, when comparing their scoring system with the well-established molecular prognostic factors, they found that their score better predicted overall survival.[68]

Circulating miRNAs (circ-miRNAs) are miRNAs detected in serum or plasma. The interest in determining the levels of circ-miRNAs differentially expressed in cancer, for example, is to identify biomarkers that are reproducible and non-invasive.[85–87] Moussay *et al.* detected the number of circ-miRNAs as being one-third higher in patients with CLL compared to normal subjects.[88] In addition, a series of 14 circ-miRNAs differentiated CLL from normal, multiple myeloma and hairy cell lymphoma patients. When analysing expression of the same 14 miRNAs in B-cells, these researchers found the expression pattern was quite different from plasma, suggesting they were released by cell types other than B-cells. In the same study, different circ-miRNAs were detected in patients with higher levels of ZAP70 than in patients with normal ZAP70 levels. In another study, increased levels of circ-mir-155 were detected in plasma of patients with incomplete disease remission compared to complete remission, indicating a possible use of circ-miRNAs to predict response to therapy.[89]

Resistance to therapy and relapse are constant features that afflict CLL patients and are of concern to doctors. To date, the standard of care of CLL is shifting from the combination of chemotherapeutic drugs, being the most common regimen fludarabine, cyclophosphamide and rituximab (FCR) to the treatment with ibrutinib with or without venetoclax.[90] Fludarabine is a purine analogue and cyclophosphamide is an alkylating agent, both interfering in the process of DNA synthesis, while rituximab is a monoclonal antibody against CD20 in B-cells, triggering cell death.[90] Ibrutinib is a small molecule inhibitor of Bruton's tyrosine kinase (BTK), present in B-cells, that blocks BCR signalling, inducing apoptosis in leukaemic cells. Venetoclax is also a small molecule which blocks BCL2, leading to apoptosis of CLL cells.[91] In *in vivo* experiments, the combination of these two drugs, more specific for at their targets than the chemotherapeutic drugs, promised synergistic effects with reduced toxicity.[92] Clinical trials have been taking place with this new drug combination and results show increased patient response to therapy, reduced treatment resistance and refractory disease, especially in high-risk CLL cases.[93] It is reported that in aggressive CLL cases, patients show recurrent resistance to therapy and relapse after initial treatment, with a 5-year disease-free rate of <50%.[94] Some 30–40% of chemo-refractory CLL cases have mutated or deleted p53, and another 30% are del11q or have mutated ATM;[95] for the remaining cases the causes of refraction remain unclear.

From when the downregulation of mir-15a/16-1 cluster was first identified in del13q cases of CLL, and its tumour suppressive function associated with the pathogenesis of the disease, there has been a great interest in the development of miRNA-based therapy. While having several targets can be an advantage as resistance to miRNA-based therapy due to target site mutation would be unlikely, it can also cause off-target effects. Also, the supra-physiological conditions of miRNA reintroduction for therapy are still incompletely understood.

mir-34 was found to have tumour-suppressive roles in several types of cancer, including CLL. mir-34 levels were downregulated in del17p or mutated p53 cases, associated with a poor response to therapy,[96] and extremely downregulated in fludarabine-refractory patients.[95] It was found to be part of the chemoresistance network in CLL, in which high levels of mir-34 were found in patients with a functional p53, while very low levels were found in patients with mutated p53 and fludarabine-refractory disease.[95] mir-34 is a direct target of p53, so patients with mutated or deleted p53 and downregulated mir-34 had an impaired DNA damage response.

In an *in vitro* experiment, reintroduction of mir-34 *via* miRNA mimics to restore expression levels increased levels of BCL-2-associated X (BAX) and p21, downstream effectors of p53 to induce apoptosis.[95] Therefore, mir-34 presented itself as a very promising target for therapeutics. Unfortunately, the first clinical trial in patients with liver cancer with mir-34 reintroduction had very adverse side-effects, which caused the company to bring the trial to a halt (information available at https://clinicaltrials.gov/ct2/show/NCT01829971). Hopefully, after reassessment of pre-clinical tests, a new form of mir-34 mimic can be used in different clinical trials, perhaps even for CLL.

The mir-181 family has also been associated with treatment resistance. Low levels of miR181a/b were found in refractory CLL patients and predicted treatment-free survival.[68] It also helped with prognosis, since patients with a progressive disease have decreasing levels of mir-181b compared to stable levels in patients with stable disease.[73,97] Reintroduction of mir-181a/b enhances drug sensitivity in primary cultured CLL B-cells through targeting of BLC2, MCL1, and XIAP, promoting apoptosis.[98] Perhaps mir-181a/b mimic therapy, just like mir-34, could be an option for treatment of refractory CLL cases. Surely, assessment of all possible side-effects is of the highest importance, taking into consideration the lessons learned with mir-34.

mir-155 is an oncogenic miRNA overexpressed and associated with poor prognosis in several types of cancer[99] and almost all haematological malignancies and involved in chemoresistance in osteosarcoma[100] and breast cancer.[101] Toll-like receptor 9 (TLR9) stimulation in B-cells leads to fludarabine resistance in patients with adverse prognostic factors. In the same patients, there was an increase in mir-155-3p levels, indicating that it could be binding to TLR9 and stimulating cell proliferation and survival.[102]

Ferracin *et al.* evaluated the expression of more than 700 miRNAs before and after fludarabine therapy.[103] High levels of mir-148a, mir-221, and mir-21 were found in refractory CLL. *In vitro* experiments with knockdown of mir-221 and mir-21 showed an increase in caspase activity followed by cell apoptosis.

Interestingly, Epstein–Barr virus (EBV) genome or proteins have been identified in malignant B-cells and have been associated with the pathogenesis of CLL.[104,105] The EBV is a human-specific type of herpes virus with subclinical symptoms and latent infection of B-cells.[106] In EBV-associated lymphomas, >90% of malignant B-cells had EBV genome identified using FISH.[107] Since then, researchers have been trying to find the role of EBV in the progression of CLL. mir-155 in EBV-positive FISH status was overexpressed in malignant B-cells.[108] In this context, our group tried to assess the pattern of exogenously expressed miRNAs and their contribution to disease progression in CLL,[106] using as a detection method qPCR of plasma viral miRNA levels, as it was shown by our group that this methodology is more accurate to identify viral infections.[109] Bam HI fragment H rightward open reading frame 1 (BHRF1-1) expression levels were higher in plasma of CLL patients compared to healthy individuals and were positively associated with advanced CLL clinical stages.[106] Moreover, patients with the highest plasma levels of BHRF1-1 had reduced levels of p53, indicating that this EBV-miRNA might have a role in the disease progression of CLL.[106]

These studies present a good promise for miRNA-based therapy. The biggest challenge, however, is the development of a successful *in vivo* delivery system, meeting with efficiency, stability, safety and specificity requirements. Naked miRNA molecules cannot be introduced in the body due to their rapid degradation, so their encapsulation is necessary. Nanocarriers present an interesting approach, but more research is necessary to assess safety and efficiency for these molecules.

11.5.2 miRNA Implications in AML

Acute myeloid leukaemia is characterized by accumulation of incompletely differentiated haematopoietic cell lines of diverse myeloid origins and reduced production of healthy haematopoietic elements.[110] The majority of patients present leukaemic cells of various stages of incomplete maturation, with different subclones of cell populations. Improvements in AML therapy have been developed over the past 30 years, mainly through supportive care and stem-cell transplantation. For decades, the standard of care for AML has been anthracyclines (usually daunorubicin) combined with cytarabine for induction of remission, and intermediate doses of cytarabine with allogenic stem-cell transplantation for consolidation of remission. Although a few parameters vary, and some patients receive autologous instead of allogeneic stem-cell transplantation, the regimen for AML treatment is as described above, achieving survival rates of as much as 70% in patients age <60 years.[110]

Despite these advances, AML remains the deadliest type of leukaemia among the most common types (AML, ALL, CML, and CLL), with mortality rates of >50% in all diagnosed cases.[30] The main obstacles in AML treatment and cure are refractory disease during initial treatment and, more frequently, relapse after complete remission. Clonal diversity and the different cytogenetic types and mutations between patients make it difficult to stratify the disease to predict outcome and delineate proper treatment.

A few chromosomal abnormalities associated with AML pathogenesis have been described (Table 11.2). Translocations of chromosome 8 and 21 (t[8;21]) and cytogenetic abnormalities in the long arm of chromosome 11 (abn11q23) are among the most studied aberrations in AML. T[8;21] confers fusions of myeloid transcription factors and Runt-related transcription factor 1 (*RUXN1*) mutations, culminating in impaired haematopoietic differentiation. Abn11q23 affects multiple lineage leukaemia (*MLL*) genes and its partners and this subtype is often associated with poor prognosis and relapse. Other mutations have been identified and associated with AML. Mutations in the FMS-like tyrosine kinase 3 *FLT3* gene activate sustained signalling of this tyrosine-kinase receptor and its downstream

Table 11.2 The chromosomal aberrations found in AML to date, with listing of affected genes, frequency of occurrence and estimated prognosis.[a]

Alteration	Affected genes	Incidence in *de novo* AML	Prognosis
t[8;21]	*RUNX1/RUNX1T1*	5–7%	Favourable
t[15;17]	*PML/RARA*	5–8%	Favourable
t[11;17]	*ZBTB16/RARA*	<1%	Poor
abn[16q22]	*CBFB/MYH11*	5%	Favourable
abn[11q23]	*MLL* and partners	3%	Poor
+8		3%	Poor
del5q, del7q		15–20%	Poor
inv 3	*RPN1/MECOM*	<1%	Poor
abn(17p)	*TP53*	5%	Poor
+13		1–2%	Poor
t[6;9][2p34q]	*DEK/NUP214*	<1%	Poor
t[9;22]	*BCR/ABL1*	1%	Poor
t[1;22]	*RBM15/MKL1*	<1%	Poor

[a]AML: acute myeloid leukaemia. Adapted from ref. 110 with permission from Elsevier, Copyright 2013.

effectors, which include RAS-Rapidly accelerated fibrosarcoma (RAS-RAF), Janus kinase 2–signal transducer and activator of transcription (JAK-STAT) and phosphatidylinositol-4,5-bisphosphate 3-kinase–RAC-alpha serine/threonine-protein kinase (PI3K-AKT) proteins, all drivers of cell proliferation. Nucleophosmin 1 (*NPM1*) mutations impair its correct cell localization, along with its binding proteins. Mutations in proteins forming the spliceosome complex, deregulating RNA processing, and the cohesin complex, impairing chromosomal segregation and transcriptional regulation, have also been identified. Mutations in genes involved in epigenetic homeostasis and p53 were also present in AML.[48,110] Patients with normal cytogenetics (cytogenetically normal AML, CN-AML) are also present.

Although several chromosomal and genetic alterations have been described, there is no homogeneity in the frequency and occurrence of these abnormalities, and patients have a very diverse phenotype. Besides, many subclones with different patterns of molecular abnormalities are present at diagnosis, so different subclones might become predominant under the selective pressure of different chemotherapeutic agents.[110] A patient might be identified with a specific type of translocation, say t[8;21], and be treated with high-dose cytarabine to reach a complete remission of >60%. However, if subclones are present, which they usually are, cytarabine alone might not eliminate all leukaemic cells, leading to a therapy resistant relapse. Therefore, stratification of AML and development of new therapies have been a major challenge.

The most recent miRNAs reported to be involved in AML pathogenesis (see review[111]) had either oncogenic or tumor-suppressive functions (Figure 11.3). mir-9 was appointed to act as an oncogeneoncomiR in MLL-rearranged AML, where its overexpression induces proliferation and survival, and leukaemogenesis in mice.[112] However, in t[8;21] and EVI1-induced AML, it acted as a tumour suppressor, where its downregulation increased cell proliferation[113] and its upregulation induced monocytic differentiation (Figure 11.3).[114] mir-22 was also found to have a dual role in AML, acting as an oncomiR in myelodysplastic syndrome-derived AML (MDS-derived AML) and a tumour suppressor in *de novo* AML (Figure 11.3). Its targets, which include the Tet methylcytosine dioxygenase 2 (*TET2*) gene, are mainly responsible for epigenetic regulation of several other genes involved in cell proliferation and survival. Therefore, when acting as an oncomiR, its overexpression leads to increase in cell proliferation, survival and self-renewal and myeloid malignancy in mice, while reducing cell differentiation.[115] As a tumour suppressor, it reduces AML blast cell growth and leukaemic progression in mice and increases cell differentiation.[116,117]

The mir-17-92 cluster has been reported to have an oncogenic role in several types of haematological malignancies, as well as solid tumours. In AML, especially in the MLL-rearranged subtype, its upregulation increases cell proliferation, survival, self-renewal, and colony-forming capacity and leukaemogenesis in mice, while reducing cell differentiation[118,119] (Figure 11.3). mir-125b acts as an oncomiR in several subtypes of AML, specifically in t[2;11][p21;q23],[120] MDS-derived AML[121,122] and in paediatric

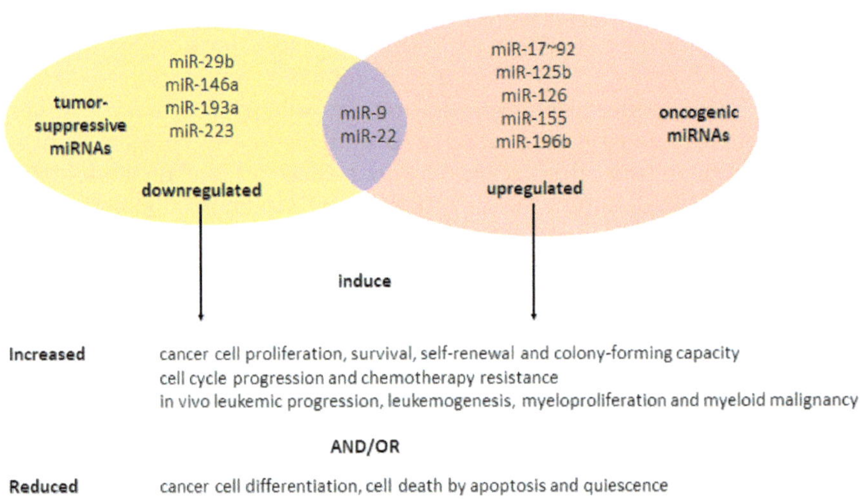

Figure 11.3 In acute myeloid leukaemia (AML), miRNAs have an oncogenic role, a tumor-suppressive role, or a dual role, context-dependent, represented in between the oncogenic miRNAs and the tumor-suppressive miRNAs. Tumor-suppressive miRNAs are usually downregulated, while oncogenic miRNAs are overexpressed. Together, these expression variations induce increased cancer cell proliferation, survival, self-renewal and colony-forming capacity, cell cycle progression and chemotherapy resistance, and *in vivo* leukaemic progression, leukaemogenesis, myeloproliferation and myeloid malignancy; and reduced cancer cell differentiation, cell death by apoptosis and quiescence.

AML[123] (Figure 11.3). It acts by increasing cancer cell proliferation, self-renewal, and myeloid progenitors, as well as its overexpression leading to development of AML in mice.[124] Upregulated in t[8;21] AML and in CN-AML, mir-126 has been reported to act as an oncomiR.[125] Its overexpression increases proliferation, survival colony-forming ability, cell-cycle progression, and chemotherapy resistance, and decreases differentiation and quiescence[126–129] (Figure 11.3).

Reported overall as an oncogenic miRNA in several types of haematological cancers, mir-155 is also upregulated in several subtypes of AML, being the highest in CN-AML[130] (Figure 11.3). It acts by increasing cancer cell proliferation and survival, and its overexpression leads to myeloproliferative neoplasms in mice.[131–134] Usually overexpression of mir-155 confers negative prognosis in CN-AML. mir-196b was reported to act as an oncomiR in MLL-rearranged AML, increasing cancer cell proliferation and survival, and reducing differentiation[135,136] (Figure 11.3).

mir-29b is downregulated in nearly all AML subtypes investigated, acting as a tumour suppressor, as reported in other haematological malignancies, such as CLL (Figure 11.3). It inhibits cell growth, leukaemic progression *in vivo* and global DNA hypermethylation,[137] and induces cell death by apoptosis.[138–141] mir-146a has been shown to be downregulated in several types of AML, acting mainly as a tumour suppressor by decreasing cell proliferation and survival and nuclear factor-κB (NF-κB) signalling, while its deletion leads to myeloproliferation in mice[142–147] (Figure 11.3).

mir-193a acts mainly as a tumour suppressor in several subtypes of AML, reducing cell growth, cell cycle progression, and global DNA hypermethylation, while increasing cell differentiation and apoptosis[148,149] (Figure 11.3). Finally, mir-223 is a tumour suppressor, downregulated in several subtypes of AML, especially in t[8;21]. It acts by reducing cancer cell proliferation and increasing cell death by apoptosis and cell differentiation[150–153] (Figure 11.3).

With all these miRNAs differentially expressed involved in AML, it is evident to think of them as important tools in the clinic, such as potential biomarkers or therapeutic targets. Indeed, there have been advances in this field. For example, Mi *et al.* performed a large-scale genome-wide miRNA expression profiling assay and found 27 miRNAs that were differentially expressed between ALL and AML.[154] Among them, mir-128a/b was overexpressed, while *let-7* and mir-223 were downregulated in ALL compared to AML. According to the authors, analysing the expression of at least two of these miRNAs was able to distinguish ALL from AML in >95% of cases.

Another study performed miRNA expression profiling of more than 230 AML patients to identify potential markers correlated with clinical outcome. They found that high levels of mir-196b and mir-644 were independently associated with shorter overall survival, and low levels of mir-135a and mir-409-3p with a higher risk of relapse.[155]

Sun *et al.* tried to identify miRNAs that were consistent significant prognostic determinants, independent from other known prognostic factors.

They reported mir-212 as an independent prognostic factor, significantly associated with a prolonged overall survival and relapse-free survival in a discovery cohort that was subsequently confirmed in an independent validation cohort.[156] Guo *et al.* identified five small non-coding RNAs that predicted myelodysplastic syndrome (MDS) patients that were likely to progress to AML, among them three members of the mir-181 family.[157]

Other studies focused on identifying biomarkers that were sensitive enough to identify minimal residual disease (MRD). MRD is the presence of a small number of leukaemic blasts in the bone marrow after chemotherapy, largely responsible for refractory disease. Circ-miRNAs were proposed as early identifiers of these blasts before they reach widespread bloodstream, since the pattern of expression of circ-miRNAs in plasma of AML patients is different from healthy donors.[158,159] Unfortunately, there is some divergence between publications, making further investigation needed.

miRNA-based therapy of targeted delivery of mir-29b has recently shown success *in vitro* and in mice engrafted with human AML cell lines. mir-29b was delivered through transferrin-conjugated lipid nanoparticles into the bloodstream, leading to decreased leukaemic cell growth and survival improvement of the AML-engrafted mice. The success of the therapy *in vivo* is attributed to mir-29b downregulating a series of genes involved in a variety of cellular processes in AML, such as FLT3, and the efficient delivery method used.[157]

mir-22 mimics delivered by polyamidoamide nanoparticles treatment was shown to increase survival in xenotransplantation mouse models and in primary AML transplantation mouse models, the latter having a 40% cure rate.[116] mir-126 antagomiR therapy, delivered *via* lipid nanoparticles, increased survival in mice engrafted with primary human AML blasts.[155] mir-181a mimics, also delivered through lipid nanoparticles, increased survival and decreased splenomegaly in mice xenografted with human AML cell lines.[158] mir-21 and mir-196b antagomiR therapy, delivered *via* implanted osmotic pumps, were shown to cure primary AML transplantation mouse models and increase survival in combination with induction chemotherapy in xenotransplantation mouse models.[159] The idea of combining chemotherapy with miRNA-based therapy, with the possibility of repurposing existing drugs, could influence miRNA levels by targeting the pathways that regulate miRNA expression, powering its effect (if mimics) or downregulating its levels (if antagomiR).

Regardless of the type of haematological malignancy, miRNAs have proven to be attractive targets for disease stratification and differentiation, biomarkers for prognosis or targeted-therapy development (Figure 11.4). Their reduced size could make them rapidly bioavailable, and their array of targets enable them to regulate several signalling pathways, reducing development of resistance to therapy strategies by cancer cells. Nevertheless, unknown off-targets of miRNA-based therapy could result in a series of unwanted treatment side-effects. Recently, scientists have used mathematical approaches to predict and understand miRNA network regulation.[160] Building miRNA networks is an *in silico* method to predict with more precision the several targets

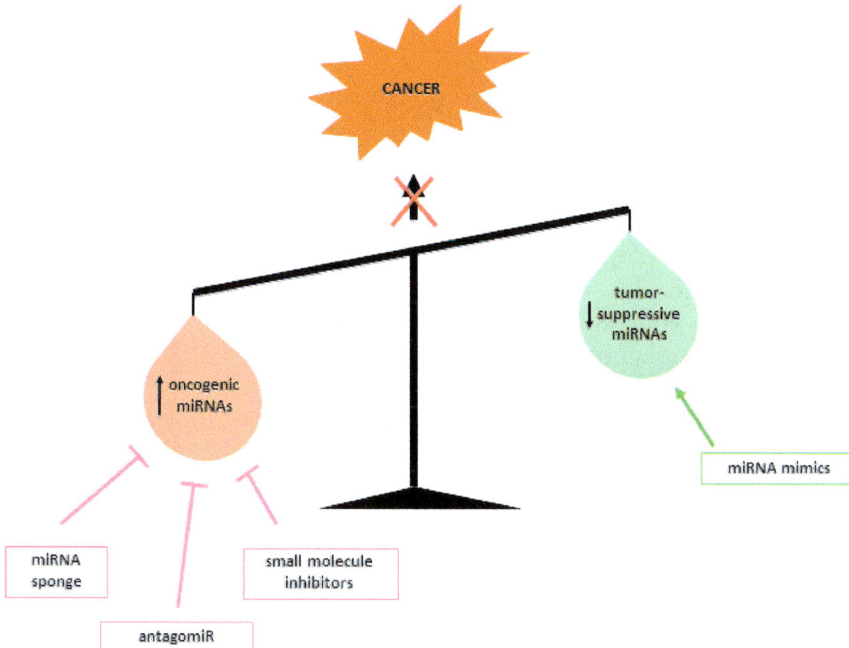

Figure 11.4 Based on the functional role of miRNAs in cancer as oncogenic miRNAs (oncomiRs) and tumor-suppressive miRNAs, therapy approaches that seek a balance between expression levels of the main miRNAs involved in cancer pathogenesis seems to be the best approach. Molecules that function as miRNA sponges, high-jacking the molecule and impeding its proper function; small molecule inhibitors; or antagomiRs, which bind to the target miRNA and repress its function, are the current possible approaches. For downregulated tumor-suppressive miRNAs, miRNA mimics is the technique developed to re-establish miRNA levels.

of a miRNA or analogous miRNAs, followed by experimental validation. It might serve as an interesting approach, given the complexity and intricacy of miRNA networks in cancer.

miRNA sponges, which high-jack the miRNA target impeding its proper function; small molecule inhibitors; and antagomiRs, antagonic small sequences that specifically target miRNAs, are approaches that are being tested *in vitro*, *in vivo* and in the clinic for targeted therapy of oncogenic miRNAs (Figure 11.4). Along the same lines, miRNA mimics have also been used to re-establish tumor-suppressive miRNA levels (Figure 11.4).

Nevertheless, the biggest challenge faced by targeted therapy in cancer to date is effective delivery methods that have specificity, efficacy, and safety, the latter including avoidance of unwanted immune reactions. Nanoparticles have been tested in pre-clinical models and show promising results, especially in the examples presented for AML. Notwithstanding this, miRNA-based therapy and the development of efficient biomarkers are beginning to enter into the clinic, with a promising future ahead for greatly improved cancer diagnosis and therapy.[161]

Acknowledgement

Dr Calin is the Felix L. Haas Endowed Professor in Basic Science. Work in Dr Calin's laboratory is supported by National Institutes of Health (NIH/NCATS) grant UH3TR00943-01 through the NIH Common Fund, Office of Strategic Coordination (OSC), the NCI grants 1R01 CA182905-01 and 1R01CA222007-01A1, an NIGMS 1R01GM122775-01 grant, a U54 grant #CA096297/CA096300 – UPR/MDACC Partnership for Excellence in Cancer Research 2016 Pilot Project, a Team DOD (CA160445P1) grant, a Ladies Leukemia League grant, a Chronic Lymphocytic Leukemia Moonshot Flagship project, a Sister Institution Network Fund (SINF) 2017 grant, and the Estate of C. G. Johnson, Jr. Ana C. P. Mafra is supported by Fundação de Amparo a Pesquisa do Estado de São Paulo (FAPESP), process #2016/06625-0, while conducting research at LNBio/CNPEM, Brazil; and process #2016/23301-4 while conducting research as part of an internship at Dr Calin's laboratory.

References

1. V. Ambros and X. Chen, *Development*, 2007, **134**, 1635–1641.
2. L. A. O'Neill, F. J. Sheedy and C. E. McCoy, *Nat. Rev. Immunol.*, 2011, **11**, 163–175.
3. R. C. Lee, R. L. Feinbaum and V. Ambros, *Cell*, 1993, **75**, 843–854.
4. B. J. Reinhart, F. J. Slack, M. Basson, A. E. Pasquinelli, J. C. Bettinger, A. E. Rougvie, H. R. Horvitz and G. Ruvkun, *Nature*, 2000, **403**, 901–906.
5. M. Pichler and G. A. Calin, *Br. J. Cancer*, 2015, **113**, 569–573.
6. C. Sevignani, G. A. Calin, S. C. Nnadi, M. Shimizu, R. V. Davuluri, T. Hyslop, P. Demant, C. M. Croce and L. D. Siracusa, *Proc. Natl. Acad. Sci. U. S. A.*, 2007, **104**, 8017–8022.
7. G. A. Calin, C. D. Dumitru, M. Shimizu, R. Bichi, S. Zupo, E. Noch, H. Aldler, S. Rattan, M. Keating, K. Rai, L. Rassenti, T. Kipps, M. Negrini, F. Bullrich and C. M. Croce, *Proc. Natl. Acad. Sci. U. S. A.*, 2002, **99**, 15524–15529.
8. Y. Lee, M. Kim, J. Han, K. H. Yeom, S. Lee, S. H. Baek and V. N. Kim, *EMBO J.*, 2004, **23**, 4051–4060.
9. L. A. Macfarlane and P. R. Murphy, *Curr. Genomics*, 2010, **11**, 537–561.
10. Y. Peng and C. M. Croce, *Signal Transduction Targeted Ther.*, 2016, **1**, 15004.
11. D. P. Bartel, *Cell*, 2009, **136**, 215–233.
12. A. Helwak, G. Kudla, T. Dudnakova and D. Tollervey, *Cell*, 2013, **153**, 654–665.
13. M. P. Dragomir, E. Knutsen and G. A. Calin, *Cell*, 2018, **174**, 1038–1038.e1.
14. S. Djuranovic, A. Nahvi and R. Green, *Science*, 2012, **336**, 237–240.
15. A. A. Bazzini, M. T. Lee and A. J. Giraldez, *Science*, 2012, **336**, 233–237.
16. D. Hanahan and R. A. Weinberg, *Cell*, 2011, **144**, 646–674.
17. H. Tagawa and M. Seto, *Leukemia*, 2005, **19**, 2013–2016.

18. Y. Hayashita, H. Osada, Y. Tatematsu, H. Yamada, K. Yanagisawa, S. Tomida, Y. Yatabe, K. Kawahara, Y. Sekido and T. Takahashi, *Cancer Res.*, 2005, **65**, 9628–9632.
19. K. J. Mavrakis, A. L. Wolfe, E. Oricchio, T. Palomero, K. de Keersmaecker, K. McJunkin, J. Zuber, T. James, A. A. Khan, C. S. Leslie, J. S. Parker, P. J. Paddison, W. Tam, A. Ferrando and H. G. Wendel, *Nat. Cell Biol.*, 2010, **12**, 372–379.
20. B. P. Lewis, I. H. Shih, M. W. Jones-Rhoades, D. P. Bartel and C. B. Burge, *Cell*, 2003, **115**, 787–798.
21. K. A. O'Donnell, E. A. Wentzel, K. I. Zeller, C. V. Dang and J. T. Mendell, *Nature*, 2005, **435**, 839–843.
22. P. S. Eis, W. Tam, L. Sun, A. Chadburn, Z. Li, M. F. Gomez, E. Lund and J. E. Dahlberg, *Proc. Natl. Acad. Sci. U. S. A.*, 2005, **102**, 3627–3632.
23. M. Metzler, M. Wilda, K. Busch, S. Viehmann and A. Borkhardt, *Genes, Chromosomes Cancer*, 2004, **39**, 167–169.
24. S. A. Ciafre, S. Galardi, A. Mangiola, M. Ferracin, C. G. Liu, G. Sabatino, M. Negrini, G. Maira, C. M. Croce and M. G. Farace, *Biochem. Biophys. Res. Commun.*, 2005, **334**, 1351–1358.
25. J. A. Chan, A. M. Krichevsky and K. S. Kosik, *Cancer Res.*, 2005, **65**, 6029–6033.
26. A. Cimmino, G. A. Calin, M. Fabbri, M. V. Iorio, M. Ferracin, M. Shimizu, S. E. Wojcik, R. I. Aqeilan, S. Zupo, M. Dono, L. Rassenti, H. Alder, S. Volinia, C. G. Liu, T. J. Kipps, M. Negrini and C. M. Croce, *Proc. Natl. Acad. Sci. U. S. A.*, 2005, **102**, 13944–13949.
27. U. Klein, M. Lia, M. Crespo, R. Siegel, Q. Shen, T. Mo, A. Ambesi-Impiombato, A. Califano, A. Migliazza, G. Bhagat and R. Dalla-Favera, *Cancer Cell*, 2010, **17**, 28–40.
28. J. Takamizawa, H. Konishi, K. Yanagisawa, S. Tomida, H. Osada, H. Endoh, T. Harano, Y. Yatabe, M. Nagino, Y. Nimura, T. Mitsudomi and T. Takahashi, *Cancer Res.*, 2004, **64**, 3753–3756.
29. S. M. Johnson, H. Grosshans, J. Shingara, M. Byrom, R. Jarvis, A. Cheng, E. Labourier, K. L. Reinert, D. Brown and F. J. Slack, *Cell*, 2005, **120**, 635–647.
30. American Cancer Society, *Cancer Facts & Figures 2018*, available at https://www.cancer.org/content/dam/cancer-org/research/cancer-facts-and-statistics/annual-cancer-facts-and-figures/2018/cancer-facts-and-figures-2018.pdf, accessed in July 2018.
31. R. W. Georgantas, R. Hildreth, S. Morisot, J. Alder, C. G. Liu, S. Heimfeld, G. A. Calin, C. M. Croce and C. I. Civin, *Proc. Natl. Acad. Sci. U. S. A.*, 2007, **104**, 2750–2755.
32. F. Fazi, A. Rosa, A. Fatica, V. Gelmetti, M. L. De Marchis, C. Nervi and I. Bozzoni, *Cell*, 2005, **123**, 819–831.
33. L. Fontana, E. Pelosi, P. Greco, S. Racanicchi, U. Testa, F. Liuzzi, C. M. Croce, E. Brunetti, F. Grignani and C. Peschle, *Nat. Cell Biol.*, 2007, **9**, 775–787.
34. H. Bruchova, D. Yoon, A. M. Agarwal, J. Mendell and J. T. Prchal, *Exp. Hematol.*, 2007, **35**, 1657–1667.

35. N. Felli, L. Fontana, E. Pelosi, R. Botta, D. Bonci, F. Facchiano, F. Liuzzi, V. Lulli, O. Morsilli, S. Santoro, M. Valtieri, G. A. Calin, C. G. Liu, A. Sorrentino, C. M. Croce and C. Peschle, *Proc. Natl. Acad. Sci. U. S. A.*, 2005, **102**, 18081–18086.
36. R. Garzon, F. Pichiorri, T. Palumbo, R. Iuliano, A. Cimmino, R. Aqeilan, S. Volinia, D. Bhatt, H. Alder, G. Marcucci, G. A. Calin, C. G. Liu, C. D. Bloomfield, M. Andreeff and C. M. Croce, *Proc. Natl. Acad. Sci. U. S. A.*, 2006, **103**, 5078–5083.
37. S. Monticelli, K. M. Ansel, C. Xiao, N. D. Socci, A. M. Krichevsky, T. H. Thai, N. Rajewsky, D. S. Marks, C. Sander, K. Rajewsky, A. Rao and K. S. Kosik, *Genome Biol.*, 2005, **6**, R71.
38. B. Zhou, S. Wang, C. Mayr, D. P. Bartel and H. F. Lodish, *Proc. Natl. Acad. Sci. U. S. A.*, 2007, **104**, 7080–7085.
39. C. Xiao, D. P. Calado, G. Galler, T. H. Thai, H. C. Patterson, J. Wang, N. Rajewsky, T. P. Bender and K. Rajewsky, *Cell*, 2016, **165**, 1027.
40. S. Costinean, N. Zanesi, Y. Pekarsky, E. Tili, S. Volinia, N. Heerema and C. M. Croce, *Proc. Natl. Acad. Sci. U. S. A.*, 2006, **103**, 7024–7029.
41. T. H. Thai, D. P. Calado, S. Casola, K. M. Ansel, C. Xiao, Y. Xue, A. Murphy, D. Frendewey, D. Valenzuela, J. L. Kutok, M. Schmidt-Supprian, N. Rajewsky, G. Yancopoulos, A. Rao and K. Rajewsky, *Science*, 2007, **316**, 604–608.
42. A. Rodriguez, E. Vigorito, S. Clare, M. V. Warren, P. Couttet, D. R. Soond, S. van Dongen, R. J. Grocock, P. P. Das, E. A. Miska, D. Vetrie, K. Okkenhaug, A. J. Enright, G. Dougan, M. Turner and A. Bradley, *Science*, 2007, **316**, 608–611.
43. C. Z. Chen, L. Li, H. F. Lodish and D. P. Bartel, *Science*, 2004, **303**, 83–86.
44. J. R. Neilson, G. X. Zheng, C. B. Burge and P. A. Sharp, *Genes Dev.*, 2007, **21**, 578–589.
45. S. H. Ramkissoon, L. A. Mainwaring, Y. Ogasawara, K. Keyvanfar, J. P. McCoy, E. M. Sloand, S. Kajigaya and N. S. Young, *Leuk. Res.*, 2006, **30**, 643–647.
46. M. Fabbri, R. Garzon, M. Andreeff, H. M. Kantarjian, G. Garcia-Manero and G. A. Calin, *Leukemia*, 2008, **22**, 1095–1105.
47. K. Van Roosbroeck and G. A. Calin, *Semin. Oncol.*, 2016, **43**, 209–214.
48. H. Döhner, D. J. Weisdorf and C. D. Bloomfield, *N. Engl. J. Med.*, 2015, **373**, 1136–1152.
49. M. Ciccone, A. Ferrajoli, M. J. Keating and G. A. Calin, *Cancer Cell*, 2014, **26**, 770–770.e1.
50. G. A. Calin and C. M. Croce, *Semin. Oncol.*, 2006, **33**, 167–173.
51. G. A. Calin and C. M. Croce, *Blood*, 2009, **114**, 4761–4770.
52. H. Döhner, S. Stilgenbauer, A. Benner, E. Leupolt, A. Kröber, L. Bullinger, K. Döhner, M. Bentz and P. Lichter, *N. Engl. J. Med.*, 2000, **343**, 1910–1916.
53. F. Bullrich, H. Fujii, G. Calin, H. Mabuchi, M. Negrini, Y. Pekarsky, L. Rassenti, H. Alder, J. C. Reed, M. J. Keating, T. J. Kipps and C. M. Croce, *Cancer Res.*, 2001, **61**, 6640–6648.

54. D. Mertens, S. Wolf, P. Schroeter, C. Schaffner, H. Döhner, S. Stilgenbauer and P. Lichter, *Blood*, 2002, **99**, 4116–4121.
55. A. Migliazza, F. Bosch, H. Komatsu, E. Cayanis, S. Martinotti, E. Toniato, E. Guccione, X. Qu, M. Chien, V. V. Murty, G. Gaidano, G. Inghirami, P. Zhang, S. Fischer, S. M. Kalachikov, J. Russo, I. Edelman, A. Efstratiadis and R. Dalla-Favera, *Blood*, 2001, **97**, 2098–2104.
56. G. Rondeau, I. Moreau, S. Bézieau, J. L. Petit, R. Heilig, S. Fernandez, E. Pennarun, J. S. Myers, M. A. Batzer, J. P. Moisan and M. C. Devilder, *Mutat. Res.*, 2001, **458**, 55–70.
57. G. A. Calin, A. Cimmino, M. Fabbri, M. Ferracin, S. E. Wojcik, M. Shimizu, C. Taccioli, N. Zanesi, R. Garzon, R. I. Aqeilan, H. Alder, S. Volinia, L. Rassenti, X. Liu, C. G. Liu, T. J. Kipps, M. Negrini and C. M. Croce, *Proc. Natl. Acad. Sci. U. S. A.*, 2008, **105**, 5166–5171.
58. M. Negrini, G. Cutrona, C. Bassi, S. Fabris, B. Zagatti, M. Colombo, M. Ferracin, L. D'Abundo, E. Saccenti, S. Matis, M. Lionetti, L. Agnelli, M. Gentile, A. G. Recchia, S. Bossio, D. Reverberi, G. Rigolin, G. A. Calin, S. Sabbioni, G. Russo, P. Tassone, F. Morabito, M. Ferrarini and A. Neri, *Clin. Cancer Res.*, 2014, **20**, 4141–4153.
59. T. Stankovic and A. Skowronska, *Leuk. Lymphoma*, 2014, **55**, 1227–1239.
60. D. C. Corney, A. Flesken-Nikitin, A. K. Godwin, W. Wang and A. Y. Nikitin, *Cancer Res.*, 2007, **67**, 8433–8438.
61. V. Balatti, Y. Pekarky, L. Rizzotto and C. M. Croce, *Adv. Exp. Med. Biol.*, 2013, **792**, 309–325.
62. S. Deneberg, M. Kanduri, D. Ali, S. Bengtzen, M. Karimi, Y. Qu, E. Kimby, L. Mansouri, R. Rosenquist, A. Lennartsson and S. Lehmann, *Epigenetics*, 2014, **9**, 910–917.
63. L. Q. Wang, Y. L. Kwong, K. F. Wong, C. S. Kho, D. Y. Jin, E. Tse, A. Rosèn and C. S. Chim, *J. Transl. Med.*, 2014, **12**, 52.
64. M. Fabbri, A. Bottoni, M. Shimizu, R. Spizzo, M. S. Nicoloso, S. Rossi, E. Barbarotto, A. Cimmino, B. Adair, S. E. Wojcik, N. Valeri, F. Calore, D. Sampath, F. Fanini, I. Vannini, G. Musuraca, M. Dell'Aquila, H. Alder, R. V. Davuluri, L. Z. Rassenti, M. Negrini, T. Nakamura, D. Amadori, N. E. Kay, K. R. Rai, M. J. Keating, T. J. Kipps, G. A. Calin and C. M. Croce, *JAMA*, 2011, **305**, 59–67.
65. K. Suzuki and H. Matsubara, *J. Biomed. Biotechnol.*, 2011, **2011**, 978312.
66. M. Mraz, K. Malinova, J. Kotaskova, S. Pavlova, B. Tichy, J. Malcikova, K. Stano Kozubik, J. Smardova, Y. Brychtova, M. Doubek, M. Trbusek, J. Mayer and S. Pospisilova, *Leukemia*, 2009, **23**, 1159–1163.
67. L. He, X. He, L. P. Lim, E. de Stanchina, Z. Xuan, Y. Liang, W. Xue, L. Zender, J. Magnus, D. Ridzon, A. L. Jackson, P. S. Linsley, C. Chen, S. W. Lowe, M. A. Cleary and G. J. Hannon, *Nature*, 2007, **447**, 1130–1134.
68. S. Rossi, M. Shimizu, E. Barbarotto, M. S. Nicoloso, F. Dimitri, D. Sampath, M. Fabbri, S. Lerner, L. L. Barron, L. Z. Rassenti, L. Jiang, L. Xiao, J. Hu, P. Secchiero, G. Zauli, S. Volinia, M. Negrini, W. Wierda, T. J. Kipps, W. Plunkett, K. R. Coombes, L. V. Abruzzo, M. J. Keating and G. A. Calin, *Blood*, 2010, **116**, 945–952.

69. V. Balatti, L. Rizzotto, C. Miller, A. Palamarchuk, P. Fadda, R. Pandolfo, L. Z. Rassenti, E. Hertlein, A. S. Ruppert, A. Lozanski, G. Lozanski, T. J. Kipps, J. C. Byrd, C. M. Croce and Y. Pekarsky, *Proc. Natl. Acad. Sci. U. S. A.*, 2015, **112**, 2169–2174.
70. A. Bottoni and G. A. Calin, *Microrna*, 2014, **2**, 158–164.
71. T. L. Parker and M. P. Strout, *Discovery Med.*, 2011, **11**, 115–123.
72. E. Porpaczy, M. Bilban, G. Heinze, M. Gruber, K. Vanura, I. Schwarzinger, S. Stilgenbauer, B. Streubel, C. Fonatsch and U. Jaeger, *Eur. J. Clin. Invest.*, 2009, **39**, 568–575.
73. R. Visone, L. Z. Rassenti, A. Veronese, C. Taccioli, S. Costinean, B. D. Aguda, S. Volinia, M. Ferracin, J. Palatini, V. Balatti, H. Alder, M. Negrini, T. J. Kipps and C. M. Croce, *Blood*, 2009, **114**, 3872–3879.
74. N. Chiorazzi, K. R. Rai and M. Ferrarini, *N. Engl. J. Med.*, 2005, **352**, 804–815.
75. G. A. Calin, M. Ferracin, A. Cimmino, G. Di Leva, M. Shimizu, S. E. Wojcik, M. V. Iorio, R. Visone, N. I. Sever, M. Fabbri, R. Iuliano, T. Palumbo, F. Pichiorri, C. Roldo, R. Garzon, C. Sevignani, L. Rassenti, H. Alder, S. Volinia, C. G. Liu, T. J. Kipps, M. Negrini and C. M. Croce, *N. Engl. J. Med.*, 2005, **353**, 1793–1801.
76. S. Marton, M. R. Garcia, C. Robello, H. Persson, F. Trajtenberg, O. Pritsch, C. Rovira, H. Naya, G. Dighiero and A. Cayota, *Leukemia*, 2008, **22**, 330–338.
77. N. Papakonstantinou, S. Ntoufa, E. Chartomatsidou, G. Papadopoulos, A. Hatzigeorgiou, A. Anagnostopoulos, K. Chlichlia, P. Ghia, M. Muzio, C. Belessi and K. Stamatopoulos, *Mol. Med.*, 2013, **19**, 115–123.
78. V. Fulci, S. Chiaretti, M. Goldoni, G. Azzalin, N. Carucci, S. Tavolaro, L. Castellano, A. Magrelli, F. Citarella, M. Messina, R. Maggio, N. Peragine, S. Santangelo, F. R. Mauro, P. Landgraf, T. Tuschl, D. B. Weir, M. Chien, J. J. Russo, J. Ju, R. Sheridan, C. Sander, M. Zavolan, A. Guarini, R. Foà and G. Macino, *Blood*, 2007, **109**, 4944–4951.
79. D. X. Zhu, K. R. Miao, C. Fang, L. Fan, W. Zhu, H. Y. Zhu, Y. Zhuang, M. Hong, P. Liu, W. Xu and J. Y. Li, *Leuk. Res.*, 2011, **35**, 730–734.
80. S. Li, H. F. Moffett, J. Lu, L. Werner, H. Zhang, J. Ritz, D. Neuberg, K. W. Wucherpfennig, J. R. Brown and C. D. Novina, *PLoS One*, 2011, **6**, e16956.
81. B. Stamatopoulos, N. Meuleman, B. Haibe-Kains, P. Saussoy, E. Van Den Neste, L. Michaux, P. Heimann, P. Martiat, D. Bron and L. Lagneaux, *Blood*, 2009, **113**, 5237–5245.
82. K. Zhou, S. Yi, Z. Yu, Z. Li, Y. Wang, D. Zou, J. Qi, Y. Zhao and L. Qiu, *Leuk. Lymphoma*, 2012, **53**, 1155–1161.
83. M. Mraz, L. Chen, L. Z. Rassenti, E. M. Ghia, H. Li, K. Jepsen, E. N. Smith, K. Messer, K. A. Frazer and T. J. Kipps, *Blood*, 2014, **124**, 84–95.
84. B. Stamatopoulos, M. Van Damme, E. Crompot, B. Dessars, H. E. Housni, P. Mineur, N. Meuleman, D. Bron and L. Lagneaux, *Mol. Med.*, 2015, **21**, 123–133.

85. T. Kawaguchi, S. Komatsu, D. Ichikawa, M. Tsujiura, H. Takeshita, S. Hirajima, M. Miyamae, W. Okajima, T. Ohashi, T. Imamura, J. Kiuchi, H. Konishi, A. Shiozaki, K. Okamoto and E. Otsuji, *Int. J. Mol. Sci.*, 2016, **17**(9), 1459.
86. M. Dragomir and G. A. Calin, *Front. Oncol.*, 2018, **8**, 179.
87. S. Anfossi, A. Babayan, K. Pantel and G. A. Calin, *Nat. Rev. Clin. Oncol.*, 2018, **15**(9), 541–563.
88. E. Moussay, K. Wang, J. H. Cho, K. van Moer, S. Pierson, J. Paggetti, P. V. Nazarov, V. Palissot, L. E. Hood, G. Berchem and D. J. Galas, *Proc. Natl. Acad. Sci. U. S. A.*, 2011, **108**, 6573–6578.
89. A. Ferrajoli, T. D. Shanafelt, C. Ivan, M. Shimizu, K. G. Rabe, N. Nouraee, M. Ikuo, A. K. Ghosh, S. Lerner, L. Z. Rassenti, L. Xiao, J. Hu, J. M. Reuben, S. Calin, M. J. You, J. T. Manning, W. G. Wierda, Z. Estrov, S. O'Brien, T. J. Kipps, M. J. Keating, N. E. Kay and G. A. Calin, *Blood*, 2013, **122**, 1891–1899.
90. N. Jain and S. O'Brien, *Blood Rev.*, 2016, **30**, 233–244.
91. G. Del Poeta, M. Postorino, L. Pupo, M. I. Del Principe, M. Dal Bo, T. Bittolo, F. Buccisano, B. Mariotti, E. Iannella, L. Maurillo, A. Venditti, V. Gattei, P. de Fabritiis, M. Cantonetti and S. Amadori, *Drugs Today*, 2016, **52**, 249–260.
92. F. Cervantes-Gomez, B. Lamothe, J. A. Woyach, W. G. Wierda, M. J. Keating, K. Balakrishnan and V. Gandhi, *Clin. Cancer Res.*, 2015, **21**, 3705–3715.
93. N. Jain, P. A. Thompson, A. Ferrajoli, J. A. Burger, G. Borthakur, K. Takahashi, P. Bose, Z. Estrov, E. J. Jabbour, M. Konopleva, Y. Alvarado, T. Kadia, M. Yilmaz, C. D. DiNardo, M. Ohanian, J. E. Cortes, R. Kanagal-Shamanna, K. Patel, N. Garg, N. X. Wang, N. Fru, N. Cruz, V. Gandhi, W. Plunkett, H. M. Kantarjian, M. J. Keating and W. G. Wierda, Blood Abstracts and Meeting Program - 59th ASH Annual Meeting, 2017, p. 469.
94. X. C. Badoux, M. J. Keating, X. Wang, S. M. O'Brien, A. Ferrajoli, S. Faderl, J. Burger, C. Koller, S. Lerner, H. Kantarjian and W. G. Wierda, *Blood*, 2011, **117**, 3016–3024.
95. T. Zenz, J. Mohr, E. Eldering, A. P. Kater, A. Bühler, D. Kienle, D. Winkler, J. Dürig, M. H. van Oers, D. Mertens, H. Döhner and S. Stilgenbauer, *Blood*, 2009, **113**, 3801–3808.
96. D. Asslaber, J. D. Piñón, I. Seyfried, P. Desch, M. Stöcher, I. Tinhofer, A. Egle, O. Merkel and R. Greil, *Blood*, 2010, **115**, 4191–4197.
97. R. Visone, A. Veronese, V. Balatti and C. M. Croce, *Oncotarget*, 2012, **3**, 195–202.
98. D. X. Zhu, W. Zhu, C. Fang, L. Fan, Z. J. Zou, Y. H. Wang, P. Liu, M. Hong, K. R. Miao, W. Xu and J. Y. Li, *Carcinogenesis*, 2012, **33**, 1294–1301.
99. D. Jurkovicova, M. Magyerkova, L. Kulcsar, M. Krivjanska, V. Krivjansky, A. Gibadulinova, I. Oveckova and M. Chovanec, *Neoplasma*, 2014, **61**, 241–251.
100. L. Chen, K. Jiang, H. Jiang and P. Wei, *Exp. Ther. Med.*, 2014, **8**, 527–532.

101. D. D. Yu, M. M. Lv, W. X. Chen, S. L. Zhong, X. H. Zhang, L. Chen, T. F. Ma, J. H. Tang and J. H. Zhao, *Tumor Biol.*, 2015, **36**, 1395–1401.
102. E. Fonte, B. Apollonio, L. Scarfò, P. Ranghetti, C. Fazi, P. Ghia, F. Caligaris-Cappio and M. Muzio, *Clin. Cancer Res.*, 2013, **19**, 367–379.
103. M. Ferracin, B. Zagatti, L. Rizzotto, F. Cavazzini, A. Veronese, M. Ciccone, E. Saccenti, L. Lupini, A. Grilli, C. De Angeli, M. Negrini and A. Cuneo, *Mol. Cancer*, 2010, **9**, 123.
104. A. M. Tsimberidou, M. J. Keating, C. E. Bueso-Ramos and R. Kurzrock, *Leuk. Lymphoma*, 2006, **47**, 827–836.
105. J. J. Tarrand, M. J. Keating, A. M. Tsimberidou, S. O'Brien, R. P. LaSala, X. Y. Han and C. E. Bueso-Ramos, *Cancer*, 2010, **116**, 880–887.
106. A. Ferrajoli, C. Ivan, M. Ciccone, M. Shimizu, Y. Kita, M. Ohtsuka, L. D'Abundo, J. Qiang, S. Lerner, N. Nouraee, K. G. Rabe, L. Z. Rassenti, K. Van Roosbroeck, J. T. Manning, Y. Yuan, X. Zhang, T. D. Shanafelt, W. G. Wierda, S. Sabbioni, J. J. Tarrand, Z. Estrov, M. Radovich, H. Liang, M. Negrini, T. J. Kipps, N. E. Kay, M. Keating and G. A. Calin, *EBioMedicine*, 2015, **2**, 572–582.
107. H. J. Delecluse, R. Feederle, B. O'Sullivan and P. Taniere, *J. Clin. Pathol.*, 2007, **60**, 1358–1364.
108. S. D. Linnstaedt, E. Gottwein, R. L. Skalsky, M. A. Luftig and B. R. Cullen, *J. Virol.*, 2010, **84**, 11670–11678.
109. E. Fuentes-Mattei, D. E. Giza, M. Shimizu, C. Ivan, J. T. Manning, S. Tudor, M. Ciccone, O. A. Kargin, X. Zhang, P. Mur, N. S. do Amaral, M. Chen, J. J. Tarrand, F. Lupu, A. Ferrajoli, M. J. Keating, C. Vasilescu, S. J. Yeung and G. A. Calin, *EBioMedicine*, 2017, **20**, 182–192.
110. F. Ferrara and C. A. Schiffer, *Lancet*, 2013, **381**, 484–495.
111. J. A. Wallace and R. M. O'Connell, *Blood*, 2017, **130**, 1290–1301.
112. P. Chen, C. Price, Z. Li, Y. Li, D. Cao, A. Wiley, C. He, S. Gurbuxani, R. B. Kunjamma, H. Huang, X. Jiang, S. Arnovitz, M. Xu, G. M. Hong, A. G. Elkahloun, M. B. Neilly, M. Wunderlich, R. A. Larson, M. M. Le Beau, J. C. Mulloy, P. P. Liu, J. D. Rowley and J. Chen, *Proc. Natl. Acad. Sci. U. S. A.*, 2013, **110**, 11511–11516.
113. S. Emmrich, J. E. Katsman-Kuipers, K. Henke, M. E. Khatib, R. Jammal, F. Engeland, F. Dasci, C. M. Zwaan, M. L. den Boer, L. Verboon, J. Stary, A. Baruchel, V. de Haas, A. A. Danen-van Oorschot, M. Fornerod, R. Pieters, D. Reinhardt, J. H. Klusmann and M. M. van den Heuvel-Eibrink, *Leukemia*, 2014, **28**, 1022–1032.
114. V. Senyuk, Y. Zhang, Y. Liu, M. Ming, K. Premanand, L. Zhou, P. Chen, J. Chen, J. D. Rowley, G. Nucifora and Z. Qian, *Proc. Natl. Acad. Sci. U. S. A.*, 2013, **110**, 5594–5599.
115. S. J. Song, K. Ito, U. Ala, L. Kats, K. Webster, S. M. Sun, M. Jongen-Lavrencic, K. Manova-Todorova, J. Teruya-Feldstein, D. E. Avigan, R. Delwel and P. P. Pandolfi, *Cell Stem Cell*, 2013, **13**, 87–101.
116. X. Jiang, C. Hu, S. Arnovitz, J. Bugno, M. Yu, Z. Zuo, P. Chen, H. Huang, B. Ulrich, S. Gurbuxani, H. Weng, J. Strong, Y. Wang, Y. Li, J. Salat, S. Li, A. G. Elkahloun, Y. Yang, M. B. Neilly, R. A. Larson, M. M. Le Beau, T.

Herold, S. K. Bohlander, P. P. Liu, J. Zhang, Z. Li, C. He, J. Jin, S. Hong and J. Chen, *Nat. Commun.*, 2016, **7**, 11452.
117. C. Shen, M. T. Chen, X. H. Zhang, X. L. Yin, H. M. Ning, R. Su, H. S. Lin, L. Song, F. Wang, Y. N. Ma, H. L. Zhao, J. Yu and J. W. Zhang, *PLoS Genet.*, 2016, **12**, e1006259.
118. P. Wong, M. Iwasaki, T. C. Somervaille, F. Ficara, C. Carico, C. Arnold, C. Z. Chen and M. L. Cleary, *Cancer Res.*, 2010, **70**, 3833–3842.
119. S. Mi, Z. Li, P. Chen, C. He, D. Cao, A. Elkahloun, J. Lu, L. A. Pelloso, M. Wunderlich, H. Huang, R. T. Luo, M. Sun, M. He, M. B. Neilly, N. J. Zeleznik-Le, M. J. Thirman, J. C. Mulloy, P. P. Liu, J. D. Rowley and J. Chen, *Proc. Natl. Acad. Sci. U. S. A.*, 2010, **107**, 3710–3715.
120. M. Bousquet, C. Quelen, R. Rosati, V. Mansat-De Mas, R. La Starza, C. Bastard, E. Lippert, P. Talmant, M. Lafage-Pochitaloff, D. Leroux, C. Gervais, F. Viguié, J. L. Lai, C. Terre, B. Beverlo, C. Sambani, A. Hagemeijer, P. Marynen, G. Delsol, N. Dastugue, C. Mecucci and P. Brousset, *J. Exp. Med.*, 2008, **205**, 2499–2506.
121. R. M. O'Connell, A. A. Chaudhuri, D. S. Rao, W. S. Gibson, A. B. Balazs and D. Baltimore, *Proc. Natl. Acad. Sci. U. S. A.*, 2010, **107**, 14235–14240.
122. A. A. Chaudhuri, A. Y. So, A. Mehta, A. Minisandram, N. Sinha, V. D. Jonsson, D. S. Rao, R. M. O'Connell and D. Baltimore, *Proc. Natl. Acad. Sci. U. S. A.*, 2012, **109**, 4233–4238.
123. H. Zhang, X. Q. Luo, P. Zhang, L. B. Huang, Y. S. Zheng, J. Wu, H. Zhou, L. H. Qu, L. Xu and Y. Q. Chen, *PLoS One*, 2009, **4**, e7826.
124. A. Y. So, R. Sookram, A. A. Chaudhuri, A. Minisandram, D. Cheng, C. Xie, E. L. Lim, Y. G. Flores, S. Jiang, J. T. Kim, C. Keown, P. Ramakrishnan and D. Baltimore, *Blood*, 2014, **124**, 1502–1512.
125. Z. Li, J. Lu, M. Sun, S. Mi, H. Zhang, R. T. Luo, P. Chen, Y. Wang, M. Yan, Z. Qian, M. B. Neilly, J. Jin, Y. Zhang, S. K. Bohlander, D. E. Zhang, R. A. Larson, M. M. Le Beau, M. J. Thirman, T. R. Golub, J. D. Rowley and J. Chen, *Proc. Natl. Acad. Sci. U. S. A.*, 2008, **105**, 15535–15540.
126. A. M. Dorrance, P. Neviani, G. J. Ferenchak, X. Huang, D. Nicolet, K. S. Maharry, H. G. Ozer, P. Hoellarbauer, J. Khalife, E. B. Hill, M. Yadav, B. N. Bolon, R. J. Lee, L. J. Lee, C. M. Croce, R. Garzon, M. A. Caligiuri, C. D. Bloomfield and G. Marcucci, *Leukemia*, 2015, **29**, 2143–2153.
127. Z. Li, P. Chen, R. Su, Y. Li, C. Hu, Y. Wang, S. Arnovitz, M. He, S. Gurbuxani, Z. Zuo, A. G. Elkahloun, S. Li, H. Weng, H. Huang, M. B. Neilly, S. Wang, E. N. Olson, R. A. Larson, M. M. Le Beau, J. Zhang, X. Jiang, M. Wei, J. Jin, P. P. Liu and J. Chen, *Blood*, 2015, **126**, 2005–2015.
128. E. R. Lechman, B. Gentner, S. W. K. Ng, E. M. Schoof, P. van Galen, J. A. Kennedy, S. Nucera, F. Ciceri, K. B. Kaufmann, N. Takayama, S. M. Dobson, A. Trotman-Grant, G. Krivdova, J. Elzinga, A. Mitchell, B. Nilsson, K. G. Hermans, K. Eppert, R. Marke, R. Isserlin, V. Voisin, G. D. Bader, P. W. Zandstra, T. R. Golub, B. L. Ebert, J. Lu, M. Minden, J. C. Y. Wang, L. Naldini and J. E. Dick, *Cancer Cell*, 2016, **29**, 602–606.

129. D. C. de Leeuw, F. Denkers, M. C. Olthof, A. P. Rutten, W. Pouwels, G. J. Schuurhuis, G. J. Ossenkoppele and L. Smit, *Cancer Res.*, 2014, **74**, 2094–2105.
130. G. Marcucci, K. S. Maharry, K. H. Metzeler, S. Volinia, Y. Z. Wu, K. Mrózek, D. Nicolet, J. Kohlschmidt, S. P. Whitman, J. H. Mendler, S. Schwind, H. Becker, A. K. Eisfeld, A. J. Carroll, B. L. Powell, J. E. Kolitz, R. Garzon, M. A. Caligiuri, R. M. Stone and C. D. Bloomfield, *J. Clin. Oncol.*, 2013, **31**, 2086–2093.
131. D. Gerloff, R. Grundler, A. A. Wurm, D. Bräuer-Hartmann, C. Katzerke, J. U. Hartmann, V. Madan, C. Müller-Tidow, J. Duyster, D. G. Tenen, D. Niederwieser and G. Behre, *Leukemia*, 2015, **29**, 535–547.
132. E. Schneider, A. Staffas, L. Röhner, K. Krowiorz, M. Heuser, K. Döhner, L. Bullinger, H. Döhner, L. Fogelstrand, A. Rouhi, F. Kuchenbauer and L. Palmqvist, *Exp. Hematol.*, 2016, **44**, 1166–1171.
133. J. A. Wallace, D. A. Kagele, A. M. Eiring, C. N. Kim, R. Hu, M. C. Runtsch, M. Alexander, T. B. Huffaker, S. H. Lee, A. B. Patel, T. L. Mosbruger, W. P. Voth, D. S. Rao, R. R. Miles, J. L. Round, M. W. Deininger and R. M. O'Connell, *Blood*, 2017, **129**, 3074–3086.
134. R. M. O'Connell, D. S. Rao, A. A. Chaudhuri, M. P. Boldin, K. D. Taganov, J. Nicoll, R. L. Paquette and D. Baltimore, *J. Exp. Med.*, 2008, **205**, 585–594.
135. Z. Li, H. Huang, P. Chen, M. He, Y. Li, S. Arnovitz, X. Jiang, C. He, E. Hyjek, J. Zhang, Z. Zhang, A. Elkahloun, D. Cao, C. Shen, M. Wunderlich, Y. Wang, M. B. Neilly, J. Jin, M. Wei, J. Lu, P. J. M. Valk, R. Delwel, B. Lowenberg, M. M. Le Beau, J. Vardiman, J. C. Mulloy, N. J. Zeleznik-Le, P. P. Liu and J. Chen, *Nat. Commun.*, 2018, **9**, 16192.
136. R. Popovic, L. E. Riesbeck, C. S. Velu, A. Chaubey, J. Zhang, N. J. Achille, F. E. Erfurth, K. Eaton, J. Lu, H. L. Grimes, J. Chen, J. D. Rowley and N. J. Zeleznik-Le, *Blood*, 2009, **113**, 3314–3322.
137. R. Garzon, S. Liu, M. Fabbri, Z. Liu, C. E. Heaphy, E. Callegari, S. Schwind, J. Pang, J. Yu, N. Muthusamy, V. Havelange, S. Volinia, W. Blum, L. J. Rush, D. Perrotti, M. Andreeff, C. D. Bloomfield, J. C. Byrd, K. Chan, L. C. Wu, C. M. Croce and G. Marcucci, *Blood*, 2009, **113**, 6411–6418.
138. J. N. Gong, J. Yu, H. S. Lin, X. H. Zhang, X. L. Yin, Z. Xiao, F. Wang, X. S. Wang, R. Su, C. Shen, H. L. Zhao, Y. N. Ma and J. W. Zhang, *Cell Death Differ.*, 2014, **21**, 100–112.
139. R. Garzon, C. E. Heaphy, V. Havelange, M. Fabbri, S. Volinia, T. Tsao, N. Zanesi, S. M. Kornblau, G. Marcucci, G. A. Calin, M. Andreeff and C. M. Croce, *Blood*, 2009, **114**, 5331–5341.
140. S. Liu, L. C. Wu, J. Pang, R. Santhanam, S. Schwind, Y. Z. Wu, C. J. Hickey, J. Yu, H. Becker, K. Maharry, M. D. Radmacher, C. Li, S. P. Whitman, A. Mishra, N. Stauffer, A. M. Eiring, R. Briesewitz, R. A. Baiocchi, K. K. Chan, P. Paschka, M. A. Caligiuri, J. C. Byrd, C. M. Croce, C. D. Bloomfield, D. Perrotti, R. Garzon and G. Marcucci, *Cancer Cell*, 2010, **17**, 333–347.
141. M. Eyholzer, S. Schmid, L. Wilkens, B. U. Mueller and T. Pabst, *Br. J. Cancer*, 2010, **103**, 275–284.

142. D. T. Starczynowski, F. Kuchenbauer, B. Argiropoulos, S. Sung, R. Morin, A. Muranyi, M. Hirst, D. Hogge, M. Marra, R. A. Wells, R. Buckstein, W. Lam, R. K. Humphries and A. Karsan, *Nat. Med.*, 2010, **16**, 49–58.
143. D. T. Starczynowski, R. Morin, A. McPherson, J. Lam, R. Chari, J. Wegrzyn, F. Kuchenbauer, M. Hirst, K. Tohyama, R. K. Humphries, W. L. Lam, M. Marra and A. Karsan, *Blood*, 2011, **117**, 595–607.
144. M. E. Varney, M. Niederkorn, H. Konno, T. Matsumura, J. Gohda, N. Yoshida, T. Akiyama, S. Christie, J. Fang, D. Miller, A. Jerez, A. Karsan, J. P. Maciejewski, R. A. Meetei, J. Inoue and D. T. Starczynowski, *J. Exp. Med.*, 2015, **212**, 1967–1985.
145. J. L. Zhao, D. S. Rao, M. P. Boldin, K. D. Taganov, R. M. O'Connell and D. Baltimore, *Proc. Natl. Acad. Sci. U. S. A.*, 2011, **108**, 9184–9189.
146. J. Fang, B. Barker, L. Bolanos, X. Liu, A. Jerez, H. Makishima, S. Christie, X. Chen, D. S. Rao, H. L. Grimes, K. Komurov, M. T. Weirauch, J. A. Cancelas, J. P. Maciejewski and D. T. Starczynowski, *Cell Rep.*, 2014, **8**, 1328–1338.
147. M. P. Boldin, K. D. Taganov, D. S. Rao, L. Yang, J. L. Zhao, M. Kalwani, Y. Garcia-Flores, M. Luong, A. Devrekanli, J. Xu, G. Sun, J. Tay, P. S. Linsley and D. Baltimore, *J. Exp. Med.*, 2011, **208**, 1189–1201.
148. Y. Li, L. Gao, X. Luo, L. Wang, X. Gao, W. Wang, J. Sun, L. Dou, J. Li, C. Xu, M. Zhou, M. Jiang, J. Zhou, M. A. Caligiuri, C. Nervi, C. D. Bloomfield, G. Marcucci and L. Yu, *Blood*, 2013, **121**, 499–509.
149. X. N. Gao, J. Lin, Y. H. Li, L. Gao, X. R. Wang, W. Wang, H. Y. Kang, G. T. Yan, L. L. Wang and L. Yu, *Oncogene*, 2011, **30**, 3416–3428.
150. J. A. Pulikkan, V. Dengler, P. S. Peramangalam, A. A. Peer Zada, C. Müller-Tidow, S. K. Bohlander, D. G. Tenen and G. Behre, *Blood*, 2010, **115**, 1768–1778.
151. Y. Xiao, C. Su and T. Deng, *Oncol. Lett.*, 2016, **12**, 3531–3536.
152. B. Gentner, N. Pochert, A. Rouhi, F. Boccalatte, T. Plati, T. Berg, S. M. Sun, S. M. Mah, M. Mirkovic-Hösle, J. Ruschmann, A. Muranyi, S. Leierseder, B. Argiropoulos, D. T. Starczynowski, A. Karsan, M. Heuser, D. Hogge, F. D. Camargo, S. Engelhardt, H. Döhner, C. Buske, M. Jongen-Lavrencic, L. Naldini, R. K. Humphries and F. Kuchenbauer, *Exp. Hematol.*, 2015, **43**, 858–868.e7.
153. F. Fazi, S. Racanicchi, G. Zardo, L. M. Starnes, M. Mancini, L. Travaglini, D. Diverio, E. Ammatuna, G. Cimino, F. Lo-Coco, F. Grignani and C. Nervi, *Cancer Cell*, 2007, **12**, 457–466.
154. S. Mi, J. Lu, M. Sun, Z. Li, H. Zhang, M. B. Neilly, Y. Wang, Z. Qian, J. Jin, Y. Zhang, S. K. Bohlander, M. M. Le Beau, R. A. Larson, T. R. Golub, J. D. Rowley and J. Chen, *Proc. Natl. Acad. Sci. U. S. A.*, 2007, **104**, 19971–19976.
155. M. Díaz-Beyá, S. Brunet, J. Nomdedéu, R. Tejero, T. Díaz, M. Pratcorona, M. Tormo, J. M. Ribera, L. Escoda, R. Duarte, D. Gallardo, I. Heras, M. P. Queipo de Llano, J. Bargay, M. Monzo, J. Sierra, A. Navarro, J. Esteve and Cooperative AML group CETLAM (Grupo Cooperativo Para el Estudio y Tratamiento de las Leucemias Agudas y Mielodisplasias), *Leukemia*, 2014, **28**, 804–812.

156. S. M. Sun, V. Rockova, L. Bullinger, M. K. Dijkstra, H. Döhner, B. Löwenberg and M. Jongen-Lavrencic, *Leukemia*, 2013, **27**, 100–106.
157. Y. Guo, S. A. Strickland, S. Mohan, S. Li, A. Bosompem, K. C. Vickers, S. Zhao, Q. Sheng and A. S. Kim, *Leuk. Lymphoma*, 2017, **58**, 1–15.
158. C. S. Hourigan, R. P. Gale, N. J. Gormley, G. J. Ossenkoppele and R. B. Walter, *Leukemia*, 2017, **31**, 1482–1490.
159. J. A. Yin, M. A. O'Brien, R. K. Hills, S. B. Daly, K. Wheatley and A. K. Burnett, *Blood*, 2012, **120**, 2826–2835.
160. M. Dragomir, A. C. P. Mafra, S. M. G. Dias, C. Vasilescu and G. A. Calin, *Int. J. Mol. Sci.*, 2018, **19**(7), 1871.
161. M. Y. Shah, A. Ferrajoli, A. K. Sood, G. Lopez-Berestein and G. A. Calin, *EBioMedicine*, 2016, **12**, 34–42.

CHAPTER 12

Discovery and Therapeutic Relevance of Small Molecule Modifiers of MicroRNA Function

NICHOLAS ANKENBRUCK, YUTA NARO, ROHAN KUMBHARE AND ALEXANDER DEITERS*

Department of Chemistry and University of Pittsburgh Cancer Institute, Pittsburgh, Pennsylvania 15260, USA
*E-mail: deiters@pitt.edu

12.1 Introduction

RNA is involved in many essential cellular functions, thus making it an attractive therapeutic target for modulation by small molecules.[1,2] Despite being considered largely 'undruggable', recent studies have identified small molecules targeting RNAs that adopt specific conformations, contain repeat sequences, or display three-way junctions.[1–12] MicroRNAs (miRNAs) are short, non-coding RNAs that function in the post-transcriptional regulation of protein-coding genes.[13–15] In the canonical pathway, miRNAs are transcribed from the genome into primary miRNAs (pri-miRNAs). The pri-miRNA is processed into a ~60–70-nucleotide precursor miRNA

Drug Discovery Series No. 69
MicroRNAs in Diseases and Disorders: Emerging Therapeutic Targets
Edited by Philip V. Peplow, Bridget Martinez, George A. Calin and Aurora Esquela-Kerscher

Published by the Royal Society of Chemistry, www.rsc.org

(pre-miRNA) hairpin by Drosha before being exported from the nucleus. Once in the cytoplasm, the pre-miRNA is cleaved by Dicer, generating a ~22-nucleotide mature miRNA duplex comprised of a guide strand and passenger strand. The guide strand is then loaded into the RNA-induced silencing complex (RISC) which suppresses gene expression by binding to a complementary sequence within the 3′ untranslated region (UTR) of messenger RNAs (mRNAs), leading to translational repression. Because of the critical role they play in the regulation of gene expression, dysregulation of miRNAs has been implicated in various human diseases,[16] most notably cancer.[17–20] Specifically, oncogenic miRNAs (oncomiRs) target tumor suppressor genes or genes that promote apoptosis and are frequently upregulated in cancer, while tumor suppressor miRNAs target oncogenes and are often downregulated.[21] Traditional approaches such as antisense oligonucleotides and miRNA sponges have been investigated towards targeting miRNAs for therapeutic purposes, including in clinical trials;[22] however, these technologies are limited by potential immunogenic effects, poor cellular uptake, and limited chemical stability *in vivo*. Furthermore, antisense oligonucleotides and miRNA sponges are only capable of directly blocking their nucleic acid target. Limitations imposed by oligonucleotide-based therapies, together with recent success in targeting RNA and RNA-processing proteins with small molecules, has prompted the investigation and discovery of compounds that modulate nearly all stages of miRNA biogenesis (Figure 12.1).

In this review, we discuss recent approaches employed towards the discovery of new small molecules controlling miRNA function. We describe the development, potential mechanism of action, and pre-therapeutic evaluation of small molecule modulators of disease-relevant miRNAs. We also describe key design rules and potential pitfalls of methodologies used to discover miRNA-targeting small molecules.

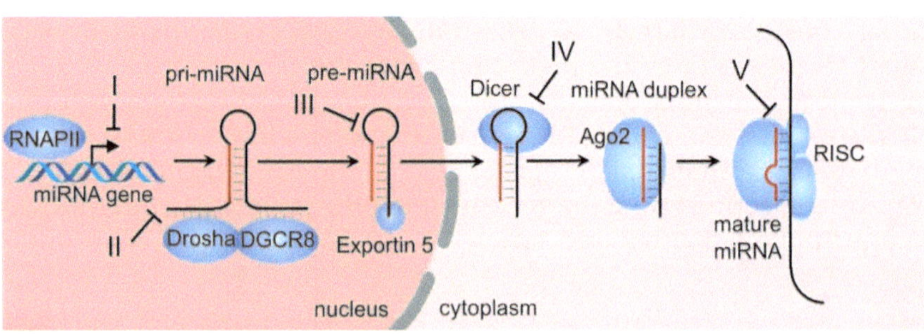

Figure 12.1 Different modes of action of small molecule modifiers of miRNA function: (I) inhibition of transcription; (II) inhibition of pri-miRNA processing; (III) direct binding to pre-miRNA, blocking further processing; (IV) inhibition of Dicer processing; (V) modulation of mature miRNA function *via* unknown mechanisms.

12.2 miRNA Inhibitors Discovered Through Biochemical Assays

The development of *in vitro* assays to monitor miRNA maturation has proven instrumental in the discovery of novel inhibitors of miRNA function (for an overview see Figure 12.2). Advantages of biochemical assays are a robust read-out, scalability for use in high-throughput screens, and the limited number of components that obviates target identification for hit compounds, as is frequently the case in cell-based phenotypic screens. Many *in vitro* assays that have been reported were designed to probe direct binding of small molecules to primary and precursor miRNAs or other protein machinery involved in miRNA biogenesis. A second group of assays has been constructed around monitoring the enzymatic function of the machinery involved in miRNA biogenesis. Taken together, these pursuits have led to the development of microarray platform assays, fluorescence-based assays, and more recently, catalytic chemiluminescence-based assays. This section describes the development and application of each of these different technologies, while also discussing the distinct advantages and disadvantages that each approach offers.

Microarray technology has provided an efficient approach to screen for specific ligand–RNA interactions in a high-throughput fashion.[23] A typical microarray involves the covalent immobilization of a ligand library onto a glass slide, which is then incubated with a fluorescently labeled RNA molecule. Following washing steps, the molecules that bind and immobilize the

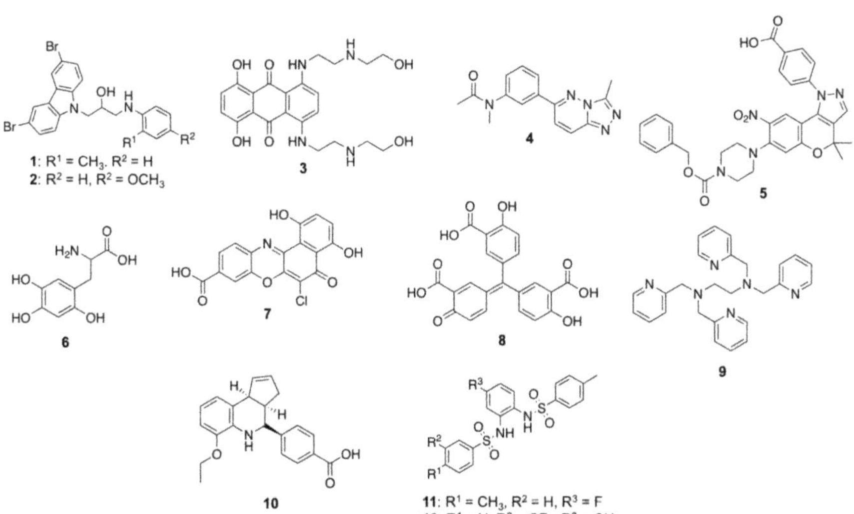

Figure 12.2 Structures of small molecule miRNA inhibitors identified through *in vitro* assays.

labeled RNA can be identified using fluorescence imaging. Conveniently, the slides can also be re-probed with control oligonucleotides to identify interactions that provide specificity for the RNA of interest. As such, the high-throughput capability, design specificity, and robust fluorescence read-out renders microarrays a valuable approach for the identification of ligand–miRNA interactions. The first successful applications of microarray assays for the identification of miRNA inhibitors was the screening of peptoids[24] and peptides.[25] Both peptoids and peptides are ideal for use in a microarray format because large, diverse libraries can be synthesized rapidly *via* automated systems and the ligands can be readily immobilized onto slides *via* the C-terminus. A recent approach described by the Schneekloth lab utilized a small molecule microarray to identify low molecular weight compounds capable of directly binding to the precursor miR-21 (pre-miR-21) hairpin.[26] A Cy5-labeled pre-miR-21 hairpin was incubated with a printed array of 20 000 small molecules, while as a negative control, a Cy5-labeled HIV TAR hairpin was employed to reveal non-specific interactions (Figure 12.3a). The screen produced 19 hits, of which 11 were selected for additional binding assays. Initially, differential scanning fluorimetry was used as a qualitative approach to monitor the binding interaction

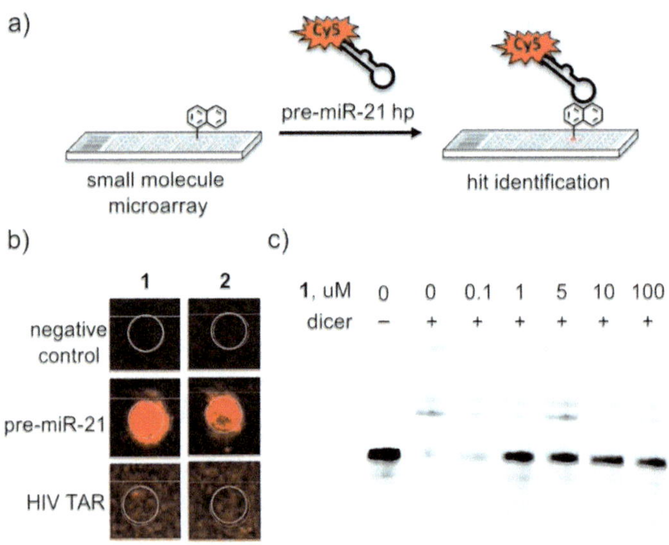

Figure 12.3 (a) Cartoon depiction of a small molecule microarray screen against the pre-miR-21 hairpin. (b) Raw images from the screen for hit compound **1** and **2** following treatment with negative control (buffer), Cy5-labeled pre-miR-21 hairpin, or a Cy5-labeled HIV TAR hairpin. (c) In-gel fluorescence of AlexFluor647-labeled pre-miR-21 cleaved with Dicer in the presence of increasing concentration of **1**. Reproduced from ref. 26 with permission from American Chemical Society, Copyright 2017.

between compound and RNA by measuring the melting temperature (T_m) of the pre-miR-21 hairpin in the presence and absence of compound, where changes in T_m indicate a direct interaction between the small molecule and the RNA. Of the 11 selected compounds, **1** and **2** (Figure 12.2) led to reductions in T_m indicating that small molecule binding to the RNA had a destabilizing effect on the hairpin structure. This mechanism contrasts with many previously reported small molecule–RNA binding interactions, such as aminoglycosides,[27] which typically stabilize the hairpin structure resulting in an increased T_m. To quantify these interactions and to provide equilibrium dissociation constants (K_d) for the two compounds, two orthogonal fluorescence titration assays were employed. A 1-aminopurine fluorescence titration assay resulted in K_d values of 2.3 μM and 800 nM for compounds **1** and **2**, respectively, while a fluorescence intensity assay produced similar results, with K_d values of 3.2 μM and 700 nM, respectively. Furthermore, both compounds demonstrated specificity for the pre-miR-21 hairpin and showed no binding to the negative control HIV TAR hairpin (Figure 12.3b); however, no negative control pre-miRNA hairpins, *e.g.*, of similar sequence, were tested offering limited insight into miRNA specificity of compounds **1** and **2**. Compound **1** was also investigated for its ability to inhibit Dicer processing *in vitro* and was found to abrogate Dicer-mediated cleavage of fluorophore labeled pre-miR-21 in a dose-dependent manner with an IC_{50} of <1 μM (Figure 12.3c). The effect of **1** on other unrelated precursor miRNAs was not explored and the biological activity of either compound has not been confirmed in cells.

To design a unique platform for probing RNA-motifs and chemical interactions simultaneously, the Disney lab developed a microarray-based two-dimensional combinatorial screening (2DCS) approach (Figure 12.4).[28,29] Using 2DCS, both RNA motifs and ligands are varied and analyzed simultaneously to identify selective interactions between small molecules and specific RNA motifs. First, a collection of azide-bearing small molecules is immobilized onto an alkyne-modified agarose microarray slide. The immobilized library is then probed for binding to a ^{32}P-labeled RNA library displaying a random 3 × 3 nucleotide internal loop pattern. Competitor RNA oligonucleotides are included to ensure that RNA–ligand interactions are confined to the random region, while competitor DNA oligonucleotides are added to ensure that the interactions are RNA-specific. After incubation and washing, the microarray slide is imaged using a phosphorimager and bound RNAs are excised from the agarose. The excised RNAs are then amplified and sequenced to determine the identity of the bound RNAs. In the original report, this approach was applied to determine RNA-binding preferences for four aminoglycosides: kanamycin A, tobramycin, neamine, and neomycin B. Disney's results demonstrated that specific RNA motifs have discreet binding preferences between the aminoglycosides screened. Neomycin preferentially interacted with GA pairs, tobramycin preferred internal loops with GG pairs, kanamycin A preferred internal loops containing pyrimidine–pyridimine pairs, and neamine preferred various internal loops, especially GA pairs.

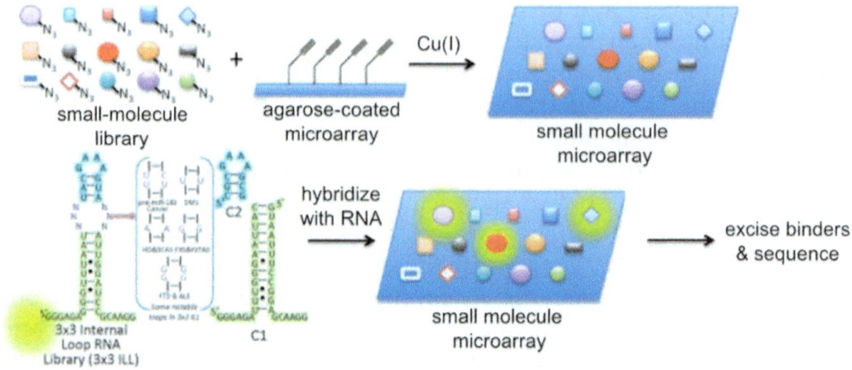

Figure 12.4 Identification of small molecules binding specific RNA motifs and structural determination of RNA motifs by two-dimensional combinatorial screening (2DCS). In 2DCS, a small molecule library is conjugated to an agarose-coated microarray and hybridized with an RNA motif library. The binding RNAs are excised from the microarray, amplified, and finally sequenced to identify small molecule–RNA binding partners. Reproduced from ref. 29 with permission from American Chemical Society, Copyright 2016.

As a proof-of-concept, this innovative study demonstrated the robust and high-throughput capabilities of 2DCS. A large amount of RNA–ligand interaction data was produced using this two-dimensional approach, prompting the development of additional bioinformatics and statistical analysis tools for larger screens. To meet this need and expand upon this technology, the Disney lab generated a more comprehensive discovery platform, which combines the screening approach of 2DCS with a method to statistically analyze selection data to identify and score RNA motif–small molecule interactions. This approach, termed 'Inforna', was successfully applied to identify several small molecule miRNA inhibitors and is described in more detail in the rational design section of this chapter (Section 12.4).

A critical limitation of the previously described small molecule microarray technologies is the requirement that the compounds be immobilized onto slides *via* a covalent tag. This requirement significantly restricts library diversity and accessibility, while also running the risk that covalent modification of the small molecule may render it inactive towards potential targets. To overcome this challenge, the Disney lab developed an approach termed 'Absorb-Array', which allows unmodified compound libraries to be non-covalently adhered onto the surface of microarray plates and probed for binding to RNA libraries in 2DCS fashion, as was described above with Inforna.[30] They found that small molecules could be successfully absorbed onto hydrated agarose-coated surfaces and demonstrated that the molecules were retained following washing steps. To validate the platform, an Inforna screen was carried out using a library of 727 pharmacologically active compounds that have been used in human clinical trials to identify small molecules that can bind to the pre-miR-21 hairpin. Three topoisomerase inhibitors were identified to bind the adenosine bulge (A-bulge) in the Dicer site of pre-miR-21 with K_d values between 24 and 58 nM, while the most potent hit, compound **3** (Figure 12.2), also bound the uracil bulge (U-bulge) adjacent to the Dicer site with a K_d of 46 nM. Compound **3** inhibited biogenesis of mature miR-21 in MDA-MB-231 breast cancer cells with an IC_{50} value of 1 μM, as revealed by quantitative reverse transcription polymerase chain reaction (RT-qPCR) (Figure 12.5a). As expected, inhibition of mature miR-21 was accompanied by an increase in pre-miR-21 levels, supporting a mechanism in which the compound inhibits Dicer processing of pre-miR-21. To determine the selectivity of **3**, nine additional miRNAs were quantified using RT-qPCR with only miR-21 showing a significant decrease in mature miRNA levels (Figure 12.5b). Compound **3** was further shown to rescue expression of a miR-21 target, tumor suppressor phosphatase and tensin homolog (PTEN). Using a Boyden chamber assay, reduction of miR-21 levels by **3** inhibited the invasive phenotype of MDA-MB-231

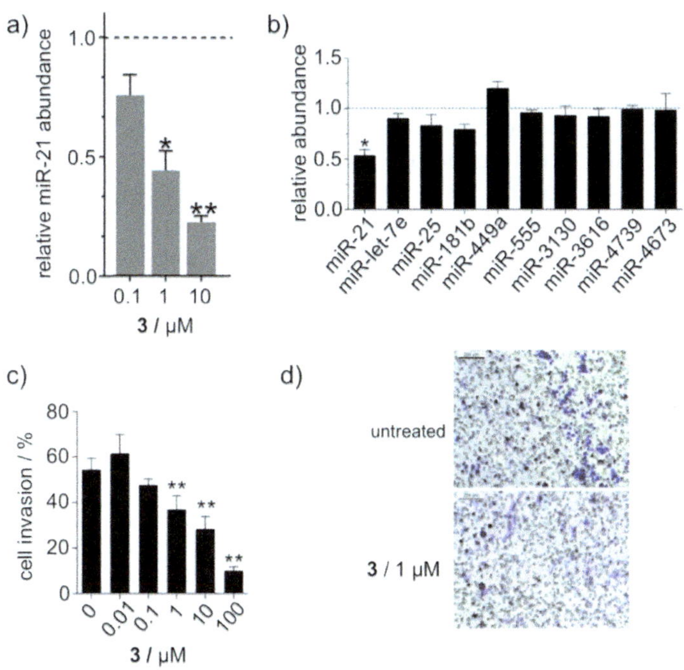

Figure 12.5 (a) Reverse transcription quantitative PCR (RT-qPCR) analysis demonstrates treatment of MDA-MB-231 cells with **3** inhibits mature miR-21 in a dose-dependent fashion. (b) Treatment with **3** has no effect on mature miRNA levels of several additional miRNA targets. (c) Treatment of MDA-MB-231 cells with **3** inhibits cell invasion in a dose-dependent fashion using a Boyden chamber assay. (d) Representative images of the reduction in cell invasion observed in MDA-MB-231 cells treated with **3**. Reproduced from ref. 30 with permission from Elsevier, Copyright 2018.

cells in a dose-dependent fashion (Figure 12.5c and d). To confirm that **3** binds pre-miR-21 in cells, a small molecule-RNA profiling approach called Chemical Cross-Linking and Isolation by Pull-Down (Chem-CLIP) was employed. The Chem-CLIP probe consists of the parent compound **3** modified with a biotin and a chlorambucil crosslinking moiety capable of covalently modifying its cellular RNA targets.[31] Cells were treated with the probe and the resulting conjugates were isolated by biotin capture followed by analysis by RT-qPCR. Indeed, the Chem-CLIP probe successfully enriched pre-miR-21 by >10-fold, confirming that **3** directly interacts with miR-21 in cells. Furthermore, Chem-CLIP experiments using excess parent compound **3** demonstrated competitive inhibition of labeling in a dose-dependent manner. Lastly, to further probe the selectivity of **3**, the Chem-CLIP approach was expanded to the nine additional miRNAs, which contain the same A-bulge RNA motif found in miR-21 and targeted by **3**. Only two miRNAs, *let-7e* and miR-25, exhibited binding; however, these interactions were not sufficient to reduce mature miRNA levels in cells. The authors believe the selectivity of **3** for miR-21 over the other miRNAs containing the same A-bulge may be due to the existence of the adjacent U-bulge, which is also capable of binding **3** and is only present in pre-miR-21. These results exemplify the promising potential that microarray-based screens have for identifying novel small molecule modulators of miRNA function and the development of AbsorbArray represents a significant solution to a well-known limitation of microarray technology.

While these applications demonstrate the advantages of microarray-based screens, they also highlight key limitations. For instance, the assay formats only monitor ligand–RNA interactions and do not assess the ability of compounds to inhibit miRNA maturation or function. In the approaches highlighted above, many ligands were identified that were capable of binding to the RNA with high affinity, but were unable to inhibit their function. The development of multiplexed microarray approaches that simultaneously monitor both ligand binding and perturbation of function may further facilitate identification of high-quality inhibitors.

With the goal of identifying inhibitors capable of perturbing miRNA biogenesis, fluorescence resonance energy transfer (FRET) assays have been developed to identify small molecule inhibitors of miRNA maturation. The Arenz lab was the first to develop FRET assays using fluorophore-labeled precursor miRNAs, similar to molecular beacons.[32,33] Installation of a 5′ fluorophore and a 3′ fluorescence quencher at the termini of a pre-miRNA hairpin enabled efficient FRET monitoring of Dicer-mediated cleavage. Importantly, this approach identifies compounds capable of inhibiting miRNA processing rather than identifying compounds that may merely bind the miRNA of interest. This technology has been extensively applied to identify and develop aminoglycoside derivatives as miRNA maturation inhibitors;[34–37] however, the broad RNA-binding ability of many aminoglycosides[38] has hampered their ability to distinguish between different miRNA sequences, defining most aminoglycosides as non-specific inhibitors of miRNA biogenesis.[37,39] Tailored FRET assays have also been developed for targeting specific miRNA–protein interactions, such as the approach developed by Hall and co-workers to identify inhibitors of the abnormal cell lineage protein 28 (LIN28)-pre-*let-7* interaction.[40] The LIN28 protein inhibits *let-7* maturation *via* binding to pri- or pre-*let-7* RNA, resulting in significant inhibition of Drosha- or Dicer-mediated processing.[41] Utilizing a green fluorescent protein (GFP)-LIN28 fusion donor and a black-hole quencher (BHQ)-labeled pre-*let-7* acceptor, Hall was able to detect inhibition of the LIN28-pre-*let-7* interaction.[40] Screening of 16 000 drug-like molecules led to the identification of 14 initial hits, which were subsequently subjected to a secondary luciferase reporter assay in cells to confirm their activity. Indeed, seven of the hit compounds demonstrated an increase in mature *let-7*, which corresponds to inhibition of the LIN28-pre-*let-7* interaction. The small molecule **4** was shown to directly bind LIN28 protein as its principal mechanism of action. Treatment of HeLa cells with **4** led to increased levels of *let-7a*, *let-7g*, and *let-7f*, while having no effect on a control miRNA, miR-15; however, additional miRNAs were not evaluated and thus the extent of selectivity is not yet known. Lastly, **4** inhibited proliferation and stem-like properties in several human cancer cell lines with IC_{50} values between 20 and 42 μM.

In an alternative FRET approach, Park and co-workers employed unnatural amino acid mutagenesis to site-specifically fluorescently label the LIN28 protein (Figure 12.6a).[42] Compared to the approach undertaken by Hall and co-workers, this design significantly improved FRET efficiency, imparted by placement of the donor and acceptor in optimal proximity to each other to obtain maximal overlap of the emission spectra of the donor and the absorption spectra of the acceptor. Indeed, FRET quenching efficiencies as high as 85% were obtained, marking a significant improvement over previous assays (Figure 12.6b). Using this assay, the authors screened 4500 drug-like compounds and identified the benzopyranylpyrazole **5** as a hit compound (Figure 12.2). Compound **5** elicited a dose-dependent increase in fluorescence in the FRET assay yielding an IC_{50} of 4 μM (Figure 12.6c). Mechanism of action studies indicated that **5** directly binds LIN28 and inhibits the LIN28-*let-7* interaction in JAR cells leading to increased levels of all *let-7* family miRNAs without affecting two control miRNAs, miR-20b and miR-214, as revealed by RT-qPCR. This increase in *let-7* miRNA also decreased the endogenous levels of oncogenic target genes of *let-7*, including cellular myelocytomatosis (c-MYC), high mobility group AT-hook 2 (HMGA2), and rat sarcoma viral oncogene homolog (RAS) (Figure 12.6d).

As an alternative to FRET-based approaches, the Balasubramanian lab developed a high-throughput fluorescence polarization assay to identify inhibitors of the LIN28–pre-*let-7* interaction.[43] Fluorophore-labeled pre-*let-7* was titrated with varying concentrations of recombinant LIN28 in the presence of small molecules and fluorescence polarization was measured. After removing false positives and confirming activity through secondary assays, five of the confirmed hits were identified and subjected to an *in vitro* Dicer processing assay which revealed two compounds **6** and **7** (Figure 12.2) that were capable of rescuing pre-*let-7* processing, demonstrating the ability to

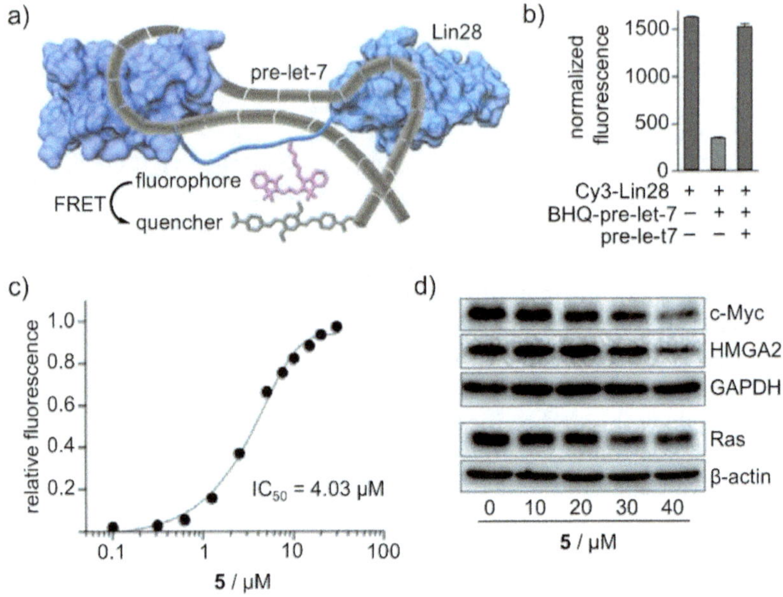

Figure 12.6 (a) Model of LIN28-pre-*let-7* binding and the design of fluorescence resonance energy transfer (FRET)-based probes. (b) Complex of Cy3-Lin28 and BHQ-pre-*let-7* leads to decreased fluorescence intensity, while addition of excess unlabeled pre-*let-7* rescues fluorescent signal. (c) Dose–response rescue of fluorescence signal in the FRET-assay following treatment with **5**. (d) Western blot analysis of the effect of **5** on expression of endogenous *let-7* target gene products in JAR cells. Reproduced from ref. 42 with permission from American Chemical Society, Copyright 2016.

inhibit the LIN28–pre-*let-7* interaction in a productive fashion. Compound **6** was capable of fully rescuing Dicer processing of pre-*let-7* at a concentration of 20 μM; however, these results were only demonstrated *in vitro* and the selectivity of the compound was not investigated. Also employing a fluorescence polarization-based approach, Kiriakidou and co-workers developed a screen to identify small molecule inhibitors of miRNA–RISC loading using tetramethylrhodamine (TAMRA)-labeled RNAs and recombinant human Argonaute 2 protein (AGO2).[44] Screening of small molecule collections identified aurintricarboxylic acid **8** (Figure 12.2) as an inhibitor of AGO2 loading, with an IC_{50} of 470 nM. Furthermore, **8** inhibited siRNA function in cells and thus likely represents a general inhibitor of AGO2 loading. The ability of **8** to inhibit protein–nucleic acid interactions non-discriminately has been established previously and explains its lack of selectivity.[45a]

In a similar design, Sliz and co-workers used FAM-labeled pre-*let-7* and recombinantly-expressed LIN28 protein to carry out a fluorescence polarization-based high-throughput screen of 101 017 compounds.[45b] Of the compounds screened, six were found to inhibit LIN28-*let-7* binding and impair LIN28-mediated *let-7* oligouridylation. Inhibitors **9** and **10** (Figure 12.2) blocked the LIN28-*let-7* interaction with an IC_{50} of 2.5 μM and 7 μM, respectively. Using a heteronuclear single-quantum coherence (HSQC) approach, the authors identified that **9** interacted with the zinc-knuckle domain of LIN28, inducing significant perturbations and destabilizing its structure. Additionally, HSQC analysis and saturation transfer difference (STD) spectroscopy revealed that **10** binds to the cold shock domain of LIN28, inhibiting oligouridylation activity of LIN28 with an IC_{50} of 27 μM. To test whether **10** was capable of inhibiting LIN28 function in cells, a dual luciferase reporter assay containing eight tandem *let-7* recognition sites in the 3′ UTR of *Renilla* luciferase was constructed and stably transduced into HeLa cells ectopically expressing LIN28A and LIN28B, two paralogs of LIN28 expressed in mammals. Indeed, compound **10** was capable of significantly reducing *Renilla* luciferase expression, demonstrating inhibition of LIN28 and activation of *let-7* in cells. Supporting this finding, RT-qPCR analysis of *let-7* family miRNAs in cells treated with **10** showed significant enhancement of all *let-7* miRNA levels, while a control miRNA, miR16, showed little change in expression. The overall success of these approaches may point to the benefit of designing highly customized approaches to identifying inhibitors for a specific protein–RNA interaction; however, the limited potential for translating these assays to additional protein–RNA targets should not be ignored and may require significant modification and optimization.

As a unique approach to identify inhibitors of pre-miRNA processing, the Garner lab has developed an assay technology termed 'cat-ELCCA' (catalytic enzyme-linked click chemistry assay).[46] In this assay, biotin-labeled pre-miRNAs containing a click chemistry handle (*e.g.*, *trans*-cyclooctene) within the loop-region are immobilized onto streptavidin-coated microtiter plates. Treatment of the wells with Dicer in the absence of inhibitor results in Dicer-mediated cleavage of the loop-region, and loss of the click handle. However, in the presence of an inhibitor, Dicer function is blocked and the click handle remains intact. Subsequent treatment with horseradish peroxidase (HRP) containing a complementary click chemistry reaction partner (*e.g.*, tetrazine) results in efficient covalent labeling of the intact pre-miRNA and enables robust detection *via* catalytic HRP-based chemiluminescence signal amplification (Figure 12.7a). To identify inhibitors with high substrate specificity, wells containing additional alternative pre-miRNAs can be analyzed in parallel to allow discrimination between general inhibitors and substrate-specific inhibitors. This assay design offers: (1) increased sensitivity and improved signal-to-noise ratios due to signal amplification; (2) negligible compound fluorescence interference due to a luminescence read-out; and (3) a functional read-out identifying only ligands capable of inhibiting Dicer processing of the pre-miRNAs. As an implementation of this assay format, the Garner lab applied this approach to a two-dimensional screen identifying inhibitors of pre-miR-21 and pre-*let-7d* Dicer processing.[47] In total, 50 000 small molecules and 33 000 natural product extracts (NPE) were

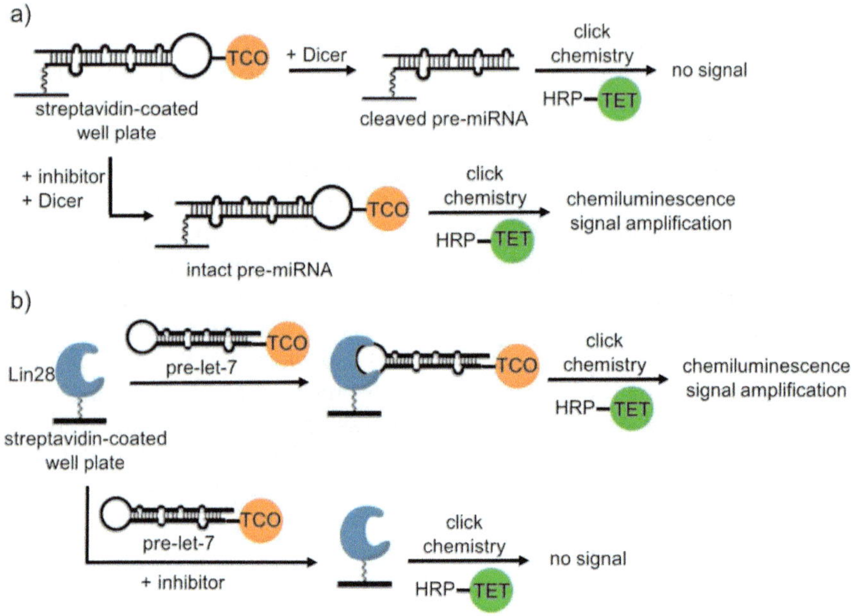

Figure 12.7 Two different cat-ELCCA (catalytic enzyme-linked click chemistry assay) assays for high-throughput screening. (a) cat-ELCCA for Dicer-mediated pre-miRNA processing and the identification of inhibitors. (b) cat-ELCCA for identification of small molecule inhibitors of the LIN28–pre-*let-7d* interaction.

tested, with the small molecules producing 170 confirmed hits and the NPEs producing 47 confirmed hits. Comparing the hit lists from both pre-miR-21 and pre-*let-7d* against each other, the selectivity of the identified hits was examined. While most of the small molecule hits exhibited little to no selectivity for pre-miR-21 over pre-*let-7d*, several NPEs exhibited significant selectivity ratios skewed towards pre-miR-21. The authors are still following up on these NPEs, but they may contain promising candidates for selective inhibition of pre-miR-21 processing. More recently, the Garner lab has also described an adaptation of the cat-ELCCA approach toward identification of small molecule inhibitors of the LIN28–pre-*let-7d* RNA protein interaction (Figure 12.7b).[48] In this assay, biotin-labeled LIN28 was immobilized on streptavidin-coated microtiter plates and bound with pre-*let-7d* bearing a *trans*-cyclooctene, which in turn was reacted with a tetrazine-modified HRP. This led to efficient and robust detection of the RNA–protein complex *via* a catalytic HRP-based chemiluminescence signal. However, in the presence of a small molecule inhibitor, protein–RNA complex formation is inhibited and no chemiluminescence signal is generated. Using this assay, a high-throughput screen was carried out with a total of 127 007 small molecules screened at a concentration of 25 μM. Initially, 1468 hits were identified and after follow-up confirmation screens including dose–response assays and pan assay interference compounds (PAINS) filtering to remove false positives attributed to chemical motifs that are notoriously identified in screens without a defined and target-specific mode of action,[49] 20 hits were selected for further investigation. From these hits, two structurally similar compounds (**11** and **12**) were identified which displayed dose-dependent activity in both the cat-ELCCA assay and an electrophoretic mobility shift assay. Furthermore, both compounds did not inhibit Dicer-mediated processing of pre-*let-7* nor an additional unrelated cat-ELCCA-based protein–protein interaction assay indicating that the compounds are specific for the LIN28–pre-*let-7d* interaction. While a brief structure–activity relationship (SAR) study was conducted in hopes of improving their potency, the use of these compounds in cells and the selectivity against other precursor miRNAs was not investigated. Thus, their potential use as chemical probes is still unknown. The application of cat-ELCCA for the identification of novel inhibitors of miRNAs appears promising and, as mentioned above, comes with several advantages over existing assay technologies.

In summary, the use of *in vitro* assays to identify inhibitors of miRNA has proven successful *via* several approaches. While direct binding assays may benefit from high-throughput capabilities, they can often result in the identification of ligands that bind, but do not inhibit miRNA activity or miRNA maturation. On the other hand, while assays that directly monitor function tend to produce active inhibitors, obtaining high substrate specificity remains challenging. Future *in vitro* screening platforms should combine aspects of both direct binding assays and functional inhibition assays to identify more robust hits that have both strong efficacy and substrate specificity. Lastly, but importantly, inhibitors identified through *in vitro* assays run the risk of not eliciting biological activity in a cellular environment due to several factors, including cell permeability, stability, and toxicity, as well as other biophysical and pharmacological determinants, which need to be evaluated separately.

12.3 miRNA Modifiers Discovered Through Cell-based Assays

Cellular assays for the discovery of small molecule modulators of miRNA function and biogenesis have distinct advantages over biochemical screening approaches. For example, biochemical assays do not account for factors that may impact the efficacy of a potential hit compound in cells, such as toxicity, cell permeability, and stability. Furthermore, cell-based assays are not limited to known targets, but can be conducted in an unbiased way, thereby enabling the discovery of novel biological interactions and pathways. This section discusses the implementation of cell-based reporter assays for the discovery of small molecule modulators of miRNA function (Figure 12.8).

In a typical cell-based assay approach, the miRNA target sequence (either a natural sequence or a full complement) is cloned downstream of a reporter gene, such as luciferase or GFP, thus placing expression of the reporter

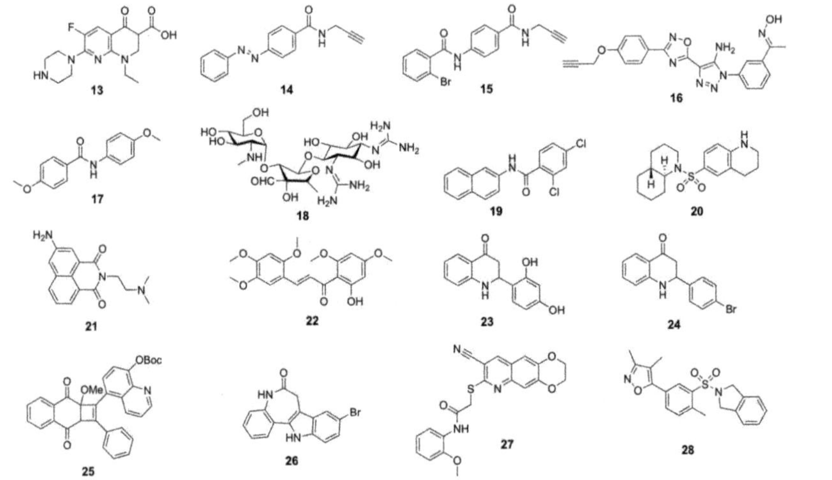

Figure 12.8 Structures of small molecule modifiers of miRNA function identified in cell-based reporter assays.

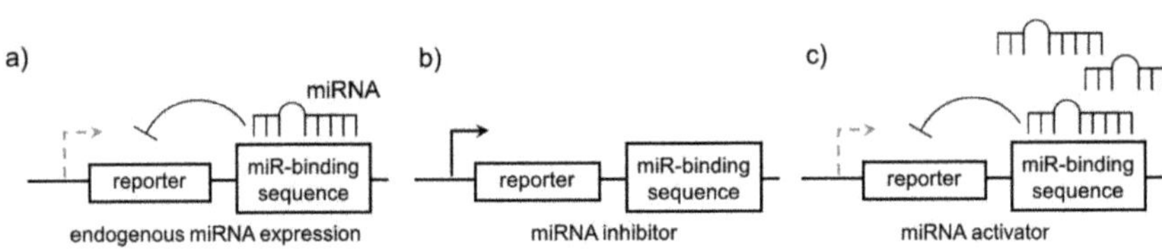

Figure 12.9 Cell-based reporter assay design. A miRNA target sequence is cloned in the 3′ untranslated region (UTR) of a luciferase or enhanced green fluorescent protein (EGFP) reporter gene. (a) Binding of a complementary miRNA to the target site results in repression of the reporter. (b) Treatment with a small molecule inhibitor of miRNA function leads to an increase in reporter expression. (c) Induction of miRNA expression or stabilization of miRNA function elicits further reduction in reporter expression.

gene under repression by the endogenous miRNA (Figure 12.9a). Selection of appropriate secondary assays for validation of potential hit compounds is important to rule out false-positives, as can be the case with luciferase-based reporter systems,[50–52] and to assist in determination of mechanisms of action. Additionally, because cell-based assays only indirectly measure alterations in miRNA function or biogenesis, minor changes in activity may be muted, resulting in potential hit compounds being overlooked. This is especially problematic in high-throughput screens where small molecules are commonly tested only at a single concentration. Cell-based reporter assays that result in an increase in signal, rather than a decrease, are more robust since the latter can possibly stem from compound-induced changes in cell health unrelated to reporter gene expression.

Jin and co-workers utilized a HEK293 cell line expressing an enhanced green fluorescent protein (EGFP) reporter together with a stably expressed short hairpin RNA (shRNA) targeting the EGFP mRNA in order to identify small molecule modifiers of the RNA interference (RNAi) pathway (which shares many components with the miRNA pathway).[53] Stable clones expressing low levels of EGFP, a consequence of efficient shRNA repression, were isolated to serve as a platform for the identification of small molecule modifiers of RNAi (Figure 12.9c). Following a screen of 2000 compounds from the United States Food and Drug Administration-approved compounds and natural products collection, **13** (Figure 12.8) was found to reduce EGFP expression, indicative of an enhancement in shRNA repression, with an EC_{50} of 30 μM. Additionally, treatment with **13** improved RNAi-mediated inhibition of several additional genes in multiple cell lines, suggesting that it is a general enhancer of the RNAi pathway. A two-fold increase in AGO2-associated small interfering RNAs (siRNAs) was observed following treatment with **13**, suggesting a mechanism by which **13** enhances siRNA loading into RISC. Compound **13** also induced a two-fold increase in the mature forms of 13 selected miRNAs, while decreasing their corresponding precursor and primary miRNAs, as shown by Northern blot analysis. Further investigations into a potential mechanism of action including immunoprecipitation and siRNA knockdown experiments suggest that **13**-mediated enhancement of RNAi is dependent on modulation of *trans*-activation-responsive region RNA-binding protein 2 (TRBP), a critical component required for the formation of the RISC.[54]

A follow-up study confirmed the binding interaction between **13** and TRBP and drew a correlation between the proposed interaction and the enhancement of tumor suppressor miRNAs.[55] To expand upon these initial findings, the Jerónimo lab performed a comprehensive expression profiling study covering 742 miRNAs in prostate cancer cell lines following treatment with **13**.[56] They discovered that while >50% of miRNAs were upregulated in both LNCaP and DU145 prostate cancer cells, decreased levels of several oncogenic miRNAs was also observed, suggesting additional mechanisms of action to the **13**–TRBP interaction. This discovery highlights the importance of evaluating a potential hit compound's activity in the context of multiple miRNAs to develop a more complete picture of its mode of action. To assist in identifying additional cellular targets of **13**, an affinity-based probe was developed.[57] MCF-7 breast cancer cells were treated with the alkyne-modified probe, followed by pull-down with biotin-azide and streptavidin beads. Mass spectrometry analysis of the isolated proteins identified P-element induced wimpy testis (Piwi)-like RNA-mediated gene silencing 3 (PIWIL3), an AGO2 protein, to be the most abundant protein and likely target of **13**. This was further supported when siRNA-mediated knockdown of PIWIL3 led to similar changes in both miRNA levels and cell proliferation as that of **13**-treated cells.

Deiters and Huang were the first to develop a cell-based reporter assay to identify small molecule inhibitors of miRNA function, specifically miR-21.[58]

miR-21 is upregulated in numerous cancers and has been implicated in carcinogenesis and drug resistance,[59] highlighting its importance as a therapeutic target. To this end, a reporter in which the miR-21 binding sequence was cloned downstream of the luciferase gene was stably expressed in HeLa cervical cancer cells (HeLa-miR-21).[58] High endogenous miR-21 levels in HeLa cells[60] resulted in suppression of luciferase expression and decreased luminescence signal; however, in the presence of a miR-21 inhibitor, rescue of luciferase expression (Figure 12.9b) results in an increase in luminescence signal. While the authors demonstrated that the reporter construct containing the miR-21 target site exhibited reduced luciferase expression in HeLa cells, to further validate that the reporter system would respond to miRNA inhibition, an antagomir specific to miR-21 could have been delivered. The HeLa-miR-21 cell line was employed in a pilot screen of >1000 small molecules from an in-house compound collection and the Library of Pharmacologically Active Compounds.[58] An initial hit from the pilot screen was optimized through SAR studies leading to the discovery of **14** (Figure 12.8), which elicited two-fold higher potency than the initial hit and an EC_{50} of 2 μM. Both mature and primary miR-21 levels were found to be reduced by 78% and 87%, respectively, indicating that **14** likely acts at the transcriptional or pre-transcriptional level. Treatment with **14** in a control HeLa cell line stably expressing a miR-30 reporter showed no reduction in luminescence signal, suggesting the small molecule is not a general inhibitor of the miRNA pathway; however, testing the compound against a more comprehensive list of miRNAs was not completed and may reveal additional targets. Despite the absence of a clear biological target, the Yao lab formulated cell-penetrating poly(disulfide)-coated mesoporous silica nanoparticles containing both **14** and a miR-21 antagomir and delivered them into HeLa cells.[61] Upon treatment with the nanoparticles, pronounced reduction of miR-21 levels coinciding with inhibition of cell migration and invasion was observed.

Lu and co-workers performed further SAR studies of **14**, synthesizing a set of 4-benzoylamino-*N*-(prop-2-yn-1-yl)benzamide derivatives,[62] and compound **15** (Figure 12.8) was identified as the most potent inhibitor, eliciting a ~90% reduction in mature miR-21 levels at 10 μM—a modest but significant improvement over the parent compound. In addition, **15** did not inhibit levels of *let-7a*, miR-25, miR-26, miR-126, miR-133, miR-143, miR-145, miR-146, miR-148, miR-152, or miR-195, providing further support for the initial claim that this compound class is not a general miRNA pathway inhibitor, but instead shows some specificity for miR-21.

After these pilot screen evaluations of the HeLa-miR-21 reporter cell line, a high-throughput screen of 333 519 compounds was performed,[63] and after a panel of confirmatory secondary assays, inhibitors were identified and optimized *via* SAR to yield the oxadiazole **16** (Figure 12.8), eliciting a 4.3-fold increase in activity relative to the parent compounds and an EC_{50} of 5.3 μM. This compound showed no effect in an *in vitro* luciferase assay indicating that it did not bind directly to the luciferase enzyme, and no inhibition of miR-122 was observed in response to treatment with **16** in a Huh7-miR-122

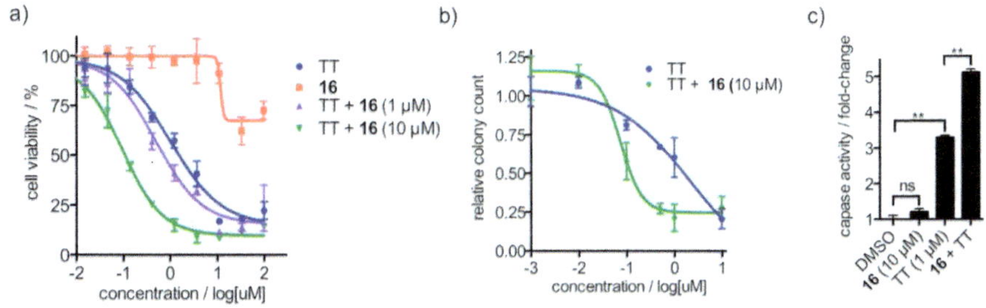

Figure 12.10 Evaluation of small molecule inhibitors of miR-21 function. (a) Treatment of A498 cells with the inhibitor **16** (10 μM) in combination with topotecan leads to an 11-fold reduction in the IC_{50}. (b) Colony formation in A498 cells is reduced 27-fold after co-treatment with **16** (10 μM) and topotecan. (c) Combination treatment of **16** (10 μM) and topotecan (1 μM) induces caspase activation by 1.6-fold. Reproduced from ref. 63 with permission from American Chemical Society, Copyright 2018.

reporter cell line,[64] suggesting that **16** is not a general miRNA inhibitor. Furthermore, **16** had little effect in cells transfected with additional reporter constructs for miR-125b, miR-182, or miR-221, providing further support for its miR-21 specificity. Interestingly, compound **16** had no effect on primary or mature miR-21 levels, suggesting a mode of action by which the small molecule inhibits miRNA function and not its biogenesis. In pre-therapeutic assays, **16** elicited low micromolar IC_{50} values in cell viability assays conducted in MIA PaCa (pancreatic cancer), A172 (breast cancer), U87 (glioblastoma), MCF7 (breast cancer), and HCT-116 (colon cancer) cell lines; however, an IC_{50} of >100 μM was observed in A498 renal cell carcinoma. Interestingly, co-treatment of **16** (10 μM) and topotecan elicited an 11-fold reduction in the IC_{50} of the chemotherapeutic agent toward A498 cells (Figure 12.10a). A similar response was observed in a clonogenic assay where co-treatment enhanced the ability of topotecan to inhibit colony formation by 27-fold (Figure 12.10b). Furthermore, the combination of **16** (10 μM) and topotecan (1 μM) induced caspase-3/7 activity two-fold compared to topotecan alone (Figure 12.10c). Taken together, the pre-therapeutic assays suggest that small molecule-mediated miR-21 knockdown may hold potential in the treatment of renal cancer by alleviating chemoresistance.

Similarly, SAR optimization of an aryl amide inhibitor identified in the high-throughput screen led to the discovery of **17** (Figure 12.8).[65] The inhibitor was subjected to numerous secondary assays analogous to those described above. Treatment of HeLa cells with **17** following transfection with a reporter plasmid in which the luciferase gene was placed under the control of the miR-21 promoter sequence had no effect on luminescence signal. This is in good agreement with a RT-qPCR assay in which the compound showed no effect on pri-miR-21 levels, suggesting that it also acts downstream of transcription.

Maiti and co-workers developed a similar luciferase-based reporter plasmid by cloning the 3′ UTR of programmed cell death protein 4 (PDCD4), a known target of miR-21,[66] downstream of a luciferase gene to screen aminoglycosides as potential inhibitors of miR-21.[67] After transfection of the reporter into MCF-7 cells, **18** (Figure 12.8) was identified as the most potent miR-21 inhibitor among the 15 aminoglycosides screened, eliciting a 1.4-fold increase in luminescence. Confirmatory transfection of a miR-21 antagomir may shed more light on how meaningful this increase is and provide much needed validation for the assay. Additionally, limited compound numbers and confining the screen to a single compound class limited diversity in this study and reduced chances of identifying inhibitors with desired characteristics. While the small increase in luminescence signal suggests that **18** is not a potent miRNA inhibitor, treatment with **18** (5 μM) led to a ~90% reduction in miR-21 levels as observed by RT-qPCR. In addition to miR-21, expression of nine other miRNAs was evaluated and while most of the miRNAs remained unaffected, miR-27a was downregulated by ~60%. Western blot analysis showed that **18** led to a minor dose-dependent increase in PDCD4 expression while a structurally similar analog elicited no change. Binding of **18** to pre-miR-21 was confirmed through melting experiments and a docking study showed a consensus structure in which **18** forms several hydrogen bonds with residues between G28 and G44 in pre-miR-21. Because **18** is also known to bind the phosphate backbone of 16S ribosomal RNA (rRNA) in the 30S ribosomal subunit *via* hydrogen bonds with four nucleotides,[68] it may be prone to promiscuous binding to other RNAs with similar sequence motifs, limiting its potential as a biochemical probe or a therapeutic. Incubation with **18** led to ~0.5-fold reduced Dicer processing of pre-miR-21, while no effect on the processing of pre-miR-103 or -95 was observed, suggesting that the compound is not a general pre-miRNA inhibitor.

A cell-based luciferase reporter plasmid was developed by the Deiters lab in order to identify small molecule modifiers of miR-122 function.[64] In contrast to the miR-21 reporter cell line,[58] the miR-122 reporter bears a second luciferase gene not under control of any miRNA. Use of a constitutively active internal control in cell-based reporter assays accounts for experimental variations including cell number, transfection efficiency, and cell viability. Transfection of the reporter plasmid into Huh7 hepatocellular carcinoma (HCC) cells, which overexpress miR-122,[69] led to a >90% reduction in luciferase activity compared to miR-122 deficient HeLa cells.[70] Furthermore, co-transfection with a miR-122 antagomir led to an increase in luciferase signal indicating the reporter was amenable for discovery of miR-122 inhibitors. A pilot screen of 1364 small molecules from the National Cancer Institute (NCI) Diversity Set II[64] yielded compounds **19** and **20** (Figure 12.8) which induced a 7.7- and 12.5-fold increase in luminescence, respectively (similar to the response observed through antagomir transfection). Additionally, amonafide (**21**) (Figure 12.8) was found to be a miR-122 activator, eliciting a seven-fold reduction in luminescence. After verifying that the compounds did not inhibit miR-21 in HeLa–miR21 cells,[58] SAR studies were performed for **19**, **20**, and **21**. Unfortunately, none of the structural modifications led to an improvement in activity relative to the parent molecules. EC_{50} values of 3 μM and 0.6 μM were observed for compounds **19** and **20**, respectively, while **21** elicited an IC_{50} of 3 μM. Treatment with compounds **19** and **20** led to 45% and 72% reductions in mature miR-122 levels and 78% and 97% reductions in pri-miR-122 levels, respectively, while treatment with **21** induced a four-fold increase of both mature and pri-miR-122 levels. Taken together, these results suggest that all three compounds perturb miR-122 function by acting at the transcriptional or pre-transcriptional level. The psiCHECK-miR122 reporter was also converted into a stable cell line that is amenable to high-throughput screening.[71] miR-122 has been implicated in the hepatitis C virus (HCV) replication cycle[72] and long-term HCV infection may contribute to the development of HCC.[73] To this end, compounds **19** and **20** were evaluated for their ability to inhibit HCV replication in Huh7 cells and both were found to reduce HCV RNA levels by almost 50%. In HCC, miR-122 is frequently down-regulated and has been found to inactivate caspase-3.[74] Treatment of HepG2 liver cancer cells with **21** led to an 80% reduction in cell viability coinciding with a 20-fold increase in caspase-3/7, in contrast to Huh7 cells, in which more modest responses were observed, suggesting the activator may have therapeutic potential towards selective treatment of HCC *via* induction of miR-122. Amonafide is a topoisomerase II inhibitor[75] that recently entered phase III clinical trials as a combination therapy towards the treatment of acute myeloid leukemia (AML).[76] Interestingly, several other known chemotherapeutics have also been shown to elicit changes in miRNA expression patterns including cisplatin,[77] 5-fluorouracil, rituximab,[78] bortezomib, perifosine, retinoic acid,[79] and decitabine.[80,81] Furthermore, the Disney lab recently demonstrated that several topoisomerase inhibitors were capable of binding pre-miR-21, as described above.[30] Taken together, these reports suggest that regulation of miRNA expression may be a secondary mode of action for these drugs and could also play a role in their therapeutic effects.

In an approach to continue exploring the therapeutic potential of **20**, the Yao lab generated mesoporous silica nanoparticles coated in a peptide for simultaneous cellular uptake of the small molecule and a miR-122 antagomir.[82] Treatment of Huh7 cells with the nanoparticles led to a ~20% greater reduction in miR-122 levels relative to **20** or miR-122 antagomir alone, confirming that the system could be used for controlled release of oligonucleotides and small molecules.

The Chen lab developed a reporter plasmid to identify small molecule modifiers of miR-34a function in HCC.[83] miR-34a is a known tumor suppressor miRNA and has been implicated in cell-cycle regulation, differentiation, migration, invasion, and senescence.[84] After confirming activity in HepG2 and Huh7 cells *via* co-transfection of a miR-34a mimic or antagomir, the miR-34a reporter was utilized in a small molecule screen of 640 compounds from a natural product collection.[83] Among the molecules screened, **22** (Figure 12.8), a plant chalcone derivative,[85] inhibited luciferase activity in a dose-dependent fashion with an IC_{50} of 4 μM. Furthermore, mature miR-34a levels increased by ~3- and ~2-fold in Huh7 and HepG2 cells, respectively, in the presence of **22** (10 μM). Under the same conditions, pri-miR-34a levels increased by ~4- and ~10-fold. Conversely, mature and pri-miR-34a levels remained unchanged in Hep3B cells. A global analysis of miRNA expression showed that the small molecule affects numerous other miRNAs as well, suggesting the compound is not specific to miR-34a activation. The pre-therapeutic potential of **22** was evaluated in five hepatocyte cell lines and growth inhibition was observed in HepG2, Bel-7404, Huh7, and PLC HCC cell lines, while the compound had no effect on cell growth of non-carcinogenic MIHA cells. In HepG2 xenografted mice, tumor growth was inhibited by up to 90% following treatment and **22** induced apoptosis and inhibited proliferation, further validating the therapeutic potential of small molecule modifiers of miRNA function.

An analogous luciferase reporter assay for small molecule modulators of miR-4644 function stably expressed in MCF-7 cells was employed in a screen of nine aza-flavanones. As mentioned above, utilizing a small compound library severely limits chemical diversity and may impede identification of lead compounds since hit rates from small molecule screens are typically low.[86] Two compounds, **23** and **24** (Figure 12.8), elicited a four-fold increase in luminescence at 4 μM. Both compounds reduced mature and pri-miR-4644 levels by 20–25%, but had no effect on miR-30a levels, suggesting that the compounds were likely not general miRNA pathway inhibitors; however, additional miRNAs would need to be screened to further elucidate specificity. The aza-flavanones **23** and **24** reduced cell viability by 40% and were found to elicit G1 phase arrest. To evaluate the efficacy of the two hit compounds *in vivo*, a transgenic fly strain expressing a reporter construct with two copies of the miR-14 (a miR-4644 homolog) binding sequence in the 3′ UTR of an EGFP gene under control of a tubulin promoter was developed. The reporter is primarily expressed in the wings and endogenous miR-14 suppresses expression of EGFP leading to reduced fluorescence. Treatment with **23** and **24** at 50 μM led to an increase in EGFP fluorescence due to miR-14 inhibition. Furthermore, western blot analysis demonstrated that EGFP expression was upregulated by ~2.5-fold and ~3-fold in the presence of **23** and **24**, respectively. Because of the homology between the two miRNAs, **23** and **24** could be used to elucidate novel miR-4644 or miR-14 regulatory mechanisms in human cells or *Drosophila melanogaster*, while also shedding light on potential evolutionary links.

Following the identification of a universal miRNA activator from the photoreaction products of naphthalene-1,4-dione and acetylenes,[87] Zhang and co-workers sought to employ a similar strategy towards discovery of selective miRNA-targeting small molecules.[88] A series of photocycloaddition reactions between 2-methoxy-1,4-naphthalenequinone and substituted aryl acetylenes led to the generation of 10 analogs that were subsequently screened in parallel cell-based assays utilizing luciferase reporters against miR-1, miR-214, miR-25, and miR-150, miRNAs that have been implicated in the development and proliferation of various diseases.[89–92] The small molecule **25** (10 µM; Figure 12.8) selectively inhibited miR-1, a myogenic miRNA (myomiR) that is enriched in cardiac and skeletal muscle and is implicated in muscle and heart development,[93] leading to a ~7-fold increase in luminescence in C2C12 cells.[88] RT-qPCR analysis showed that intracellular levels of miR-1 were reduced ~4-fold following treatment with **25**. A small SAR study of 15 analogs was performed; however, most analogs showed similar activity to **25**, providing no further insight into important structural features required for miRNA inhibition.

In a subsequent report, compound **25** was used to investigate cell signaling pathways downstream of miR-1 to better understand its role in myogenesis.[94] After discovering that compound **25** inhibited differentiation and proliferation of C2C12 cells, expression of myoblast differentiation markers muscle creatine kinase (MCK), myosin heavy chain (MHC), and myoGenin were evaluated. All three transcripts were downregulated (up to 90%), while a proliferation marker, proliferating cell nuclear antigen (PCNA), showed no reduction. While **25** was initially hypothesized to be selective for miR-1,[88] it was found to inhibit myomiRs miR-133a and -206, eliciting 10-fold and six-fold increases in luminescence in C2C12 cells transfected with their respective reporter plasmids—a similar response to the miR-1 reporter assay. Furthermore, pri-miRNA levels of miR-1, miR-133a, and miR-206 were found to be downregulated by 60–90%, indicating that the compound acts at the transcriptional or pre-transcriptional level and suggesting it may be a general regulator of myomiRs. To determine if **25** regulates myomiRs *via* myoblast determination protein (myoD), a major transcriptional regulator of myomiRs[95] and regulator of cell differentiation,[96] myoD expression levels were evaluated following compound treatment. Interestingly, while mRNA levels remained unchanged, myoD protein expression was found to be reduced, suggesting that **25** inhibits myoD translation. The miRNAs miR-221 and miR-222 have been identified as inhibitors of myoD translation and treatment of C2C12 cells with compound **25** led to three-fold increases in miR-221/222 expression. Taken together, these results suggest an interesting mechanism of action by which compound **25** inhibits a regulatory pathway by inducing miR-221/222 expression, leading to repression of the transcription factor myoD, and subsequent downregulation of myomiRs.

Similar to the reporter assay developed by the Maiti lab discussed above,[67] Gorospe and co-workers developed a luciferase-based reporter assay system bearing the 3′ UTR of KRAS or HMGA2, oncoprotein targets of *let-7*,[97,98] in order to further probe the mechanisms underlying *let-7* biogenesis.[99] An additional luciferase reporter containing two copies of *let-7f* complementary sequences was also constructed. Following co-transfection of the reporters with pre-*let-7f* or control miRNA into BG-1 or UCI-101 ovarian cancers cells for evaluation of activity, the plasmids containing KRAS or HMGA2 3′ UTRs elicited only limited reductions in luminescence, while the construct bearing complementary *let-7* sequences showed almost complete loss of luminescence and thus was used for all subsequent assays. This suggests that, in this case, a fully complementary binding sequence is necessary in cell-based reporter assays to reach an optimal dynamic range. A luciferase assay performed after co-transfection of the luciferase reporter and precursors of nine *let-7* miRNAs confirmed the plasmid was sufficient for examining overall *let-7* expression levels in cells. A screen of the Screen-Well™ Kinase Inhibitor Library led to the identification of 10 compounds (two activators and eight inhibitors) that affected *let-7* function. The compounds elicited no activity in the presence of reporter constructs bearing target sites for five other miRNAs, indicating that the compounds were not general pathway inhibitors. However, a more comprehensive analysis of miRNA expression levels in response to treatment with the compounds should be employed to confirm specificity for *let-7*. Among the 10 hit compounds, **26** (Figure 12.8), a glycogen synthase kinase 3β (GSK3β) inhibitor, was found to induce *let-7* expression, eliciting a 49% reduction in luminescence and was selected for follow-up experiments. Because p53 is known to be a downstream target of GSK3β signaling,[100] as well as a key transcription factor in *let-7* biogenesis,[101] Gorospe and co-workers hypothesized that **26** induced *let-7* levels in a p53-dependent mechanism. siRNA-mediated silencing of GSK3β in p53-null HCT-116 cells led to only a <20% reduction in luciferase activity of the *let-7* reporter while silencing in wild-type colon cancer cells led to a 40% reduction, suggesting that GSK3β inhibits *let-7* production by repressing p53. Because **26** inhibits several other kinases, its therapeutic potential may be limited; however, it holds promise as a chemical probe to better understand the interaction between GSK3β and p53, as well as signaling networks upstream of *let-7*.

While the majority of the reported cell-based reporter constructs utilize a single complementary miRNA sequence for activity, Lowry and co-workers inserted eight repeats of the *let-7* target site in the 3′ UTR of a *Renilla* luciferase gene in a psiCHECK plasmid, hypothesizing that it would lead to a higher rate of mRNA degradation.[102] Following stable expression of the construct in a Huh7.5.1 cell line, a high-throughput screen of 36 480 small molecules from commercial collections yielded 60 potential hit compounds. As an additional control to rule out potential luciferase inhibitors, the parent psiCHECK plasmid bearing no miRNA target sequence was used. After confirming activity of 29 of the 60 potential hit compounds in the stable Huh7.5.1 cell line and ruling out false positive hits, several remaining compounds demonstrated variable effects on downstream targets. The 60 initial hit compounds were retested in a RT-qPCR experiment monitoring their effect on expression of HMGA2 mRNA in order to identify molecules that specifically and robustly inhibit the repressive function of *let-7* on downstream targets. Treatment with compound **27** (Figure 12.8) elicited a reduction in HMGA2 levels by ~60%, and led to suppression of *let-7* targets v-myc myelocytomatosis viral related oncogene, neuroblastoma derived (NMYC) and insulin-like growth factor 2 mRNA-binding protein 2 (IMP2). Upon evaluation of the small molecule using SEA viewer structure-target prediction software,[103] phosphodiesterase 10A (PDE10) was identified as a potential target. Because PDE10 is known to regulate cyclic-AMP (cAMP), expression of the downstream target cAMP responsive element binding protein (CREB) was monitored. Consistent with inhibition of PDE10 and a subsequent increase in cAMP signaling, compound **27** induced activation of phosphorylated CREB. In Huh7 and TSU, 686, and UM-SSC-22B squamous cell carcinomas, treatment with **27** (1 μM) led to growth inhibition. Furthermore, compound **27** inhibited cell viability in several lung, liver, and AML cancer cell lines with a low IC_{50} of 0.1 μM.

In contrast to the approaches described above, Min and co-workers developed a reporter assay system in which the miR-21, miR-31, miR-92a-1-3p, or miR-223-3p binding sequences were cloned into the 3′ UTR of a secreted alkaline phosphatase (SEAP) gene.[104] Secreted reporter assays can be advantageous over other approaches as they enable monitoring of gene expression without the need for cell lysis, allowing for time-resolved studies and additional secondary assays to be performed on the same cells, such as RT-qPCR; however, secondary assays to account for small molecules that may inhibit secretion of the reporter should also be included. Initial screening performed by co-transfecting a miRNA expression plasmid and a reporter plasmid for the miRNA of interest into HEK293T cells followed by treatment with compounds from an in-house library led to the identification of an active compound with an aminosulfonylarylisoxazole scaffold. Ectopic expression of miRNAs may lead to exclusion of proteins or other targets present in the normal regulatory pathway of the miRNA thereby potentially lowering the hit rate of a small molecule screen. Additionally, this system could lead to an increase in false positive hits because the mechanisms governing artificial upregulation of the miRNAs differs from the endogenous mechanisms. Inhibition of endogenous miR-31 as well as six other miRNAs by compound **28** (Figure 12.8) was evaluated *via* RT-qPCR in A549 lung adenocarcinoma cells. Compound **28** was found to selectively inhibit miR-31 expression by ~50%. Treatment with **28** was also found to increase expression of downstream miR-31 targets serine/threonine kinase 40 (STK40) and early region 2 promoter binding factor (E2F) transcription factor 2 (E2F2), but had no effect on protein phosphatase 2 regulatory subunit Balpha (PPP2R2A) expression.

Rather than relying on a reporter system to monitor changes in miRNA function, the Hwang lab directly measured miRNA expression levels *via* RT-qPCR following treatment with small molecules.[105] After an initial screening of eight miRNAs that target B-cell lymphoma 2 (BCL2)/adenovirus E1B 19 kDa protein-interacting protein 3 (BNIP3), a regulator of cardiac cell death,[106] in hypoxic cardiomyocytes, it was determined that miR-182 was most effective at reducing expression of the protein following treatment with a miRNA mimic. In an effort to identify potential small molecules capable of inducing miR-182 expression to further investigate this interaction, compounds from an in-house library, including receptor agonists and antagonists, kinase inhibitors, and ion channel inhibitors and activators, were screened in hypoxic cardiomyocytes followed by RT-qPCR analysis for changes in miR-182 levels. Among the small molecules screened, the GSK3β inhibitor **26** (Figure 12.8) induced miR-182 expression by two-fold and inhibited expression of BNIP3 in hypoxic cardiomyocytes. Furthermore, treatment with **26** reduced apoptosis by ~25% relative to a DMSO control and led to a decrease in the production of hypoxia-induced reactive oxygen species (Figure 12.11a). The discovery that **26** induces expression of miR-182 and *let-7*, as discussed above,[99] may indicate that it is a non-specific enhancer of miRNAs or may reveal potential biological networks or gene regulatory pathways that link miR-182 and *let-7*. An SAR study of nine derivatives to identify modifications that would enhance the ability of **26** to induce miR-182 expression led

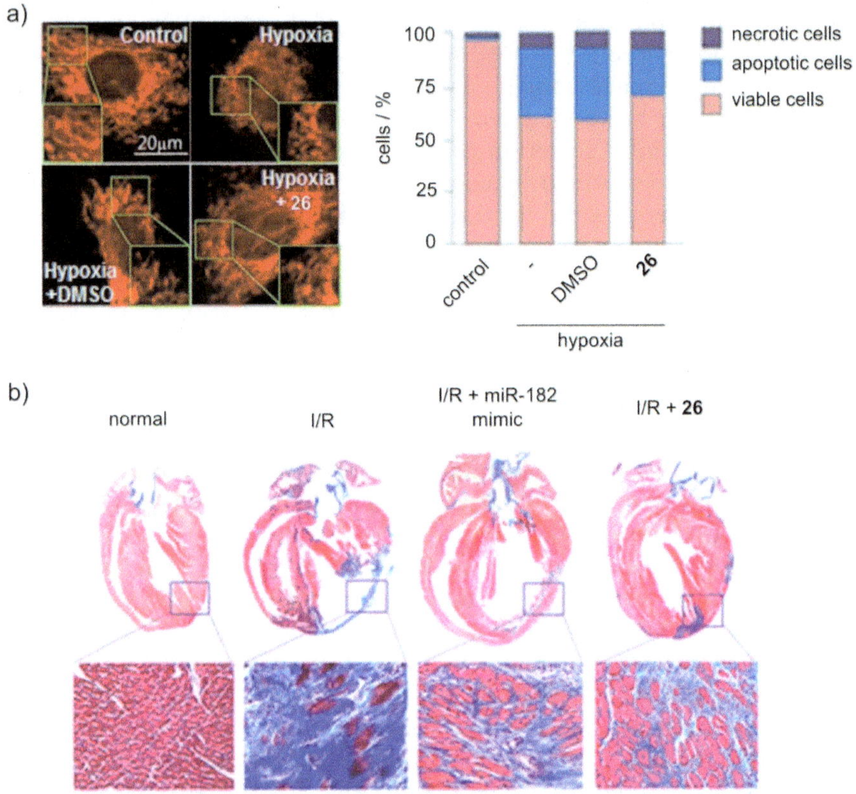

Figure 12.11 Effect of **26** on cardiomyocyte survival. (a) Treatment with small molecule **26** (10 μM) in cardiomyocytes inhibited hypoxia-induced mitochondrial fission (yellow box) and reduced apoptotic cell numbers by ~25%. Cells were stained with MitoTracker dye and imaged *via* confocal microscopy. (b) Delivery of miR-182 mimic or **26** after ischaemia/reperfusion (I/R) injury inhibits cardiac fibrosis in a rat heart model. Images were captured at 3 weeks post-injury following Trichrome staining. Reproduced from ref. 105 [https://doi.org/10.1038/srep23472], under the terms of a CC BY 4.0 license [https://creativecommons.org/licenses/by/4.0/].

to the identification of an analog bearing a nitro substituent, which elicited a ~1.5-fold improvement in activity relative to the parent compound and was also shown to inhibit BNIP3 expression. To determine if miR-182 levels were increased *via* a β-catenin-dependent pathway as a result of GSK3β inhibition,[107] miR-182 expression was examined in the presence of a β-catenin inhibitor. Subsequent treatment with compound **26** led to an induction of miR-182, suggesting that the GSK3β/β-catenin pathway is not an underlying mechanism of miR-182 expression. Despite the lack of a clear correlation between GSK3β and miR-182, the therapeutic potential of the analog **26** was evaluated in a rat ischemia–reperfusion (I/R) injury model. Delivery of a miR-182 mimic or compound **26** immediately after I/R injury led to a reduction in cardiac fibrosis (Figure 12.11b),[108] suggesting that induction of miR-182 *via* small molecules may hold therapeutic potential in the treatment of cardiovascular disease.

In summary, cell-based approaches have afforded the discovery of small molecule modulators of miRNAs from a variety of compound classes that may not have been identified through biochemical methods, since the presence of the entire miRNA biogenesis pathway, including upstream regulatory mechanisms, was required. The insertion of a miRNA target sequence in the 3′ UTR of a luciferase gene remains the most common reporter design as it provides high sensitivity and a positive readout (luminescence increase), which reduces false positive hits. Installation of a fully complementary binding sequence has been demonstrated to deliver improved dynamic range compared to the use of the 3′ UTR of a target gene, which may contain mismatches in the target site and increases background reporter activity. A key consideration in the use of luciferase as a reporter gene in cell-based assays is the necessity of employing secondary assays to rule out false positives, such as luciferase stabilizing molecules. Most of the compounds discovered act at the transcriptional level, pre-transcriptional level, or have no known mode of action. In order to better utilize these molecules as biological probes and potential therapeutics, future efforts should focus on identifying their protein targets.

12.4 Computational and Rational Design Approaches to miRNA Small Molecule Inhibitors

Rational design of miRNA-targeting small molecule inhibitors relies on structural information of both the target and the potential small molecule candidates.[109] This is in stark contrast to traditional phenotypic screening assays where the mode of action of potential hit compounds is unknown, necessitating additional assays for target identification. The ability to computationally predict RNA secondary and tertiary motifs accurately from sequences affords an accelerated approach to small molecule drug design.[110,111] However, the requirement for detailed structural information

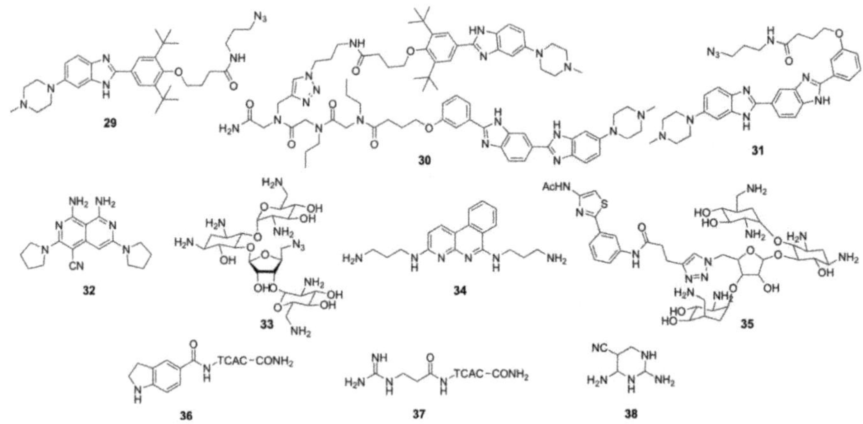

Figure 12.12 Structures of small molecules identified through rational design and computational approaches.

of a potential target can also be a limitation to rational drug design because structures of many disease-relevant RNAs still remain to be elucidated. Additionally, small molecule inhibitors that bind secondary RNA sequence motifs often display low micromolar binding affinity at best, rendering them less appealing as potential drug candidates. Here, reports involving clever rational design and computational approaches to targeting miRNAs with small molecules (Figure 12.12) are summarized.[109]

The Disney lab developed a new approach for small molecule lead identification, termed Inforna, which integrates experimentally determined RNA motif–small molecule interactions with an algorithm for scoring such interactions based on statistical analysis (Figure 12.13).[2] The initial identification of RNA motif–small molecule interactions is carried out by 2DCS, as described in Section 12.2.[28] Following selection of RNA motifs, these sequences are analyzed by SAR studies through sequencing (StARTS). StARTS is a statistical approach that enables identification of specific features in RNAs by comparing their rate of occurrence in a specific RNA to the rate of occurrence in the entire RNA library.[112,113] Combination of 2DCS data and StARTS analysis within the Inforna platform serves as a database that facilitates the identification of specific small molecule–miRNA pairs based upon tight binding interactions.

The Inforna approach was utilized in the discovery of small molecules that bind the Dicer and Drosha sites in disease-relevant pre-miRNAs and pri-miRNAs, respectively.[2] A compound set of 11 benzimidazoles and aminoglycosides, previously identified to bind RNAs,[114,115] were screened against 794 RNA motifs. Compound **29** (Figure 12.12) targeted the Drosha cleavage site of pre-miR-96 and was predicted to have the highest affinity among the 29 small molecule–pre-miRNA binding interactions identified through 2DCS and StARTS analysis, and thus was selected for further evaluation. In MCF-7 cells, compound **29** (40 μM) reduced the levels of mature miR-96 by 90% and

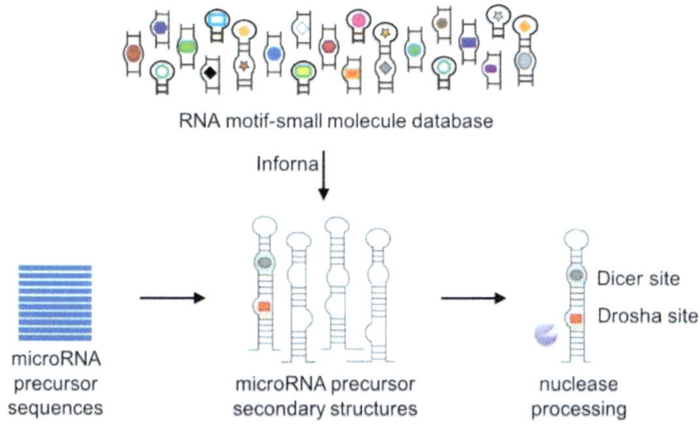

Figure 12.13 Identified sequences from two-dimensional combinatorial screening (2DCS) are analyzed through a computational approach StARTS to identify lead small molecule–RNA motif interactions. Inforna analysis extracts the secondary structural motifs from each RNA and compares them to the experimental results from 2DCS. Reproduced from ref. 2 with permission from Springer Nature, Copyright 2014.

pre-miR-96 levels by 50% while eliciting a 2.5-fold increase in pri-miR-96 expression, consistent with compound **29**-mediated inhibition observed in an *in vitro* Drosha processing assay. Pre-miR-96, pre-miR-182, and pre-miR-183 belong to a polycistronic miRNA cluster, and thus are transcriptionally co-regulated.[116] Interestingly, **29** was found to specifically downregulate only miR-96 expression, while having no effect on mature miR-182 and miR-183 levels, in good agreement with the hypothesis that **29** inhibits miR-96 by direct RNA binding and not by targeting of an upstream regulatory protein. Furthermore, RT-qPCR profiling of 149 additional miRNAs following treatment with **29** revealed miR-96 to be the only downregulated target; however, transfection with a miR-96 antagomir led to >2.5-fold variations in the expression levels of 11 other miRNAs along with miR-96, suggesting that **29** shows higher specificity towards miR-96 biogenesis. miR-96 is upregulated in cancer cells and is known to repress the pro-apoptotic transcription factor Forkhead box protein O1 (FOXO1).[117–119] Endogenous expression of FOXO1 increased 2.5-fold upon treatment with **29** (40 μM) in MCF-7 cells consistent with an increase in apoptosis detected *via* a TUNEL assay.

An additional bis-benzimidazole identified *via* Inforna to target a 1 × 1 nucleotide internal loop close to the Drosha site on pri-miR-96[120] was conjugated to **29** by a peptoid backbone yielding the dimeric compound **30** (Figure 12.12), in an attempt to generate a more potent inhibitor. After optimizing the length of the peptoid backbone, compound **30** (with a spacer of two glycine units) was found to increase pri-miR-96 levels by ~2-fold while reducing pre- and miR-96 levels by almost 50% at a concentration of 50 nM, indicating that **30** maintains a high affinity for pri-miR-96. Compound **30** binds pri-miR-96 with a K_d of 85 nM, a ~40-fold increase compared to **29**, and elicited a reduction in mature miR-96 levels with an IC_{50} of 50 nM. FOXO1 levels increased two-fold, consistent with an 80% induction of apoptosis following treatment of MDA-MB-231 cells with **30** (50 nM). Furthermore, **30** (10 mg kg^{-1}) inhibited tumor growth in immunodeficient mice injected with MDA-MB-231 cells (Figure 12.14a) concurrent with changes in miR-96, pri-miR-96, and FOXO1 levels (Figure 12.14b).

In contrast to **29** and **30**[121] binding the Drosha processing site of pre-miR-96, compound **31** (Figure 12.12) was identified *via* Inforna to bind the Dicer

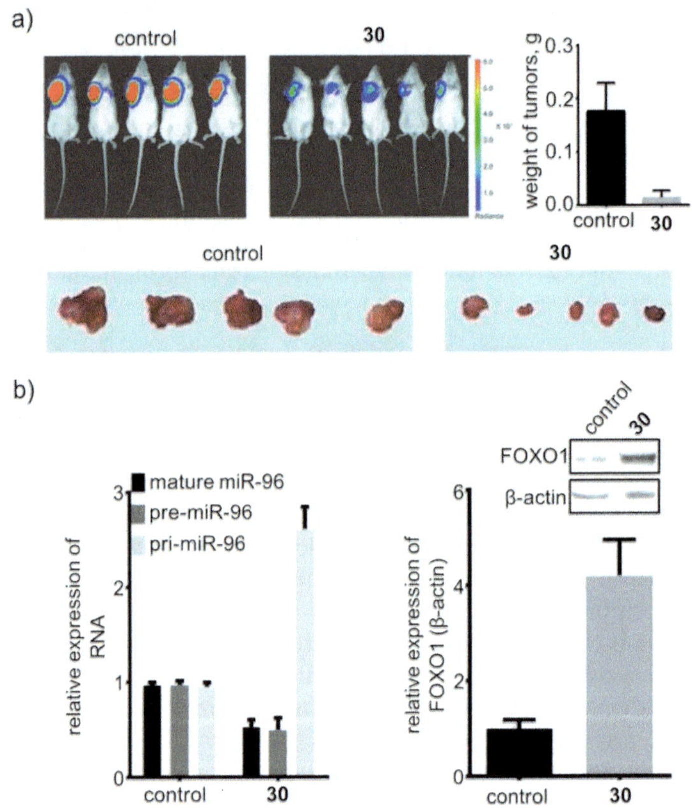

Figure 12.14 Effect of the small molecule miRNA inhibitor **30** on immunodeficient mice implanted with MDA-MB-231 cells expressing constitutively active luciferase. Tumor masses were resected and monitored 21 days post-implantation. (a) Treatment with **30** resulted in reduced tumor masses as seen by luciferase expression. Representative images of excised tumors relative to untreated control are shown. (b) Reverse transcription quantitative PCR (RT-qPCR) profiling of the excised tumors revealed increased levels of pri-miR-96 while pre-miR-96 and mature miR-96 levels were reduced. FOXO1 expression increased following treatment with **30**. Reproduced from ref. 120 with permission from PNAS.

cleavage site on pre-miR-210 with a K_d of 200 nM. Inforna analysis identified 15 other miRNAs that contain motifs that can also bind to **31** with varying affinities; however, in a RT-qPCR analysis in MDA-MB-231 cells only miR-210 maturation was affected, as indicated by a ~50% reduction in mature miR-210 levels. miR-210 acts as a central regulator of hypoxic response in solid tumors[122] by inhibiting the expression of glycerol-3-phosphate dehydrogenase 1-like enzyme (GPD1L) and subsequent repression of prolyl hydroxylase (PHD). Under normoxic conditions, PHD hydroxylates prolines on hypoxia-inducible factor 1-alpha (HIF-1α), leading to proteolytic degradation of HIF-1α.[123] However, under hypoxic conditions, miR-210 is upregulated, inhibiting expression of GPD1L.[122] In MDA-MB-231 cells cultured under hypoxic conditions, miR-210 and HIF-1α protein levels were upregulated by 15-fold and 10-fold, respectively, while GPD1L protein levels were decreased by 10-fold compared to cells cultured under normoxic conditions. Following treatment with **31** (200 nM), HIF-1α expression levels were reduced by ~75% and GPD1L levels were increased four-fold. Immunodeficient mice injected with MDA-MB-231 cells displayed reduced tumor sizes upon treatment with **31** relative to untreated control, suggesting that **31** may hold some therapeutic potential. As expected, RT-qPCR analysis of the resected tumor masses showed mature miR-210 and HIF-1α levels decreased to ~10% and ~25% of levels in untreated controls, respectively, while GPD1L expression was doubled in **31**-treated mice compared to untreated mice.

Haga *et al.* employed Inforna to identify small molecule binders of a 1 × 1 nucleotide UU internal loop in pre-miR-544 and discovered **32** (Figure 12.12) to interact with the Dicer cleavage site.[124] miR-544 was previously found to induce expression of ataxia telangiectasia mutated (ATM) protein.[125] Under hypoxic conditions, induction of ATM protein leads to reduced expression of mammalian target of rapamycin complex 1 (mTORC1), thereby restricting cell growth.[126] Additionally, two potential miR-544 binding sites within the 3′-UTR of mTOR have been identified by a bioinformatics analysis.[124] RT-qPCR analysis confirmed that ATM and miR-544 were upregulated while mTOR was downregulated under hypoxic conditions in MDA-MB-231 and MCF-7 cells. Upon treatment with **32** (20 nM), miR-544 levels were reduced by 80%, with a corresponding reduction in ATM levels, while mTOR expression was increased 10-fold. Annexin V staining revealed that apoptosis (~70% of cells) in hypoxic MDA-MB-231 cells was more pronounced upon exposure to **32** compared to transfection of a miR-544 antagomir (~60% apoptotic cells). Furthermore, drug resistant MDA-MB-231 and MCF-7 cells were found to be sensitized to the chemotherapeutic agent 5-fluorouracil upon co-treatment with **32**. Similar to **31** discussed above, treatment of an immunodeficient mouse model with **32** (40 μM) led to decreases in tumor size concurrent with reductions in expression levels of hypoxia-associated factors ATM and HIF-1α isolated from the excised tumors.

Inforna was also utilized in the identification of a potential interaction between azido-neomycin B **33** (Figure 12.12) and miR-525.[127] Treatment with **33** (25 μM) decreased both mature and pre-miR-525 levels by 40% in HepG2, while pri-miR-525 levels were increased two-fold, suggesting that **33** binds the Drosha site on pri-miR-525. The invasive and migratory properties of HCC cells is, in part, the result of zinc finger protein 395 (ZNF395) repression by miR-525.[128] Western blotting analysis showed a two-fold increase in ZNF395 expression in HepG2 cells exposed to **33**.[127] Expression levels of 19 other miRNAs predicted to bind the ZNF395 3′ UTR were unaffected following treatment with **33**;[127] however, a previous study showed decreased levels of miR-517c, miR-518e, miR-519d, and miR-525 in response to **33**,[2] suggesting that while **33** is likely not a general miRNA pathway inhibitor it is not specific for miR-525.

Targeting bulges with small molecules is advantageous compared to double-stranded base-paired RNA due to a larger binding pocket and exposed unpaired nucleobase(s).[4] Small molecules such as 2-acylamino-1,8-naphthyridine[129] and *N*,*N*-bis(3-aminopropyl)-2,7-diamino-1,8-naphthyridine (DANP)[130,131] have been reported to bind a single purine or pyrimidine bulge *via* hydrogen bonding, mimicking natural base pairing interactions. Furthermore, molecular modeling of the RNA–DANP complex suggested π-π stacking interactions of DANP with the neighboring base pairs to be a stabilizing interaction in the B-form RNA duplex, but may be insufficient to promote binding to the A-form RNA duplex (Figure 12.15a).[132] The Nakatani lab developed the benzo-DANP **34** (Figure 12.12) to bind a single cytosine bulge near the Dicer cleavage site in pre-miR-29a with the hypothesis that expansion to a three-ring system would enable formation of π-π stacking interactions and a similar hydrogen bond network as DANP, but with improved binding affinity (Figure 12.15a and b). miR-29a is known to enhance invasion and cell migration in nasopharyngeal carcinoma.[133] *In vitro* processing of pre-miR-29a by recombinant Dicer was reduced to ~50% in the presence of **34** (200 μM), with an IC_{50} of 70 μM (Figure 12.15c and d). As expected, **34** (200 μM) did not significantly affect the Dicer processing of pre-miR-29a without the C-bulge. However, Hela cells exposed to **34** (100 μM) elicited a two-fold increase in miR-29a levels, in stark contrast to previous examples of direct binding between small molecules and pre-miRNA.[121,124] Additionally, expression levels of five other miRNAs (miR-21, miR-92a-1, miR-16-1, miR-15a, and miR-34a) exhibited increased expression following treatment with **34**. The authors rationalized that the high concentration of **34** elicited off-target effects contributing to this unexpected result, but this was not experimentally confirmed.[132]

miR-372 and miR-373, which share the same seed sequence, act as oncomiRs and are upregulated in gastric cancer, as well as testicular germ cell tumors, esophageal tumors, and thyroid adenomas.[134–137] The Duca lab developed multimodal ligands which bind to the stem-loop region of pre-miR-372 and pre-miR-373 and inhibit Dicer processing.[34] The ligands were based on two RNA-binding motifs: (1) an aminoglycoside that exhibits general high-affinity binding to the stem–loop region of RNAs;[138–141] and (2) an artificial nucleobase capable of forming specific Hoogsteen bond interactions, enabling formation of triplex structures with the double-stranded region of the pre-miRNA.[142,143] In an analogous approach to that employed by other

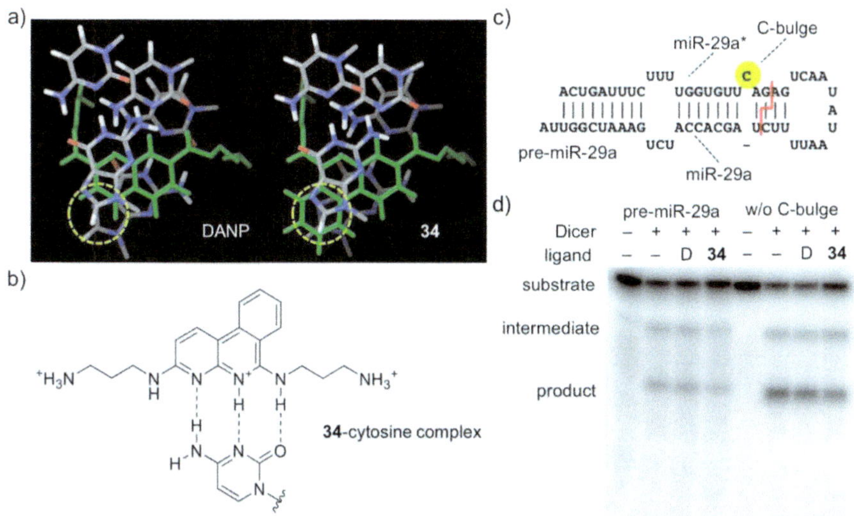

Figure 12.15 (a) Molecular modeling simulation shows the position of DANP (green, left) and **34** (green, right) together with the neighboring bases. Stabilizing π-π stacking interactions in **34** are highlighted by dashed yellow circles. RNA phosphate backbone and ribose sugar are not shown for clarity. (b) Hydrogen bond network in the **34**-cytosine complex. (c) Secondary structure of pre-miR-29a with the C-bulge (highlighted in yellow). The red line indicates Dicer cleavage site. (d) Denaturing polyacrylamide gel electrophoresis (PAGE) analysis of *in vitro* Dicer processing assay following DANP (D) and treatment with **34**. Reproduced from ref. 132 with permission from American Chemical Society, Copyright 2016.

labs discussed in Section 12.2, a FRET-based assay with a fluorophore (fluorescein) and quencher (dabcyl) on the 3′ and 5′ termini of pre-miR-372 was used to assess Dicer activity. Upon screening of aminoglycosides, neomycin was found to completely inhibit Dicer cleavage at 250 μM and thus was selected for stem–loop binding. The multimodal ligands were synthesized by conjugating several artificial and natural nucleobases bearing alkynes to an azide-modified analog of neomycin. After screening the ligands in the FRET assay, **35** (Figure 12.12) was identified as the best inhibitor with an IC_{50} of 2 μM and a K_d of 16 nM. The artificial nucleobase in **35** is known to form three hydrogen bonds with the A-T base pair (Figure 12.16a).[144,145] Importantly, neomycin alone showed a dramatically higher IC_{50} of 125 μM and a K_d of 7.4 μM, highlighting the synergistic effect of both motifs on RNA binding.[34] Both miR-372 and miR-373 are highly upregulated in gastric cancers, while virtually undetectable in normal gastric epithelial cells. Additionally, these miRNAs repress the translation of large tumor suppressor 2 (LATS2), a serine–threonine kinase involved in cell cycle regulation.[146] To evaluate the pre-therapeutic potential of **35**, AGS gastric cancer cells overexpressing miR-372 and miR-373 were treated with the ligand at 100 μM. After 5 days, the cells exhibited dose-dependent growth inhibition, whereas no significant

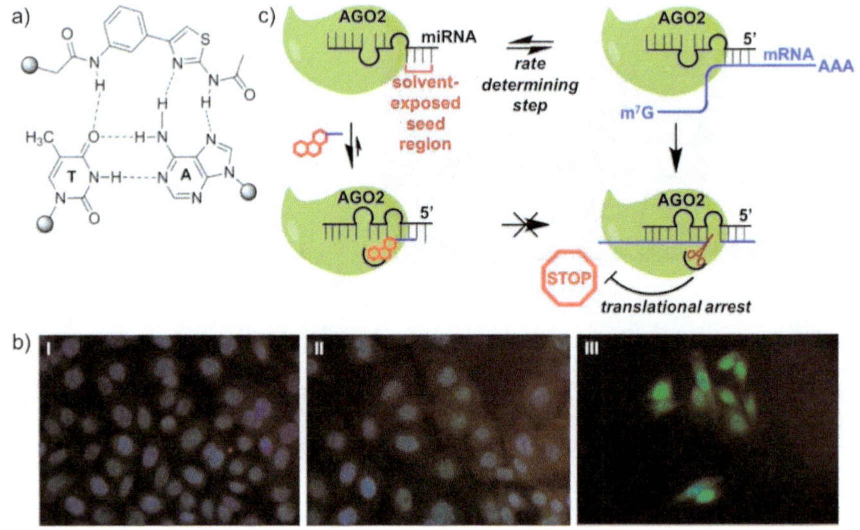

Figure 12.16 (a) Hoogsteen-type base pairing interactions of A-T Watson–Crick base pair with the artificial nucleobase motif of **35**. (b) Immunofluorescence imaging reveals LATS2 protein localization (green) in the DAPI-stained nuclei (blue) of AGS cells treated with **35**. (I) Untreated cells; (II) cells treated with neomycin (50 µM); (III) cells treated with **35** (50 µM). (c) Representation of the multimodal inhibitors of miR-122 consisting of AGO2 active site inhibitor– peptide nucleic acid PNA sequence hybrids (red–blue) which bind AGO2-miRNA complex. Adapted from ref. 34 and ref. 148 with permission from American Chemical Society, Copyright 2014 and 2013.

inhibition was observed in a neomycin control, suggesting cytostatic rather than cytotoxic effects. Immunolabeling of LATS2 protein in AGS cells after exposure to **35** (50 µM) for 5 days revealed LATS2 protein to be upregulated and localized in the nuclei and cytoplasm (Figure 12.16b), due to reduction of miR-372 and miR-373 levels. Furthermore, pri-miR-372 and pri-miR-373 levels remained almost unchanged in **35**-treated AGS cells, suggesting the mode of action of **35** to be post-transcriptional. However, this result was not confirmed as mature miR-372 and miR-373 levels were not quantified in these cells. Additionally, treatment with the ligand led to downregulation of miR-17-5p, miR-21, and miR-200b, indicating the binding interaction between **35** and pre-miR-372/373 to be non-specific.

In addition to targeting the Dicer cleavage site, efforts have also been made towards identifying small molecule inhibitors of AGO2. The binding of the AGO2–mature miRNA complex to the 3′-UTR of a target mRNA proceeds with the solvent-exposed seed region (an 8-nucleotide region from the miRNA 5′ end) interacting with the mRNA as a rate-dependent step.[147] The Schmidt lab reported a class of multimodal inhibitors consisting of a short PNA sequence complementary to the bases in the miR-122 seed region covalently attached to an AGO2 active site inhibitor (Figure 12.16c).[148] An *in silico* approach was adopted to identify a small molecule ligand for AGO2 by screening the ZINC fragment subset[149] of 627 000 compounds against the AGO2 active site using genetic optimization for ligand docking (GOLD) software.[150] GOLD predicts binding modes of ligands in various conformations and a target protein. The Chemscore piecewise linear potential (ChemPLP) function, an empirical fitness function that scores interactions between ligands and proteins by modeling steric complementarity between structures,[151] was utilized to rank the interactions revealing 1-(4-phenoxybenzyl)guanidine as the most promising candidate. The GOLD program was also used to identify a tetrameric PNA sequence (TCAC) by superimposing the PNA onto the AGO2-miRNA bound structure.[148,152] The *N*-acetylated-PNA sequence exhibited an IC_{50} of 1 μM, indicating that the tetrameric PNA alone was not sufficient for inhibition. However, when the identified AGO2 inhibitor and the PNA were combined, docking studies were unable to predict a plausible binding conformation, prompting evaluation of different linkers and heterocyclic motifs commonly found in medicinal chemistry.[153] The selected analogs were then synthesized and evaluated in an activity assay where the AGO2 protein was incubated with the compounds, mature miR-122, and a fluorescently labeled anti-miR-122 substrate. Cleavage products were then visualized by denaturing polyacrylamide gel electrophoresis. Two inhibitors, **36** and **37** (Figure 12.12), exhibited an IC_{50} of 100 nM. The tetrameric PNA sequence was proposed to confer specificity to these inhibitors towards miR-122; however, their effect on other miRNAs was not investigated. Additionally, no cell-based assays were reported, which may be attributed to poor cellular uptake of PNAs.

The Zhang group developed an *in silico* screening method based on the tertiary structure of the Dicer binding site of pre-miR-21.[154] The authors constructed a three-dimensional model of the pre-miR-21 by utilizing the MC-Fold/MC-Sym pipeline, a web-hosted service for prediction of RNA secondary and tertiary structures.[155] The pipeline consists of two programs, MC-Fold and MC-Sym, working in tandem. The RNA sequence is uploaded to MC-Fold, yielding a secondary structure that is directly fed into MC-Sym to generate the tertiary structure. Energy optimization was carried out on the RNA tertiary structure with TINKER molecular modeling software.[156] The optimized RNA structure was then utilized in a molecular docking study using Autodock.[154] A library of 1990 structures from the NCI diversity database was virtually screened to reveal 48 compounds with high binding affinity towards the Dicer binding site on pre-miR-21. Without performing *in vitro* Dicer processing or RNA binding assays, miR-21 inhibition in response to treatment with the top five compounds was directly evaluated in four human epithelial cancer cell lines (U87, LN229, SGC7901, and MCF-7). It is important to note that *in silico* predictive approaches are limited by the quality and amount of data provided to them, thus *in vitro* assays should be employed to confirm activity. Despite further validation beyond the virtual screen, compound **38** (30 μM; Figure 12.12) was identified as the most potent with ~50% inhibition of miR-21 as determined by RT-qPCR. Levels of 11 other miRNAs remained unchanged following treatment with **38**, suggesting that **38** has specificity toward miR-21. To confirm that **38** inhibits pre-miR-21 processing, Dicer expression was inhibited in U87 and MCF-7 cells by siRNA treatment.

Following siRNA knockdown of Dicer, miR-21 levels were decreased to ~50% and subsequent treatment with **38** did not affect miR-21 levels further, supporting that **38** binds to the Dicer cleavage site of pre-miR-21.[154] Additionally, exposure to **38** (30 μM) led to increased expression levels of miR-21 targets, including PTEN, PDCD4 and reversion-inducing-cysteine-rich protein with kazal motifs (RECK); however, no further pre-therapeutic evaluation of **38** was performed.

Taken together, rational design approaches have mainly targeted primary and precursor miRNAs, presumably due to the availability of more complex secondary structures for small molecule targeting. Although the approaches vary broadly, Drosha and Dicer cleavage sites in pri- and pre-miRNAs are attractive features for small molecule targeting due to the ease of high-throughput screening and computational prediction of secondary motifs. However, rigorous biochemical and cell-based secondary assays need to be performed on potential hits to confirm predicted target specificity.

12.5 Summary and Outlook

Based on the success stories summarized here, the continued discovery of small molecule reagents capable of selectively perturbing miRNA function will be highly valuable for elucidating miRNA biology and for providing new therapeutic modalities for miRNA-related diseases. Some small molecules have shown enhanced potency and selectivity compared to antagomirs in cell-based assays and *in vivo* studies and could serve as drug candidates after structural modifications to improve pharmacological and pharmacokinetic properties are being performed. Specifically, many miRNA-targeting small molecules must be delivered at high concentrations to achieve a phenotypic response, which may lead to off target effects in cellular environments. Current approaches have primarily focused on identifying miRNA inhibitors, but few have explored upregulation of specific miRNAs—an important area of investigation for additional studies. Future efforts may also be directed at targeting RNA-binding protein–RNA interactions in the miRNA biogenesis pathway with small molecules as a current focus has been placed on targeting RNA only. Aminoglycoside and benzimidazole scaffolds possess structural features that facilitate interactions with RNA targets and are common in many rational design approaches. However, the identification of novel scaffolds and pharmacophores that target miRNAs with high specificity requires continued efforts.

The various applications of *in vitro*, cell-based, and computational approaches described above have all achieved success in discovering novel probes, with each approach having distinct advantages and disadvantages. More recently, the most innovative work has combined multiple aspects from each approach in the hopes of capitalizing on the successes and overcoming the pitfalls of any one individual approach. Future studies are likely to continue this trend in the hopes of overcoming existing challenges in targeting miRNAs with small molecules.

Acknowledgments

Work in our laboratory on the discovery of miRNA inhibitors has been supported, in part, by a Research Scholar Grant (120130-RSG-11-066-01-RMC) from the American Cancer Society, the Teva USA Scholars Grants Program (50181-TEV), and the NIH (R21 NS073068).

References

1. C. M. Connelly, M. H. Moon and J. S. Schneekloth Jr, *Cell Chem. Biol.*, 2016, **23**, 1077–1090.
2. S. P. Velagapudi, S. M. Gallo and M. D. Disney, *Nat. Chem. Biol.*, 2014, **10**, 291.
3. L. Guan and M. D. Disney, *ACS Chem. Biol.*, 2012, **7**, 73–86.
4. J. R. Thomas and P. J. Hergenrother, *Chem. Rev.*, 2008, **108**, 1171–1224.
5. A. C. Stelzer, A. T. Frank, J. D. Kratz, M. D. Swanson, M. J. Gonzalez-Hernandez, J. Lee, I. Andricioaei, D. M. Markovitz and H. M. Al-Hashimi, *Nat. Chem. Biol.*, 2011, **7**, 553.
6. J. Parsons, M. P. Castaldi, S. Dutta, S. M. Dibrov, D. L. Wyles and T. Hermann, *Nat. Chem. Biol.*, 2009, **5**, 823.
7. A. Davidson, T. C. Leeper, Z. Athanassiou, K. Patora-Komisarska, J. Karn, J. A. Robinson and G. Varani, *Proc. Natl. Acad. Sci. U. S. A.*, 2009, **106**, 11931–11936.
8. J. F. Arambula, S. R. Ramisetty, A. M. Baranger and S. C. Zimmerman, *Proc. Natl. Acad. Sci. U. S. A.*, 2009, **106**, 16068–16073.
9. J. L. Childs-Disney, J. Hoskins, S. G. Rzuczek, C. A. Thornton and M. D. Disney, *ACS Chem. Biol.*, 2012, 7, 856–862.
10. L. O. Ofori, J. Hoskins, M. Nakamori, C. A. Thornton and B. L. Miller, *Nucleic Acids Res.*, 2012, **40**, 6380–6390.
11. A. Kumar, R. Parkesh, L. J. Sznajder, J. L. Childs-Disney, K. Sobczak and M. D. Disney, *ACS Chem. Biol.*, 2012, 7, 496–505.
12. S. A. Barros, I. Yoon and D. M. Chenoweth, *Angew. Chem., Int. Ed. Engl.*, 2016, **128**, 8398–8401.
13. R. C. Friedman, K. K.-H. Farh, C. B. Burge and D. P. Bartel, *Genome Res.*, 2009, **19**, 92–105.
14. S. M. Hammond, *Adv. Drug Delivery Rev.*, 2015, **87**, 3–14.
15. M. Hafner, M. Landthaler, L. Burger, M. Khorshid, J. Hausser, P. Berninger, A. Rothballer, M. Ascano Jr, A.-C. Jungkamp, M. Munschauer, A. Ulrich, G. S. Wardle, S. Dewell, M. Zavolan and T. Tuschl, *Cell*, 2010, **141**, 129–141.
16. M. Tatsuguchi, H. Y. Seok, T. E. Callis, J. M. Thomson, J.-F. Chen, M. Newman, M. Rojas, S. M. Hammond and D.-Z. Wang, *J. Mol. Cell. Cardiol.*, 2007, **42**, 1137–1141.
17. Y. Peng and C. M. Croce, *Signal Transduction Targeted Ther.*, 2016, **1**, 15004.
18. D. Jin and H. Lee, *Sci. Rep.*, 2016, **6**, 35350.

19. M. A. Jafri, M. H. Al-Qahtani and J. W. Shay, *Semin. Cancer Biol.*, 2017, **44**, 117–131.
20. S. M. Hammond, *Curr. Opin. Genet. Dev.*, 2006, **16**, 4–9.
21. A. A. Svoronos, D. M. Engelman and F. J. Slack, *Cancer Res.*, 2016, **76**, 3666–3670.
22. M. Y. Shah, A. Ferrajoli, A. K. Sood, G. Lopez-Berestein and G. A. Calin, *EBioMedicine*, 2016, **12**, 34–42.
23. C. M. Connelly, F. A. Abulwerdi and J. S. Schneekloth, in *Small Molecule Microarrays: Methods and Protocols*, ed. M. Uttamchandani and S. Q. Yao, Springer New York, New York, NY, 2017, pp. 157–175.
24. S. Chirayil, R. Chirayil and K. J. Luebke, *Nucleic Acids Res.*, 2009, **37**, 5486–5497.
25. J. Pai, S. Hyun, J. Y. Hyun, S.-H. Park, W.-J. Kim, S.-H. Bae, N.-K. Kim, J. Yu and I. Shin, *J. Am. Chem. Soc.*, 2016, **138**, 857–867.
26. C. M. Connelly, R. E. Boer, M. H. Moon, P. Gareiss and J. S. Schneekloth, *ACS Chem. Biol.*, 2017, **12**, 435–443.
27. M. Maiti, K. Nauwelaerts and P. Herdewijn, *Bioorg. Med. Chem. Lett.*, 2012, **22**, 1709–1711.
28. M. D. Disney, L. P. Labuda, D. J. Paul, S. G. Poplawski, A. Pushechnikov, T. Tran, S. P. Velagapudi, M. Wu and J. L. Childs-Disney, *J. Am. Chem. Soc.*, 2008, **130**, 11185–11194.
29. M. D. Disney and A. J. Angelbello, *Acc. Chem. Res.*, 2016, **49**, 2698–2704.
30. S. P. Velagapudi, M. G. Costales, B. R. Vummidi, Y. Nakai, A. J. Angelbello, T. Tran, H. S. Haniff, Y. Matsumoto, Z. F. Wang, A. K. Chatterjee, J. L. Childs-Disney and M. D. Disney, *Cell Chem. Biol.*, 2018, **25**, 1086–1094.e7.
31. A. Sunters, C. J. Springer, K. D. Bagshawe, R. L. Souhami and J. A. Hartley, *Biochem. Pharmacol.*, 1992, **44**, 59–64.
32. B. P. Davies and C. Arenz, *Bioorg. Med. Chem.*, 2008, **16**, 49–55.
33. P. Davies Brian and C. Arenz, *Angew. Chem., Int. Ed. Engl.*, 2006, **45**, 5550–5552.
34. D. D. Vo, C. Staedel, L. Zehnacker, R. Benhida, F. Darfeuille and M. Duca, *ACS Chem. Biol.*, 2014, **9**, 711–721.
35. D. Bose, G. G. Jayaraj, S. Kumar and S. Maiti, *ACS Chem. Biol.*, 2013, **8**, 930–938.
36. M. Klemm Claudine, A. Berthelmann, S. Neubacher and C. Arenz, *Eur. J. Org. Chem.*, 2009, **2009**, 2788–2794.
37. D. Vo Duc, A. Tran Thi Phuong, C. Staedel, R. Benhida, F. Darfeuille, A. Di Giorgio and M. Duca, *Chem. - Eur. J.*, 2016, **22**, 5350–5362.
38. M. Chittapragada, S. Roberts and Y. W. Ham, *Perspect. Med. Chem.*, 2009, **3**, 21–37.
39. T. P. A. Tran, D. D. Vo, A. Di Giorgio and M. Duca, *Bioorg. Med. Chem.*, 2015, **23**, 5334–5344.
40. M. Roos, U. Pradère, R. P. Ngondo, A. Behera, S. Allegrini, G. Civenni, J. A. Zagalak, J.-R. Marchand, M. Menzi, H. Towbin, J. Scheuermann, D. Neri, A. Caflisch, C. V. Catapano, C. Ciaudo and J. Hall, *ACS Chem. Biol.*, 2016, **11**, 2773–2781.

41. Y. Nam, C. Chen, R. I. Gregory, J. J. Chou and P. Sliz, *Cell*, 2011, **147**, 1080–1091.
42. D. Lim, W. G. Byun, J. Y. Koo, H. Park and S. B. Park, *J. Am. Chem. Soc.*, 2016, **138**, 13630–13638.
43. H. L. Lightfoot, E. A. Miska and S. Balasubramanian, *Org. Biomol. Chem.*, 2016, **14**, 10208–10216.
44. G. S. Tan, C.-H. Chiu, B. G. Garchow, D. Metzler, S. L. Diamond and M. Kiriakidou, *ACS Chem. Biol.*, 2012, **7**, 403–410.
45. (a) R. G. Gonzalez, R. S. Haxo and T. Schleich, *Biochemistry*, 1980, **19**, 4299–4303; (b) L. Wang, R. G. Rowe, A. Jaimes, C. Yu, Y. Nam, D. S. Pearson, J. Zhang, X. Xie, W. Marion, G. J. Heffron, G. Q. Daley and P. Sliz, *Cell Rep.*, 2018, **23**, 3091–3101.
46. D. A. Lorenz, J. M. Song and A. L. Garner, *Bioconjugate Chem.*, 2015, **26**, 19–23.
47. D. A. Lorenz, S. Vander Roest, M. J. Larsen and A. L. Garner, *SLAS Discovery*, 2017, **23**, 47–54.
48. D. Lorenz, T. Kaur, S. A. Kerk, E. Gallagher, J. Sandoval and A. L. Garner, *ACS Med. Chem. Lett.*, 2018, **9**, 517–521.
49. J. B. Baell and G. A. Holloway, *J. Med. Chem.*, 2010, **53**, 2719–2740.
50. N. Thorne, J. Inglese and D. S. Auld, *Chem. Biol.*, 2010, **17**, 646–657.
51. N. Thorne, D. S. Auld and J. Inglese, *Curr. Opin. Chem. Biol.*, 2010, **14**, 315–324.
52. D. S. Auld, N. Thorne, D.-T. Nguyen and J. Inglese, *ACS Chem. Biol.*, 2008, **3**, 463–470.
53. G. Shan, Y. Li, J. Zhang, W. Li, K. E. Szulwach, R. Duan, M. A. Faghihi, A. M. Khalil, L. Lu, Z. Paroo, A. W. S. Chan, Z. Shi, Q. Liu, C. Wahlestedt, C. He and P. Jin, *Nat. Biotechnol.*, 2008, **26**, 933.
54. S. M. Daniels, C. E. Melendez-Peña, R. J. Scarborough, A. Daher, H. S. Christensen, M. El Far, D. F. J. Purcell, S. Lainé and A. Gatignol, *BMC Mol. Biol.*, 2009, **10**, 38.
55. S. Melo, A. Villanueva, C. Moutinho, V. Davalos, R. Spizzo, C. Ivan, S. Rossi, F. Setien, O. Casanovas, L. Simo-Riudalbas, J. Carmona, J. Carrere, A. Vidal, A. Aytes, S. Puertas, S. Ropero, R. Kalluri, C. M. Croce, G. A. Calin and M. Esteller, *Proc. Natl. Acad. Sci. U. S. A.*, 2011, **108**, 4394–4399.
56. E. J. Sousa, I. Graça, T. Baptista, F. Q. Vieira, C. Palmeira, R. Henrique and C. Jerónimo, *Epigenetics*, 2013, **8**, 548–558.
57. N. S. Abell, M. Mercado, T. Cañeque, R. Rodriguez and B. Xhemalce, *J. Am. Chem. Soc.*, 2017, **139**, 1400–1403.
58. K. Gumireddy, D. D. Young, X. Xiong, J. B. Hogenesch, Q. Huang and A. Deiters, *Angew. Chem., Int. Ed. Engl.*, 2008, **47**, 7482–7484.
59. Y.-H. Feng and C.-J. Tsao, *Biomed. Rep.*, 2016, **5**, 395–402.
60. T. D. Schmittgen, J. Jiang, Q. Liu and L. Yang, *Nucleic Acids Res.*, 2004, **32**, e43.
61. Y. Changmin, Q. Linghui, G. Jingyan, F. Jiaqi, Y. Peiyan, S. C. L. Yao and S. Q. Yao, *Angew. Chem., Int. Ed. Engl.*, 2016, **55**, 9272–9276.
62. C.-S. Jiang, X.-M. Wang, S.-Q. Zhang, L.-S. Meng, W.-H. Zhu, J. Xu and S.-M. Lu, *Bioorg. Med. Chem.*, 2015, **23**, 6510–6519.

63. Y. Naro, N. Ankenbruck, M. Thomas, Y. Tivon, C. M. Connelly, L. Gardner and A. Deiters, *J. Med. Chem.*, 2018, **61**, 5900–5909.
64. D. D. Young, C. M. Connelly, C. Grohmann and A. Deiters, *J. Am. Chem. Soc.*, 2010, **132**, 7976–7981.
65. Y. Naro, M. Thomas, M. D. Stephens, C. M. Connelly and A. Deiters, *Bioorg. Med. Chem. Lett.*, 2015, **25**, 4793–4796.
66. L. B. Frankel, N. R. Christoffersen, A. Jacobsen, M. Lindow, A. Krogh and A. H. Lund, *J. Biol. Chem.*, 2008, **283**, 1026–1033.
67. D. Bose, G. Jayaraj, H. Suryawanshi, P. Agarwala, S. K. Pore, R. Banerjee and S. Maiti, *Angew. Chem., Int. Ed. Engl.*, 2012, **51**, 1019–1023.
68. A. P. Carter, W. M. Clemons, D. E. Brodersen, R. J. Morgan-Warren, B. T. Wimberly and V. Ramakrishnan, *Nature*, 2000, **407**, 340.
69. C. Esau, S. Davis, S. F. Murray, X. X. Yu, S. K. Pandey, M. Pear, L. Watts, S. L. Booten, M. Graham, R. McKay, A. Subramaniam, S. Propp, B. A. Lollo, S. Freier, C. F. Bennett, S. Bhanot and B. P. Monia, *Cell Metab.*, 2006, **3**, 87–98.
70. A. M. Cheng, M. W. Byrom, J. Shelton and L. P. Ford, *Nucleic Acids Res.*, 2005, **33**, 1290–1297.
71. C. M. Connelly, M. Thomas and A. Deiters, *J. Biomol. Screening*, 2012, **17**, 822–828.
72. J. A. Wilson and S. M. Sagan, *Curr. Opin. Virol.*, 2014, **7**, 11–18.
73. J. Mitchell, S. M. Lemon and D. R. McGivern, *Curr. Opin. Virol.*, 2015, **14**, 101–108.
74. C. J.-F. Lin, H.-Y. Gong, H.-C. Tseng, W.-L. Wang and J.-L. Wu, *Biochem. Biophys. Res. Commun.*, 2008, **375**, 315–320.
75. Y.-H. Hsiang, J. B. Jiang and L. F. Liu, *Mol. Pharmacol.*, 1989, **36**, 371–376.
76. R. M. Stone, E. Mazzola, D. Neuberg, S. L. Allen, A. Pigneux, R. K. Stuart, M. Wetzler, D. Rizzieri, H. P. Erba and L. Damon, *J. Clin. Oncol.*, 2015, **33**, 1252–1257.
77. R. Hummel, T. Wang, D. I. Watson, M. Z. Michael, M. Van der Hoek, J. Haier and D. J. Hussey, *Oncol. Rep.*, 2011, **26**, 1011–1017.
78. A. M. Roccaro, A. Sacco, C. Chen, J. Runnels, X. Leleu, F. Azab, A. K. Azab, X. Jia, H. T. Ngo, M. R. Melhem, N. Burwick, L. Varticovski, C. D. Novina, B. J. Rollins, K. C. Anderson and I. M. Ghobrial, *Blood*, 2009, **113**, 4391–4402.
79. R. Garzon, F. Pichiorri, T. Palumbo, M. Visentini, R. Aqeilan, A. Cimmino, H. Wang, H. Sun, S. Volinia, H. Alder, G. A. Calin, C. G. Liu, M. Andreeff and C. M. Croce, *Oncogene*, 2007, **26**, 4148.
80. Y. Saito, G. Liang, G. Egger, J. M. Friedman, J. C. Chuang, G. A. Coetzee and P. A. Jones, *Cancer Cell*, 2006, **9**, 435–443.
81. A. Lujambio, S. Ropero, E. Ballestar, M. F. Fraga, C. Cerrato, F. Setién, S. Casado, A. Suarez-Gauthier, M. Sanchez-Cespedes, A. Gitt, I. Spiteri, P. P. Das, C. Caldas, E. Miska and M. Esteller, *Cancer Res.*, 2007, **67**, 1424.
82. Y. Changmin, Q. Linghui, U. Mahesh, L. Lin and S. Q. Yao, *Angew. Chem., Int. Ed. Engl.*, 2015, **127**, 10720–10724.
83. Z. Xiao, C. H. Li, S. L. Chan, F. Xu, L. Feng, Y. Wang, J.-D. Jiang, J. J. Y. Sung, C. H. K. Cheng and Y. Chen, *Cancer Res.*, 2014, **74**, 6236.

84. E. Slabáková, Z. Culig, J. Remšík and K. Souček, *Cell Death Dis.*, 2017, **8**, e3100.
85. S. P. Velagapudi, M. G. Costales, B. R. Vummidi, Y. Nakai, A. J. Angelbello, T. Tran, H. S. Haniff, Y. Matsumoto, Z. F. Wang, A. K. Chatterjee, J. L. Childs-Disney and M. D. Disney, *Cell Chem. Biol.*, 2015, **25**, 1086–1094.e7.
86. T. Zhu, S. Cao, P.-C. Su, R. Patel, D. Shah, H. B. Chokshi, R. Szukala, M. E. Johnson and K. E. Hevener, *J. Med. Chem.*, 2013, **56**, 6560–6572.
87. X. Chen, C. Huang, W. Zhang, Y. Wu, X. Chen, C.-Y. Zhang and Y. Zhang, *Chem. Commun.*, 2012, **48**, 6432–6434.
88. S.-B. Tan, C. Huang, X. Chen, Y. Wu, M. Zhou, C. Zhang and Y. Zhang, *Bioorg. Med. Chem.*, 2013, **21**, 6124–6131.
89. J. Ai, R. Zhang, X. Gao, H.-F. Niu, N. Wang, Y. Xu, Y. Li, N. Ma, L.-H. Sun, Z.-W. Pan, W.-M. Li and B.-F. Yang, *Cardiovasc. Res.*, 2012, **95**, 385–393.
90. F. Petrocca, A. Vecchione and C. M. Croce, *Cancer Res.*, 2008, **68**, 8191.
91. H. Yang, W. Kong, L. He, J.-J. Zhao, J. D. O'Donnell, J. Wang, R. M. Wenham, D. Coppola, P. A. Kruk and S. V. Nicosia, *Cancer Res.*, 2008, **68**, 425–433.
92. J. Contreras and D. S. Rao, *Leukemia*, 2011, **26**, 404.
93. T. E. Callis, J.-F. Chen and D.-Z. Wang, *DNA Cell Biol.*, 2007, **26**, 219–225.
94. S.-B. Tan, J. Li, X. Chen, W. Zhang, D. Zhang, C. Zhang, D. Li and Y. Zhang, *Chem. Biol.*, 2014, **21**, 1265–1270.
95. P. K. Rao, R. M. Kumar, M. Farkhondeh, S. Baskerville and H. F. Lodish, *Proc. Natl. Acad. Sci. U. S. A.*, 2006, **103**, 8721.
96. C. A. Berkes and S. J. Tapscott, *Semin. Cell Dev. Biol.*, 2005, **16**, 585–595.
97. C. Mayr, M. T. Hemann and D. P. Bartel, *Science*, 2007, **315**, 1576.
98. S. M. Johnson, H. Grosshans, J. Shingara, M. Byrom, R. Jarvis, A. Cheng, E. Labourier, K. L. Reinert, D. Brown and F. J. Slack, *Cell*, 2005, **120**, 635–647.
99. R. Guo, K. Abdelmohsen, P. J. Morin and M. Gorospe, *PLoS One*, 2013, **8**, e66330.
100. J. C. Ghosh and D. C. Altieri, *Clin. Cancer Res.*, 2005, **11**, 4580.
101. A. D. Saleh, J. E. Savage, L. Cao, B. P. Soule, D. Ly, W. DeGraff, C. C. Harris, J. B. Mitchell and N. L. Simone, *PLoS One*, 2011, **6**, e24429.
102. J. Cinkornpumin, M. Roos, L. Nguyen, X. Liu, X. Gaeta, S. Lin, D. N. Chan, A. Liu, R. I. Gregory, M. Jung, J. Chute, H. Zhu and W. E. Lowry, *Sci. Rep.*, 2017, 7, 15973.
103. M. J. Keiser, B. L. Roth, B. N. Armbruster, P. Ernsberger, J. J. Irwin and B. K. Shoichet, *Nat. Biotechnol.*, 2007, **25**, 197.
104. K. Im, J. Song, Y. T. Han, S. Lee, S. Kang, K. W. Hwang, H. Min and K. H. Min, *PLoS One*, 2017, **12**, e0182331.
105. S.-Y. Lee, S. Lee, E. Choi, O. Ham, C. Y. Lee, J. Lee, H.-H. Seo, M.-J. Cha, B. Mun, Y. Lee, C. Yoon and K.-C. Hwang, *Sci. Rep.*, 2016, **6**, 23472.
106. L. A. Kubasiak, O. M. Hernandez, N. H. Bishopric and K. A. Webster, *Proc. Natl. Acad. Sci. U. S. A.*, 2002, **99**, 12825–12830.
107. C. Liu, Y. Li, M. Semenov, C. Han, G.-H. Baeg, Y. Tan, Z. Zhang, X. Lin and X. He, *Cell*, 2002, **108**, 837–847.
108. H. Hara, N. Takeda and I. Komuro, *Inflammation Regener.*, 2017, **37**, 13.

109. A. Todd, R. Anderson and P. W. Groundwater, *Pharm. J.*, 2009, **283**, 19–20.
110. D. H. Mathews, M. D. Disney, J. L. Childs, S. J. Schroeder, M. Zuker and D. H. Turner, *Proc. Natl. Acad. Sci. U. S. A.*, 2004, **101**, 7287–7292.
111. R. R. Breaker, R. Gesteland, T. Cech and J. Atkins, *The RNA World*, Cold Spring Harbor Laboratory Press, New York, 2006.
112. S. P. Velagapudi, S. J. Seedhouse and M. D. Disney, *Angew. Chem., Int. Ed. Engl.*, 2010, **49**, 3816–3818.
113. D. J. Paul, S. J. Seedhouse and M. D. Disney, *Nucleic Acids Res.*, 2009, **37**, 5894–5907.
114. M. D. Disney and O. J. Barrett, *Biochemistry*, 2007, **46**, 11223–11230.
115. S. P. Velagapudi, S. J. Seedhouse, J. French and M. D. Disney, *J. Am. Chem. Soc.*, 2011, **133**, 10111–10118.
116. S. Xu, P. D. Witmer, S. Lumayag, B. Kovacs and D. Valle, *J. Biol. Chem.*, 2007, **282**, 25053–25066.
117. H. Lu and H. Huang, *Curr. Drug Targets*, 2011, **12**, 1235–1244.
118. L. Xie, A. Ushmorov, F. Leithäuser, H. Guan, C. Steidl, J. Färbinger, C. Pelzer, M. J. Vogel, H. J. Maier and R. D. Gascoyne, *Blood*, 2012, **119**, 3503–3511.
119. I. K. Guttilla and B. A. White, *J. Biol. Chem.*, 2009, **284**, 23204–23216.
120. S. P. Velagapudi, M. D. Cameron, C. L. Haga, L. H. Rosenberg, M. Lafitte, D. R. Duckett, D. G. Phinney and M. D. Disney, *Proc. Natl. Acad. Sci. U. S. A.*, 2016, **113**, 5898–5903.
121. M. G. Costales, C. L. Haga, S. P. Velagapudi, J. L. Childs-Disney, D. G. Phinney and M. D. Disney, *J. Am. Chem. Soc.*, 2017, **139**, 3446–3455.
122. T. J. Kelly, A. L. Souza, C. B. Clish and P. Puigserver, *Mol. Cell. Biol.*, 2011, **31**, 2696–2706.
123. P. H. Maxwell, M. S. Wiesener, G.-W. Chang, S. C. Clifford, E. C. Vaux, M. E. Cockman, C. C. Wykoff, C. W. Pugh, E. R. Maher and P. J. Ratcliffe, *Nature*, 1999, **399**, 271.
124. C. L. Haga, S. P. Velagapudi, J. R. Strivelli, W.-Y. Yang, M. D. Disney and D. G. Phinney, *ACS Chem. Biol.*, 2015, **10**, 2267–2276.
125. C. Haga and D. Phinney, *J. Biol. Chem.*, 2012, **287**, 42695–42707.
126. H. Cam, J. B. Easton, A. High and P. J. Houghton, *Mol. Cell*, 2010, **40**, 509–520.
127. J. L. Childs-Disney and M. D. Disney, *ACS Chem. Biol.*, 2015, **11**, 375–380.
128. F. Pang, R. Zha, Y. Zhao, Q. Wang, D. Chen, Z. Zhang, T. Chen, M. Yao, J. Gu and X. He, *PLoS One*, 2014, **9**, e90867.
129. K. Nakatani, S. Sando and I. Saito, *J. Am. Chem. Soc.*, 2000, **122**, 2172–2177.
130. H. Suda, A. Kobori, J. Zhang, G. Hayashi and K. Nakatani, *Bioorg. Med. Chem.*, 2005, **13**, 4507–4512.
131. J. Zhang, F. Takei and K. Nakatani, *Bioorg. Med. Chem.*, 2007, **15**, 4813–4817.
132. A. Murata, T. Otabe, J. Zhang and K. Nakatani, *ACS Chem. Biol.*, 2016, **11**, 2790–2796.

133. F. Qiu, R. Sun, N. Deng, T. Guo, Y. Cao, Y. Yu, X. Wang, B. Zou, S. Zhang and T. Jing, *PLoS One*, 2015, **10**, e0120969.
134. W. J. Cho, J. M. Shin, J. S. Kim, M. R. Lee, K. S. Hong, J.-H. Lee, K. H. Koo, J. W. Park and K.-S. Kim, *Mol. Cells*, 2009, **28**, 521–527.
135. P. M. Voorhoeve, C. Le Sage, M. Schrier, A. J. Gillis, H. Stoop, R. Nagel, Y.-P. Liu, J. Van Duijse, J. Drost and A. Griekspoor, *Cell*, 2006, **124**, 1169–1181.
136. K.-H. Lee, Y.-G. Goan, M. Hsiao, C.-H. Lee, S.-H. Jian, J.-T. Lin, Y.-L. Chen and P.-J. Lu, *Exp. Cell Res.*, 2009, **315**, 2529–2538.
137. V. Rippe, L. Dittberner, V. N. Lorenz, N. Drieschner, R. Nimzyk, W. Sendt, K. Junker, G. Belge and J. Bullerdiek, *PLoS One*, 2010, **5**, e9485.
138. M. Duca, V. Malnuit, F. Barbault and R. Benhida, *Chem. Commun.*, 2010, **46**, 6162–6164.
139. H.-Y. Mei, D. P. Mack, A. A. Galan, N. S. Halim, A. Heldsinger, J. A. Loo, D. W. Moreland, K. A. Sannes-Lowery, L. Sharmeen and H. N. Truong, *Bioorg. Med. Chem.*, 1997, **5**, 1173–1184.
140. M. L. Zapp, S. Stern and M. R. Green, *Cell*, 1993, **74**, 969–978.
141. S. Kumar, P. Kellish, W. E. Robinson Jr, D. Wang, D. H. Appella and D. P. Arya, *Biochemistry*, 2012, **51**, 2331–2347.
142. M. Duca, P. Vekhoff, K. Oussedik, L. Halby and P. B. Arimondo, *Nucleic Acids Res.*, 2008, **36**, 5123.
143. V. Malnuit, M. Duca and R. Benhida, *Org. Biomol. Chem.*, 2011, **9**, 326–336.
144. D. Guianvarc'h, R. Benhida, J.-L. Fourrey, R. Maurisse and J.-S. Sun, *Chem. Commun.*, 2001, 1814–1815.
145. D. Guianvarc'h, J.-L. Fourrey, R. Maurisse, J.-S. Sun and R. Benhida, *Bioorg. Med. Chem.*, 2003, **11**, 2751–2759.
146. C. Belair, J. Baud, S. Chabas, C. M. Sharma, J. Vogel, C. Staedel and F. Darfeuille, *Silence*, 2011, **2**, 7.
147. D. P. Bartel, *Cell*, 2009, **136**, 215–233.
148. M. F. Schmidt, O. Korb and C. Abell, *ACS Chem. Biol.*, 2013, **8**, 2122–2126.
149. J. J. Irwin, T. Sterling, M. M. Mysinger, E. S. Bolstad and R. G. Coleman, *J. Chem. Inf. Model.*, 2012, **52**, 1757–1768.
150. G. Jones, P. Willett, R. C. Glen, A. R. Leach and R. Taylor, *J. Mol. Biol.*, 1997, **267**, 727–748.
151. O. Korb, T. Stutzle and T. E. Exner, *J. Chem. Inf. Model.*, 2009, **49**, 84–96.
152. G. Jones, P. Willett and R. C. Glen, *J. Mol. Biol.*, 1995, **245**, 43–53.
153. M. F. Schmidt, O. Korb, N. I. Howard, M. V. Dias, T. L. Blundell and C. Abell, *ChemMedChem*, 2013, **8**, 54–58.
154. Z. Shi, J. Zhang, X. Qian, L. Han, K. Zhang, L. Chen, J. Liu, Y. Ren, M. Yang and A. Zhang, *Cancer Res.*, 2013, **73**, 5519–5531.
155. M. Parisien and F. Major, *Nature*, 2008, **452**, 51.
156. J. W. Ponder, *TINKER: Software Tools for Molecular Design*, Washington University School of Medicine, Saint Louis, MO, 2004, vol. 3.

CHAPTER 13

MicroRNA-specific Argonaute 2 Protein Inhibitors and AGO Loading Inhibitors

KATRIEN VAN ROOSBROECK[†]*

The University of Texas MD Anderson Cancer Center, Department of Experimental Therapeutics – Unit 1950, 1515 Holcombe Blvd, Houston, TX 77030, USA
*E-mail: kvan1@mdanderson.org

13.1 Introduction

Small non-coding microRNAs (miRNAs) are involved in virtually all physiological processes and deregulated in many human diseases. They exert their gene silencing function (mainly messenger RNA (mRNA) degradation and translational repression) through association with Argonaute (AGO) proteins in the RNA-induced silencing complex (RISC), an essential component in the miRNA biogenesis process, in which mature miRNAs are generated from primary miRNA transcripts through two processing steps. Due to their involvement in physiological processes and many human diseases, including cancer, miRNAs are very attractive targets for novel therapeutics. Several miRNA-based therapeutics are currently being evaluated in phase 1 clinical

[†]Current address: The University of Texas MD Anderson Cancer Center, Department of Investigational Cancer Therapeutics – Unit 455, 1515 Holcombe Blvd, Houston, TX 77030, USA.

Drug Discovery Series No. 69
MicroRNAs in Diseases and Disorders: Emerging Therapeutic Targets
Edited by Philip V. Peplow, Bridget Martinez, George A. Calin and Aurora Esquela-Kerscher

Published by the Royal Society of Chemistry, www.rsc.org

trials (some examples are represented in Table 13.1) and many more are in the preclinical developmental stage.[1–3] Despite the development of some promising compounds, there are still some challenges associated with direct targeting of miRNAs, mainly related to *in vivo* delivery, and off-target effects and toxicities. The problem of off-target toxicities can be illustrated by the miR-34 mimic, MRX34, which was evaluated in a phase I clinical study for the treatment of advanced solid tumors.[4] However, the study was prematurely terminated due to the occurrence of five immune-related serious adverse events as patients developed severe cytokine release syndrome[3] (ClinicalTrials.gov identifier NCT01829971).

There are several ways to overcome the challenges related to miRNA-based therapeutics, one being the targeting of proteins associated with miRNAs, such as Argonaute proteins, which are necessary to exert the miRNAs' gene silencing function. The AGO family of proteins consists of eight family members belonging to two subfamilies ("clades"), based on the small non-coding RNA (ncRNA) they are partnering with. The AGO subfamily, whose four members AGO1-4 are ubiquitously expressed, bind mainly to miRNAs and small interfering RNAs (siRNAs).[5,6] AGO subfamily members, especially AGO1 and AGO2, have also been found to be deregulated in cancer in human diseases, including cancer.[7] AGO2 is a main effector of miRNA function, and thus a very feasible candidate for miRNA-based therapeutic interventions. AGO2-based drugs can either be antagonists of the miRNA–AGO2 binding complex (miRNA-specific AGO2 protein inhibitors), or inhibitors of miRNA binding to AGO2 (AGO2 loading inhibitors). In contrast, AGO2 can also be used to promote RNA interference (RNAi).

This chapter first discusses the miRNA biogenesis process and the Argonaute family of proteins (AGO). Next, the mechanisms of AGO-mediated gene silencing will be explained. A last section describes miRNA-based therapeutics, where we focus on therapeutic strategies of targeting AGO proteins for the treatment of human disease, with an emphasis on cancer.

13.2 miRNA Biogenesis

miRNAs are small non-coding RNAs, which originate from their primary miRNA transcript, named pri-miRNAs. Pri-miRNAs are ~1000–3000-nucleotide-long molecules that contain one or more hairpin structures[8] and are transcribed in the nucleus by RNA polymerase II. They then undergo a first processing step by the protein complex Microprocessor in which the RNase endonuclease III Drosha, in complex with DGCR8, cleaves the pri-miRNAs into ~70 nucleotide sequences or precursor microRNAs (pre-miRNAs) containing one hairpin loop.[6] Pre-miRNAs are then shuttled from the nucleus to the cytoplasm by Exportin 5 and RanGTPase (the nuclear export machinery),[9,10] where they are further processed into ~19–25-nucleotide double-stranded RNA (dsRNA) duplexes by the RNA-specific nuclease Dicer in cooperation with TRBP.[11] These duplexes are then loaded into RISC, where they are unwound by Argonaute 2 (AGO2). The guide strand is retained, while the passenger strand is removed, giving rise to the mature miRNA–RISC complex[6] which binds to

Table 13.1 Overview of miRNA-based drugs in clinical studies.[a]

miRNA	Drug name	Mode of action	Indication	Company	Developmental stage	Adverse reactions	Outcome	ClinicalTrials.gov /EudraCT identifier	Reference
miR-122	SPC3649 (miravirsen)	LNA-modified DNA phosphorothioate antisense oligonucleotide	HCV infection	Santaris Pharma A/S	Phase 2a	No DLTs and subject discontinuations due to adverse events	Prolonged dose-dependent reductions in HCV RNA levels without evidence of viral resistance	NCT01200420	107, 119
miR-34a	MRX34	Liposomal miR-34 mimic	Primary liver cancer, small-cell lung cancer lymphoma, melanoma, multiple myeloma, renal cell carcinoma, non-small cell lung cancer	Mirna Therapeutics	Phase 1	Grade 4 cytokine release syndrome	Terminated prematurely due to five immune-related serious adverse events	NCT01829971	4
miR-21	RG-012 (lademirsen)	ss, chemically modified oligo	Alport syndrome	Regulus Therapeutics/Genzyme	Phase 2	Not yet known	Recruitment suspended due to strategic and corporate restructuring	NCT02855268	regulusrx.com
miR-17	RGLS4326	ss chemically modified oligo that preferentially targets kidney	ADPKD	Regulus Therapeutics	Phase 1	Not yet known, being tested in healthy volunteers	Paused due to unexpected observations in a 27-week mouse chronic toxicity study	Not available	regulusrx.com
miR-103/miR-107	RG-125/AZD4076	GalNAc-conjugated anti-miRNA	NAFLD in patients with type 2 diabetes mellitus	Regulus Therapeutics/AstraZeneca	Phase 1/2	Unknown	Company discontinued clinical development of RG-125/AZD4076	NCT02826525	regulusrx.com

miR-122	RG-101	GalNAc-conjugated anti-miRNA	HCV infection	Regulus Therapeutics	Phase 1	Serious adverse events of jaundice	Viral load reduction; clinical hold due to jaundice;company discontinued clinical development of RG-101	2013-002978-49	133
miR-155	MRG-106(Cobomarsen)	LNA antimiR	Cutaneous T-cell lymphoma, mycosis fungoides, chronic lymphocytic leukemia, diffuse large B-cell lymphoma, adult T-cell leukemia/lymphoma	miRagen Therapeutics	Phase 1	Well tolerated, two DLTs (grade 3 pruritis, grade 3 tumor flare)	Well-tolerated, has clinical activity, potential to impact mycosis fungoides, quality of life	NCT02580552	miragen.com;[125]
miR-155	MRG-106 (Cobomarsen)	LNA antimiR	Cutaneous T-cell lymphoma, mycosis fungoides subtype	miRagen Therapeutics	Phase 2	Not yet known	Not yet known, study active and recruiting	NCT03713320	miragen.com
miR-29b	MRG-201 (Remlarsen)	Synthetic miRNA mimic (promiR)	Keloid	miRagen Therapeutics	Phase 2	Not yet known	Not yet known, study active and recruiting	NCT03601052	miragen.com
miR-16	TangomiRs	Targeted minicells containing a microRNA mimic	Recurrent malignant pleural mesothelioma, non-small cell lung cancer	Asbestos Diseases Research Foundation/EnGeneIC Limited	Phase 1	Infusion-related inflammatory symptoms, coronary ischemia, anaphylaxis, cardiomyopathy, non-cardiac pain	Acceptable safety profile and early signs of activity of TargomiRs	NCT02369198	134

[a]ADPKD: autosomal dominant polycystic kidney disease; DLTs: dose-limiting toxicities; HCV: hepatitis C virus; LNA: locked nucleic acid; NAFLD: non-alcoholic fatty liver disease; ss: single stranded.

its target, primarily 3′ untranslated regions (UTRs) of protein-coding genes, resulting in degradation of the target mRNA and reduced protein levels. Besides 3′ UTRs, miRNAs also target protein-coding genes internally, resulting in translation inhibition and decreased levels of the target protein as well. The miRNA biogenesis process is illustrated in Figure 13.1.

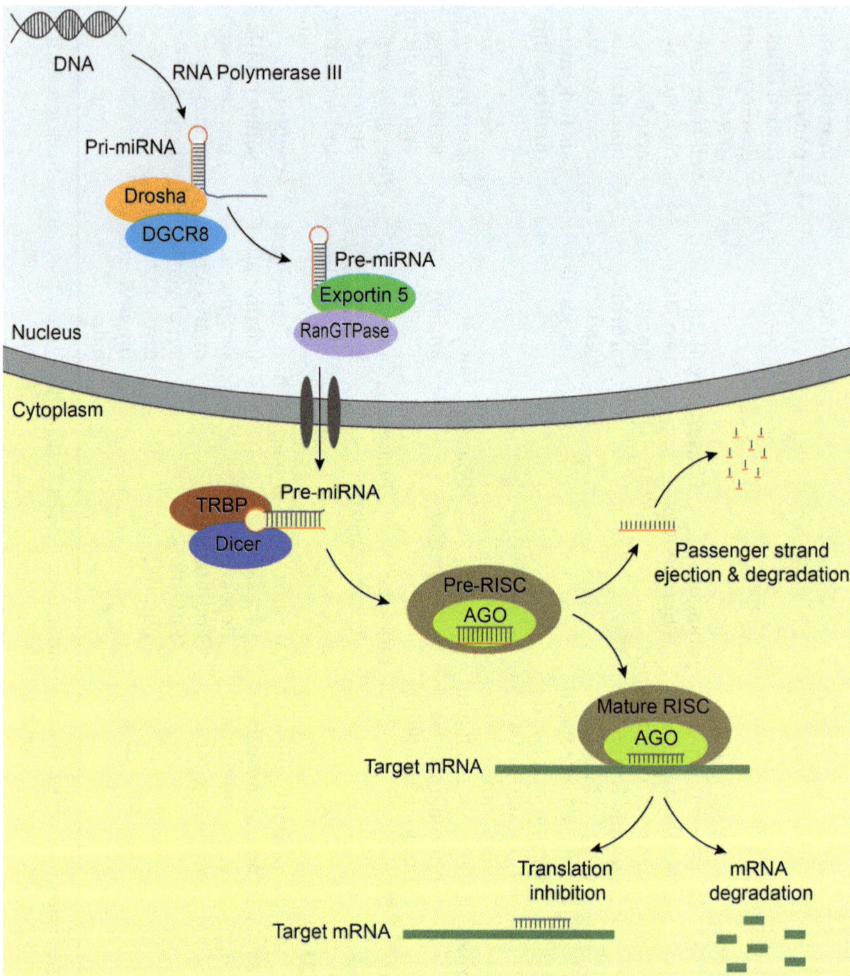

Figure 13.1 miRNA biogenesis process. Long primary miRNAs (pri-miRNAs) are transcribed in the nucleus by RNA polymerase III, and processed by Drosha and DGCR8 to shorter precursor miRNAs (pre-miRNAs). Pre-miRNAs are then shuttled from the nucleus to the cytoplasm by Exportin 5 and RanGTPase, and are further processed by Dicer and TRBP to double-stranded (ds) RNA duplexes, which are loaded into the RNA-induced silencing complex (RISC) where they are unwound by Argonaute 2 (AGO2). The guide strand is retained, while the passenger strand is removed, giving rise to the mature miRNA–RISC complex, which binds to its target mRNA to induce mRNA degeneration or inhibition of translation.

13.2.1 RNA-induced Silencing Complex

During the miRNA biogenesis process, small RNA duplexes are formed; however, miRNAs exert their function as single-stranded RNAs (ssRNAs), a process that is facilitated by the RISC complex. During the first "loading" step, RNA duplexes are transferred from Dicer onto AGO proteins, which is mediated by the heat shock protein 90 (HSP90) chaperone through an ATP-dependent mechanism (Figure 13.2). HSP90 uses ATP to induce an open AGO conformation which is receptive for the small RNA duplex.[5,12,13] This complex is called the "pre-RISC" complex. The way duplexes are loaded into AGO proteins defines which strand will become the guide strand, and which strand will be discarded as the passenger strand. This loading is not a random process, but depends on several factors, including the duplex structure and the relative thermodynamic stabilities of the duplex ends.[14,15] Generally, the strand with the less stable base pairing at its 5′ end will function as the guide strand, while the other strand (passenger strand) will be removed.

The second step of RISC assembly is the unwinding of the dsRNA or "wedging" (Figure 13.2). The exact mechanisms by which dsRNAs are unwound are not yet fully understood, but it was suggested that the N-terminal domain of AGO proteins disrupts base pairs at the 3′ end of the loaded duplex through active wedging.[16]

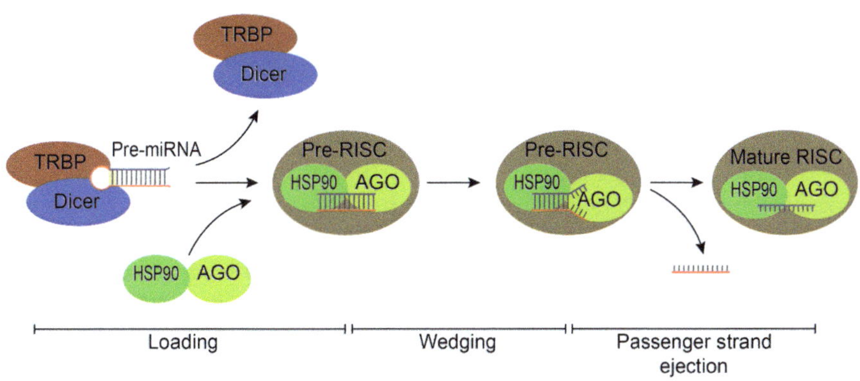

Figure 13.2 RNA-induced silencing complex (RISC) assembly. During the first "loading" step of the RISC assembly process, RNA duplexes are transferred from Dicer onto Argonaute (AGO) proteins. This is mediated by the heat shock protein 90 (HSP90) chaperone through an ATP-dependent mechanism. The second step of RISC assembly is the unwinding of the dsRNA or "wedging", where both strands of the duplex are separated at the 3′ end of the double-stranded RNA (dsRNA) duplex. This "wedged state" of the dsRNA is necessary for the final step of RISC maturation, the ejection of the passenger strand. After the passenger strand has been removed, the RISC complex is "mature" and ready to target RNAs for gene silencing through mRNA degradation or inhibition of translation.

This "wedged state" of the dsRNA is necessary for the final step of RISC maturation, the ejection of the passenger strand (Figure 13.2), which can occur either in a slicer-dependent or slicer-independent manner. The slicer-dependent ejection of the passenger strand requires perfectly complementary base pairing to cleave the passenger strand of the dsRNA duplex, which is then released from the AGO protein. Therefore, this mechanism of passenger strand removal is typically used for the processing of siRNA, rather than miRNAs.[17,18] As miRNA dsRNA duplexes often contain mismatches, the slicer-independent passenger ejection is more suitable for the processing of miRNAs, especially mismatches at the miRNA seed region (positions 2–7 or 2–8 on the guide strand) and at the 3′ end of the guide strand (positions 12–16), which correspond to the regions essential for AGO-mediated target recognition, significantly enhance the efficiency of slicer-independent target injection. It has been suggested that the PIWI-AGO-ZWILLE (PAZ) domain of AGO is critical for this process, as it anchors to the 3′ end of the guide strand after wedging, and may result in further unwinding from the 3′ end and ejection of the passenger strand.[19]

After the passenger strand has been removed, the RISC complex is "mature" and ready to target RNAs for gene silencing. The guide strand will lead the RISC complex to its target RNAs, where AGO acts as a scaffold onto which additional proteins necessary for mRNA degradation and translational repression are recruited.

13.3 Argonaute Family of Proteins

Argonaute proteins play a vital role in RISC complex assembly and gene silencing, as they are the direct binding partners of mature, small ncRNAs, such as miRNAs, siRNAs and piwi-interacting RNAs (piRNAs). They are members of a family with a molecular weight ~100 kDa, and can be divided into two subfamilies (or "clades") based on the small ncRNA they are partnering with: the P-element induced wimpy testis (PIWI) subfamily, which are only expressed in germline cells and whose members bind exclusively to piRNAs, and the AGO subfamily, whose members are ubiquitously expressed and bind mainly to miRNAs and siRNAs.[5,6] From a functional point of view, AGO proteins are mainly implicated in cytoplasmic post-transcriptional gene silencing,[20,21] while PIWI proteins are mainly involved in the silencing of transposable genetic elements.[22] Eight Argonaute family members have been identified in the human genome so far, four in each subfamily: human piwi 1 (HIWI1), HIWI2, HIWI3 and piwi-like RNA-mediated gene silencing 2 (PIWIL2, also known as HILI) (PIWI subfamily), and AGO1, AGO2, AGO3 and AGO4 (AGO subfamily).[23] Their functions are diverse and include germ cell maintenance and division, transcriptional and translational regulation, alternative splicing and heterochromatin formation.[24–28]

AGO proteins contain four structural domains, which form two lobes. The first lobe consists of the N-terminal domain and the PAZ domain, while the second lobe consists of the MID (middle) and PIWI domains. It has been shown that the N-terminal domain is essential for loading and unwinding of small RNA duplexes in the RISC complex.[16] The PAZ and MID domains are found to be important for providing binding pockets for the 3′ and 5′ ends of the small RNA duplexes, respectively.[29–31] The PIWI domain has a structure similar to RNase H, has endonuclease activity, and is responsible for slicing small RNAs.[30,32] Even though all four AGO proteins have a PIWI domain, only AGO2 has retained the sequences necessary for endolytic activity, while AGO1, AGO3 and AGO4 have lost their slicer capabilities. The catalytic activity of AGO2 is necessary for cleavage of target mRNAs that are perfectly complementary to the small RNA; however, when cleavage is not possible (for example with miRNAs that are only partially complementary), the AGO proteins recruit additional proteins to the RISC complex to aid in gene silencing through mRNA degradation.[33,34]

All four human AGO clade members are able to bind miRNAs as well as siRNAs; they do not discriminate between siRNA and miRNA duplexes for loading and processing.[6,20,35,36] AGO1, AGO2 and AGO3 are broadly expressed in cell lines and tissues, whereas the expression of AGO4 is more restricted.[37]

13.3.1 Regulation of Argonaute Proteins

Like the majority of proteins, AGO proteins can be functionally regulated through post-translational modifications, such as phosphorylation, ubiquitylation, sumoylation, hydroxylation and poly(ADP-ribose) modification (summarized in ref. 38). A few examples are mentioned below. For instance, multi-site phosphorylation followed by rapid dephosphorylation of AGO2 (AGO2 phosphorylation cycle), triggered by target mRNA engagement, was recently found to be essential for the regulation of miRNA::target interactions to maintain the global efficiency of miRNA-mediated silencing.[39] In addition, phosphorylation of AGO proteins has been shown to affect cellular localization,[40,41] regulation of small RNA binding,[42–44] target mRNA cleavage and translational repression,[44–46] and to be essential for AGO function and miRNA RISC (miRISC) assembly *in vivo*.[47] Sumoylation of AGO2 at lysine-402 (Lys402) has been shown to negatively regulate AGO2 stability[48] and influence its RNA interference (RNAi) activity.[49]

13.3.2 AGO Proteins and Their Roles in Cancer

Given the fact that miRNAs are widely involved in the development and progression of many diseases, including cancer, it is of no surprise that AGO proteins, which are highly important for the exertion of miRNAs' gene silencing function, also have a role in tumorigenesis. Although evidence of deregulation of AGO3 and AGO4 in tumors is limited, and of AGO1 is evolving, AGO2 has been found to be more widely deregulated in cancer. Of interest, Li *et al.* demonstrated overexpression of all four human AGO proteins in colon cancer.[50] A recent study by Jeong *et al.* showed that *Dicer1*, *AGO3*, and *AGO4* genes are overexpressed during ovarian carcinogenesis in chickens.[51]

AGO1 has been found to be overexpressed in a few cancers, including epithelial skin cancers,[52] urothelial carcinomas of the bladder,[53] ovarian cancer[54] and clear renal cell carcinoma.[55] Its overexpression *in vitro* has been shown to cause cell cycle arrest, decrease cellular motility and increase apoptosis upon UV irradiation in neuroblastoma cells, suggesting a tumor-suppressive role for AGO1.[56] In contrast, in hepatocellular carcinoma, downregulation of AGO1 resulted in reduced cell proliferation, migration and invasion, indicative of an oncogenic role of AGO1,[57] which was further supported by findings that intratumoral AGO1 expression was associated with shorter overall and recurrence-free survival.[57] Moreover, AGO1 was shown to be targeted by the hypoxia-responsive miRNAs *let-7* and miR-103/miR-107 to promote angiogenesis, and translational derepression of vascular endothelial growth factor mediated by AGO1 may be important for tumor angiogenesis and associated with poor prognosis.[58] These data suggest that the role of AGO1 in cancer may be dependent on cell and tumor type. In addition, AGO1 directly interacts with RNA polymerase II, is broadly associated with promoters of transcriptionally active genes and regulates the expression of AGO1-bound genes involved in cell cycle progression, cell growth and survival.[59] Furthermore, two AGO1 single nucleotide polymorphisms (SNPs), rs595961 and rs636832, were found to be associated with lung cancer risk,[60,61] suggesting that these variants may be a useful genetic marker to predict lung cancer susceptibility, and that AGO1 may be implicated in the development of lung cancer.

AGO2 has been found to be deregulated in a variety of human cancers[7,62] including overexpression in colorectal cancer,[63] head and neck squamous cell carcinoma,[64] bladder carcinoma,[53,65,66] ovarian cancer,[54] gastric carcinoma,[67] breast cancer,[68] melanoma,[69] prostate cancer,[70,71] high-grade glioma,[55] clear renal cell carcinoma[55] and hepatocellular carcinoma,[72] and downregulation in lung adenocarcinoma[73] and melanoma.[74] *In vitro*, AGO2 was found to be involved in many processes which are affected in cancer, such as cell growth, proliferation, migration and invasion of cancer cells, angiogenesis and apoptosis.[71,75–79] *In vivo*, it was demonstrated that AGO2 is associated with proliferation, colony formation, migration, tumorigenicity and metastasis of hepatocellular carcinoma cells.[72] AGO2 expression was also found to correlate with clinical factors, including progression-free survival, overall survival and prognosis.[55,65,80] In addition, AGO2 was shown to be targeted by miRNAs, such as miR-99a, resulting in translational repression in hepatocellular carcinoma.[81] AGO2 is targeted by miR-346 and miR-100 as well, leading to upregulation of AGO2 and an increase in migration, invasion, epithelial-to-mesenchymal transition and/or stemness properties.[76,82] Furthermore, the AGO2 SNP rs3928672 was found to be associated with the risk for advanced lymph node metastasis of Chinese nasopharyngeal carcinoma patients[79] and the AGO2 SNPs rs11786030 and rs2292779 were shown to be significantly associated with disease-free survival, progression and overall survival of breast cancer patients.[83]

Very recently, a circular RNA derived from AGO2 (circAGO2) was identified and proven to drive cancer progression.[84] The authors found circAGO2 to be overexpressed in several types of cancer, including gastric cancer, colon cancer, prostate cancer and neuroblastoma. They showed both *in vitro* and *in vivo* that increased levels of circAGO2 were associated with increased tumor cell growth, invasion and metastasis.[84] This indicates that AGO2 deregulation can be caused by several factors, including the well characterized mechanisms of amplification or copy number gains[69] and mutation,[85] as well as newly identified mechanisms, such as circAGO2.[84]

13.4 Mechanisms of AGO-mediated Gene Silencing

AGO proteins can associate with both siRNAs and miRNAs to induce gene silencing. In this section, we primarily focus on gene regulation through miRNA inhibition.

13.4.1 mRNA Degradation

Gene silencing through mRNA degradation occurs in three steps: deadenylation, decapping and mRNA decay.[33,34,86] During the deadenylation step, AGO first associates with GW182, which consists of an AGO-binding domain at its N-terminus and a C-terminal silencing domain.[5,33,86] Poly(A) binding protein cytoplasmic (PABPC), which is bound to the poly-(A) tail of the target mRNA, is then engaged to the complex through binding with GW182.[87,88] This triggers recruitment of the deadenylases poly(A) specific ribonuclease subunit (PAN2) and PAN3 (PAN2-PAN3)[89] and carbon catabolite repressor 4 (CCR4)-negative regulator of transcription 1 (NOT1) complexes,[87,88] resulting in deadenylation of the target mRNA. Next, the decapping protein 2 (DCP2) and various additional decapping proteins, such as DCP1, enhancer of decapping 3 and 4 (EDC3 and EDC4), DEAD box protein 6 (DDX6) and protein associated with topoisomerase II homolog 1-like protein 1 (PATL1), interact with the complex and initiate decapping of the mRNA target.[90,91] Finally, 5′-to-3′ exoribonuclease 1 (XRN1) is recruited and initiates target mRNA degradation.[92,93]

13.4.2 Translational Repression

The exact mechanism(s) in which miRNAs and AGO mediate translational repression are not yet fully elucidated. Increasing evidence points to inhibition at the translation initiation step, rather than post-initiation, elongation or termination, in a cap-dependent manner.[5,34,94] The 5′ end of target mRNAs is known to have a 7-methylguanosine (m7GpppN) cap structure,[95] which is recognized by the eukaryotic translation initiation factor 4E (eIF4E), resulting in recruitment of additional translation initiation factors and initiation of translation.[96] This 5′ end cap has been proved important for miRISC-mediated translational inhibition. It is believed that miRNAs interfere with the assembly of the eIF4F complex[97,98] consisting of RNA helicase eIF4A, cap-binding protein eIF4E and adaptor protein eIF4G, which is important for the binding of the 43S pre-initiation complex to the target mRNA and translation initiation.[99] It was recently shown that eIF4E2 is essential for miRNA-mediated translational repression by two potential mechanisms, either through competition with eIF4E for binding to the EIF4E binding protein 4E-T,[100] or through direct interaction with trinucleotide repeat containing 6 (TNRC6) proteins.[94] Even though several pieces of the puzzle have been put together, the precise mechanism(s) of miRNA/AGO-mediated translational repression still remain(s) incompletely understood.

13.5 miRNA-based Therapeutics

Due to their functional involvement in physiological processes and their widespread implication in many human diseases, including cancer, miRNAs are extremely attractive targets for novel therapeutics. In cancer, they can act as tumor suppressors or oncogenes, hence, therapies focused on targeting of miRNAs are based on reintroduction of downregulated miRNAs or inhibition of overexpressed oncogenic miRNAs (or oncomiRs).[101–104] Their suitability as targets for a novel class of drugs is further supported by some specific characteristics: miRNAs are short sequences, which can easily be chemically modified, and one miRNA has the potential to target multiple protein-coding genes.[105] One type of strategy for the development of miRNA-based therapeutics is the use of antisense oligonucleotides.[105,106] However, when used as naked single-stranded (ss) or double-stranded (ds) RNA or DNA, oligonucleotides complementary to miRNA sequences are quickly degraded by serum proteases, and cannot easily penetrate cell membranes to be delivered intracellularly.[1] Therefore, chemical modifications, such as 2′-*O*-methyl (2′-OMe) modifications, phosphorothioate (PS) bonds, addition of methoxyethyl group at the 2′-sugar position (2′-MOE), introduction of a fluorine atom substituent at the C2′-ribose position (2′-F) and locked nucleic acids (LNAs, in which the ribose unit is "locked" in the 3′ endo confirmation due to an extra methylene bridge connecting the 2′ oxygen and 4′ carbon atoms) have been used to overcome the challenges of serum nuclease degradation (reviewed in ref. 1). The best known and most successful LNA-PS-modified antisense oligonucleotide is miravirsen (SPC3649, Santaris Pharma A/S, now Hoffmann-La Roche), which was successfully evaluated in phase 2 clinical trials for the treatment of hepatitis C virus (HCV) infection[107] (ClinicalTrials.gov). Miravirsen is discussed in more detail in Section 13.5.1. LNA-based anti-miRs against miR-155 (cobomarsen, MRG-106 from miRagen Therapeutics[108]) and miR-92 (MRG-110, S95010 from miRagen Therapeutics in collaboration with Servier) are currently being tested in phase 1/2 clinical trials for the treatment of cancers (cutaneous T-cell lymphoma, mucosis fungoides, chronic lymphocytic leukemia, diffuse large B-cell lymphoma and adult T-cell leukemia/lymphoma; ClinicalTrials.gov identifiers NCT02580552 and NCT03713320) and in healthy volunteers with an objective of evaluating in ischemic conditions (including heart failure, incisional complications and wound healing; clinicaltrials.gov identifier; ClinicalTrials.gov identifier NCT03603431), respectively. Another challenge of developing anti-sense oligonucleotides as miRNA-based drugs, is efficient *in vivo* delivery, as their negatively charged backbone hampers cell permeability and intracellular uptake. To overcome this challenge, delivery vehicles, such as covalent conjugation with uptake enhancers, nanoparticle-based delivery systems (lipid-based, polymer-based or others, for example exosomes and high-density lipoproteins), and targeted delivery systems (for instance aptamers, peptides and antibodies, or folate) are currently being developed and evaluated (reviewed in ref. 109).

The obstacle of *in vivo* delivery can also be addressed by using another approach for miRNA-based drug development: the use of small molecule inhibitors of miRNAs.[110,111] Small molecule modulators are widely used for the modulation (mainly inhibition) of proteins, and are efficiently delivered into cells through diffusion. Due to their low molecular weight (<1000 Da), small RNAs can easily cross cell membranes and travel within the cell to their target protein.[1] However, as mature miRNAs are only short sequences (~19–25 bp), they do not provide the necessary binding pockets for small molecule inhibitors to exert their function. Reports have been published on targeting of miRNAs by small molecule modulators. For example, a recent study reports that the small molecule inhibitor linifanib is able to specifically reduce miR-10b levels.[112] It is not clear whether inhibition occurs through direct interaction with pre-miR-10b, or *via* indirect mechanisms likely involving direct binding to proteins or other macromolecules that are involved in the miRNA regulation process. Pre-miRNAs are the longer precursors of mature miRNAs and consist of ~70 bp, including one stem–loop or hairpin structure that can form the basis for small molecule binding and modulation. Monroig-Bosque *et al.* attempted to elucidate the molecular mechanisms underlying linifanib-mediated inhibition of miR-10b; however, technological limitations because of the large size of pre-miR-10b and its low stability prevented them from unravelling the exact mechanisms.[112]

Another approach could be to not directly target miRNAs, but to target proteins associated with miRNAs, which are necessary to exert their function, such as Argonaute proteins. AGO2 is a main effector of miRNA function, and thus, a very viable candidate for miRNA-based therapeutic interventions. AGO2-based drugs can either be antagonists of the miRNA–AGO2 binding complex (miRNA-specific AGO2 protein inhibitors) or inhibitors of miRNA binding to AGO2 (AGO2 loading inhibitors). In contrast, AGO2 can also be used to promote RNA interference (RNAi) (Figure 13.3).

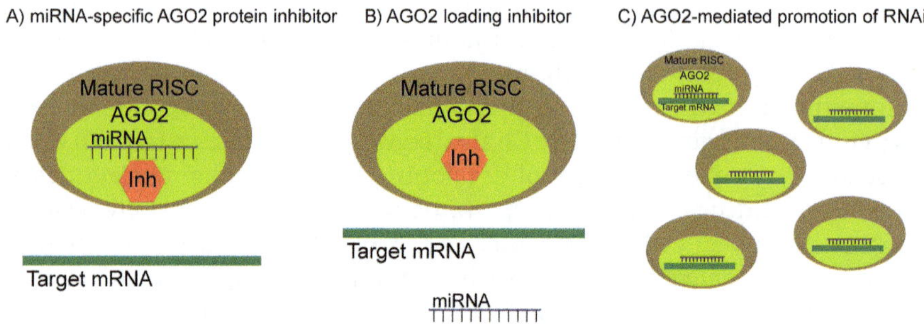

Figure 13.3 Argonaute (AGO)-based therapeutics. AGO2 is a main effector of miRNA function, and thus, a very viable candidate for miRNA-based therapeutic interventions. AGO2-based drugs can either be antagonists of (A) the miRNA-AGO2 binding complex (miRNA-specific AGO2 protein inhibitors), or (B) inhibitors of miRNA binding to AGO2 (AGO2 loading inhibitors). (C) In contrast, AGO2 can also be used to promote RNA interference (RNAi).

13.5.1 miRNA-specific Argonaute 2 Protein Inhibitors

Antagonists of the miR–AGO2 protein complex (known as anti-miR-AGOs) have a dual inhibitor function, as they bind both the miRNA seed region as well as the active site of AGO2.[113] The binding to the miRNA seed region occurs through a short oligonucleotide sequence (such as a short LNA), while the active site of AGO2 is blocked by a small molecule moiety.[113] It was shown that binding of a target mRNA to the RISC complex containing the miRNA guide and AGO2 occurs in two steps: the slow, rate-limiting interaction of the inhibitor with the seed region, followed by a rapid, full binding of the substrate.[114] Therefore, only blocking the seed region by the oligonucleotide is sufficient to prohibit gene silencing. Schmidt *et al.* used a peptide nucleic acid (PNA) as oligo to bind the miRNA sequence and mimic the target mRNA. PNAs have been shown to form stable duplexes with their target miRNA, which can be efficiently synthesized in solid phase, and monomers are commercially available.[113] A short PNA sequence of only four base pairs in length is still able to be miRNA-specific (due to the limited number of human miRNAs). PNA-based anti-miR-AGOs have a low molecular weight (<1500 kDa), and this, in combination with a more hydrophobic nature of the molecule (when compared to oligonucleotide inhibitors) leads to the fact that anti-miR-AGOs may show better pharmacokinetic characteristics and may be easier to deliver than sole oligonucleotides or small molecule inhibitors.[113]

Although not designed as such, the LNA miravirsen acts as a miRNA-specific AGO2 protein inhibitor, as it binds to miR-122 in the liver, preventing its binding to the HCV RNA for viral replication,[115] while impeding AGO2 to exert its cleavage function.[1,116,117] As mentioned, miravirsen is being tested in advanced clinical trials as a miR-122 inhibitor for the treatment of HCV infection. It is the miRNA-based therapeutic closest to clinical implementation and has shown very promising results so far. Even though it is an example of the promising potential of miRNA modulation for the treatment of human diseases, it has to be mentioned that miravirsen is one of a kind. Due to its PS modification, miravirsen specifically locates to and accumulates in the liver, where it targets the liver-expressed miR-122.[118] Effective delivery in tissues and organs besides the liver remains challenging, despite the tremendous amount of research and advances that were made over the past decade or so in optimizing (anti-)miRNA delivery systems.[109] In addition, miravirsen has shown to have a relatively high barrier to resistance; however, nucleotide changes have been detected in HCV patients treated with miravirsen, providing a rationale for combination treatments.[119]

Another LNA-modified oligonucleotide inhibitor, currently evaluated in a first-in-human phase 1 clinical trial for the treatment of cutaneous T-cell lymphoma, mycosis fungoides, chronic lymphocytic leukemia, diffuse large B-cell lymphoma and adult T-cell leukemia/lymphoma, is the miR-155 inhibitor cobomarsen (MRG-106).[108] miR-155 is a well-known oncogene, and we and others have demonstrated its potential as anti-cancer treatment.[120–124] The phase 1 study is currently recruiting patients to assess safety, tolerability and pharmacokinetics of MRG-106. Preliminary data in 24 patients with cutaneous T-cell lymphoma suggest that the compound is well tolerated and has clinical activity, associated with reductions in T-cell clonality and gene expression changes consistent with known miR-155 mechanisms.[125]

13.5.2 AGO Loading Inhibitors

Another way of interfering with AGO function is by blocking the binding of miRNAs to the AGO protein and RISC complex through so-called AGO loading inhibitors (known as miRISC loading inhibitors). Small molecule inhibitors of RISC loading interfering with miRNA binding to AGO2 efficiently block miRNA loading in AGO2[126] and produce effects similar to inhibition of AGO2 by short hairpin RNA (shRNA).[127] Tan *et al.*[126] used purified recombinant AGO2 to screen panels of small molecule inhibitors, and identified aurintricarboxylic acid as a potent inhibitor of RISC loading by using a newly developed *in vitro* method based on fluorescence polarization of TAMRA-labeled small RNAs loaded to human AGO2. The groups of Del Rio and Fazi used structure-based (on the crystallographic structure of human AGO2) high-throughput docking screening[128] to identify BCI-137 as a cell-permeable small molecule that competes with AGO2 miRNA binding.[127] The pharmacophore of BCI-137 was found to mimic the 5′ end of miRNAs, which is responsible for miRNA binding into the AGO2 MID domain. However, this method of blocking AGO2 function is more appropriate for research purposes rather than for clinical applications, as it results in a non-miRNA-specific general inhibition of AGO2, which possibly results in more systemic effects and increased toxicities.

13.5.3 Use of AGO2 to Promote RNAi

Ectopic expression of AGO2 can be used to enhance RNAi and reduce the expression of oncogenic proteins. For example, Li *et al.* coencapsulated siRNA for signal transducer and activator of transcription 3 (STAT3) and AGO2 in cationic lipids or structurally defined synthetic polypeptides to silence oncogenic STAT3.[129] The main rationale behind this was that miRNA/AGO2 or siRNA/AGO2 complexes are more stable than naked small RNAs and it has been shown that microvesicles contain miRNA/AGO2 complexes that are able to efficiently silence target mRNAs in recipient cells.[130,131] Using this combined siRNA/AGO2 delivery approach, the authors were able to successfully silence oncogenic STAT3 *in vitro* and *in vivo*, and reduce tumor burden and prolong survival in an *in vivo* mouse model of melanoma.[129] This proof of principle for siRNA/AGO2 demonstrated its potential for RNAi boosting through increased siRNA function, but can easily be extended towards delivery of tumor suppressor miRNAs to promote inhibition of oncoproteins. These data demonstrate that AGO2 can not only be used as a target for inhibitor-based strategies, but also has potential as replacement therapeutics.

13.6 Conclusions

To date, most miRNA-based therapeutic approaches have been focused on direct suppression or promotion of miRNA expression. Even though great progress has been made in the recent past, for example by chemically modifying antisense oligonucleotides to increase stability, several challenges still remain. For example, as antisense oligonucleotides and miRNAs consist of short sequences, they are prone to having off-target effects due to binding to other, similar miRNAs/sequences (hybridization-dependent off-target effects), which may result in increased side-effects and toxicities. Another challenge is effective delivery of miRNA-based drugs, which may be overcome by the development of nanoparticles or other delivery vehicles,[109] or by the use of small molecule inhibitors of miRNAs.[110,111] Although small molecule inhibitors for targeting proteins have been proven very successful, their applications as inhibitors of miRNA molecules is rather limited, as the short miRNA sequences (up to ~25 bp) do not provide the necessary binding pockets for small molecule inhibitors to exert their function.

In this chapter, we describe the potential of targeting an essential component of RISC, Argonaute proteins, which are vital for miRNA-guided gene silencing. miRNA function can effectively be blocked by AGO inhibitors that either antagonize the miRNA/AGO complex by acting as a mRNA mimic (miRNA-specific AGO protein inhibitors) or block the miRNA binding site to the AGO protein (AGO loading inhibitors). The latter approach seems to have less therapeutic potential, as it is a rather non-specific method to inhibit binding of all miRNAs to AGO. In contrast, the miRNA-specific AGO2 protein inhibitors are more promising drugs, due to their dual function of direct and specific binding to the miRNA seed region, as well as blocking AGO2 through direct binding to its active site. In fact, the only miRNA-based therapy to have been successfully tested in phase 2 clinical studies, miravirsen, is an LNA against miR-122 that falls in this category and specifically binds to miR-122, while also blocking AGO2 function.

As more details are uncovered regarding the biological functions and molecular mechanisms underlying tumorigenesis, novel strategies can be designed to specifically target certain components. For example, it was recently shown that a circular RNA generated from AGO2 (circAGO2) is overexpressed in a variety of human cancers, and this circAGO2 was shown to drive cancer progression.[84] A lentivirus-mediated short hairpin RNA targeting circAGO2 was able to successfully inhibit circAGO2, resulting in reduced tumor formation and cancer aggressiveness in an *in vivo* xenograft experiment with nude mice. Another recent report showed that suppression of AGO2 through siRNA inhibition reduced cell growth, increased apoptosis and resulted in cell cycle arrest *in vitro*, as well as tumor growth suppression in a fibrosarcoma xenograft *in vivo* model.[132] However, similar concerns regarding clinical applicability arise for direct AGO2 inhibition as for the AGO loading inhibitors described in Section 13.5.2, as the expression of many miRNAs is affected simultaneously here, also.

In conclusion, similar to direct miRNA modulation, targeting of the AGO proteins in the RISC complex shows great therapeutic potential for the treatment of human diseases including cancer, and AGO-specific inhibitors may become a new generation of drugs, but some obstacles still need to be overcome before they can be generally applied as new therapies.

References

1. M. F. Schmidt, *Methods Mol. Biol.*, 2017, **1517**, 3–22.
2. T. Catela Ivkovic, G. Voss, H. Cornella and Y. Ceder, *Cancer Lett.*, 2017, **407**, 113–122.
3. K. Van Roosbroeck and G. A. Calin, *Adv. Cancer Res.*, 2017, **135**, 119–149.
4. M. S. Beg, A. J. Brenner, J. Sachdev, M. Borad, Y. K. Kang, J. Stoudemire, S. Smith, A. G. Bader, S. Kim and D. S. Hong, *Invest. New Drugs*, 2017, **35**, 180–188.
5. A. Azlan, N. Dzaki and G. Azzam, *J. Genet. Genomics*, 2016, **43**, 481–494.
6. H. Kobayashi and Y. Tomari, *Biochim. Biophys. Acta*, 2016, **1859**, 71–81.
7. Z. Ye, H. Jin and Q. Qian, *J. Cancer*, 2015, **6**, 877–882.
8. Y. Lee, M. Kim, J. Han, K. H. Yeom, S. Lee, S. H. Baek and V. N. Kim, *EMBO J.*, 2004, **23**, 4051–4060.
9. M. T. Bohnsack, K. Czaplinski and D. Gorlich, *RNA*, 2004, **10**, 185–191.
10. E. Lund, S. Guttinger, A. Calado, J. E. Dahlberg and U. Kutay, *Science*, 2004, **303**, 95–98.
11. R. I. Gregory, T. P. Chendrimada, N. Cooch and R. Shiekhattar, *Cell*, 2005, **123**, 631–640.
12. S. Iwasaki, M. Kobayashi, M. Yoda, Y. Sakaguchi, S. Katsuma, T. Suzuki and Y. Tomari, *Mol. Cell*, 2010, **39**, 292–299.

13. T. Kawamata, H. Seitz and Y. Tomari, *Nat. Struct. Mol. Biol.*, 2009, **16**, 953–960.
14. A. Khvorova, A. Reynolds and S. D. Jayasena, *Cell*, 2003, **115**, 209–216.
15. D. S. Schwarz, G. Hutvagner, T. Du, Z. Xu, N. Aronin and P. D. Zamore, *Cell*, 2003, **115**, 199–208.
16. P. B. Kwak and Y. Tomari, *Nat. Struct. Mol. Biol.*, 2012, **19**, 145–151.
17. C. Matranga, Y. Tomari, C. Shin, D. P. Bartel and P. D. Zamore, *Cell*, 2005, **123**, 607–620.
18. T. A. Rand, S. Petersen, F. Du and X. Wang, *Cell*, 2005, **123**, 621–629.
19. S. Gu, L. Jin, Y. Huang, F. Zhang and M. A. Kay, *Curr. Biol.*, 2012, **22**, 1536–1542.
20. G. Meister, *Nat. Rev. Genet.*, 2013, **14**, 447–459.
21. L. Peters and G. Meister, *Mol. Cell*, 2007, **26**, 611–623.
22. M. C. Siomi, K. Sato, D. Pezic and A. A. Aravin, *Nat. Rev. Mol. Cell Biol.*, 2011, **12**, 246–258.
23. J. Hock and G. Meister, *Genome Biol.*, 2008, **9**, 210.
24. G. Azzam, P. Smibert, E. C. Lai and J. L. Liu, *Dev. Biol.*, 2012, **365**, 384–394.
25. K. T. Gagnon, L. Li, Y. Chu, B. A. Janowski and D. R. Corey, *Cell Rep.*, 2014, **6**, 211–221.
26. V. Huang and L. C. Li, *RNA Biol.*, 2014, **11**, 18–24.
27. K. M. Michalik, R. Bottcher and K. Forstemann, *Nucleic Acids Res.*, 2012, **40**, 9596–9603.
28. W. Wei, Z. Ba, M. Gao, Y. Wu, Y. Ma, S. Amiard, C. I. White, J. M. Rendtlew Danielsen, Y. G. Yang and Y. Qi, *Cell*, 2012, **149**, 101–112.
29. A. Boland, E. Huntzinger, S. Schmidt, E. Izaurralde and O. Weichenrieder, *Proc. Natl. Acad. Sci. U. S. A.*, 2011, **108**, 10466–10471.
30. M. Jinek and J. A. Doudna, *Nature*, 2009, **457**, 405–412.
31. J. B. Ma, K. Ye and D. J. Patel, *Nature*, 2004, **429**, 318–322.
32. J. J. Song, S. K. Smith, G. J. Hannon and L. Joshua-Tor, *Science*, 2004, **305**, 1434–1437.
33. M. R. Fabian and N. Sonenberg, *Nat. Struct. Mol. Biol.*, 2012, **19**, 586–593.
34. S. Jonas and E. Izaurralde, *Nat. Rev. Genet.*, 2015, **16**, 421–433.
35. J. Liu, M. A. Carmell, F. V. Rivas, C. G. Marsden, J. M. Thomson, J. J. Song, S. M. Hammond, L. Joshua-Tor and G. J. Hannon, *Science*, 2004, **305**, 1437–1441.
36. M. Yoda, T. Kawamata, Z. Paroo, X. Ye, S. Iwasaki, Q. Liu and Y. Tomari, *Nat. Struct. Mol. Biol.*, 2010, **17**, 17–23.
37. S. Petri, A. Dueck, G. Lehmann, N. Putz, S. Rudel, E. Kremmer and G. Meister, *RNA*, 2011, **17**, 737–749.
38. D. Jee and E. C. Lai, *Trends Cell Biol.*, 2014, **24**, 546–553.
39. R. J. Golden, B. Chen, T. Li, J. Braun, H. Manjunath, X. Chen, J. Wu, V. Schmid, T. C. Chang, F. Kopp, A. Ramirez-Martinez, V. S. Tagliabracci, Z. J. Chen, Y. Xie and J. T. Mendell, *Nature*, 2017, **542**, 197–202.
40. Y. Zeng, H. Sankala, X. Zhang and P. R. Graves, *Biochem. J.*, 2008, **413**, 429–436.

41. J. Lopez-Orozco, J. M. Pare, A. L. Holme, S. G. Chaulk, R. P. Fahlman and T. C. Hobman, *RNA*, 2015, **21**, 2030–2038.
42. S. Rudel, Y. Wang, R. Lenobel, R. Korner, H. H. Hsiao, H. Urlaub, D. Patel and G. Meister, *Nucleic Acids Res.*, 2011, **39**, 2330–2343.
43. M. Yang, A. D. Haase, F. K. Huang, G. Coulis, K. D. Rivera, B. C. Dickinson, C. J. Chang, D. J. Pappin, T. A. Neubert, G. J. Hannon, B. Boivin and N. K. Tonks, *Mol. Cell*, 2014, **55**, 782–790.
44. D. Rajgor, T. M. Sanderson, M. Amici, G. L. Collingridge and J. G. Hanley, *EMBO J.*, 2018, **37**, e97943.
45. S. R. Horman, M. M. Janas, C. Litterst, B. Wang, I. J. MacRae, M. J. Sever, D. V. Morrissey, P. Graves, B. Luo, S. Umesalma, H. H. Qi, L. J. Miraglia, C. D. Novina and A. P. Orth, *Mol. Cell*, 2013, **50**, 356–367.
46. N. Paradis-Isler and J. Boehm, *J. Biol. Chem.*, 2018, **293**, 9311–9325.
47. K. S. Bridge, K. M. Shah, Y. Li, D. E. Foxler, S. C. K. Wong, D. C. Miller, K. M. Davidson, J. G. Foster, R. Rose, M. R. Hodgkinson, P. S. Ribeiro, A. A. Aboobaker, K. Yashiro, X. Wang, P. R. Graves, M. J. Plevin, D. Lagos and T. V. Sharp, *Cell Rep.*, 2017, **20**, 173–187.
48. U. Sahin, P. Lapaquette, A. Andrieux, G. Faure and A. Dejean, *PLoS One*, 2014, **9**, e102957.
49. F. Josa-Prado, J. M. Henley and K. A. Wilkinson, *Biochem. Biophys. Res. Commun.*, 2015, **464**, 1066–1071.
50. L. Li, C. Yu, H. Gao and Y. Li, *BMC Cancer*, 2010, **10**, 38.
51. W. Jeong, H. Bae, W. Lim and G. Song, *J. Anim. Sci.*, 2017, **95**, 4857–4868.
52. M. Sand, M. Skrygan, D. Georgas, C. Arenz, T. Gambichler, D. Sand, P. Altmeyer and F. G. Bechara, *Mol. Carcinog.*, 2012, **51**, 916–922.
53. A. Rabien, N. Ratert, A. Hogner, A. Erbersdobler, K. Jung, T. H. Ecke and E. Kilic, *Int. J. Mol. Sci.*, 2018, **19**(6), 1622.
54. O. Vaksman, T. E. Hetland, C. G. Trope, R. Reich and B. Davidson, *Hum. Pathol.*, 2012, **43**, 2062–2069.
55. B. Feng, P. Hu, S. J. Lu, J. B. Chen and R. L. Ge, *Asian Pac. J. Cancer Prev.*, 2014, **15**, 4079–4083.
56. C. Parisi, C. Giorgi, E. M. Batassa, L. Braccini, G. Maresca, I. D'Agnano, V. Caputo, A. Salvatore, F. Pietrolati, C. Cogoni and C. Catalanotto, *FEBS Lett.*, 2011, **585**, 2965–2971.
57. M. Wang, L. Zhang, Z. Liu, J. Zhou, Q. Pan, J. Fan, R. Zang and L. Wang, *Cell Death Dis.*, 2018, **9**, 324.
58. Z. Chen, T. C. Lai, Y. H. Jan, F. M. Lin, W. C. Wang, H. Xiao, Y. T. Wang, W. Sun, X. Cui, Y. S. Li, T. Fang, H. Zhao, C. Padmanabhan, R. Sun, D. L. Wang, H. Jin, G. Y. Chau, H. D. Huang, M. Hsiao and J. Y. Shyy, *J. Clin. Invest.*, 2013, **123**, 1057–1067.
59. V. Huang, J. Zheng, Z. Qi, J. Wang, R. F. Place, J. Yu, H. Li and L. C. Li, *PLoS Genet.*, 2013, **9**, e1003821.
60. J. S. Kim, Y. Y. Choi, G. Jin, H. G. Kang, J. E. Choi, H. S. Jeon, W. K. Lee, D. S. Kim, C. H. Kim, Y. J. Kim, J. W. Son, T. H. Jung and J. Y. Park, *Mol. Carcinog.*, 2010, **49**, 913–921.

61. X. Fang, Z. Yin, X. Li, L. Xia and B. Zhou, *Int. J. Environ. Res. Public Health*, 2016, **13**(10), 939.
62. J. T. Huang, J. Wang, V. Srivastava, S. Sen and S. M. Liu, *Front. Oncol.*, 2014, **4**, 113.
63. D. J. Papachristou, A. Korpetinou, E. Giannopoulou, A. G. Antonacopoulou, H. Papadaki, P. Grivas, C. D. Scopa and H. P. Kalofonos, *Virchows Arch.*, 2011, **459**, 431–440.
64. S. S. Chang, I. Smith, C. Glazer, P. Hennessey and J. A. Califano, *ORL*, 2010, **72**, 337–343.
65. F. Q. Yang, J. H. Huang, M. Liu, F. P. Yang, W. Li, G. C. Wang, J. P. Che and J. H. Zheng, *Int. J. Clin. Exp. Pathol.*, 2014, **7**, 340–347.
66. Z. Zhang, G. Zhang, C. Kong, J. Bi, D. Gong, X. Yu, D. Shi, B. Zhan and P. Ye, *Tumor Biol.*, 2015, **36**, 5071–5079.
67. J. Zhang, X. S. Fan, C. X. Wang, B. Liu, Q. Li and X. J. Zhou, *Med. Oncol.*, 2013, **30**, 628.
68. B. D. Adams, K. P. Claffey and B. A. White, *Endocrinology*, 2009, **150**, 14–23.
69. Y. Zhou, L. Chen, B. Barlogie, O. Stephens, X. Wu, D. R. Williams, M. A. Cartron, F. van Rhee, B. Nair, S. Waheed, M. Pineda-Roman, Y. Alsayed, E. Anaissie and J. D. Shaughnessy Jr, *Proc. Natl. Acad. Sci. U. S. A.*, 2010, **107**, 7904–7909.
70. X. Fu, C. Xue, Y. Huang, Y. Xie and Y. Li, *Mol. BioSyst.*, 2010, **6**, 2561–2572.
71. X. J. Bian, G. M. Zhang, C. Y. Gu, Y. Cai, C. F. Wang, Y. J. Shen, Y. Zhu, H. L. Zhang, B. Dai and D. W. Ye, *Tumor Biol.*, 2014, **35**, 11571–11578.
72. N. Cheng, Y. Li and Z. G. Han, *Hepatology*, 2013, **57**, 1906–1918.
73. X. Zhang, P. Graves and Y. Zeng, *Biochim. Biophys. Acta*, 2013, **1830**, 2553–2561.
74. D. Voller, J. Reinders, G. Meister and A. K. Bosserhoff, *Br. J. Cancer*, 2013, **109**, 3116–3124.
75. P. K. Naoghare, Y. K. Tak, M. J. Kim, E. Han and J. M. Song, *Basic Clin. Pharmacol. Toxicol.*, 2011, **109**, 274–282.
76. M. Wang, D. Ren, W. Guo, Z. Wang, S. Huang, H. Du, L. Song and X. Peng, *Int. J. Oncol.*, 2014, **45**, 362–372.
77. S. Wu, W. Yu, X. Qu, R. Wang, J. Xu, Q. Zhang, J. Xu, J. Li and L. Chen, *J. Hematol. Oncol.*, 2014, **7**, 40.
78. Z. L. Ye, Y. Huang, L. F. Li, H. L. Zhu, H. X. Gao, H. Liu, S. Q. Lv, Z. H. Xu, L. N. Zheng, T. Liu, J. L. Zhang, H. J. Jin and Q. J. Qian, *Acta Pharmacol. Sin.*, 2015, **36**, 1237–1245.
79. P. Li, J. Meng, Y. Zhai, H. Zhang, L. Yu, Z. Wang, X. Zhang, P. Cao, X. Chen, Y. Han, Y. Zhang, H. Chen, Y. Ling, Y. Li, Y. Cui, J. X. Bei, Y. X. Zeng, F. He and G. Zhou, *BMC Cancer*, 2015, **15**, 862.
80. H. Zhang, Y. Wang, J. Dou, Y. Guo, J. He, L. Li, X. Liu, R. Chen, R. Deng, J. Huang, R. Xie, X. Zhao and J. Yu, *Oncogene*, 2018, **38**(9), 1410–1431.
81. J. Zhang, H. Jin, H. Liu, S. Lv, B. Wang, R. Wang, H. Liu, M. Ding, Y. Yang, L. Li, J. Zhang, S. Fu, D. Xie, M. Wu, W. Zhou and Q. Qian, *Oncogenesis*, 2014, **3**, e97.

82. J. Guo, J. Lv, M. Liu and H. Tang, *J. Biol. Chem.*, 2015, **290**, 30342–30350.
83. H. Sung, S. Jeon, K. M. Lee, S. Han, M. Song, J. Y. Choi, S. K. Park, K. Y. Yoo, D. Y. Noh, S. H. Ahn and D. Kang, *BMC Cancer*, 2012, **12**, 195.
84. Y. Chen, F. Yang, E. Fang, W. Xiao, H. Mei, H. Li, D. Li, H. Song, J. Wang, M. Hong, X. Wang, K. Huang, L. Zheng and Q. Tong, *Cell Death Differ.*, 2018, DOI: 10.1038/s41418-018-0220-6ID.
85. M. S. Kim, J. E. Oh, Y. R. Kim, S. W. Park, M. R. Kang, S. S. Kim, C. H. Ahn, N. J. Yoo and S. H. Lee, *J. Pathol.*, 2010, **221**, 139–146.
86. E. Huntzinger and E. Izaurralde, *Nat. Rev. Genet.*, 2011, **12**, 99–110.
87. M. Jinek, M. R. Fabian, S. M. Coyle, N. Sonenberg and J. A. Doudna, *Nat. Struct. Mol. Biol.*, 2010, **17**, 238–240.
88. G. Kozlov, N. Safaee, A. Rosenauer and K. Gehring, *J. Biol. Chem.*, 2010, **285**, 13599–13606.
89. D. Kuzuoglu-Ozturk, E. Huntzinger, S. Schmidt and E. Izaurralde, *Nucleic Acids Res.*, 2012, **40**, 5651–5665.
90. M. Arribas-Layton, D. Wu, J. Lykke-Andersen and H. Song, *Biochim. Biophys. Acta*, 2013, **1829**, 580–589.
91. T. Nishihara, L. Zekri, J. E. Braun and E. Izaurralde, *Nucleic Acids Res.*, 2013, **41**, 8692–8705.
92. J. E. Braun, V. Truffault, A. Boland, E. Huntzinger, C. T. Chang, G. Haas, O. Weichenrieder, M. Coles and E. Izaurralde, *Nat. Struct. Mol. Biol.*, 2012, **19**, 1324–1331.
93. C. I. Jones, D. P. Grima, J. A. Waldron, S. Jones, H. N. Parker and S. F. Newbury, *RNA Biol.*, 2013, **10**, 1345–1355.
94. S. Chen and G. Gao, *Protein Cell*, 2017, **8**, 750–761.
95. G. Varani, *Structure*, 1997, **5**, 855–858.
96. A. C. Gingras, B. Raught and N. Sonenberg, *Annu. Rev. Biochem.*, 1999, **68**, 913–963.
97. A. Fukao, Y. Mishima, N. Takizawa, S. Oka, H. Imataka, J. Pelletier, N. Sonenberg, C. Thoma and T. Fujiwara, *Mol. Cell*, 2014, **56**, 79–89.
98. H. Su, S. Meng, Y. Lu, M. I. Trombly, J. Chen, C. Lin, A. Turk and X. Wang, *Mol. Cell*, 2011, **43**, 97–109.
99. R. J. Jackson, C. U. Hellen and T. V. Pestova, *Nat. Rev. Mol. Cell Biol.*, 2010, **11**, 113–127.
100. C. Chapat, S. M. Jafarnejad, E. Matta-Camacho, G. G. Hesketh, I. A. Gelbart, J. Attig, C. G. Gkogkas, T. Alain, N. Stern-Ginossar, M. R. Fabian, A. C. Gingras, T. F. Duchaine and N. Sonenberg, *Proc. Natl. Acad. Sci. U. S. A.*, 2017, **114**, 5425–5430.
101. R. S. Redis, I. Berindan-Neagoe, V. I. Pop and G. A. Calin, *J. Cell. Biochem.*, 2012, **113**, 1451–1459.
102. K. Van Roosbroeck, J. Pollet and G. A. Calin, *Expert Rev. Mol. Diagn.*, 2013, **13**, 183–204.
103. I. Berindan-Neagoe, C. Monroig Pdel, B. Pasculli and G. A. Calin, *Ca-Cancer J. Clin.*, 2014, **64**, 311–336.
104. K. Van Roosbroeck and G. A. Calin, *Semin. Oncol.*, 2016, **43**, 209–214.
105. H. Ling, M. Fabbri and G. A. Calin, *Nat. Rev. Drug Discovery*, 2013, **12**, 847–865.

106. Z. Li and T. M. Rana, *Nat. Rev. Drug Discovery*, 2014, **13**, 622–638.
107. H. L. Janssen, H. W. Reesink, E. J. Lawitz, S. Zeuzem, M. Rodriguez-Torres, K. Patel, A. J. van der Meer, A. K. Patick, A. Chen, Y. Zhou, R. Persson, B. D. King, S. Kauppinen, A. A. Levin and M. R. Hodges, *N. Engl. J. Med.*, 2013, **368**, 1685–1694.
108. A. G. Seto, X. Beatty, J. M. Lynch, M. Hermreck, M. Tetzlaff, M. Duvic and A. L. Jackson, *Br. J. Haematol.*, 2018, **183**(3), 428–444.
109. X. Chen, L. S. Mangala, C. Rodriguez-Aguayo, X. Kong, G. Lopez-Berestein and A. K. Sood, *Cancer Metastasis Rev.*, 2018, **37**, 107–124.
110. C. Monroig Pdel, L. Chen, S. Zhang and G. A. Calin, *Adv. Drug Delivery Rev.*, 2015, **81**, 104–116.
111. D. Wen, M. Danquah, A. K. Chaudhary and R. I. Mahato, *J. Controlled Release*, 2015, **219**, 237–247.
112. P. D. C. Monroig-Bosque, M. Y. Shah, X. Fu, E. Fuentes-Mattei, H. Ling, C. Ivan, N. Nouraee, B. Huang, L. Chen, V. Pileczki, R. S. Redis, E. J. Jung, X. Zhang, M. Lehrer, R. Nagvekar, A. C. P. Mafra, M. D. M. Monroig-Bosque, A. Irimie, C. Rivera, C. Dan Dumitru, I. Berindan-Neagoe, E. P. Nikonowicz, S. Zhang and G. A. Calin, *Sci. Rep.*, 2018, **8**, 13106.
113. M. F. Schmidt, O. Korb and C. Abell, *ACS Chem. Biol.*, 2013, **8**, 2122–2126.
114. D. P. Bartel, *Cell*, 2009, **136**, 215–233.
115. C. L. Jopling, M. Yi, A. M. Lancaster, S. M. Lemon and P. Sarnow, *Science*, 2005, **309**, 1577–1581.
116. S. Obad, C. O. dos Santos, A. Petri, M. Heidenblad, O. Broom, C. Ruse, C. Fu, M. Lindow, J. Stenvang, E. M. Straarup, H. F. Hansen, T. Koch, D. Pappin, G. J. Hannon and S. Kauppinen, *Nat. Genet.*, 2011, **43**, 371–378.
117. M. F. Schmidt, O. Korb and C. Abell, *Methods Mol. Biol.*, 2017, **1517**, 239–249.
118. M. Lindow and S. Kauppinen, *J. Cell Biol.*, 2012, **199**, 407–412.
119. S. Ottosen, T. B. Parsley, L. Yang, K. Zeh, L. J. van Doorn, E. van der Veer, A. K. Raney, M. R. Hodges and A. K. Patick, *Antimicrob. Agents Chemother.*, 2015, **59**, 599–608.
120. R. Bayraktar and K. Van Roosbroeck, *Cancer Metastasis Rev.*, 2018, **37**, 33–44.
121. K. Van Roosbroeck, F. Fanini, T. Setoyama, C. Ivan, C. Rodriguez-Aguayo, E. Fuentes-Mattei, L. Xiao, I. Vannini, R. S. Redis, L. D'Abundo, X. Zhang, M. S. Nicoloso, S. Rossi, V. Gonzalez-Villasana, R. Rupaimoole, M. Ferracin, F. Morabito, A. Neri, P. P. Ruvolo, V. R. Ruvolo, C. V. Pecot, D. Amadori, L. Abruzzo, S. Calin, X. Wang, M. J. You, A. Ferrajoli, R. Orlowski, W. Plunkett, T. M. Lichtenberg, R. V. Davuluri, I. Berindan-Neagoe, M. Negrini, I. I. Wistuba, H. M. Kantarjian, A. K. Sood, G. Lopez-Berestein, M. J. Keating, M. Fabbri and G. A. Calin, *Clin. Cancer Res.*, 2017, **23**, 2891–2904.
122. D. D. Yu, M. M. Lv, W. X. Chen, S. L. Zhong, X. H. Zhang, L. Chen, T. F. Ma, J. H. Tang and J. H. Zhao, *Tumor Biol.*, 2015, **36**, 1395–1401.
123. Z. Chen, T. Ma, C. Huang, T. Hu and J. Li, *J. Cell. Physiol.*, 2014, **229**, 545–550.

124. P. Poltronieri, P. I. D'Urso, V. Mezzolla and O. F. D'Urso, *Chem. Biol. Drug Des.*, 2013, **81**, 79–84.
125. C. Querfeld, F. M. Foss, Y. H. Kim, L. Pinter-Brown, B. M. William, P. Porcu, T. Pacheco, B. M. Haverkos, J. DeSimone, J. Guitart, A. S. Halwani, H. A. Eradat, A. Huen, K. Schroeder, L. A. Pestano, P. J. Williams, I. Cheronis, G. S. Gordon, D. Escolar, P. Rubin and W. S. Marshall, *Blood*, 2018, **132**, 2903.
126. G. S. Tan, C. H. Chiu, B. G. Garchow, D. Metzler, S. L. Diamond and M. Kiriakidou, *ACS Chem. Biol.*, 2012, **7**, 403–410.
127. S. Masciarelli, R. Quaranta, I. Iosue, G. Colotti, F. Padula, G. Varchi, F. Fazi and A. Del Rio, *ACS Chem. Biol.*, 2014, **9**, 1674–1679.
128. T. Bellissimo, S. Masciarelli, E. Poser, I. Genovese, A. Del Rio, G. Colotti and F. Fazi, *Methods Mol. Biol.*, 2017, **1517**, 211–221.
129. J. Li, C. Wu, W. Wang, Y. He, E. Elkayam, L. Joshua-Tor and P. T. Hammond, *Proc. Natl. Acad. Sci. U. S. A.*, 2018, **115**, E2696–E2705.
130. B. Laffont, A. Corduan, H. Ple, A. C. Duchez, N. Cloutier, E. Boilard and P. Provost, *Blood*, 2013, **122**, 253–261.
131. Z. Lv, Y. Wei, D. Wang, C. Y. Zhang, K. Zen and L. Li, *PLoS One*, 2014, **9**, e103599.
132. T. Tagami, T. Suzuki, K. Hirose, J. M. Barichello, N. Yamazaki, T. Asai, N. Oku, T. Ishida and H. Kiwada, *Drug Delivery Transl. Res.*, 2011, **1**, 277–288.
133. M. H. van der Ree, J. M. de Vree, F. Stelma, S. Willemse, M. van der Valk, S. Rietdijk, R. Molenkamp, J. Schinkel, A. C. van Nuenen, U. Beuers, S. Hadi, M. Harbers, E. van der Veer, K. Liu, J. Grundy, A. K. Patick, A. Pavlicek, J. Blem, M. Huang, P. Grint, S. Neben, N. W. Gibson, N. A. Kootstra and H. W. Reesink, *Lancet*, 2017, **389**, 709–717.
134. N. van Zandwijk, N. Pavlakis, S. C. Kao, A. Linton, M. J. Boyer, S. Clarke, Y. Huynh, A. Chrzanowska, M. J. Fulham, D. L. Bailey, W. A. Cooper, L. Kritharides, L. Ridley, S. T. Pattison, J. MacDiarmid, H. Brahmbhatt and G. Reid, *Lancet Oncol.*, 2017, **18**, 1386–1396.

CHAPTER 14

Strategies for Safe and Targeted Delivery of MicroRNA Therapeutics

SUNGHYUN (SANDER) MYOUNG[a,b,c] AND ANDREA L. KASINSKI*[a,b]

[a]Purdue University, Department of Biological Sciences, 1203 West State St., West Lafayette, IN, 47907, USA; [b]Purdue University, Center for Cancer Research, West Lafayette, IN, 47907, USA; [c]Purdue University, Purdue University Interdisciplinary Life Science Program (PULSe), West Lafayette, IN, 47907, USA

*E-mail: akasinski@purdue.edu

14.1 Introduction

From the elucidation that microRNAs (miRNAs) are involved in numerous biological processes to the knowledge that many are reduced in expression in human diseases such as cancer,[1–5] it is only logical that the next step is to explore if these 22-nucleotide entities can be utilized as therapeutic agents. Indeed, miRNAs are recognized as key modulators of cellular processes often dysregulated in disease.[6,7]

Unbeknown to the scientists involved at the time, the miRNA field began with the discovery of *lin-4*, a small RNA involved in heterochronic signaling in *Caenorhabditis elegans*.[8] Mutated in worms with a defect in developmental

Drug Discovery Series No. 69
MicroRNAs in Diseases and Disorders: Emerging Therapeutic Targets
Edited by Philip V. Peplow, Bridget Martinez, George A. Calin and Aurora Esquela-Kerscher

Published by the Royal Society of Chemistry, www.rsc.org

timing, *lin-4* does not encode a protein-coding gene but instead produces a small RNA. This small RNA binds to the 3′ untranslated region of a messenger RNA (mRNA), *lin-14* negatively regulating its translation.[8] Seven years later, a second miRNA was discovered, also in *C. elegans*. In this case, worms mutated for this miRNA exhibited a lethal phenotype, thus the mapped gene was given the name *lethal-7* (*let-7*). Like *lin-4*, *let-7* is also involved in the heterochronic pathway in *C. elegans* and, extraordinarily, is evolutionarily conserved in all metazoans,[9] including humans.[10] Currently more than 2500 miRNAs have been predicted in humans (miRBase v22).[11]

Determined through genomic sequencing, expression profiling, and functional studies, miRNAs are intimately involved in multiple cellular processes such as cell differentiation and growth, apoptosis, neurogenesis, and immune response.[12,13] Because of their contribution to nearly every biological process, it is not surprising that miRNA dysregulation alters normal cellular homeostasis, influencing various disease states. The first example of miRNAs dysregulation in disease was determined after careful examination of a region of chromosome 13q14 that is deleted in more than half of B-cell chronic lymphocytic leukemias (B-CLLs).[14] In this case, the only genes within the 13q14 region that showed consistent involvement in B-CLL were *mir-15a* and *mir-16-1*, two miRNA-encoding genes contained within a single miRNA cluster. Loss of miR-15a and miR-16-1, which function as tumor suppressors, resulted in elevations in MCL1, BCL-2, and JUN, proteins involved in cell proliferation and anti-apoptosis. As such, when miR-15a and miR-16-1 are restored in leukemic mouse models a significant decrease in tumor growth and volume occurs.[15] This extraordinary finding, followed by others, found that miRNAs are globally downregulated in tumor tissues compared to normal tissues, as opposed to being overexpressed.[16] Many of these so-called downregulated miRNAs are located in fragile regions of the genome often deleted in various cancers.[17] This evidence is highly suggestive that dysregulation of miRNAs is not only involved in disease pathogenesis, but that restoration of some of these depleted miRNAs may have therapeutic relevance.[18]

Some of the more frequently reduced miRNAs in human carcinogenesis belong to the *let-7* family of miRNAs. The *let-7* family consists of 13 family members that primarily function as tumor suppressors through repressing the translation of multiple oncogenes including *KRAS*, *MYC*, and *HMGA2*,[19] maintaining cellular differentiation and preventing uncontrollable cellular proliferation. Thus, loss of mature *let-7* predisposes the cells to a tumorigenic fate. There are multiple mechanisms that can contribute to reduced *let-7* levels including chromosomal deletion, transcriptional defects, or impaired *let-7* biogenesis,[20] the latter of which can reduce the levels of multiple *let-7* family members at the same time. Attempts to restore *let-7* by increasing the pools in various cancer models have been quite successful. For example, systemic delivery of mature *let-7* into mouse models with constitutively activated Kras ($Kras^{G12D/+}$) prevented lung tumor formation if administered at the same time as *Kras*$^{G12D/+}$ activation.[21] Importantly, *let-7* administration post *Kras*$^{G12D/+}$ activation decreased the volume of already formed lung tumors.[22]

Similar to the *let-7* family, the miR-34 family is typically downregulated in various malignancies. This is perhaps not surprising, since all the three *mir-34* family members are transcriptionally induced by the most mutated gene in human cancers, namely p53.[23–27] The bulk of p53 mutations occur in two hot-spots that alter either the DNA binding domain[28] of p53 or its conformation,[29–31] impairing p53 function. As a transcription factor, wild-type p53 regulates the expression of various protein-coding genes, and in 2007 was determined to be a major transcriptional regulator of the miR-34 family of miRNAs.[23–27] The miR-34 family consist of three isoforms: miR-34a, miR-34b, and miR-34c, which function as tumor suppressors. Loss of any isoform of miR-34 allows the target mRNAs, such as *CDK4*,[32] *BCL-2*,[33] and *MET*[26] to be expressed, leading to sustained cell growth, anti-apoptotic responses, and metastasis. Unsurprisingly, restoration of miR-34 in various murine tumor models slows tumor growth, including tumors in the lung,[34–36] prostate,[1] and liver.[37] These results, among many others, supported the advancement of miR-34a as the first miRNA mimic to enter clinical trials in 2013.[38]

Although miRNAs are globally lost in cancer, a select few upregulated miRNAs act as prominent oncogenes that tumors can become addicted to; so-called "oncomiRs". For example, the oncomiR miR-21, located in a region of chromosome 17 (17q23), is amplified in multiple tumor types.[17] Indeed, miR-21 is the most upregulated miRNA in carcinogenesis.[39] Importantly, miR-21 not only serves as a biomarker of disease, it directly contributes to disease onset and maintenance. As an oncogenic miRNA, overexpression of miR-21 promotes cell proliferation, anti-apoptosis, metastasis, and chemoresistance by inhibiting tumor-suppressive genes such as *PTEN*,[40] *PDCD4*,[41,42] *and BCL-2*.[43] Through indirect mechanisms, miR-21 also negatively regulates p53.[44] By reducing the levels of these critical tumor suppressors, miR-21 initiates a tumorigenic cascade that cells become addicted to.[45] This phenomenon, referred to as oncomiR addiction, is what makes miR-21 inhibitors attractive therapeutics. Inhibiting miR-21 using an antagomiR (a chemically engineered oligonucleotide used to sequester miRNAs) reduces cell growth and induces apoptosis in glioblastoma,[46,47] breast cancer,[42] and colorectal cancer cells.[48] Inhibiting miR-21 activity also sensitizes various tumors to chemotherapeutics, thereby expanding the clinical utility of miR-21 inhibitors.[49–52]

In addition to miR-21, miR-155 is also classified as an oncomiR. miR-155, an important regulator of immune cell development, located in the B-cell integration cluster on chromosome 21q23, is overexpressed in lymphoma.[53] Functionally, miR-155 has multiple targets, including PU.1,[54] AID,[55] C/EBPβ,[56] and SHIP1.[57] SHIP1 is a negative regulator of myeloid cell proliferation and survival, thus reduced SHIP1 through miR-155 targeting blocks these critical functions, leading to sustained growth. Inhibition of miR-155 with antagomiRs increases SHIP1 expression and reduces tumor burden in mouse models of lymphoma,[58] further supporting the use of antagomiRs as miRNA-based therapeutics.

Although many studies report promising anti-cancer effects following the delivery of miRNAs or antagomiRs into cancer cells in culture and in mouse models there are critical hurdles that still need to be addressed and overcome.[59] The importance of efficiently delivering intact miRNAs and antagomiRs to the diseased tissue without promoting toxicity in normal cells is crucial if miRNA-based agents are to be advanced as therapeutics in humans. In this review, we discuss the strengths and limitations of conventional chemotherapeutics and targeted therapeutics, the challenges and advances in developing miRNAs as effective therapeutics, and the strategies being developed to deliver miRNAs to obtain clinical utility.

14.2 Limitations of Conventional Cancer Therapeutics and the Advent of "Therapeutic-miRNAs"

The development and discovery of new drugs has impacted the lives of many patients. Many of these new drugs are designed to target disease-associated molecules. Indeed, our understanding of the genetic and molecular mechanisms of cancer has improved tremendously over the past few decades; however, safe and targeted drug development is slow to progress, with high failure rates.[60] Major reasons why current cancer therapeutics fail can be classified into the following: (1) toxicity from the drug or its carrier; (2) lack of specificity in targeting cancer cells; (3) development of acquired resistance to the drug; and (4) difficulty in targeting sites of metastatic disease that are often inaccessible to the agent.

Expectedly, many of the targeted agents currently on the market target proteins such as enzymes or receptors. For example, tyrosine kinases, which are crucial mediators of cell growth, cell proliferation, anti-apoptosis, metastasis, and angiogenesis are often constitutively activated in various tumors due to overexpression and/or mutation, and thus, have been investigated as effective therapeutic targets.[61] The majority of drugs used to target tyrosine kinases either compete with the substrate adenosine triphosphate (ATP) or block the interaction of the ligand with the receptor, both impairing downstream signaling. Epidermal growth factor receptor (EGFR), a receptor tyrosine kinase, is notorious for being overexpressed and/or mutated, primarily in lung cancer[62] and glioblastoma.[63] In wild-type cells, two EGFRs homodimerize allowing ATP to bind, which results in *trans*-phosphorylation of the two receptors. Once the receptors are phosphorylated, signaling proteins are recruited, leading to transient activation of downstream RAS and PI3K pathways. In cancer, due to overexpression and/or mutation, the receptor is constitutively activated promoting sustained stimulation of the downstream signaling proteins and subsequent tumor formation. Erlotinib (Tarceva), an EGFR inhibitor, binds to the ATP binding site of the receptor with higher affinity than ATP, preventing downstream signaling from occurring.[64–66] Erlotinib, approved by the United States Food and Drug Administration (FDA) in 2004 as a treatment for cancer, has been effective in lung cancer patients and is currently under investigation for efficacy in other cancers. Similar FDA-approved targeted agents have been developed and approved for targeting other tyrosine kinases, such as imatinib (Gleevec) targeting BCR-ABL, used for treating leukemia[67] and sunitinib (Sutent) which targets both the vascular endothelial growth factor receptor and the platelet-derived growth factor receptor for use in renal cell carcinoma.[68,69] Despite the positive effects of tyrosine kinase inhibitors (TKIs) in the clinic, a critical problem is target-cell specificity. These drugs target not only diseased cells, but also normal cells. Likewise, many of the TKIs accumulate in the bone marrow and gastrointestinal tract, where cells are rapidly dividing and depend on these pro-growth pathways.[43] An additional concern is the development of acquired resistance. Cancer cells that have initially responded to EGFR TKI can acquire a second mutation in EGFR, leading to drug resistance.[70] Beyond secondary mutations in the target, cells challenged with drugs that block one signaling pathway can respond through alternative by-pass signaling tracks that can compensate or overcome the targeted cancer treatment. A recent study has shown that TKI drug resistance is also tied to epithelial-to-mesenchymal transition (EMT). Cells that undergo EMT gain the ability to invade, migrate, and proliferate. Examining response to erlotinib, resistance was significantly more abundant in mesenchymal cells than in epithelial cells.[71] Therefore, cells that migrate to other tissue sites may acquire erlotinib resistance at a greater frequency. Due to these concerns, and many others (for a thorough review on drug resistance in cancer please refer to the review by Holohan *et al.*[72]), alternative novel therapeutic approaches are being investigated.

Apart from tyrosine kinases, for the past few decades there have been countless attempts to target RAS. RAS is a GTPase which cycles between its guanosine triphosphate (GTP)-bound active state and guanosine diphosphate (GDP)-bound inactive state. Guanine nucleotide exchange factors exchange GDP for GTP, promoting the active state, whereas GTPase-activating proteins (GAPs) induce hydrolysis of GTP, promoting the inactivate state.[73] Mutations in RAS disrupt the interaction with GAPs, leading to constitutive activation and stimulation of downstream signaling pathways. Sustained activation of RAS promotes cellular proliferation, survival, and invasion.[74] Unfortunately, RAS is considered "undruggable" due to its smooth surface, lack of a deep binding pocket, and the high binding affinity it has for GTP, which hampers the ability of a drug to act as a competitive inhibitor at a concentration that would be physiologically achievable.[75] Researchers have continued to search for new methods to target undruggable RAS. Included among them are targeting downstream pathways and the use of synthetic lethal approaches.[76] One recently reported method is based on altering the localization of RAS to endomembranes. Delocalization prevents lipid modifications on RAS, blocking the translocation of RAS to the plasma membrane, which is essential for RAS activity. The target, prenyl-binding protein phosphodiesterase δ (PDEδ), directly alters the localization and activation of RAS, facilitating its diffusion in the cytosol. A small molecule inhibitor that disrupts the PDEδ–RAS interaction blocks RAS signaling in pancreatic cancer cells and xenografts.[77] While this study provides a new method for reducing RAS activity, the challenge with targeting undruggable targets such as RAS has been realized by the National Cancer Institute and is reflected in an initiative put forth in 2013, the RAS Initiative (www.cancer.gov/research/key-initiatives/ras).

Due to lack of specificity in targeting cancer cells, development of drug resistance, and unidentifiable drug target sites on the protein structure for some targets, current drugs are limited in their use. Therefore, alternative and/or additional approaches are necessary. For many of these targets, inhibiting the production of the protein through transcriptional or translational block may represent an effective way to target diseases that are not treatable by current drugs.

miRNAs hold great therapeutic promise for these undruggable targets, including RAS, due to their ability to target the messenger RNA and reduce expression of the protein product.[78] The main advantage of using miRNAs as a replacement therapeutic is that miRNAs are small and thus presumably easy to deliver. Furthermore, because non-diseased cells are expected to still harbor the miRNA, it is probable that toxicity due to unintentional delivery to non-diseased cells would be modest in comparison to toxicity from small molecules. Moreover, miRNAs target multiple genes, and many of the genes targeted by a single miRNA can modulate several target genes in a single pathway, or in complementary pathways leading to an enhanced response.[36,79] To identify the critical miRNAs that when altered in cancer promote or maintain disease, various transcriptomic and functional studies have been conducted. In a recent study, the transcriptome of 15 of the most common tumor subtypes was analyzed.[80] The study identified differential expression of 147 genes and 95 miRNAs that were shared in most tumor types. Importantly, 36 paired interactions between the miRNA and the reported targeted transcript were deregulated together. This result indicates that a single dysregulated miRNA can alter the expression of its target genes, and that these altered mRNA–miRNA interactions are conserved in cancer pathways between various tumors.[80] Therefore, balancing the levels of miRNAs that are dysregulated in cancer, through miRNA-based therapeutic approaches, can potentially impact multiple major signaling pathways in carcinogenesis.

14.3 miRNA Therapeutic Strategies

Due to advances in RNA sequencing, profiling miRNA expression levels in normal and tumorigenic samples has become routine. Indeed, aberrant expression of miRNAs is often associated with numerous disease states,[16] and in many cases changes in miRNA abundance can directly contribute to various hallmarks of cancer.[81] These early findings have paved the way for the development of novel miRNA-based therapeutic options to alter the levels of dysregulated miRNA. Depending on how the miRNA expression level in the disease state changes, two different strategies for miRNA-based therapeutics have been developed: (1) miRNA inhibition—for cases where the miRNA is found to be overexpressed in the disease state; or (2) miRNA restoration—for cases where the miRNA is downregulated in the disease state (Figure 14.1A).

14.3.1 Antagonizing Overexpressed miRNAs

When a causal miRNA is overexpressed in the disease state, miRNA inhibition therapy is used to block miRNA function. This very common strategy has been achieved using a variety of mechanisms that collectively sequester the overly abundant miRNA. Antisense oligonucleotides (ASO) are synthetic short single-stranded RNA or DNA molecules that bind to the mature miRNA sequence *via* complementary base pairing, thus sequestering the miRNA away from the endogenous targets allowing target-gene translation. Initial attempts to introduce naked ASOs systemically were ineffective due to rapid degradation by nucleases in the bloodstream and poor uptake by the target tissue and cells.[82,83] However, chemical modifications were incorporated into the design of the ASO to improve stability and cell permeability.[84] The first oligonucleotide modification that was incorporated to prevent RNA ASOs from degradation by nucleases was 2′-*O*-methyl of the ribose sugar.[85] The addition of the 2′-*O*-methyl group provides nuclease resistance and improved binding to the miRNA target. Alternative modifications such as, 2′-*O*-methoxyethyl,[86] 2′-fluoro,[87] 2′-4′-methylene (locked nucleic acids, LNAs),[88] phosphorothioate linkages,[89] and peptide nucleic acids[90] also provide resistance to nucleases and improve overall binding affinity (Figure 14.1B). Moreover, addition of a cholesterol to the 3′ end of the ASO increases cellular delivery kinetics *in vivo*. Cholesterol promotes uptake of the ASO by facilitating ASO binding to cell surface membrane receptors.[91] ASOs synthesized with such modifications have been effective against hepatitis C virus (HCV) infection, which causes liver inflammation. Long-term illnesses associated with HCV infection can lead to liver cancer. In HCV-infected cells, miR-122 is elevated, and through binding to the 5′ non-coding region of the viral genome miR-122 stimulates viral replication.[92] Thus, silencing miR-122 through ASO targeting was hypothesized as a valid treatment option. Miravirsen (SPC3649), developed by Santaris Pharma, is a LNA that does just that. ASO inhibition of miR-122 results in reduced HCV RNA with no signs of side effects in mice and primates,[93–95] nor were there any reported short- or long-term side effects in human patients.[96] Miravirsen is currently in phase II clinical trials (ClinicalTrials.gov identifier NCT01872936).[97]

Although ASOs are currently the most clinically relevant method for miRNA sequestration, other technologies have been used in pre-clinical studies as proof of concept with the intent for future development; for example, DNA plasmids that encode multiple complimentary miRNA binding sites in tandem. These so-called miRNA sponges, which can be transfected into cells, sequester miRNAs preventing miRNA–target gene engagement.[98] Recent studies suggest that some competing endogenous RNAs (ceRNAs) such as pseudogenes,[99] long non-coding RNAs[100] or circular RNAs[101] act as endogenous sponges sequestering miRNAs leading to upregulation of miRNA target genes. Studies attempting to adopt some of these endogenous ceRNAs as therapeutics to inhibit miRNA activity are of great interest.

Not all attempts to inhibit miRNA activity are based on base pairing between the miRNA and the inhibitor. Small molecule inhibitors are also used to inhibit miRNA expression and function. These compounds primarily inhibit the miRNA biogenesis pathway or disrupt the interaction between the miRNA and its target mRNA,[102] although other small molecules have been identified that inhibit transcription of miRNAs, miR-21 or miR-122 being two examples of miRNAs that fall into this class.[103,104]

14.3.2 Restoring Downregulated miRNAs

Conversely, when a miRNA is lost or downregulated in the disease state, miRNA restoration therapy supplements miRNA levels, leading to a therapeutic effect. While miRNAs levels can be enriched through delivery of DNA-encoding miRNA vectors,[105] restoration is typically done through introducing synthetic oligonucleotides that have the same mature miRNA sequence and a fully complementary passenger strand, generating a miRNA duplex referred to as mimic.[106] Similar to ASOs, miRNA mimics are chemically modified to prevent nuclease degradation and enhance their binding affinity. However, to maintain activity, careful consideration of the composition of the RNA bases and various modifications must be taken into account. For example, duplex RNAs composed entirely of 2′-*O*-methyl modifications tend to over-wind, while RNAs duplexes composed solely of 2′-fluoro modified ribose bases are under-wound. Similarly, a duplex composed of all LNAs has decreased helical winding and an enlarged helical pitch.[107] To overcome the helical constraints, short interfering RNAs (siRNAs) have been developed where 2′-*O*-methyl residues are alternated with 2′-fluoro residues. Additionally, some of the phosphodiester groups on the ends of the two strands have been replaced with phosphorothioate internucleotide linkages to reduce the activity of extra- and intra-cellular nucleases. While these modified bases and linkages lead to reduced innate immunity, increased stability, and roughly a 100-fold increase in tissue distribution,[108] displacement of the passenger strand from the guide strand poses an issue. This rate-limiting effect is alleviated *via* shortening of a fully-modified passenger strand to 15 nucleotides, reducing the thermal melt of the duplex.[108] Several other siRNA chemistries have been developed by Alnylam Pharmaceuticals,[109] many of which show exceptional stability and targeting, and reduced off-target effects relative to unmodified

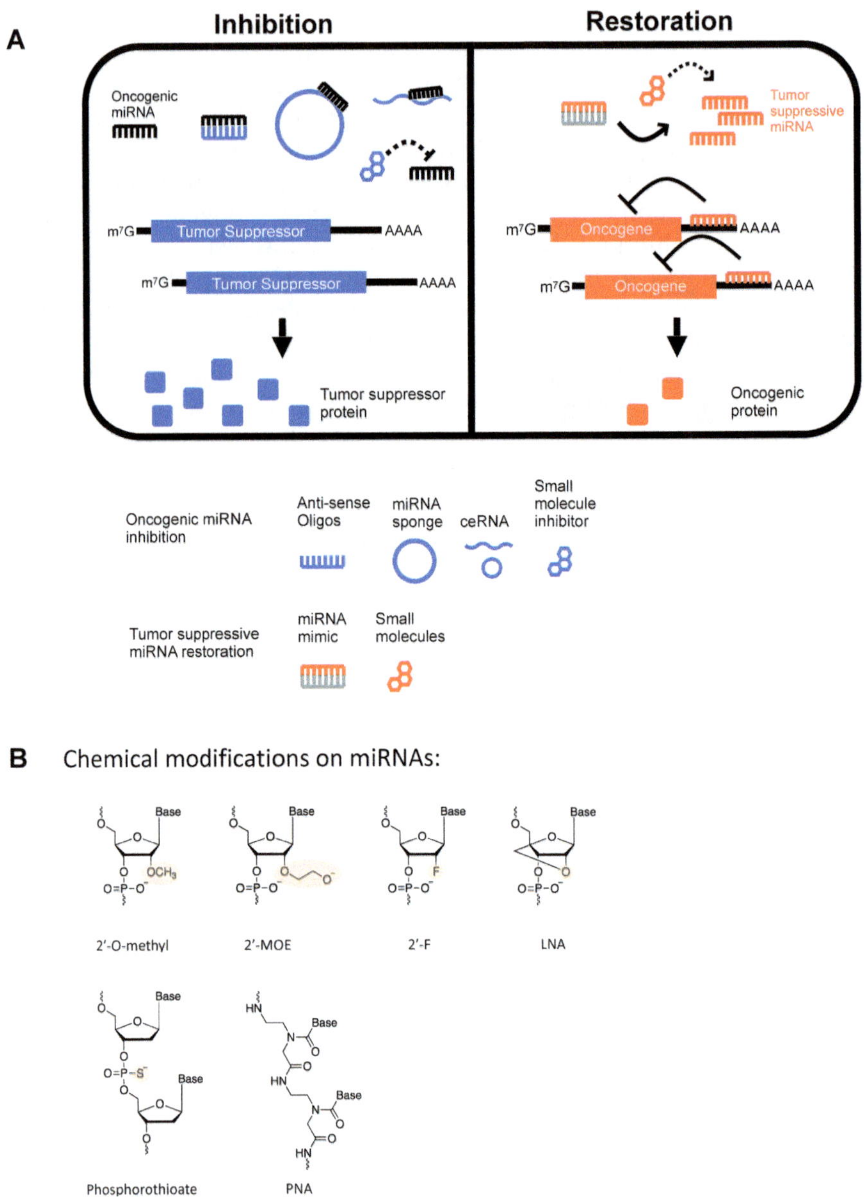

Figure 14.1 Therapeutic strategies for altering miRNA activity. (A) Two major strategies for miRNA-based therapeutics are being developed. Left: When oncogenic miRNAs are upregulated in the disease state, miRNA inhibition is performed using anti-sense oligonucleotides (ASOs), miRNA sponges, competing endogenous RNAs (ceRNAs: pseudogenes, long non-coding RNAs, circular RNAs), or small molecule inhibitors. Sequestration of oncogenic miRNAs promotes the expression of tumor suppressive proteins (blue squares). Right: When tumor-suppressive miRNAs are downregulated in the disease state, miRNA

siRNAs.[110] Importantly, while these modifications and shorter passenger strand design have been efficacious for siRNA delivery, attempts to modify miRNA mimics in a similar manner have had mixed results, requiring more careful engineering.

A minimally modified miR-34a mimic entered clinical trials; in this case the mature strand was not modified while the passenger strand contained modifications that facilitated loading of the mature strand into the RNA-induced silencing complex (RISC). As the first miRNA to enter into clinical trial, miR-34a had promising initial results, including prolonged confirmed partial response lasting 48 weeks in a patient with hepatocellular carcinoma.[111] Unfortunately, as dose escalation increased, several severe immune-related adverse events occurred, resulting in termination of the phase I study in 2016. Additional miRNAs have since entered into the clinic. Therapeutic delivery of miR-29b mimics (MRG-201), developed by miRagen Therapeutics, was tested in patients with cutaneous fibrosis and had promising effect. Previous studies from the van Rooij lab observed that restoration of miR-29b levels during pulmonary fibrosis progression decreased collagen expression and blocked pulmonary fibrosis.[112] Currently, MRG-201 has completed phase I clinical trials (ClinicalTrials.gov identifier NCT02603224); however, the future of MRG-201 is not certain at this time. MiRagen has other miRNA-based drugs in the pipeline, although the remaining candidates are all miRNA inhibitors: MRG-106, which inhibits miR-155 and is being developed for hematological malignancies; MRG-107, also a miR-155 inhibitor, although the target population in this case is patients with amyotrophic lateral sclerosis; and MRG-110*, an antagomiR to miR-92a which is predicted to enhance blood vessel growth to accelerate the healing process.

With regard to enhancing the cellular abundance of miRNA, direct administration of the miRNA mimics is the obvious mechanism; however, due to challenges with miRNA delivery (see subsequent sections), small molecules are being used to restore miRNA expression. Compounds that can activate specific transcription factors or inhibit negative regulators of miRNA transcription or biogenesis can increase the cytosolic levels of miRNA. For example, compounds that inhibit the expression of *let-7* targets using a cell-based assay were identified by the Lowry lab. In this study, three compounds with a similar structure targeted and inhibited phosphodiesterase. Inhibition of phosphodiesterase stimulated cAMP levels and activated transcription factor

restoration is performed using miRNA mimics or small molecules to induce expression and/or restore biogenesis. Increased levels of the tumor-suppressive miRNA reduce the expression of oncogenic proteins (red squares). (B) Chemical modifications of miRNA mimics and ASOs improve stability, cell permeability, and binding affinity to mRNA. Modifications are made at the 2′ position on the sugar ring such as, 2′-*O*-methyl, 2′-*O*-methyoxyethyl (2′-MOE), 2′-fluoro (2′-F), and locked nucleic acid (LNA). Additional modifications are made on the RNA backbone: phosphorothioate and peptide nucleic acid (PNA).

CREB. Activation of CREB led to a reduction of the *let-7* target genes *HMGA-2* and *MYC*.[113] An additional study by the Gregory lab identified compounds that selectively block a negative regulator of *let-7* biogenesis. LIN28, an RNA-binding protein, negatively regulates *let-7* biogenesis through recruitment of the 3′ terminal uridylyl transferases, TUT4 and TUT7. Once recruited, the TUTases add an oligo (U)-tail to pre-*let-7*. The uridylated pre-*let-7* is then targeted for degradation preventing processing to the mature tumor suppressive form. Using a biochemical assay that monitors TUT4 activity, six compounds were identified as TUT4 inhibitors impairing the TUT4-pre-*let-7* interaction *in vitro* as well as in cells.[114] Modulating miRNA levels *via* small molecules may not necessarily be the most efficient approach yet as there are anticipated challenges that need to be overcome, such as permeability of the small molecule, and potential off-target effects. Nonetheless, these studies shed light on some of the basic biology of miRNA regulation, and perhaps, in the future, may lead to miRNA-based small molecule therapeutics.

While there is substantial evidence to suggest that modulating the levels of certain miRNAs can alter the pathology of disease, the biggest challenge in the field is achieving safe, specific, and efficient delivery of the miRNAs to the target tissue. Modification of the oligos have prevented degradation and increased binding efficacy as described; however, still due to delivery vehicle associated toxicity, poor transfection efficiency, systemic clearance, and non-specific biodistribution safe and targeted delivery strategies are essential to realize the full potential of miRNA therapeutics.[115] Initially, viral vectors were used to deliver DNA encoding miRNA precursors into cells, which following transcription are subject to processing by the endogenous machinery. Viruses such as lentivirus, adenovirus, and adeno-associated viruses (AAV) were modified by genetically engineering viral capsid protein to increase the binding affinity between the virus and the cancer cell receptor for specific targeting. In the context of hepatocellular carcinoma, miR-26a is dramatically reduced in expression, which contributes to disease.[116] To revert the cancerous phenotype, modified AAV was used to deliver the gene encoding *mir-26a* specifically to the liver.[117] Exogenous miR-26a resulted in tumor regression by promoting cell cycle arrest and apoptosis without toxicity.[118] However, despite efficient delivery using viruses there are some concerns. Viruses can trigger an immune response, which can reduce the efficiency of delivery. In addition, lentiviruses, which integrate their DNA into the host genome, can activate or inhibit genes depending on the integration site. Lastly, high-quality viruses are difficult to scale, limiting their application. Although virus-based delivery is efficient for delivering miRNAs and for establishing proof of concept, due to the concerns described, improvements are necessary to utilize viruses in the clinic. Thus, other non-viral delivery vehicles are receiving more attention since they tend to be safer, and easier to manufacture and scale up for miRNA therapy.

14.4 Non-viral miRNA Delivery Systems for Cancer Therapy

Development and design of an appropriate delivery system for miRNA mimics is as important as selecting therapeutically relevant miRNAs. An ideal delivery system should be (1) biocompatible with minimum toxicity or immune response; (2) stable in circulation, but should be excreted after the payload is released; (3) efficient in loading and releasing the payloads into or out of the delivery system; (4) selective in targeting disease cell; (5) easy to penetrate the endothelial layer to reach the target tissue; and (6) cost-effective and easy to scale up. There has been ongoing research to address each of these criteria, through the development and testing of various delivery platforms (Figure 14.2). This section gives an overview of the main delivery vehicles that have been tested and developed, including lipid-based vehicles, polymer-based vehicles, inorganic material-based vehicles, and a class of novel vehicles. Due to the amount of effort that has been conducted in the cancer field, a majority of the review is dedicated to studies delivering miRNAs to tumor tissue. However, many of the same strategies can be applied to other disease models.

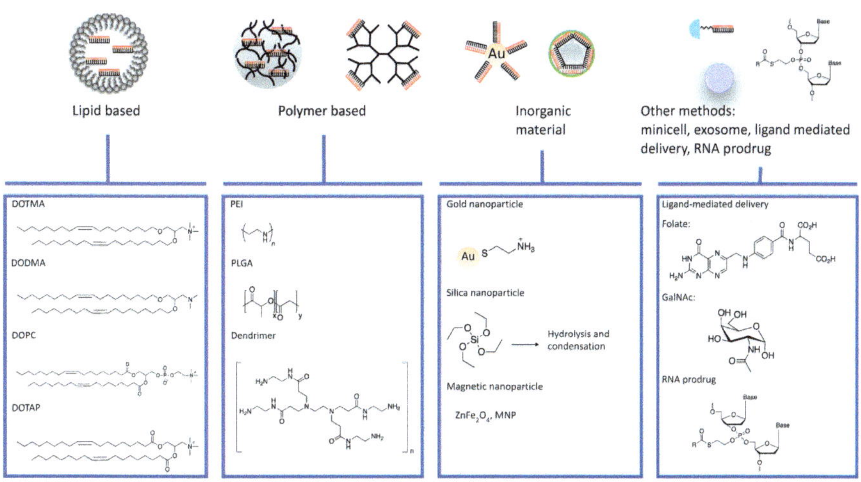

Figure 14.2 Delivery systems for miRNA therapeutics. Modified anti-sense oligonucleotides (ASOs) or miRNA mimics are encapsulated in multiple delivery vehicles to efficiently deliver miRNAs. The lipid composition is modified to create lipid vesicles with different properties using DOTMA, DODMA, DOPC, and DOTAP. Different polymer-based systems such as polyethylenimine (PEI), polylactic-*co*-glycolic acid (PLGA), and dendrimers were also developed to deliver miRNAs. Lately, inorganic gold, silica, and magnet nanoparticles have been developed as promising tools for miRNA delivery. Due to toxicity associated with some of the delivery vehicles, novel methods have been developed such as mini cells, exosomes, FolamiRs, and RNA prodrugs.

14.4.1 Lipid-based Delivery

The most well-studied vehicles used for delivering miRNAs are lipid-based vesicles. A lipid vesicle is composed of a lipid bilayer and a hydrophilic core that incorporates the payload, in this case a miRNA mimic. Cationic lipids, which facilitate packaging of negatively charged miRNAs, are typically used. The positively charged lipid–miRNA complexes interact with the negatively charged cell membrane and undergo endocytosis. However, cationic lipids often promote immunomodulatory activity. To alleviate this, the lipid composition can be modified to create lipid vesicles with different surface charges. Surface charges also help facilitate endosomal escape and stability in circulation. Although there have been many successful studies using lipid vesicles for delivery, there are some drawbacks that have prevented lipid-based miRNA delivery platforms from successfully making their way through clinical trials. These include, but are not limited to, toxicity, non-specific uptake, immune responses, and endosomal sequestration. Improvements in modifying the lipid composition and structure, and adding unique moieties to the complexes have begun to overcome some of these challenges, increasing efficacy and reducing toxicity.

A cationic lipid composed of *N*-[1-(2,3-dioleyloxy)propyl]-*N*,*N*,*N*-trimethylammonium chloride (DOTMA), cholesterol, and polyethylene glycol (PEG) has been developed as a vehicle to deliver miRNA for cancer therapy. Cholesterol facilitates the release of miRNA into the cells and reduces cytotoxicity by neutralizing some of the positive charges of DOTMA.[119] Addition of PEGylation helps to dampen the immune response and increases stability during circulation. Delivery of miR-122 mimics using DOTMA–cholesterol, with the addition of a helper lipid, oleic acid, effectively delivered miR-122 to hepatocellular carcinoma (HCC) cells *in vivo*. Delivery was more robust than commercially available liposomal transfection reagents.[120] This DOTMA–cholesterol complex, loaded with precursor miR-133b, was also tested in non-small cell lung cancer (NSCLC). When miR-133b is processed into its mature form, mature miR-133b functions as a tumor suppressor by targeting the anti-apoptotic gene *MCL-1*. Delivery of miR-133b using this delivery vehicle achieved better accumulation in the lung tissue compared to commercial siPORT NeoFX transfection reagent.[121] The same group also evaluated delivery of the tumor suppressor miR-29b to lung tumors *in vivo* using DOTMA–cholesterol as a vehicle. Delivery reduced xenograft lung tumor growth through targeting *CDK6*, *DMNT3B*, *and MCL-1*.[122]

DOTMA cationic lipids have permanently charged tertiary amine head group whereas, the 2-dioleyloxy-*N*,*N*-dimethyl-3-aminopropane (DODMA) lipids have a protonatable tertiary amine head group. DODMA allows the lipid to be ionizable depending on the pH. At physiological pH the lipid has a neutral charge, but in more acidic environments the amphoteric lipid is converted into a positively charged, cationic lipid. When DODMA, in complex with miRNA, is delivered to the cell, the complex undergoes endocytosis and is trafficked to endosomes. The acidic endosomal environment alters the charge of DODMA–miRNA from neutral to positive, resulting in fusion with the anionic endosomal membrane. This allows the encapsulated miRNA to be released into the cytoplasm. Using this method, miR-122 was delivered to HCC cells, resulting in a significant reduction in tumor burden *in vivo* by targeting *Adam10* and *Srf* genes.[123]

Although cationic lipids are effectively internalized into cells, they are rapidly cleared from circulation due to their net positive charge. To improve biodistribution in target tissues, neutral lipids were used as an alternative for generating lipid vehicles. When miR-34a was encapsulated in a neutral lipid emulsion (NLE; MaxSuppressor™ *In Vivo* RNA-LANCEr II, BIOO Scientific, Inc.) composed of 1,2-dioleoyl-*sn*-glycero-3-phosphocholine (DOPC), squalene oil, polysorbate 20, and an antioxidant, it effectively accumulated in the target lung tissue and reduced tumor burden in mouse models of NSCLC.[34,37] Other studies show therapeutic effects with miRNAs encapsulated in the same NLE. For example, delivery of miR-221/222 inhibitors using NLE promoted anti-proliferative effects in human multiple myeloma both in cells and *in vivo* by upregulating *p27Kip1*, *PUMA*, and *PTEN*.[124] In addition, NLE-delivery of miR-182 antagomiRs increased *Bcat2*, *Foxo3*, *and Adcy6* expression to regulate hypertrophy *in vivo*.[125]

Although use of neutral lipid delivery vehicles results in reduced toxicity and improved biodistribution compared to cationic lipid vehicles, there are challenges with inefficient delivery of the miRNAs into cells and loading of the miRNAs into the lipid vehicles. Therefore, to overcome these issues liposome-polycation-hyaluronic acid (LPH) was developed.[126] The hyaluronic acid-based biopolymer complexed with miRNAs and protamine improved miRNA loading and formed an anionic complex. This anionic complex is then encapsulated in cationic lipids *N*-[1-(2,3-dioleoyloxy)propyl]-*N*,*N*,*N*-trimethylammonium chloride (DOTAP), cholesterol, and PEG, to neutralize some of the positive charges, reducing delivery-associated toxicity. To improve targeting, and thus specificity, additional modifications were added to this complex. For example, LPH with the addition of a GC4 single-chain variable fragment (scFv) efficiently targeted lung metastasis in the B16F10 murine melanoma model. The GC4 scFv is a monoclonal antibody that was initially selected and isolated from a phage antibody library to target the CD133 epitope on B16F10 melanoma cells. Delivery of miR-34a using the GC4-targeted nanoparticles resulted in a reduction in lung metastasis as well as primary lung tumors *in vivo*.[126] Similarly, modification of LPH with cyclic pentapeptide c(Arg-Gly-Asp-D-Phe-Lys) (cRGD) targets $\alpha_v\beta_3$ integrin-positive angiogenic endothelial cells. Delivery of anti-miR-296 using cRGD–LPH inhibited angiogenesis through enhancing the miR-296 target hepatocyte growth factor-regulated tyrosine kinase substrate (HSG) expression. Elevated HSG promoted growth factor receptor degradation to ultimately diminish blood tube formation and endothelial cell migration in cells and *in vivo*.[127,128]

Improvements in lipid-based vehicles have resulted in their advancement into clinical trials. A pH-dependent lipid vehicle in complex with miR-34a, called MRX34, delivered the first miRNA mimic to human patients in a phase I study (Mirna Therapeutics).[129] MRX34 is composed of a double-stranded miR-34a mimic encapsulated in a lipid carrier called a Smarticle. The Smarticle, composed of amphoteric lipids, has an overall anionic charge at physiological pH, while at lower pH conditions, such as those that occur in the tumor environment, the lipids become cationic, facilitating uptake into the tumor cells. Initially, MRX34 was effective in multiple *in vivo* mouse models promoting its transition into a phase I clinical trial for patients with primary liver cancer or with primary cancers that metastasize to the liver. However, due to severe immune-related side effects resulting in the death of three patients, the clinical trial was terminated.[111] Whether the toxicity was due to miR-34a or the vehicle has not yet been reported.

Lipid vehicles have been developed as an effective delivery system for miRNA. However, despite the protected miRNAs stability in circulation and modifications of the lipids that help to reduce toxicity, further improvements are needed to translate lipid-based delivery systems into the clinic. In the meantime, other delivery methods have been developed to advance miRNA delivery with unique features.

14.4.2 Polymer-based Delivery

Synthetic or natural polymers can be tailored in size and charge to efficiently deliver miRNAs. Polyethylenimines (PEIs) are the most widely used cationic polymer for miRNA delivery. Due to the high cationic charge density, the PEI and miRNA complex retain a net positive charge. Thus, the complex targets and interacts with the negatively charged polysaccharides on the cell surface, promoting efficient endocytosis. Once the complex is inside the cell, the pH buffer effect facilitates endosomal swelling through the influx of protons and water acting as a so-called "proton sponge", ultimately leading to endosomal bursting. Release of the miRNAs into the cytoplasm facilitates miRNA-mediated target gene silencing.[130] Examples of miRNAs delivered by PEI include miR-145 and miR-33a, which were administered either locally or systemically. Local delivery of miR-145 or systemically delivered miR-33a reduced cell proliferation, increased apoptosis, and reduced tumor growth in mouse models of colon carcinoma.[131] Additionally, delivery of PEI-encapsulated miR-145 in combination with chemoradiotherapy and cisplatin-enhanced therapeutic efficacy and improved survival in xenograft mouse models of lung cancer.[132]

Although there are multiple benefits to using PEI, there are also critical limitations, such as a high charge density of the particle and poor particle biodegradation that contribute to toxicity. To circumvent this, modifications of PEIs have been made to reduce toxicity and enhance polymer biocompatibility and biodegradability. For example, by adding a disulfide linkage to the PEI the polymer complex can be reduced by enzymes such as glutathione reductase as it enters the cell forming smaller and less toxic PEI polymer fragments.[133] Further refinement of PEIs can lead to enhanced uptake in specific tissues. For example, conjugation of PEI to a polyarginine (R11) peptide enhances PEI uptake by prostate cancer cells.[134] These polyarginine-disulfide-linked PEIs efficiently delivered miR-145 to prostate tumor cells, reducing tumor growth and increasing animal survival.[135]

Polylactic-*co*-glycolic acid (PLGA) has been well characterized and widely studied due to its safe, biodegradable, and biocompatible properties. PLGA is composed of two copolymers: poly-lactic acid and poly-glycolic acid. The ratio of the two copolymers changes the particle's overall physical properties. These properties alter the degradation kinetics of the molecule and release of the payload. For example, increasing the content of glycolic acid, which is slightly more hydrophilic than lactic acid, increases the hydrolysis rate of the particle leading to faster release of the payload. To increase circulation time and retention of the particles in the tumor tissue, multiple surface modifications have been added.[136] Such surface-modified particles have successfully co-delivered anti-miR-21 and anti-miR-10b, resulting in reduced tumor growth in a breast cancer xenograft model.[137] To improve targeting, PLGA particles have also been coated with a cell-penetrating peptide, penetratin, which was used to efficiently deliver anti-miR-155 to lymphocytes, delaying tumor growth in a mouse model of lymphoma.[58] A key advantage of using PLGA as a delivery vehicle is the high loading efficiency of PLGA particles. Taking advantage of the high-loading capacity of PLGA particles and efficient cellular uptake of PEI particles, delivery using copolymers containing both PLGA and PEI was attempted. Combining the hydrophobic and neutral properties of PLGA with the cationic PEI, this micelle-like copolymer was loaded with the chemotherapeutic drug doxorubicin along with miR-542-3p. Use of this hybrid particle efficiently promoted apoptosis and cytotoxicity in breast cancer cells.[138]

Another polymer-based carrier that has been developed is composed of dendrimers. Dendrimers are highly branched polymers with multiple surface moieties. These particles typically contain cationic amine groups on their branches which facilitate association with negatively charged miRNAs. Similar to PEIs, dendrimers act as proton sponges enabling endosomal burst and release of miRNA once inside of the cells. Dendrimer vehicles are highly efficient in delivering miRNAs into cells; however, due to their charge-related toxicity there have only been a few successful deliveries reported. For example, poly(amidoamine) (PAMAM) dendrimers have been generated and loaded with anti-miR-21 and the chemotherapeutic 5-fluorouracil (5-FU). Delivery of this complex to glioblastoma cells increased chemosensitivity of 5-FU and decreased migration of tumor cells.[139,140] Similarly, a triple helix RNA molecule containing miR-205 and anti-miR-221 was conjugated to PAMAM G5 dendrimer. Delivery of this complex to a triple-negative breast cancer mouse model resulted in reduced tumor size and increased survival.[141]

In addition to synthetic polymers, natural polymers are also being developed and used as delivery vehicles for miRNAs. Chitosan is a natural cationic polymer composed of glucosamine and *N*-acetylglucosamine residues.[142] The biocompatibility, biodegradability, and low toxicity of chitosan makes it an attractive delivery vehicle. The addition of hyaluronic acid enhanced biocompatibility as well as targeting to tumor cells expressing high levels of CD44 on their surface.[143] When miR-34a and doxorubicin were co-encapsulated in hyaluronic acid–chitosan complexes, breast cancer cells expressing high CD44 selectively took up the particle and delivered miR-34a promoted anti-tumor activity through silencing anti-apoptotic *BCL-2* and suppressed cell migration through targeting the Notch signaling pathway.[144]

14.4.3 Inorganic Material-based Delivery

Lately, inorganic materials such as gold, silica, and magnetic-based nanoparticles have been developed as miRNA delivery vehicles. The advantage of inorganic-based delivery systems is that they are biocompatible, non-toxic, and non-immunogenic. For example, gold nanoparticles (AuNPs) have unique physical and chemical properties such as shape, surface area, biocompatibility, and low cytotoxicity that aids in effective delivery of miRNAs.[145–147] Specifically, the coating on gold nanoparticles are a mix of hydrophobic and hydrophilic components that can bind to scavenger receptors on the cell surface leading to efficient uptake *via* endocytosis. In addition, cysteamine-functionalized AuNP-S-PEG nanoparticles also have high miRNA loading and release capacities, and low toxicity. This is exemplified in a study that delivered two miRNAs with different functions. Using AuNP-S-PEG nanoparticles, delivery of tumor suppressor miR-31 reduced cell proliferation in neuroblastoma cell lines, whereas delivery of oncogenic miR-1323 increased cell proliferation in ovarian cancer cell lines.[148]

Silica-based nanoparticles are stable, biodegradable, and non-toxic, making these nanoparticles an attractive system for miRNA delivery. Taking advantage of neuroblastoma cells, which express high levels of antigen disialoganglioside (GD_2) on their cell surface, systemic delivery of silica nanoparticles conjugated with an anti-GD_2 antibody specifically targeted neuroblastoma cells. Delivery of these nanoparticles containing miR-34a to a tumor-bearing mouse model resulted in tumor growth inhibition through the induction of apoptosis and reduction in vascularization.[149]

Magnetic nanoparticles are an additional inorganic particle used to deliver miRNAs. The imaging capability of these magnetic nanoparticles through magnetic resonance imaging allows for visualization of the nanoparticle in the target tissues. Breast cancer mouse models treated with LNA anti-miR-10b encapsulated in ultra-small magnetic particles modified with integrin binding peptide RGD prevented breast cancer metastasis.[150] Similarly, magnetic zinc-doped iron oxide nanoparticles ($ZnFe_2O_4$) complexed with the tumor suppressor *let-7* and coated with PEI efficiently delivered the miRNA into glioblastoma cancer cell. Once delivered into the cells, and under magnetic hyperthermia, the cancer cells became sensitized to the treatment and apoptosis was enhanced.[151]

The use of inorganic materials for miRNA delivery does have some disadvantages. In general, inorganic nanoparticles lack cargo protection and are thus targeted for endosomal entrapment. In addition, other than the gold nanoparticle, most inorganic nanoparticles have low loading efficiency. Although these studies are expanding the potential vehicles that can be used to deliver miRNAs, marked improvements are needed before these vehicles would have clinical utility.

14.4.4 Novel Delivery Methods

Other novel approaches currently being developed to deliver miRNAs include the use of enucleated non-living nano-sized bacterially derived cells called mini cells or EnGenelC Delivery Vehicle (EDV). These mini cells have been modified to selectively target cells overexpressing EGFR through coating the nanoparticle with anti-EGFR antibodies. A phase I clinical trial of EDV nanoparticles containing miR-16 mimic for the treatment of malignant pleural mesothelioma and NSCLC was completed in early 2017 (ClinicalTrials.gov identifier NCT02369198).[152–154] The results of the study have yet to be released.

Delivery of miRNAs has also been achieved using endogenous miRNA delivery vehicles such as exosomes. Exosomes are 50–100 nm extracellular vesicles that aid in cell-to-cell communication through the transfer of bioactive materials such as lipids, proteins, and nucleic acids, including miRNAs. These miRNA-containing exosomes are transported to the recipient cells leading to miRNA-directed targeting in the recipient. Utilizing this mechanism, modified exosomes that express a peptide that specifically targets upregulated EGFR on cancer cells has been developed. These modified exosomes, loaded with *let-7a* mimics were delivered to a breast cancer xenograft model reducing tumor growth.[155] However, in order to advance the clinical utility of exosomes some critical hurdles need to be resolved, such as reducing unwanted accumulation of the exosomes in the liver and increasing their low miRNA loading efficiency.

While all these approaches rely on a protective delivery vehicle, the idea of abandoning the protective vehicle can prevent many of the vehicle-associated problems such as toxicity and immune response. Indeed, a vehicle-free method for delivering miRNAs has been developed. Our group has generated folate-conjugated miRNAs (also called FolamiRs) that can specifically target cancer cells expressing high levels of the folate receptor on the cell surface,[156] a feature common to some hematological malignancies, and lung,[157] breast,[158] ovarian,[159] and other solid tumors. These cancer cells internalize the bound FolamiR *via* endocytosis, ultimately releasing some of the miRNA cargo into the cytosol to induce gene silencing.[160] Therapeutic effects of miR-34a-conjugated FolamiRs were observed in a breast cancer xenograft model and in a genetically engineered model of NSCLC.[156] Although the results are promising, endosomal entrapment of the FolamiR molecules is a rate-limiting step, a feature common to most of the delivery vehicles. Various methods to promote endosomal escape and enhance efficiency of delivered miRNAs are being evaluated.[161]

Additional targeting ligands that bind to other surface receptors have been developed for RNA delivery. Using the *N*-acetylgalactosamine (GalNAc) ligand, an RNA prodrug was efficiently delivered to the liver, which highly expresses the GalNAc receptor asialoglycoprotein (ASGPR). The RNA prodrug contains a neutral phosphotriester backbone which allows the RNA to enter the cell unimpaired by the typical negative charge that prevents RNA uptake.[162] Once inside, the RNA prodrug converts to its native negative charged phosphodiester backbone by cytoplasmic thioesterases, restoring its activity. The RNA prodrug displayed cellular stability and low immune response *in vivo*.[163] In addition, an siRNA targeting transthyretin (TTR) was developed by Alnylam Pharmaceuticals for the treatment of TTR-mediated amyloidosis. When mutated, TTR is unable to carry thyroid hormones and retinol binding proteins from the liver, resulting in accumulation of amyloid protein in the tissue, subsequently leading to peripheral neuropathy and cardiomyopathy.[164] A TTR siRNA conjugated to GalNAc was efficiently delivered to the liver and knocked down both mutant and wild-type TTR.[165] This treatment, also called ALN-TTR02 or Patisiran is currently in phase III clinical trials (ClinicalTrials.gov identifier NCT01617967). While these later examples highlight siRNA-mediated delivery, based on the success of folate-conjugated miRNAs, it is expected that additional ligand-mediated miRNA delivery vehicles will be developed and validated in the near future.

14.5 Conclusion

Over the past 20 years miRNAs have emerged as important regulators of nearly every cellular process and proved to be promising entities for cancer therapeutics in pre-clinical models. Delivery of miRNA mimics to restore relevant miRNA levels or ASOs to sequester oncogenic miRNAs has led to appropriate therapeutic consequences. However, a major challenge to be met while using ASOs and miRNA mimics is to ensure their safe and targeted delivery. To accomplish this, two major factors need to be addressed: (1) the chemical and biological properties of the miRNA mimic or ASO need to be understood and improved upon; and (2) safe and effective targeted delivery strategies need to be developed, validated, and put into practice.

Previous efforts to identify specific RNA modifications that improve activity, stability, and reduce toxicity were attempted using a large-scale chemical screen.[166] In this initial screen, LNA-modified siRNAs were determined to be compatible with loading into the RISC and siRNAs synthesized with LNAs had improved targeting activity. In addition, increased stability of the RNA was found to occur with only a few modified bases rather than full modification. However, more recent reports suggest that fully modified siRNAs are more effective.[108] The discrepancy is likely due to the types of modifications and the position of the modified bases in the RNA duplex. Moreover, to alleviate the increased melting temperature of fully modified duplexes, which may have impaired activity in the earlier study, Khvorova and co-workers shortened the antisense strand to 15 nucleotides to facilitate easier strand displacement. Despite these attempts some chemical modifications that may be useful for improving activity, stability, and reducing toxicity may lead to unwanted effects such as impaired specificity and off-target effects due to sustained binding to lower-affinity targets.[167]

Indeed, the chemical properties of the delivered miRNA can significantly alter the efficacy of the miRNA; however, one cannot overlook similar challenges related to the biological properties of miRNAs. For example, one of the key features of miRNAs is that one miRNA can downregulate multiple genes that regulate diverse signaling pathways. At times, the concerted regulation of multiple pathways is an advantage—assuming the pathways converge on a desired phenotype. However, due to the broad activity of some miRNAs, therapeutic doses may lead to unwanted silencing of low-affinity targets. The solution is to selectively regulate pathways that contribute to the particular disease progression, while not affecting other pathways or wild-type genes, whose expression may be critical to certain cells. Through computational analysis, artificially generated miRNAs can be engineered that selectively target genes with single nucleotide mutations that contribute to the tumorigenic phenotype without affecting the wild-type gene.[168] This method can be expanded to designing miRNAs that selectively target genes only involved in pathology of the disease while eliminating targeting of irrelevant genes, thus preventing side-effects associated with unwanted targeting. An algorithm that may be useful for predicting such designer miRNAs is miRBooking.[169]

Although use of current miRNA delivery systems has resulted in improvements in reducing vehicle-associated toxicity and immune response, while enhancing biocompatibility, transfection efficiency, and endosomal escape, targeted delivery of miRNAs *in vivo* is still a challenge. Due to the negative charge of the miRNA, the vast majority of the delivery vehicles are engineered using cationic lipids or polymers for packaging. However, the positive charge density of the delivery vehicles can promote rapid clearance in the bloodstream and cell toxicity. To overcome this, new biocompatible materials and/or novel strategies to remove or minimize the charge will be essential. For example, neutralizing the charge of the delivery vehicle through the use of neutral lipids, polymers, or masking the charge through conjugating PEG molecules to the vehicle has reduced cationic-related toxicity. Similarly, the use of uncharged RNA pro-drugs accomplishes the same, albeit in the absence of a delivery vehicle. Together with a targeting agent, tissue specific delivery can improve biodistribution and reduce the delivered dose, further preventing delivery-associated toxicity. Many of the targeting agents currently being tested include antibodies (scFV, GD_2), peptides (cRGD, R11, penetratin), or ligands (hyaluronic acid, folate, GalNAc). The miRNA delivery methods that have been tested *in vivo* are summarized in Table 14.1.

While many of these individual challenges related to miRNA delivery are being tackled, the future involves combining these novel chemistries into a single agent. The next decade will likely be met with additional

Table 14.1 Summary of miRNA therapeutics tested *in vivo*.[a]

Delivery system	Targeted miRNA	Therapeutic strategy	Model	Administration	Reference
Lipid-based delivery					
DOTMA-cholesterol	miR-133b, miR-29b	Replacement	Lung cancer	Systemic	Wu *et al.*, 2013[122]
DODMA	miR-122	Replacement	Liver cancer	Intratumoral	Hsu *et al.*, 2013[123]
Neutral lipid emulsion	miR-34a	Replacement	Lung cancer	Systemic, intratumoral	Wiggins *et al.*, 2010[34] Trang *et al.*, 2011[37]
	miR-221/222	Inhibition	Multiple myeloma	Intratumoral	Di Martino *et al.*, 2013[124]
	miR-182	Inhibition	Myocardial hypertrophy	Systemic	Li *et al.*, 2016[125]
LPH-GC4	miR-34a	Replacement	Melanoma lung metastasis	Systemic	Chen *et al.*, 2010[126]
LPH-cRGD	miR-296	Inhibition	Angiogenesis	Subcutaneous	Liu *et al.*, 2011[127]
Polymer-based delivery					
PEI	miR-145 and miR-33a	Replacement	Colon cancer	Systemic, intratumoral	Ibrahim *et al.*, 2011[131]
	miR-145	Replacement	Lung cancer	Systemic, subcutaneous	Chiou *et al.*, 2012[132]
	miR-145	Replacement	Prostate	Systemic	Zhang *et al.*, 2015[135]
PLGA	miR-21, miR-10b	Inhibition	Breast cancer	Systemic	Devulapally *et al.*, 2015[137]
	miR-155	Inhibition	Lymphoma	Intratumoral	Babar *et al.*, 2012[58]
Dendrimer	miR-205, miR-221	Replacement and inhibition	Breast cancer	Intratumoral	Conde *et al.*, 2015[141]
Inorganic material-based delivery					
Silica-based	miR-34a	Replacement	Neuroblastoma	Systemic	Tivnan *et al.*, 2012[149]
Magnetic	miR-10b	Inhibition	Breast cancer	Systemic	Yigit *et al.*, 2013[150]
Novel delivery					
EDV nanocell (mini cell)	miR-16	Replacement	Lung cancer	Systemic	Taylor *et al.*, 2015[154]
Exosome	Let-7a	Replacement	Breast cancer	Systemic	Ohno *et al.*, 2013[155]
FolamiR (ligand-mediated)	miR-34a	Replacement	Breast cancer Lung cancer	Systemic	Orellana *et al.*, 2017[156]

[a]cRGD: cyclic (Arg-Gly-Asp-D-Phe-Lys); EDV: EnGenelC Delivery Vehicle; LPH: liposome-polycation-hyaluronic acid; PEI: polyethylenimines PLGA:

interdisciplinary approaches between chemists, biologists, engineers, and clinicians to develop these all-encompassing vehicles, to validate them in pre-clinical models, and successfully transition them into human patients. It is highly expected that the next decade will reveal the true promise of miRNA-based therapeutics as this major obstacle – delivery – is overcome.

Acknowledgements

We graciously thank members of the Kasinski Lab for editing the manuscript and for thoughtful discussion. This work was supported by NCI R01CA226259 (ALK) and NCI R01CA205420 (ALK).

References

1. C. Liu, K. Kelnar, B. Liu, X. Chen, T. Calhoun-Davis, H. Li, L. Patrawala, H. Yan, C. Jeter, S. Honorio, J. F. Wiggins, A. G. Bader, R. Fagin, D. Brown and D. G. Tang, *Nat. Med.*, 2011, **17**, 211–216.
2. L. Li, L. Yuan, J. Luo, J. Gao, J. Guo and X. Xie, *Clin. Exp. Med.*, 2013, **13**, 109–117.
3. C. D. Johnson, A. Esquela-kerscher, G. Stefani, M. Byrom, K. Kelnar, D. Ovcharenko, M. Wilson, X. Wang, J. Shelton, J. Shingara, L. Chin, D. Brown and F. J. Slack, *Cancer Res.*, 2007, **67**, 7713–7722.
4. P. A. Gregory, A. G. Bert, E. L. Paterson, S. C. Barry, A. Tsykin, G. Farshid, M. A. Vadas, Y. Khew-Goodall and G. J. Goodall, *Nat. Cell Biol.*, 2008, **10**, 593–601.
5. Y. Pekarsky and C. M. Croce, *Cell Death Differ.*, 2015, **22**, 6–11.
6. B. D. Adams, A. L. Kasinski and F. J. Slack, *Curr. Biol.*, 2014, **24**, R762–R776.
7. E. A. Orellana and A. L. Kasinski, *Cancers*, 2015, **7**, 1388–1405.
8. R. C. Lee, R. L. Feinbaum and V. Ambros, *Cell*, 1993, **75**, 843–854.
9. A. E. Pasquinelli, A. McCoy, E. Jiménez, E. Saló, G. Ruvkun, M. Q. Martindale and J. Baguñà, *Evol. Dev.*, 2003, **5**, 372–378.
10. B. J. Reinhart, F. J. Slack, M. Basson, A. E. Pasquinelli, J. C. Bettinger, A. E. Rougvie, H. R. Horvitz and G. Ruvkun, *Nature*, 2000, **403**, 901–906.
11. A. Kozomara and S. Griffiths-Jones, *Nucleic Acids Res.*, 2014, **42**, 68–73.
12. Y. Huang, X. J. Shen, Q. Zou, S. P. Wang, S. M. Tang and G. Z. Zhang, *J. Physiol. Biochem.*, 2011, **67**, 129–139.
13. A. S. Pal and A. L. Kasinski, *Adv. Cancer Res.*, 2017, **135**, 53–118.
14. G. A. Calin, C. D. Dumitru, M. Shimizu, R. Bichi, S. Zupo, E. Noch, H. Aldler, S. Rattan, M. Keating, K. Rai, L. Rassenti, T. Kipps, M. Negrini, F. Bullrich and C. M. Croce, *Proc. Natl. Acad. Sci. U. S. A.*, 2002, **99**, 15524–15529.
15. G. A. Calin, A. Cimmino, M. Fabbri, M. Ferracin, S. E. Wojcik, M. Shimizu, C. Taccioli, N. Zanesi, R. Garzon, R. I. Aqeilan, H. Alder, S. Volinia, L. Rassenti, X. Liu, C.-g. Liu, T. J. Kipps, M. Negrini and C. M. Croce, *Proc. Natl. Acad. Sci. U. S. A.*, 2008, **105**, 5166–5171.

16. J. Lu, G. Getz, E. A. Miska, E. Alvarez-Saavedra, J. Lamb, D. Peck, A. Sweet-Cordero, B. L. Ebert, R. H. Mak, A. A. Ferrando, J. R. Downing, T. Jacks, H. R. Horvitz and T. R. Golub, *Nature*, 2005, **435**, 834–838.
17. G. A. Calin, C. Sevignani, C. D. Dumitru, T. Hyslop, E. Noch, S. Yendamuri, M. Shimizu, S. Rattan, F. Bullrich, M. Negrini and C. M. Croce, *Proc. Natl. Acad. Sci. U. S. A.*, 2004, **101**, 2999–3004.
18. A. L. Kasinski and F. J. Slack, *Nat. Rev. Cancer*, 2011, **11**, 849–864.
19. S. Roush and F. J. Slack, *Trends Cell Biol.*, 2008, **18**, 505–516.
20. Y. Peng and C. M. Croce, *Signal Transduction Targeted Ther.*, 2016, **1**, 15004.
21. A. Esquela-Kerscher, P. Trang, J. F. Wiggins, L. Patrawala, A. Cheng, L. Ford, B. Joanne, D. Brown, A. G. Bader and F. J. Slack, *Cell Cycle*, 2008, **7**, 759–764.
22. P. Trang, P. P. Medina, J. F. Wiggins, L. Ruffino, K. Kelnar, M. Omotola, R. Homer, D. Brown, A. G. Bader, J. B. Weidhaas and F. J. Slack, *Oncogene*, 2010, **29**, 1580–1587.
23. V. Tarasov, P. Jung, B. Verdoodt, D. Lodygin, A. Epanchintsev, A. Menssen, G. Meister and H. Hermeking, *Cell Cycle*, 2007, **6**, 1586–1593.
24. G. T. Bommer, I. Gerin, Y. Feng, A. J. Kaczorowski, R. Kuick, R. E. Love, Y. Zhai, T. J. Giordano, Z. S. Qin, B. B. Moore, O. A. MacDougald, K. R. Cho and E. R. Fearon, *Curr. Biol.*, 2007, **17**, 1298–1307.
25. T. C. Chang, E. A. Wentzel, O. A. Kent, K. Ramachandran, M. Mullendore, K. H. Lee, G. Feldmann, M. Yamakuchi, M. Ferlito, C. J. Lowenstein, D. E. E. Arking, M. A. Beer, A. Maitra and J. T. Mendell, *Mol. Cell*, 2007, **26**, 745–752.
26. L. He, X. He, L. P. Lim, E. de Stanchina, Z. Xuan, Y. Liang, W. Xue, L. Zender, J. Magnus, D. Ridzon, A. L. Jackson, P. S. Linsley, C. Chen, S. W. Lowe, M. A. Cleary and G. J. Hannon, *Nature*, 2007, **447**, 1130–1134.
27. N. Raver-Shapira, E. Marciano, E. Meiri, Y. Spector, N. Rosenfeld, N. Moskovits, Z. Bentwich and M. Oren, *Mol. Cell*, 2007, **26**, 731–743.
28. S. Kato, S.-Y. Han, W. Liu, K. Otsuka, H. Shibata, R. Kanamaru and C. Ishioka, *Proc. Natl. Acad. Sci. U. S. A.*, 2003, **100**, 8424–8429.
29. E. A. Medcalf and J. Milner, *Cell*, 1991, **65**, 765–774.
30. J. O. Milner, E. A. Medcalf and A. C. Cook, *Mol. Cell. Biol.*, 1991, **11**, 12–19.
31. A. Sigal and V. Rotter, *Cancer Res.*, 2000, **60**, 6788–6793.
32. K. B. Spurgers, D. L. Gold, K. R. Coombes, N. L. Bohnenstiehl, B. Mullins, R. E. Meyn, C. J. Logothetis and T. J. McDonnell, *J. Biol. Chem.*, 2006, **281**, 25134–25142.
33. F. Yang, Q. Li, Z. Gong, L. Zhou, N. You, S. Wang, X. Li, J. Li, J. An, D. Wang, Y. He and K. Dou, *Technol. Cancer Res. Treat.*, 2014, **13**, 77–86.
34. J. F. Wiggins, L. Ruffino, K. Kelnar, M. Omotola, L. Patrawala, D. Brown and A. G. Bader, *Cancer Res.*, 2010, **70**, 5923–5930.
35. A. L. Kasinski and F. J. Slack, *Cancer Res.*, 2012, **72**, 5576–5587.
36. A. L. Kasinski, K. Kelnar, C. Stahlhut, E. Orellana, J. Zhao, E. Shimer, S. Dysart, X. Chen, A. G. Bader and F. J. Slack, *Oncogene*, 2015, **34**, 1–9.

37. P. Trang, J. F. Wiggins, C. L. Daige, C. Cho, M. Omotola, D. Brown, J. B. Weidhaas, A. G. Bader and F. J. Slack, *Mol. Ther.*, 2011, **19**, 1116–1122.
38. A. Bouchie, *Nat. Biotechnol.*, 2013, **31**, 577.
39. S. Volinia, G. A. Calin, C.-G. Liu, S. Ambs, A. Cimmino, F. Petrocca, R. Visone, M. Iorio, C. Roldo, M. Ferracin, R. L. Prueitt, N. Yanaihara, G. Lanza, A. Scarpa, A. Vecchione, M. Negrini, C. C. Harris and C. M. Croce, *Proc. Natl. Acad. Sci. U. S. A.*, 2006, **103**, 2257–2261.
40. J. g. Zhang, J. j. Wang, F. Zhao, Q. Liu, K. Jiang and G. hai Yang, *Clin. Chim. Acta*, 2010, **411**, 846–852.
41. I. A. Asangani, S. A. K. Rasheed, D. A. Nikolova, J. H. Leupold, N. H. Colburn, S. Post and H. Allgayer, *Oncogene*, 2008, **27**, 2128–2136.
42. L. B. Frankel, N. R. Christoffersen, A. Jacobsen, M. Lindow, A. Krogh and A. H. Lund, *J. Biol. Chem.*, 2008, **283**, 1026–1033.
43. F. Sicard, M. Gayral, H. Lulka, L. Buscail and P. Cordelier, *Mol. Ther.*, 2013, **21**, 986–994.
44. X. Ma, S. N. Choudhury, X. Hua, Z. Dai and Y. Li, *Carcinogenesis*, 2013, **34**, 1216–1223.
45. P. P. Medina, M. Nolde and F. J. Slack, *Nature*, 2010, **467**, 86–90.
46. J. A. Chan, A. M. Krichevsky, K. S. Kosik, J. A. Chan, A. M. Krichevsky and K. S. Kosik, *Cancer Res.*, 2005, **65**, 6029–6033.
47. M. F. Corsten, R. Miranda, R. Kasmieh, A. M. Krichevsky, R. Weissleder and K. Shah, *Cancer Res.*, 2007, **67**, 8994–9000.
48. R. Nedaeinia, M. Sharifi, A. Avan, M. Kazemi, L. Rafiee, M. Ghayour-Mobarhan and R. Salehi, *Cancer Gene Ther.*, 2016, **23**, 246–253.
49. G. H. Shi, D. W. Ye, X. D. Yao, S. L. Zhang, B. Dai, H. L. Zhang, Y. J. Shen, Y. Zhu, Y. P. Zhu, W. J. Xiao and C. G. Ma, *Acta Pharmacol. Sin.*, 2010, **31**, 867–873.
50. Y. Li, W. Li, Y. Yang, Y. Lu, C. He, G. Hu, H. Liu, J. Chen, J. He and H. Yu, *Brain Res.*, 2009, **1286**, 13–18.
51. K. Wang and P. F. Li, *J. Biol. Chem.*, 2010, **285**, 16958–16966.
52. J. K. Park, E. J. Lee, C. Esau and T. D. Schmittgen, *Pancreas*, 2009, **38**, 190–199.
53. P. S. Eis, W. Tam, L. Sun, A. Chadburn, Z. Li, M. F. Gomez, E. Lund and J. E. Dahlberg, *Proc. Natl. Acad. Sci. U. S. A.*, 2005, **102**, 3627–3632.
54. E. Vigorito, K. L. Perks, C. Abreu-Goodger, S. Bunting, Z. Xiang, S. Kohlhaas, P. P. Das, E. A. Miska, A. Rodriguez, A. Bradley, K. G. C. Smith, C. Rada, A. J. Enright, K. M. Toellner, I. C. M. MacLennan and M. Turner, *Immunity*, 2007, **27**, 847–859.
55. G. Teng, P. Hakimpour, P. Landgraf, A. Rice, T. Tuschl, R. Casellas and F. N. Papavasiliou, *Immunity*, 2008, **28**, 621–629.
56. R. M. O'Connell, D. S. Rao, A. A. Chaudhuri, M. P. Boldin, K. D. Taganov, J. Nicoll, R. L. Paquette and D. Baltimore, *J. Exp. Med.*, 2008, **205**, 585–594.
57. R. M. O'Connell, A. A. Chaudhuri, D. S. Rao and D. Baltimore, *Proc. Natl. Acad. Sci. U. S. A.*, 2009, **106**, 7113–7118.
58. I. A. Babar, C. J. Cheng, C. J. Booth, X. Liang, J. B. Weidhaas, W. M. Saltzman and F. J. Slack, *Proc. Natl. Acad. Sci. U. S. A.*, 2012, **109**, E1695–E1704.

59. M. Kim, A. L. Kasinski and F. J. Slack, *Lancet Oncol.*, 2011, **12**, 319–321.
60. J. A. DiMasi and H. G. Grabowski, *J. Clin. Oncol.*, 2007, **25**, 209–216.
61. A. Arora and E. M. Scholar, *J. Pharmacol. Exp. Ther.*, 2005, **315**, 971–979.
62. T. J. Lynch, D. W. Bell, R. Sordella, S. Gurubhagavatula, R. A. Okimoto, B. W. Brannigan, P. L. Harris, S. M. Haserlat, J. G. Supko, F. G. Haluska, D. N. Louis, D. C. Christiani, J. Settleman and D. A. Haber, *N. Engl. J. Med.*, 2011, **350**, 2187–2198.
63. T. E. Taylor, F. B. Furnari and W. K. Cavenee, *Curr. Cancer Drug Targets*, 2012, **12**, 197–209.
64. J. D. Moyer, E. G. Barbacci, K. K. Iwata, L. Arnold, B. Boman, A. Cunningham, C. DiOrio, J. Doty, M. J. Morin, M. P. Moyer, M. Neveu, V. A. Pollack, L. R. Pustilnik, M. M. Reynolds, D. Sloan, A. Theleman and P. Miller, *Cancer Res.*, 1997, **57**, 4838–4848.
65. M. Ranson, *Br. J. Cancer*, 2004, **90**, 2250–2255.
66. E. Raymond, S. Faivre and J. P. Armand, *Drugs*, 2000, **60**, 15–23.
67. T. O'Hare, C. A. Eide and M. W. Deininger, *Blood*, 2007, **110**, 2242–2249.
68. R. J. Motzer, M. D. Michaelson, B. G. Redman, G. R. Hudes, G. Wilding, R. A. Figlin, M. S. Ginsberg, S. T. Kim, C. M. Baum, S. E. DePrimo, J. Z. Li, C. L. Bello, C. P. Theuer, D. J. George and B. I. Rini, *J. Clin. Oncol.*, 2006, **24**, 16–24.
69. R. J. Motzer, T. E. Hutson, P. Tomczak, M. D. Michaelson, R. M. Bukowski, O. Rixe, S. Oudard, S. Negrier, C. Szczylik, S. T. Kim, I. Chen, P. W. Bycott, C. M. Baum and R. A. Figlin, *N. Engl. J. Med.*, 2007, **356**, 115–124.
70. K. Suda, R. Onozato, Y. Yatabe and T. Mitsudomi, *J. Thorac. Oncol.*, 2009, **4**, 1–4.
71. L. A. Byers, L. Diao, J. Wang, P. Saintigny, L. Girard, M. Peyton, L. Shen, Y. Fan, U. Giri, P. K. Tumula, M. B. Nilsson, J. Gudikote, H. Tran, R. J. G. Cardnell, D. J. Bearss, S. L. Warner, J. M. Foulks, S. B. Kanner, V. Gandhi, N. Krett, S. T. Rosen, E. S. Kim, R. S. Herbst, G. R. Blumenschein, J. J. Lee, S. M. Lippman, K. K. Ang, G. B. Mills, W. K. Hong, J. N. Weinstein, I. I. Wistuba, K. R. Coombes, J. D. Minna and J. V. Heymach, *Clin. Cancer Res.*, 2013, **19**, 279–290.
72. C. Holohan, S. Van Schaeybroeck, D. B. Longley and P. G. Johnston, *Nat. Rev. Cancer*, 2013, **13**, 714–726.
73. C. Herrmann, G. Horn, M. Spaargaren and A. Wittinghofer, *J. Biol. Chem.*, 1996, **271**, 6794–6800.
74. K. H. Lim, A. T. Baines, J. J. Fiordalisi, M. Shipitsin, L. A. Feig, A. D. Cox, C. J. Der and C. M. Counter, *Cancer Cell*, 2005, **7**, 533–545.
75. A. D. Cox, S. W. Fesik, A. C. Kimmelman, J. Luo and C. J. Der, *Nat. Rev. Drug Discovery*, 2014, **13**, 828–851.
76. A. Takashima and D. V. Faller, *Expert Opin. Ther. Targets*, 2013, **17**, 507–531.
77. G. Zimmermann, B. Papke, S. Ismail, N. Vartak, A. Chandra, M. Hoffmann, S. A. Hahn, G. Triola, A. Wittinghofer, P. I. H. Bastiaens and H. Waldmann, *Nature*, 2013, **497**, 638–642.

78. D. P. Bartel, R. Lee and R. Feinbaum, *Cell*, 2004, **116**, 281–297.
79. I. Ivanovska and M. A. Cleary, *Cell Cycle*, 2008, **7**, 3137–3142.
80. E. Andrés-León, I. Cases, S. Alonso and A. M. Rojas, *Sci. Rep.*, 2017, **7**, 1–13.
81. N. Momi, S. Kaur, S. Rachagani, A. K. Ganti and S. K. Batra, *Trends Mol. Med.*, 2014, **20**, 36–47.
82. A. Boutla, C. Delidakis and M. Tabler, *Nucleic Acids Res.*, 2003, **31**, 4973–4980.
83. G. Hutvágner, M. J. Simard, C. C. Mello and P. D. Zamore, *PLoS Biol.*, 2004, **2**, 465–475.
84. J. Weiler, J. Hunziker and J. Hall, *Gene Ther.*, 2006, **13**, 496–502.
85. J. Krützfeldt, N. Rajewsky, R. Braich, K. G. Rajeev, T. Tuschl, M. Manoharan and M. Stoffel, *Nature*, 2005, **438**, 685–689.
86. S. T. Crooke, M. J. Graham, J. E. Zuckerman, B. S. Conklin and H. Griffey, *J. Pharmacol. Exp. Ther.*, 1996, **277**, 923–937.
87. P. S. Pallan, E. M. Greene, P. A. Jicman, R. K. Pandey, M. Manoharan, E. Rozners and M. Egli, *Nucleic Acids Res.*, 2011, **39**, 3482–3495.
88. C. Wahlestedt, P. Salmi, L. Good, J. Kela, T. Johnsson, T. Hokfelt, C. Broberger, F. Porreca, J. Lai, K. Ren, M. Ossipov, A. Koshkin, N. Jakobsen, J. Skouv, H. Oerum, M. H. Jacobsen and J. Wengel, *Proc. Natl. Acad. Sci. U. S. A.*, 2000, **97**, 5633–5638.
89. J. M. Campbell, T. A. Bacon and E. Wickstrom, *J. Biochem. Biophys. Methods*, 1990, **20**, 259–267.
90. B. Hyrup and P. E. Nielsen, *Bioorg. Med. Chem.*, 1996, **4**, 5–23.
91. J. Soutschek, A. Akinc, B. Bramlage, K. Charisse, R. Constein, M. Donoghuw, S. Elbashir, A. Geick, P. Hadwiger, J. Harborth, M. John, V. Kesavan, G. Lavine, R. K. Pandey, T. Racie, K. G. Rajeev, I. Rohl, I. Toudjarska, G. Wang, S. Wuschko, D. Bumcrot, V. Koteliansky, S. Limmer, M. Manoharan and H. Vornlocher, *Nature*, 2004, **432**, 173–178.
92. C. L. Jopling, M. Yi, A. M. Lancaster, S. M. Lemon and P. Sarnow, *Science*, 2005, **1**, 1577–1581.
93. J. Elmén, M. Lindow, A. Silahtaroglu, M. Bak, M. Christensen, A. Lind-Thomsen, M. Hedtjärn, J. B. Hansen, H. F. Hansen, E. M. Straarup, K. McCullagh, P. Kearney and S. Kauppinen, *Nucleic Acids Res.*, 2008, **36**, 1153–1162.
94. R. E. Lanford, E. S. Hildebrandt-Eriksen, A. Petri, R. Persson, M. Lindow, M. E. Munk, S. Kauppinen and H. Ørum, *Science*, 2010, **327**, 198–201.
95. E. S. Hildebrandt-Eriksen, V. Aarup, R. Persson, H. F. Hansen, M. E. Munk and H. Ørum, *Nucleic Acid Ther.*, 2012, **22**, 152–161.
96. H. L. A. Janssen, H. W. Reesink, E. J. Lawitz, S. Zeuzem, M. Rodriguez-Torres, K. Patel, A. J. van der Meer, A. K. Patick, A. Chen, Y. Zhou, R. Persson, B. D. King, S. Kauppinen, A. A. Levin and M. R. Hodges, *N. Engl. J. Med.*, 2013, **368**, 1685–1694.
97. R. Titze-de-Almeida, C. David and S. S. Titze-de-Almeida, *Pharm. Res.*, 2017, **34**, 1339–1363.
98. M. S. Ebert, J. R. Neilson and P. A. Sharp, *Nat. Methods*, 2007, **4**, 721–726.

99. Y. Tay, L. Kats, L. Salmena, D. Weiss, S. M. Tan, U. Ala, F. Karreth, L. Poliseno, P. Provero, F. Di Cunto, J. Lieberman, I. Rigoutsos and P. P. Pandolfi, *Cell*, 2011, **147**, 344–357.
100. M. Cesana, D. Cacchiarelli, I. Legnini, T. Santini, O. Sthandier, M. Chinappi, A. Tramontano and I. Bozzoni, *Cell*, 2011, **147**, 358–369.
101. T. B. Hansen, T. I. Jensen, B. H. Clausen, J. B. Bramsen, B. Finsen, C. K. Damgaard and J. Kjems, *Nature*, 2013, **495**, 384–388.
102. P. d. C. Monroig, L. Chen, S. Zhang and G. A. Calin, *Adv. Drug Delivery Rev.*, 2015, **81**, 104–116.
103. K. Gumireddy, D. D. Young, X. Xiong, J. B. Hogenesch, Q. Huang and A. Deiters, *Angew. Chem., Int. Ed.*, 2008, **47**, 7482–7484.
104. D. D. Young, C. M. Connelly, C. Grohmann and A. Deiters, *J. Appl. Chem. Sci.*, 2010, **132**, 7976–7981.
105. K. Terasawa, K. Shimizu and G. Tsujimoto, *J. Nucleic Acids*, 2011, 1–7.
106. A. G. Bader, D. Brown, J. Stoudemire and P. Lammers, *Gene Ther.*, 2011, **18**, 1121–1126.
107. A. Eichert, K. Behling, C. Betzel, V. A. Erdmann, J. P. Fürste and C. Förster, *Nucleic Acids Res.*, 2010, **38**, 6729–6736.
108. M. R. Hassler, A. A. Turanov, J. F. Alterman, R. A. Haraszti, A. H. Coles, M. F. Osborn, D. Echeverria, M. Nikan, W. E. Salomon, L. Roux, B. M. D. C. Godinho, S. M. Davis, D. V. Morrissey, P. D. Zamore, S. A. Karumanchi, M. J. Moore, N. Aronin and A. Khvorova, *Nucleic Acids Res.*, 2018, **46**, 2185–2196.
109. D. J. Foster, C. R. Brown, S. Shaikh, C. Trapp, M. K. Schlegel, K. Qian, A. Sehgal, K. G. Rajeev, V. Jadhav, M. Manoharan, S. Kuchimanchi, M. A. Maier and S. Milstein, *Mol. Ther.*, 2018, **26**, 708–717.
110. M. M. Janas, M. K. Schlegel, C. E. Harbison, V. O. Yilmaz, Y. Jiang, R. Parmar, I. Zlatev, A. Castoreno, H. Xu, S. Shulga-Morskaya, K. G. Rajeev, M. Manoharan, N. D. Keirstead, M. A. Maier and V. Jadhav, *Nat. Commun.*, 2018, **9**, 1–10.
111. M. S. Beg, A. J. Brenner, J. Sachdev, M. Borad, Y. K. Kang, J. Stoudemire, S. Smith, A. G. Bader, S. Kim and D. S. Hong, *Invest. New Drugs*, 2017, **35**, 180–188.
112. R. L. Montgomery, G. Yu, P. A. Latimer, C. Stack, K. Robinson, C. M. Dalby, N. Kaminski and E. van Rooij, *EMBO Mol. Med.*, 2014, **6**, 1347–1356.
113. J. Cinkornpumin, M. Roos, L. Nguyen, X. Liu, X. Gaeta, S. Lin, D. N. Chan, A. Liu, R. I. Gregory, M. Jung, J. Chute, H. Zhu and W. E. Lowry, *Sci. Rep.*, 2017, **7**, 1–10.
114. S. Lin and R. I. Gregory, *RNA Biol.*, 2015, **12**, 792–800.
115. H. Wang, Y. Jiang, H. Peng, Y. Chen, P. Zhu and Y. Huang, *Adv. Drug Delivery Rev.*, 2015, **81**, 142–160.
116. J. Ji, J. Shi, A. Budhu, Z. Yu, M. Forgues, S. Roessler, S. Ambs, Y. Chen, P. S. Meltzer, C. M. Croce, L. X. Qin, K. Man, C. M. Lo, J. Lee, I. O. Ng, J. Fan, Z. Y. Tang, H. C. Sun and X. W. Wang, *N. Engl. J. Med.*, 2009, **361**, 1437–1447.
117. H. Nakai, S. Fuess, T. A. Storm, Y. Nara, M. A. Kay and S. Muramatsu, *Society*, 2005, **79**, 214–224.

118. J. Kota, R. R. Chivukula, K. A. O'Donnell, E. A. Wentzel, C. L. Montgomery, H. W. Hwang, T. C. Chang, P. Vivekanandan, M. Torbenson, K. R. Clark, J. R. Mendell and J. T. Mendell, *Cell*, 2009, **137**, 1005–1017.
119. C. Lorenz, P. Hadwiger, M. John, H. P. Vornlocher and C. Unverzagt, *Bioorg. Med. Chem. Lett.*, 2004, **14**, 4975–4977.
120. X. Wang, B. Yu, W. Ren, X. Mo, C. Zhou, H. He, H. Jia, L. Wang, S. T. Jacob, R. J. Lee, K. Ghoshal and L. J. Lee, *J. Controlled Release*, 2013, **172**, 690–698.
121. Y. Wu, M. Crawford, B. Yu, Y. Mao, S. P. Nana-Sinkam and L. J. Lee, *Mol. Pharm.*, 2011, **8**, 1381–1389.
122. Y. Wu, M. Crawford, Y. Mao, R. J. Lee, I. C. Davis, T. S. Elton, L. J. Lee and S. P. Nana-Sinkam, *Mol. Ther.--Nucleic Acids*, 2013, **2**, 1–10.
123. S. h. Hsu, B. Yu, X. Wang, Y. Lu, C. R. Schmidt, R. J. Lee, L. J. Lee, S. T. Jacob and K. Ghoshal, *Nanomedicine*, 2013, **9**, 1169–1180.
124. M. T. Di Martino, A. Gullà, M. E. G. Cantafio, M. Lionetti, E. Leone, N. Amodio, P. H. Guzzi, U. Foresta, F. Conforti, M. Cannataro, A. Neri, A. Giordano, P. Tagliaferri and P. Tassone, *Oncotarget*, 2013, **4**, 242–255.
125. N. Li, C. Hwangbo, I. M. Jaba, J. Zhang, I. Papangeli, J. Han, N. Mikush, B. Larrivée, A. Eichmann, H. J. Chun, L. H. Young and D. Tirziu, *Sci. Rep.*, 2016, **6**, 1–15.
126. Y. Chen, X. Zhu, X. Zhang, B. Liu and L. Huang, *Mol. Ther.*, 2010, **18**, 1650–1656.
127. X.-Q. Liu, W.-J. Song, T.-M. Sun and P.-Z. Zhang, *Mol. Pharm.*, 2011, **8**, 250–259.
128. R. Haubner, R. Gratias, B. Diefenbach, S. L. Goodman, A. Jonczyk and H. Kessler, *J. Am. Chem. Soc.*, 1996, **118**, 7461–7472.
129. D. S. Hong, Y.-K. Kang, A. J. Brenner, J. C. Sachdev, S. Ejadi, M. J. Borad, T.-Y. Kim, H. Y. Lim, K. Park, C. Becerra, A. G. Bader, J. Stoudemire, S. Smith, S. Kim and M. S. Beg, *J. Clin. Oncol.*, 2016, **34**, 2508.
130. O. Boussif, F. Lezoualc'h, M. A. Zanta, M. D. Mergny, D. Scherman, B. Demeneix and J. P. Behr, *Proc. Natl. Acad. Sci. U. S. A.*, 1995, **92**, 7297–7301.
131. A. F. Ibrahim, U. Weirauch, M. Thomas, A. Grünweller, R. K. Hartmann and A. Aigner, *Cancer Res.*, 2011, **71**, 5214–5224.
132. G. Y. Chiou, J. Y. Cherng, H. S. Hsu, M. L. Wang, C. M. Tsai, K. H. Lu, Y. Chien, S. C. Hung, Y. W. Chen, C. I. Wong, L. M. Tseng, P. I. Huang, C. C. Yu, W. H. Hsu and S. H. Chiou, *J. Controlled Release*, 2012, **159**, 240–250.
133. S. Son, D. W. Hwang, K. Singha, J. H. Jeong, T. G. Park, D. S. Lee and W. J. Kim, *J. Controlled Release*, 2011, **155**, 18–25.
134. J. T. Hsieh, J. Zhou, C. Gore and P. Zimmern, *BJU Int.*, 2011, **108**, 1666–1671.
135. T. Zhang, X. Xue, D. He and J. T. Hsieh, *Cancer Lett.*, 2015, **365**, 156–165.
136. L. E. Van Vlerken, T. K. Vyas and M. M. Amiji, *Pharm. Res.*, 2007, **24**, 1405–1414.
137. R. Devulapally, N. M. Sekar, T. V. Sekar, K. Foygel, T. F. Massoud, J. K. Willmann and R. Paulmurugan, *ACS Nano*, 2015, **9**, 2290–2302.
138. S. Wang, J. Zhang, Y. Wang and M. Chen, *Nanomedicine*, 2016, **12**, 411–420.
139. Y. Ren, X. Zhou, M. Mei, X. B. Yuan, L. Han, G. X. Wang, Z. F. Jia, P. Xu, P. Y. Pu and C. S. Kang, *BMC Cancer*, 2010, **10**, 1–13.

140. Y. Ren, C. S. Kang, X. B. Yuan, X. Zhou, P. Xu, L. Han, G. X. Wang, Z. Jia, Y. Zhong, S. Yu, J. Sheng and P. Y. Pu, *J. Biomater. Sci., Polym. Ed.*, 2010, **21**, 303–314.
141. J. Conde, N. Oliva, M. Atilano, H. S. Song and N. Artzi, *Nat. Mater.*, 2016, **15**, 353–363.
142. H. Ragelle, G. Vandermeulen and V. Préat, *J. Controlled Release*, 2013, **172**, 207–218.
143. N. Duceppe and M. Tabrizian, *Biomaterials*, 2009, **30**, 2625–2631.
144. X. Deng, M. Cao, J. Zhang, K. Hu, Z. Yin, Z. Zhou, X. Xiao, Y. Yang, W. Sheng, Y. Wu and Y. Zeng, *Biomaterials*, 2014, **35**, 4333–4344.
145. M. R. Papasani, G. Wang and R. A. Hill, *Nanomedicine*, 2012, **8**, 804–814.
146. R. Shukla, V. Bansal, M. Chaudhary, A. Basu, R. R. Bhonde and M. Sastry, *Langmuir*, 2005, **21**, 10644–10654.
147. E. E. Connor, J. Mwamuka, A. Gole, C. J. Murphy and M. D. Wyatt, *Small*, 2005, **1**, 325–327.
148. R. Ghosh, L. C. Singh, J. M. Shohet and P. H. Gunaratne, *Biomaterials*, 2013, **34**, 807–816.
149. A. Tivnan, W. S. Orr, V. Gubala, R. Nooney, D. E. Williams, C. McDonagh, S. Prenter, H. Harvey, R. Domingo-Fernández, I. M. Bray, O. Piskareva, C. Y. Ng, H. N. Lode, A. M. Davidoff and R. L. Stallings, *PLoS One*, 2012, **7**, 1–12.
150. M. V. Yigit, S. K. Ghosh, M. Kumar, V. Petkova, A. Kavishwar, A. Moore and Z. Medarova, *Oncogene*, 2013, **32**, 1530–1538.
151. P. T. Yin, B. P. Shah and K. B. Lee, *Small*, 2014, **10**, 4106–4112.
152. J. A. MacDiarmid, N. B. Mugridge, J. C. Weiss, L. Phillips, A. L. Burn, R. P. P. Paulin, J. E. Haasdyk, K. A. Dickson, V. N. Brahmbhatt, S. T. Pattison, A. C. James, G. Al Bakri, R. C. Straw, B. Stillman, R. M. Graham and H. Brahmbhatt, *Cancer Cell*, 2007, **11**, 431–445.
153. J. A. MacDiarmid and H. Brahmbhatt, *Curr. Opin. Biotechnol.*, 2011, **22**, 909–916.
154. K. Taylor, C. B. Howard, M. L. Jones, I. Sedliarou, J. MacDiarmid, H. Brahmbhatt, T. P. Munro and S. M. Mahler, *mAbs*, 2015, **7**, 53–65.
155. S. I. Ohno, M. Takanashi, K. Sudo, S. Ueda, A. Ishikawa, N. Matsuyama, K. Fujita, T. Mizutani, T. Ohgi, T. Ochiya, N. Gotoh and M. Kuroda, *Mol. Ther.*, 2013, **21**, 185–191.
156. E. A. Orellana, S. Tenneti, L. Rangasamy, L. T. Lyle, P. S. Low and A. L. Kasinski, *Sci. Transl. Med.*, 2017, **9**, 1–11.
157. P. T. Cagle, Q. J. Zhai, L. Murphy and P. S. Low, *Arch. Pathol. Lab. Med.*, 2013, **137**, 241–244.
158. Z. Zhang, J. Wang, D. E. Tacha, P. Li, R. E. Bremer, H. Chen, B. Wei, X. Xiao, J. Da, K. Skinner, D. G. Hicks, H. Bu and P. Tang, *Arch. Pathol. Lab. Med.*, 2014, **138**, 890–895.
159. K. R. Kalli, A. L. Oberg, G. L. Keeney, T. J. H. Christianson, P. S. Low, K. L. Knutson and L. C. Hartmann, *Gynecol. Oncol.*, 2008, **108**, 619–626.

160. L. Rangasamy, V. Chelvam, A. K. Kanduluru, M. Srinivasarao, N. A. Bandara, F. You, E. A. Orellana, A. L. Kasinski and P. S. Low, *Bioconjugate Chem.*, 2018, **29**, 1047–1059.
161. E. A. Orellana, L. Rangasamy, S. Tenneti, A. M. Abdelaal, P. S. Low and A. L. Kasinski, *bioRxiv*, 2018, **367672**, 1–29.
162. R. Breslow and R. Xu, *Proc. Natl. Acad. Sci. U. S. A.*, 1993, **90**, 1201–1207.
163. B. R. Meade, K. Gogoi, A. S. Hamil, C. Palm-Apergi, A. Van Den Berg, J. C. Hagopian, A. D. Springer, A. Eguchi, A. D. Kacsinta, C. F. Dowdy, A. Presente, P. Lönn, M. Kaulich, N. Yoshioka, E. Gros, X. S. Cui and S. F. Dowdy, *Nat. Biotechnol.*, 2014, **32**, 1256–1261.
164. W. Colon and J. W. Kelly, *Biochemistry*, 1992, **31**, 8654–8660.
165. J. K. Nair, J. L. S. Willoughby, A. Chan, K. Charisse, M. R. Alam, Q. Wang, M. Hoekstra, P. Kandasamy, A. V. Kelin, S. Milstein, N. Taneja, J. Oshea, S. Shaikh, L. Zhang, R. J. Van Der Sluis, M. E. Jung, A. Akinc, R. Hutabarat, S. Kuchimanchi, K. Fitzgerald, T. Zimmermann, T. J. C. Van Berkel, M. A. Maier, K. G. Rajeev and M. Manoharan, *J. Am. Chem. Soc.*, 2014, **136**, 16958–16961.
166. J. B. Bramsen, M. B. Laursen, A. F. Nielsen, T. B. Hansen, C. Bus, N. Langkjoør, B. R. Babu, T. Højland, M. Abramov, A. Van Aerschot, D. Odadzic, R. Smicius, J. Haas, C. Andree, J. Barman, M. Wenska, P. Srivastava, C. Zhou, D. Honcharenko, S. Hess, E. Müller, G. V. Bobkov, S. N. Mikhailov, E. Fava, T. F. Meyer, J. Chattopadhyaya, M. Zerial, J. W. Engels, P. Herdewijn, J. Wengel and J. Kjems, *Nucleic Acids Res.*, 2009, **37**, 2867–2881.
167. A. L. Jackson, J. Burchard, D. Leake, A. Reynolds, J. Schelter, J. Guo, J. M. Johnson, L. Lim, J. Karpilow, K. Nichols, W. Marshall, A. Khvorova and P. S. Linsley, *RNA*, 2006, **12**, 1197–1205.
168. M. Acunzo, G. Romano, G. Nigita, D. Veneziano, L. Fattore, A. Laganà, N. Zanesi, P. Fadda, M. Fassan, L. Rizzotto, R. Kladney, V. Coppola and C. M. Croce, *Proc. Natl. Acad. Sci. U. S. A.*, 2017, **114**, E4203–E4212.
169. N. Weill, V. Lisi, N. Scott, P. Dallaire, J. Pelloux and F. Major, *Nucleic Acids Res.*, 2015, **43**, 6730–6738.

CHAPTER 15

Modification of Drug Response in Cancer by MicroRNAs

LAURA L. MEIJER[a], JISCE R. PUIK[a], CATERINA VIVALDI[b], MJRIAM CAPULA[c], ENRICO VASILE[b], GEERT KAZEMIER[a] AND ELISA GIOVANNETTI*[c,d]

[a]Department of Surgery, Cancer Center Amsterdam, VU University Medical Center, Amsterdam, The Netherlands; [b]Department of Medical Oncology, University Hospital of Pisa, Pisa, Italy; [c]Department of Medical Oncology, Cancer Center Amsterdam, VU University Medical Center, Amsterdam, The Netherlands; [d]Cancer Pharmacology Lab, AIRC-Start-Up Unit, University Hospital of Pisa, Pisa, Italy
*E-mail: e.giovannetti@vumc.nl

15.1 Introduction

Cancer is the leading cause of death worldwide, with an estimated 1 688 780 new cancer cases diagnosed in the United States in 2017.[1] Despite advances in cancer therapeutics, current therapies fail to cure the majority of patients, since many patients exhibit or develop resistance to therapy. Therefore, drug resistance remains a major obstacle in the treatment of patients with cancer and is the main cause for tumor progression and death.

Drug resistance is a well-known phenomenon and constitutes the ability of diseases to become tolerant to pharmaceuticals.[2] Drug resistance can broadly be divided into intrinsic and acquired resistance. Intrinsic resistance

Drug Discovery Series No. 69
MicroRNAs in Diseases and Disorders: Emerging Therapeutic Targets
Edited by Philip V. Peplow, Bridget Martinez, George A. Calin and Aurora Esquela-Kerscher

Published by the Royal Society of Chemistry, www.rsc.org

consists of the resistance of the tumor caused by a pre-existing phenotype, while acquired resistance develops over time during treatment in tumors that were initially sensitive to therapy.[3] Most research has focused on drug resistance mechanisms in cytotoxic chemotherapies. A broad range of molecular mechanisms has been implicated in drug resistance, including increased drug efflux, alterations in drug metabolism and mutations of drug targets.[4] Resistance mechanisms can be genomically encoded (*e.g.*, mutations or amplifications), non-genetic (*e.g.*, gene expression changes) and influenced by the tumor microenvironment.[5] More recently, drug development has shifted towards agents that target specific molecular alterations, so-called "targeted therapies". In general, resistance mechanisms in targeted therapies lead to reactivation of the target or its critical downstream signaling pathways.[6] Even though cytotoxic chemotherapies and targeted therapies have distinct mechanisms of action, they share molecular features of drug resistance.[4]

Understanding the cellular and biochemical mechanisms behind drug resistance might help overcome resistance and develop effective therapies. The major challenge is the multifactorial etiology of drug resistance, depending both on therapy- and disease-specific mechanisms, and the heterogeneity of tumor resistance, making these complex mechanisms poorly understood.[5] Overcoming these challenges is difficult, but pattern analysis suggests the possibility to apply overlapping strategies against drug resistance.

As our understanding of cancer biology has advanced, interest has increased to overcome drug resistance by post-transcriptional regulation by microRNAs (miRNAs). miRNAs have been studied extensively to investigate their association with cancer. They are a class of short, non-coding RNAs, regulating gene expression in the post-transcriptional phase.[7] The biogenesis of miRNAs involves a complex process starting as large precursor RNAs, so called pri-miRNAs.[8] These pri-miRNAs are processed in the nucleus by the RNase III enzyme, Drosha, and the double-stranded RNA-binding protein, Pasha, into pre-miRNAs. Subsequently, they are exported into the cytoplasm by Exportin 5 and undergo additional processing to form the mature miRNA strand incorporated in miRISC (miRNA RNA-induced silencing complex).[9] miRNAs exert their function by binding to complementary sequences in the 3′ untranslated regions of their target messenger RNA (mRNA). Thereby, they either induce mRNA degradation or disrupt protein translation of their target mRNA.[10,11] The major advantage of miRNAs is their possibility to target multiple mRNAs; a single miRNA can bind hundreds of different target genes. Therefore, miRNAs potentially control approximately half of the protein-coded genes within the human genome.[11] Due to the plethora of effects on gene regulation, miRNAs can regulate a large range of biological processes, which include development, differentiation, proliferation and apoptosis.[12–14] This is of particular interest when discussing the role of miRNAs in the regulation of cancer-related pathways.[15]

15.1.1 miRNAs and Cancer

The association between miRNA expression and cancer was first described in chronic lymphocytic leukemia (CLL).[16] Since this discovery, the link between different miRNA expression profiles when comparing tumor and normal tissues has accumulated and miRNAs have been closely associated with tumor initiation and disease progression for a variety of tumor types, including lung, breast, ovarian, colon, hepatocellular, bile duct, and pancreatic cancers.[17–25]

miRNAs can either function as tumor suppressors or oncogenes.[26] The reduction of a miRNA that functions as a tumor suppressor or the increase of an oncogenic miRNA could both lead to tumor formation, by deregulated expression of the miRNA-target protein.[27] Moreover, several studies have shown the oncogenic involvement of miRNAs by overexpression of a miRNA gene in mouse models.[28–30] These studies confirm that deregulation of specific oncogenic miRNA could accelerate proliferation, while overexpression of tumor-suppressive miRNAs could prevent progression. Interestingly, a large number of miRNA genes are located in fragile regions of cancer-related sites, such as regions of deletions, amplifications and translocation breakpoints.[27] Indeed, the genomic abnormalities involved in miRNA expression are similar to the mechanisms described for protein-coded genes, such as rearrangements, amplifications, deletions, polymorphisms or mutations.[31]

Profiling of cell lines and tissue samples shows that miRNA expression is often cell-type and tissue-type specific.[32] miRNAs are aberrantly expressed in tumors and expression profiles of human tumor tissue show distinct patterns compared to normal tissue of the same organ.[26,33,34] However, overlap in miRNA expression profiles between distinct tumor types is common and this indicates a common set of miRNAs which can facilitate carcinogenesis. As such, several miRNAs have been suggested as tissue-independent oncomiRs, such as miR-21, which is most often overexpressed in cancer cells of several tissues, including cells in the gastrointestinal tract and known to be involved in cell migration and invasion.[35,36] In addition, various miRNAs have been shown to interact with mRNAs acting in essential cancer pathways, such as Bcl-2, Myc and RAS.[27,37] These findings underline the significance of miRNAs in the modulation of oncogenic signaling pathways.

The role of miRNAs in controlling essential pathways has led to the exploration of the use of miRNAs in blood as novel class of biomarkers for the diagnosis and prognosis of cancer in various tumor types.[38–41] miRNAs are released as a result of apoptosis and necrotic cell death, but are also actively excreted from cells into the circulation, either in microvesicles or in conjunction with RNA-binding proteins (*e.g.*, Argonaute–RISC complex).[10] In this way, they remain stably expressed due to protection against RNase digestion. This supports their possible use as minimally or non-invasive biomarkers in several types of biofluids, including blood, saliva, urine and bile.[39,42–45] In addition, miRNAs are easy to detect by high-throughput sequencing, microarray profiling and real-time quantitative polymerase chain reaction.[46,47] Although this discovery has led to the identification of several tumor-specific circulating miRNA expression profiles, these processes can be influenced by both biological and technical variation, making the identification more complex and delicate for clinical implementation.[48]

15.1.2 miRNA and Anticancer Drug Resistance

Many of the pathways in which miRNAs contribute to carcinogenesis and cancer progression are the targets of anticancer drugs. Understanding these mechanisms and those involved in resistance to anticancer drugs is fundamental to overcome drug resistance and improve cancer outcomes. Therefore, miRNAs are also investigated for their roles in influencing and maintaining drug resistance through the modulation of protein expression.[49] A strong correlation between miRNA expression patterns and sensitivity to anticancer drugs has been found in a panel of cell lines used for anticancer drug discovery.[50] In particular, miR-21, which is most commonly reported as an oncogenic miRNA, significantly induced chemoresistance to specific compounds.[50] Interestingly, the same study reported opposite effects for other compounds, suggesting that modulation of miRNA expression might overcome resistance mechanisms and improve sensitivity to drugs.[51] This emphasizes the critical and controversial role of miRNAs, not only in cancer development, but also in drug resistance. The following sections further discuss the role of miRNAs in mechanisms underlying chemotherapy resistance, including apoptosis and drug targets. The possibility to modulate miRNA expression and enhance sensitivity to anticancer drugs by miRNA-based therapeutics is also discussed. Practical implementations of miRNA dysregulation and modulation of drug response focus more specifically on two common tumors in which resistance mechanisms are a major obstacle, pancreatic ductal adenocarcinoma (PDAC) and lung cancer.

15.2 Interaction of miRNA Expression and Drug Response

15.2.1 General Function of miRNAs in Drug Response

Cancer drug resistance can be broadly divided into two categories: intrinsic and acquired resistance. Tumors can be insensitive to therapeutic agents before the treatment, which refers to intrinsic resistance, while other tumors are initially sensitive and gradually acquire resistance during treatment.[3] The most common reason for the acquisition of resistance is expression of one or more energy-dependent transporters that detect and eject anticancer drugs from cells, resulting in multidrug resistance (MDR).[52] However, several other mechanisms are involved in tumor resistance. In particular, cytotoxic drugs inflict cell death by affecting processes that are commonly overactive or enhanced in tumors compared with normal cells, such as DNA synthesis, whereas biological agents interact with receptors, ligands, signaling molecules or genes that are pivotal in tumor growth and development. They can inhibit tumor cell proliferation, induce programmed cell death, inhibit angiogenesis or enhance antitumor immune response.[53] All these determinants of drug activity might have aberrations that make the cancer cells resistant to the different therapeutic approaches.

Recent data demonstrated that drug resistance is regulated not only by genetic (*e.g.*, mutations, amplifications, deletions and translocations) and epigenetic (*e.g.*, DNA hypermethylation and histone post-translational modification) changes, but also by miRNAs.[54] Indeed, recent studies showed that deregulation of miRNAs influences the expression of multiple target mRNAs, and therefore multiple proteins, leading to variations in the chemosensitivity of cancer cells *via* various cellular processes. Most of the miRNAs have been demonstrated to alter cellular response to anticancer agents *via* modulation of survival pathways and/or apoptotic response.[55] Additional miRNAs are noted to affect mechanisms other than survival and apoptotic signaling, such as the targets of several anticancer drugs and/or the DNA repair systems, which play an important role in resistance to many chemotherapeutic regimens.[56,57] Finally, miRNAs have been implicated in the regulation of drug metabolism *via* targeting drug transporters and drug metabolizing enzymes.[54] These mechanisms are discussed here in more detail.

15.2.2 Alterations of the Cell Cycle and Evasion of Apoptosis

Various anticancer drugs induce intrinsic or extrinsic apoptotic response in tumor cells.[58] The balance between pro-apoptotic and survival signals determines the fate of cancer cells. Therefore, this modulation of survival, cell cycle and/or apoptosis pathways in malignant tumors can be an important mechanism for chemoresistance. miRNAs target diverse molecules involved in these signaling pathways and can alter the regulation of anticancer agent-induced apoptosis. For example, increased expression of miR-21 following transfection with a specific precursor led to a significant reduction of apoptotic index in both control cells and in cancer cells treated with gemcitabine.[59,60] This might be explained by the fact that overexpression of miR-21 can lead to down-regulation of PTEN and a more active signaling through the PI3-kinase–Akt pathway, which renders the cancer cells less susceptible to apoptosis (Figure 15.1).[61] PTEN is an important tumor-suppressor gene, which inhibits the activation of PI3K–Akt pathway and mediates cell apoptosis and therefore can contribute to the resistant profiles of tumor cells. Several studies showed that miR-21 regulates expression of PTEN and can affect phosphorylation of its downstream kinase Akt, whose reduction correlated

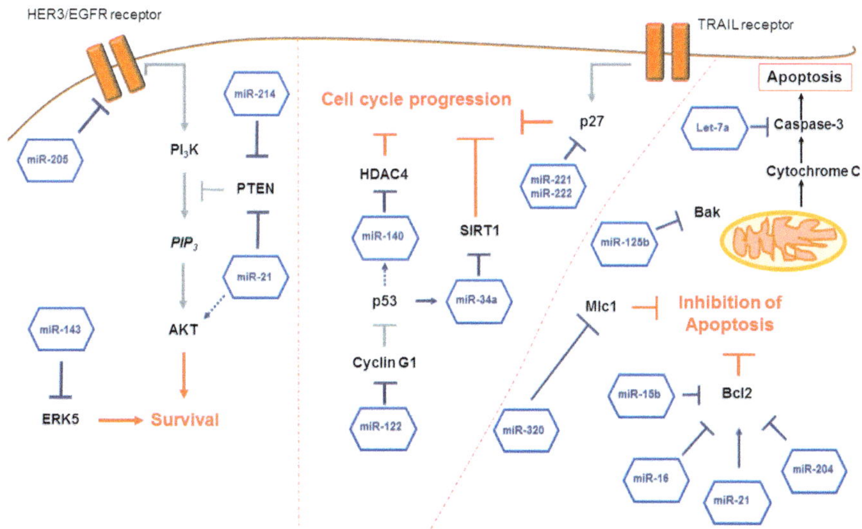

Figure 15.1 Schematic overview of regulation of survival, cell cycle, apoptosis and pathways by specific miRNAs.

with the enhancement of gemcitabine-induced apoptosis and antitumor activity *in vitro* and *in vivo*.[59,60]

Interestingly, the role of miR-21 may be tumor-specific, as inhibition of miR-21 increased sensitivity and induction of apoptosis by gemcitabine in cholangiocarcinoma and PDAC, but not in colon cancer cells.[59,61,62] In prostate cancer cells, miR-21 knockdown was not sufficient to affect the proliferative and invasive potential as well as the chemo- and radio-sensitivity profiles or to modulate the expression of PTEN.[63] Moreover, several other studies reported that miR-21 expression was correlated with resistance to several anticancer agents, suggesting that its oncogenic properties could be cell- and tissue-dependent. Therefore, the potential role of miR-21 in chemoresistance should be contextualized with respect to the tumor type and the treatment.[50,59,60,64]

Another study demonstrated that miR-199a-3p is down-regulated in hepatocellularcarcinoma (HCC) and several other tumor types. This miRNA targets the mTOR pathway and c-Met, which are subsequently up-regulated in these tumors.[65,66] The up-regulation of these oncogenic pathways enhanced proliferation and invasive capability, reduced susceptibility to hypoxia and prevention of doxorubicin-induced apoptosis in HCC cells.[65]

Furthermore, the development of resistance in various cancer cells has been linked to abnormalities in the expression of Bcl-2 family proteins and Bcl-2 expression is also directly inhibited by several miRNAs (Figure 15.1).[67] For example, miR-15b and miR-16 were revealed to directly regulate Bcl-2 expression and they are down-regulated in the gastric cell line SGC7901/VCR compared to the parental cells SGC7901. These different expression levels modulate the susceptibility of gastric cancer cells to anticancer agent-induced apoptosis.[68] Similarly, miR-204 and miR-320 are down-regulated in cholangiocarcinoma cells compared with normal cholangiocytes and the exogenous expression of miR-204 and miR-320 reduced the protein level of their targets, Bcl-2 and Mcl-1 (an anti-apoptotic member of Bcl-2 family), respectively.[69]

In addition, Bak1, a pro-apoptotic member of Bcl-2 family, is a direct target of miR-125b (Figure 15.1), and the suppression of Bak1 led to increased resistance towards taxol-induced apoptosis. This was demonstrated in taxol-resistant breast cancer cells, where a miRNA array showed the up-regulation of several miRNAs, including miR-125b, in comparison to their taxol-sensitive parental cell lines.[70]

p53 is a key tumor suppressor gene, with a pivotal role in apoptosis induction, and it is activated in response to various chemotherapeutic drugs.[71] miR-140 modulates chemoresistance by reducing proliferation and inducing cell cycle arrest, partly through suppression of HDAC4 (Figure 15.1), only in cells containing wild-type p53. Indeed, miR140 is highly expressed in osteosarcoma tumor xenografts treated with doxorubicin, cisplatin and ifosfamide and its influence on drug activity is dependent on the presence of p53.[72] Moreover, miR-140 expression is elevated in colon cancer stem cells that exhibit reduced proliferation and chemoresistance, while inhibition of this miRNA partially abrogated the resistance to 5-fluorouracil (5-FU). Furthermore, p53 was found to induce cell cycle arrest and apoptosis *via* modulation of miR-34a expression.[73] Indeed, this miRNA was down-regulated in prostate cancer cells expressing null or mutated p53, compared to cells with wild-type p53.[74]

Other studies showed that cyclin G1 was one of the several targets of miR-122 in HCC.[75] This miRNA increases the stability of p53 protein, through suppression of cyclin G1, and restoration of miR-122 expression was associated with increased sensitivity of HCC cells to doxorubicin and to doxorubicin-induced apoptosis.[76] Another study was performed in 17 patients with CLL, either responding or refractory to fludarabine and revealed that the expression levels of miR-21, miR-148a and miR-222 were significantly higher in non-responding patients. All of these patients had a dysfunctional p53 pathway.[77] However, resistance to fludarabine in CLL patients was also associated with the overexpression of miR-181a and miR-221, and reduced expression of miR-29a.[78]

15.2.3 miRNAs Interact with Drug Targets and DNA Repair

miRNAs can alter cellular response to a specific drug or class of drugs *via* interference with drug targets and DNA repair. For example, the modulation of class III-tubulin (TUBB3) was associated with resistance to microtubule-binding agents.[79,80] TUBB3 is a target for miR-200c. Reintroduction of miR-200c resulted in down-regulation of TUBB3 and enhanced sensitivity to microtubule-targeting agents.[81] In addition, several mesenchymal markers, including ZEB1, ZEB2, fibronectin-1 and neurotrophic tyrosine kinase (NTRK2) were found to be down-regulated by miR-200c, underlining the role of miR-200c in controlling the epithelial–mesenchymal transition (EMT) and sensitivity to microtubule-targeting chemotherapeutic agents.[79] In a recent study, Maftouh *et al.* demonstrated that induction of miR-211 expression in PDAC cells increased the sensitivity to gemcitabine through reduced expression of its target ribonucleotide reductase subunit 2 (RRM2), which is an important cellular target of gemcitabine.[82]

Another study reported that miR-192 and miR-215 play a role in 5-FU resistance in colorectal cancer. These miRNAs target thymidylate synthase, which is the main drug target of the fluoropyrimidine-based therapy in colorectal cancer.[83] Down-regulation of thymidylate synthase conveys sensitivity in colorectal cancer cells to 5-FU. However, down-regulation of thymidylate synthase by miR-192 and miR-215 did not lead to an increase in 5-FU sensitivity, suggesting that the activity of miR-192 and miR-215 was not mediated by thymidylate synthase. In contrast, overexpression of both miRNAs resulted in a reduction of cell proliferation and therefore diminished the effectiveness of S-phase specific drugs such as 5-FU.[83] Similarly, dihydrofolate reductase and topoisomerase I are putative targets of miR-24, which was correlated with topotecan and irinotecan sensitivity.[84]

Another miRNA that has an important role in chemoresistance is miR-98, which targets High mobility group A2 (HMGA2).[85] HMGA2 is a non-histone architectural transcription factor and was demonstrated to enhance chemosensitivity to agents causing DNA double-strand break, such as cisplatin and doxorubicin, in various cancer cells.[86] When miR-98 levels increased during hypoxia, protein levels of HGMA2 decreased; reinstatement of miR-98 expression in normoxic head and neck squamous cell carcinoma cells reduced HGMA2 protein levels and promoted resistance to doxorubicin and cisplatin.[85]

miRNAs can also alter cellular response to several anti-cancer drugs by interfering with DNA repair. Using miRNA microarrays to compare expression levels before and after radiation treatment, Weidhaas *et al.* demonstrated that miRNAs are also involved in the alteration of cellular response after DNA damage caused by irradiation, as well as UV-induced DNA damage. The expression of each member of *let-7* family, except *let-7g*, decreased after irradiation in both cancerous and normal lung cells.[87] The *let-7* family of miRNAs regulates the critical oncogene RAS and has been shown to be down-regulated in lung tumors. However, while overexpression of *let-7a* and *let-7b* promotes radio-sensitization, overexpression of *let-7g* results in radioprotection, implying that the modulation of different members of *let-7* family before irradiation affects survival.

Moreover, miR-512 is down-regulated after irradiation in prostate cancer cells.[88] The overexpression of miR-512 sensitized prostate cancer cells to radiation treatment whereas inhibition of miR-512 expression induced radio-resistance. Furthermore, inhibition of miR-512 expression reduced the levels of repair proteins involved in radiation response, such as Cockayne syndrome protein A and manganese superoxide dismutase, suggesting that altering DNA repair proteins could be a mechanism for miR-512-mediated radio-sensitivity.

15.2.4 Regulation of Drug Transporters and Metabolism

Resistance to various anticancer agents has been associated with increased expression of drug efflux pumps,[52] such as MDR1 and ABCG2 (two of the most important members of ATP-binding cassette (ABC) transporter family) and recent evidence suggests that both drug efflux pumps are directly regulated by miRNAs in different tumor types. Zhu *et al.* explored differentially expressed miRNAs between MDR human ovarian and cervical cancer cells and their parental drug-sensitive cells, and discovered that both miR-27a and miR-451 caused drug resistance by modulation of MDR1/P-gp expression.[89] Of note, this effect of miR-451 is tissue-specific, as miR-451 negatively regulated P-gp expression specifically in breast cancer cells and ectopic expression of miR-451 restored sensitivity to doxorubicin in doxorubicin-resistant MCF-7 cells.[90]

Another study showed that resistance to paclitaxel in the ovarian cancer cell line A2780 originates from the overexpression of miR-27a. One of the targets of miR-27a is the homeodomain-interacting protein kinase 2 (HIPK2), and down-regulation of this protein plays a role in the subsequent increased expression of MDR1/P-gp.[91]

ABCG2, also known as breast cancer resistance protein (BCRP), is another efflux pump protein involved in cancer cell resistance.[92] A recent study showed that its expression was directly inhibited by miR-128 in breast cancer cells and this inhibition increased the sensitivity of cells to mitoxantrone.[93] However, ABCG2 expression was also inhibited by miR-519c in an S1 parental colon cancer cell line, while this repression was relieved in resistant colon cancer cells.[94]

Regarding the miRNA-mediated modulation of drug metabolism, miR-27b was demonstrated to directly regulate CYPIB1 in breast and cervical cancer cells. This enzyme metabolizes many carcinogens and is overexpressed in a wide range of cancers compared to normal tissues.[95,96] Tsuchiya *et al.* demonstrated an inverse association between expression level of miR-27b and CYPIB1.[95] These evidences suggested that the decreased expression of miR-27b might be responsible for the high expression of CYPIB1 protein in cancerous tissues. Another study evaluated the regulation by miR-148 of the expression level of the transcription factor pregane X receptor (PXR), which regulates the expression of a number of genes involved in drug metabolism and elimination, including CYP3A4.[97] Overexpression of miR-148a diminishes the protein levels of PXR, and miR-148a-dependent decrease of PXR protein attenuates the induction of CYP3A4.[98]

To conclude, hopefully in the near future, the expression profiles of specific miRNAs could provide information about resistance of individual tumors to different treatments before starting therapy, whereas modulation of the expression of specific miRNAs during treatment might offer a new tool for the prediction of acquired resistance.

15.3 Development of miRNA-based Therapeutics

15.3.1 Mechanisms of miRNA Therapeutics

Besides their highly promising capability to function as biomarker for diagnostic, prognostic and predictive purposes, the ability of miRNAs to target multiple mRNA transcripts makes them a new target to explore for novel therapeutics. Indeed, miRNAs have successfully completed phase I clinical trials and entered into phase II trials to challenge their value as therapeutics in cancer.[99,100] miRNAs are an intricate part of the multi-step process that leads to tumor growth and suppression. Therefore, they are also proposed as both targets and means to intervene in tumor progression and modulation of sensitivity to known anticancer drugs. Previous studies have already demonstrated that miRNAs could reduce tumor growth *in vivo*, for example by restitution of tumor-suppressor miRNA.[101] To further extrapolate miRNAs therapeutics from *in vivo* studies to the clinical setting, two main strategies have been proposed and evaluated: miRNA replacement and miRNA inhibition.

15.3.1.1 Replenishing miRNA Expression

Several deregulated miRNAs in tumor tissue are suppressed, as down-regulation of specific miRNAs can result in increased expression levels of oncogenes. Previous studies have provided evidence that restoration of the expression levels of a down-regulated tumor suppressor miRNA inhibits oncogene expression and subsequently restrains tumor growth.[102] For this purpose, miRNA mimics as well as miRNA expression vectors have been developed. Synthetic miRNA mimics are double-stranded RNAs with minor chemical modifications, enabling them to form complexes with RISC for downstream mRNA inhibition.[102] The duplex consists of a guide strand that is identical to the mature miRNA strand and a passenger strand that complements the guide strand. Tumor suppressive effects of miRNA mimics have been widely validated in both *in vitro* and *in vivo* preclinical models.[101,103,104] For instance, this has been illustrated by a recent study on lipid-based systemic delivery of miR-34 and miR-143/145.[101] These miRNAs induced apoptosis and inhibited proliferation, but also inhibited PDAC growth in mouse models. Since miRNAs regulate multiple genes simultaneously, mimics can also target multiple pathways.[105] This could lead to a synergistic therapeutic effect. Hence, miRNA mimics are considered a very suitable tool to overcome the loss of important miRNA functions in the clinical setting.[100] Of note, a miR-34 mimic, MRX34 (Mirna Therapeutics) entered a clinical trial in 2013 for treatment of various types of malignancies (clinicaltrials.gov identifier NCT01829971). A total of 155 patients had been enrolled by September 2016 when the trial was terminated due to serious immune-related adverse events, which should prompt further investigations. However, delivery systems for miRNA-based therapeutics still form a major barrier to be overcome in future clinical trials.

15.3.1.2 *Inhibition of miRNA Expression*

In contrast to miRNA mimics, inhibition of oncogenic miRNA could lead to higher expression of tumor-suppressive genes. miRNA inhibition can be achieved in various ways, including antagonistic oligonucleotides (antagomirs), miRNA sponges and small molecules.[102] Antagomirs are RNAs that precisely complement targeted miRNA or its precursors and bind it in an antisense way, averting downstream complexation with RISC.[100] Similarly, miRNA sponges make use of complementarity. These sponges are produced from transgenes and consist of mRNA. The mRNA contains multiple binding sites complementary to the target miRNA and highly specific for the miRNA seed region.[106] The latter enables inhibition of a miRNA family sharing the same seed region. Furthermore, small molecule inhibitors of specific miRNAs are organic compounds of low molecular weight that can intervene in miRNA biogenesis at different maturation stages.[102] For example, they can inhibit transcription and intron splicing, but also obstruct Dicer and interactions between RISC and mRNA.

Currently, small interfering RNAs (siRNAs) are used as novel class of miRNA inhibitors. Contrary to miRNA, siRNA is fully complementary to mRNA, allowing specific gene silencing.[107] In March 2016, a phase Ib/IIa clinical trial that combined Atu027 with gemcitabine in patients with locally advanced or metastatic PDAC was successfully completed (clinicaltrials.gov identifier NCT01808638). Atu027 is a siRNA-lipoplex that inhibits protein kinase N3 (PKN3), which is known for its role in tumor progression and metastasis. In the near future, combinations of such siRNA and miRNA will also enter clinical trials. A preclinical study on ovarian cancer reported synergistic therapeutic anti-tumor effects *in vitro* and *in vivo* after siRNA-miRNA treatment.[108]

15.3.2 Delivery of miRNA Therapeutics

Successful delivery of miRNA therapeutics to targeted cells is extremely challenging. Delivery methods have to overcome the many barriers that have evolved to protect the different parts of the human body. An effective delivery system should prevent nuclease degradation, postpone or prevent renal and hepatic clearance, selectively accumulate in target cells, efficiently internalize into cells, elude the endocytic pathway and be non-toxic and non-immunogenic.[109] In addition, techniques used to improve systemic delivery and overcome these barriers, including packaging modification, inconveniently also form potent activators of the immune system.[110] Clearly, naked single-stranded RNA cannot overcome these cellular hurdles, as it is prone to RNase degradation, activates the immune system and cannot passively cross the cell membrane due to its size and negative charge.[109] Therefore, RNA modifications as well as lipid and non-lipid nanoparticles have been developed to transport miRNA-based therapeutics. Viral vehicles and their modifications go beyond the scope of this chapter, but it is important to underline that while they are capable of delivering RNA to target cells, exploiting natural mechanisms, they are also considered very immunogenic.

15.3.2.1 RNA Modulation

An important strategy in averting RNase nuclease degradation in serum and cells is backbone stability. Stability of RNA backbone nucleotides can be enhanced with chemical changes such as methylation and locked nucleic acid (LNA) addition.[111] miRNA mimics are usually modified by methylation, whereas anti-miRNAs often contain LNA structures. LNAs are modified bicyclic nucleic acid analogues with a 2′-O,4′-C methylene bridge locking a ribose in a C′3-endo conformation.[100] Such a rigid LNA structure reduces RNase susceptibility and increases binding affinity to mRNA.[111] Another modification that has substantially contributed to backbone stability is internucleotide phosphorothioates linkage; nucleotides are linked by substituting a non-bridging oxygen atom in a phosphate group with sulfur. These linkages offer resistance against RNA degradation and reduce renal excretion.[111]

Chemical modifications also play an important role in recognition by the immune system.[110] Synthetic miRNAs can trigger the innate immune system through Toll-like receptors (TLRs) 3, 7 and 8 extracellularly, and through protein kinase R (PKR) and cytoplasmic RNA receptors such as retinoic acid inducible gene I (RIG-I) intracellularly.[110] This induces high levels of interferon-α and inflammatory cytokines. Contrary to mRNA, small RNAs such as miRNAs can be substantially modified without losing their potency.[105] Chemical modifications that limit immune system activation are also based on stabilization. They focus on modification of the ribose 2′-OH group at Uracil (U) sites in particular, as gene silencing activity does not depend on this group and U-rich sequences are associated with TLR activation.[110] Substitutions with 2′-*O*-Me and 2′-fluoro (2′F) have shown to be particularly effective in abrogating the innate immune system *in vivo*.

Other chemical modifications that have been designed to improve miRNA delivery include pH-low-insertion-peptides. These peptides carry disulfide bonds that attach phalloidin to the carboxyl terminus.[112] At acidic pH, the peptide changes conformationally from an unstructured monomer to a transmembrane alpha-helix, inserting the carboxyl terminus across the membrane upon protonation.[100] The disulfide bond is cleaved inside cell cytoplasm, releasing polar cargo molecules attached to the carboxyl terminus into the cytoplasm.

15.3.2.2 Lipid-based RNA Delivery

Lipid-based delivery systems constitute a major group of vehicles in RNA delivery. With their phospholipid membrane layer, they can easily fuse with the cell membrane and thus internalize by endocytosis. Electrostatic and hydrophobic interactions allow lipids to condense naturally and easily to a lipid nanoparticle structure.[105] Cationic lipid nanoparticles (LNPs) are promising, as they carry positively charged lipids, such as amines, that electrostatically complex with the negatively charged miRNA (lipoplexes).[105] Furthermore, cationic lipids facilitate endosomal escape by formation of ion bonds with anionic phospholipids within the endosomal membrane, destabilizing the membrane and releasing the miRNA into the cytoplasm.[105] However, electrostatic-based complexes with miRNAs and lipids have poor stability and are highly cytotoxic. Therefore, hydrophobic moieties such as a derivative of polyethylene glycol (PEG), as well as cholesterol and dioleoylphosphatidylethanolamine (DOPE) are used to stabilize LNPs. PEGs render the lipids unrecognizable for phagocyte systems in the liver and spleen.[113] Unsurprisingly, they also block electrostatic interactions between cationic lipids and anionic phospholipids.[114] This hurdle was solved by using shorter acyl chains. These enable sufficient dissociation of the secured PEG-lipids for surface interaction. Despite all these modifications, lipid vehicles still caused inherent toxicity, partly due to the positive charge of cationic lipids.

To substantially limit toxicity and enhance accurate targeting of the tumor, ionizable lipids have been developed. For example, Mirna Therapeutics uses NOV340 liposomal delivery systems for their miRNA clinical trials. These lipids are only positively charged at acidic pH, such as the acidic tumor microenvironment, enhancing interaction with the negatively charged cell membrane.[105] Examples include dioleoyl-dimethylammoniumpropane (DOPAP) and dioleoyl-trimethylammoniumpropane (DOTAP). In addition, these ionizable variants aid in endosomal escape.[105] Many other alternative synthetic lipid designs have been made on the basis of these modifications. Because of their little resemblance of actual lipids, they are often referred to as "lipoids".

15.3.2.3 Polymeric-based RNA Delivery

Biodegradable synthetic cationic polymeric nanoparticles are the most widely used nanoparticles in nucleic acid delivery. Cationic polymers were one of the first non-viral nucleic acid delivery systems. Among others, naturally derived polymers include chitosan, dextran and cellulose.[115] Synthetically derived polymers include polyethylenimine (PEI), poly(-amino esters) and poly-(α)glutamic acid (PGA), which have been developed as the basis of prospective vehicles.[103,107]

PEI is one of the earliest synthetic polymers, and is known for its low molecular weight, which improves cellular uptake and release into the cytoplasm.[115] In addition, it enables escape from the endosomal route by the proton sponge effect.[109] PEI can carry secondary and tertiary unprotonated amine groups and has high buffering capacity.[109] These weak bases only become protonated when pH falls below 6, as in the intracellular environment of endosomes.[116] After encapsulation in the endosome, the unprotonated amines of PEI lead to a flux of protons crossing the lysosomal membrane through vacuolar ATPase.[116] Chloride ions follow the protons down over the charge gradient and induce osmotic swelling, which subsequently leads to rupture and release into the cytoplasm. However, a major challenge that PEI poses is its non-biodegradable structure, which leads to intracellular accumulation and cytotoxicity.[115]

Poly(lactide-*co*-glycolide) (PLGA) is a copolymer that is currently thriving in many antisense delivery studies and is already being used in the medical setting for drug delivery.[117] It has been approved by the US Food and Drug Administration as well as European Medicines Agency.[118] PLGA offers a high safety profile due to its neutral charge and is both biocompatible and biodegradable.[118] When hydrolyzed, it breaks down into lactic and glycolic acid metabolites, which are then incorporated in the Krebs cycle and removed as carbon dioxide and water. Its physiochemical properties can be highly controlled and modified, including molecular weight, structure, loading capacity and solubility through lactide to glycolic ratio.[119] Alterations in these synthesis parameters allow controlled and variable drug delivery rates.[120] Moreover, during degradation, initially fixed properties such as molecular weight can change over time, influencing drug delivery rates.[120] In the cytoplasm, PLGA has also shown the capability of sustained drug release.[121] PLGA is widely used in combination with other polymers or liposomal structures to optimize nanocarriers. A major limitation of PLGA-based nanoparticles is its low drug loading, despite efficient encapsulation.[100]

A novel synthetic polymer nanocarrier that has shown promising results is PGA-based. It is water-soluble and is biodegraded by the cathepsin B enzyme, a cysteine protease that is highly expressed in tumor tissue.[103] Moreover, it seems to have low toxicity and immunogenicity. It also uses the enhanced permeability and retention effect for selective drug targeting tumors grow vessels with wide fenestrations and abnormal architecture, allowing nanoparticles to enter the target tissue.[118] In addition, lymphatic drainage is impaired, resulting in further accumulation.

To summarize, miRNAs hold great potential as novel class as therapeutics by either miRNA replacement or miRNA inhibition. Although successful delivery of miRNAs at their target cells remains a hurdle, novel methods have evolved to improve adequate delivery of miRNAs. Future studies on miRNA-based therapeutics, either combined with conventional chemotherapy or siRNA, will further demonstrate their potential for clinical implementation.

15.4 Modification of Resistance and Prediction of Drug Response by miRNAs in Pancreatic Cancer

PDAC is a lethal disease with a 5-year overall survival of only 8% and projected to become the second leading cause of cancer-related deaths by 2030.[122] Currently, the only curative treatment is complete surgical resection; however, most patients present at advanced disease stages, for which chemotherapy is the first-line treatment.[123] Despite multimodal treatment regimens, resistance to current anticancer drugs remains a major problem, which is reflected in poor clinical outcomes. To improve outcomes for patients with PDAC, accurate diagnosis at early stages and modulation of therapy response to increase sensitivity to anticancer drugs are two main focuses of research.

Increasing efforts have been made to evaluate miRNA expression profiles as diagnostic or prognostic biomarkers in all kinds of tumor types, for example in stomach, hepatocellular, ovarian, and nasopharyngeal carcinoma.[124–128] Previously, most studies examined the expression profiles of miRNAs in tissue, while more recent researches focused their interest on miRNA expression profiles in liquid samples, including blood and urine.[10,33] Blood-based miRNA expression profiles of patients with PDAC have demonstrated aberrant expression profiles of patients with PDAC compared to healthy individuals and patients with benign disease.[129–132] The prognostic potential of miRNA expression profiles in patients with PDAC has further supported their potential as biomarkers. Most studies investigating prognostic significance of miRNA profiles have evaluated tissue expression profiles.[133–135] Lately, the prognostic significance of circulating miRNAs was confirmed in blood samples.[136,137] However, several studies showed controversial results, as techniques differ between studies and standardized methods are lacking. These challenges need to be addressed before clinical implementation of miRNA profiles as diagnostic and prognostic biomarkers is feasible.[138] To further exploit the use of miRNAs, the association between miRNA expression and response to therapy has been investigated in many tumor types, mostly focusing on cell lines. To illustrate the potential of miRNAs to enhance sensitivity to therapy, their impact in PDAC is discussed in more detail in this section.

15.4.1 Correlation Between miRNA Expression and Chemoresistance in PDAC

The first-line treatment for advanced PDAC consists of conventional chemotherapy.[139] However, limited response to therapy and rapid disease progression enforce further research to investigate the mechanisms underlying drug resistance, including the role of miRNAs. miRNAs appear to be critical regulators of chemoresistance in PDAC. Gemcitabine is one of the most commonly administrated chemotherapies, mostly used as adjuvant therapy after surgical resection of the primary tumor and combined with albumin-bound paclitaxel particles (nab-paclitaxel) for metastatic PDAC.[140,141] However, gemcitabine-based regimens show moderate activity and continuing efforts are made to improve sensitivity.[142] Modulation of miRNA expression could help to enhance sensitivity to chemotherapy and optimize response to treatment. Several studies reported deregulated miRNA expression in gemcitabine-resistant pancreatic cancer cell lines and found enhanced sensitivity to gemcitabine when this disturbed miRNA expression was restored (Table 15.1).[82,143–160] In particular, high miR-211 expression in tissue samples was associated with enhanced survival in patients with PDAC who received adjuvant gemcitabine chemotherapy and miR-211 expression was identified as prognostic factor.[135] Interestingly, induction of miR-211 increased the sensitivity to gemcitabine in PDAC cells.[82] Another study identified miR-33a as a possible miRNA to modulate sensitivity to gemcitabine.[159] High plasma levels of miR-33a were correlated to enhanced survival in patients with PDAC. Moreover, enhanced expression of miR-33a increased sensitivity to gemcitabine, both *in vitro* and *in vivo*. In a similar experimental design, Liu *et al.* explored the role of miR-153 expression to modulate sensitivity to gemcitabine.[144] Low expression of miR-153 in tissue of patients who received gemcitabine correlated with poorer survival. Conversely, by enhancing miR-153 expression in pancreatic cancer cells and in mice, sensitivity to gemcitabine was restored.

Besides modulation of a single miRNA, combination of miRNAs might be a more complete approach to enhance sensitivity. Screening of specific miRNA expression profiles in gemcitabine-resistant pancreatic cancer cells revealed a panel of deregulated miRNAs in gemcitabine-resistant cells.[161] This profile was involved in controlling cell death and cycle, suggesting further approaches to interfere and overcome chemoresistance. In another miRNA–mRNA regulatory network analysis of gemcitabine resistant cells, 56 different miRNAs were identified as important drug resistance modulators.[162] Shen *et al.* defined a core miRNA–mRNA regulatory network which correlated with gemcitabine resistance and provided further insights into the molecular mechanisms behind gemcitabine resistance.[162] These initial investigations strengthen the role of miRNA expression in therapy response and could help to understand the biology behind drug resistance in order to develop effective therapies.

The backbone of current cytotoxic treatment of patients with metastatic and locally advanced PDAC consists of 5-FU, commonly used in combination with oxaliplatin and irinotecan.[139] Few studies investigated the possible role of miRNAs to module sensitivity to 5-FU in PDAC (Table 15.1).[163–167] A known miRNA in PDAC, miR-21, was shown to modulate sensitivity to 5-FU; suppression of miR-21 enhanced sensitivity to 5-FU in pancreatic cancer cells.[163] Others reported an up-regulation of miR-320a in 5-FU resistant pancreatic cancer cells and demonstrated that miR-320a plays an important role in the regulation of 5-FU resistance.[165] Additionally, miR-34a has been reported to modulate sensitivity to 5-FU and enhanced expression could result in sensitivity to 5-FU.[166]

Table 15.1 Modulation of candidate miRNAs is related to sensitivity to chemotherapy in pancreatic cancer.[a]

Author	Year	miRNA of interest	Deregulated expression in resistant cells	Drug	Study design	Effect of miRNA restoration on drug sensitivity	Reference
Amponsah	2017	miR-210	Down-regulation	Gemcitabine	Cell lines	Restoring miRNA expression enhanced sensitivity	147
Cai	2013	miR-181b	Up-regulation	Gemcitabine	Cells and mice	Restoring miRNA expression enhanced sensitivity	151
Chaudhary	2017	miR-205	Down-regulation	Gemcitabine	Cells and mice	Restoring miRNA expression enhanced sensitivity	149
Fan	2016	miR-101-3p	Down-regulation	Gemcitabine	Cell lines	Restoring miRNA expression enhanced sensitivity	146
Huang	2018	miR-29c	NR	Gemcitabine	Cells and mice	Restoring miRNA expression enhanced sensitivity	157
Li	2017	miR-21	NR	Gemcitabine	Cells and mice	Inhibition of miR-21 enhanced sensitivity	148
Liang	2015	miR-33a	Down-regulation	Gemcitabine	Cell lines	Restoring miRNA expression enhanced sensitivity	158
Liang	2015	miR-33a	NR	Gemcitabine	Cells and mice	Restoring miRNA expression enhanced sensitivity	159
Liu	2015	miR-218	Down-regulation	Gemcitabine	Cell lines	Restoring miRNA expression enhanced sensitivity	160
Liu	2017	miR-153	Down-regulation	Gemcitabine	Cells and mice	Restoring miRNA expression enhanced sensitivity	144
Maftouh	2014	miR-211	NR	Gemcitabine	Cell lines	Restoring miRNA expression enhanced sensitivity	82
Miyamae	2015	miR-744	NR	Gemcitabine	Cell lines	Overexpression increased resistance	154
Nagano	2013	miR-29a	NR	Gemcitabine	Cell lines	Restoring miRNA expression enhanced sensitivity	152
Park	2009	miR-21	NR	Gemcitabine	Cell lines	Inhibition of miR-21 enhanced sensitivity	155

Takiuchi	2013	miR-181b	NR	Gemcitabine	Cell lines	Overexpression increased resistance	153
Xia	2017	miR-301a-3p	Up-regulation	Gemcitabine	Cells and mice	Inhibition of miR-301a-3p enhanced sensitivity	156
Xiong	2017	miR-410-3p	Down-regulation	Gemcitabine	Mice	Restoring miRNA expression enhanced sensitivity	145
Yan	2012	miR-17-5p	NR	Gemcitabine	Cell lines	Restoring miRNA expression enhanced sensitivity	150
Yu	2017	miR-429	Down-regulation	Gemcitabine	Cell lines	Restoring miRNA expression enhanced sensitivity	143
Long	2017	miR-34a	Down-regulation	5-FU	Cell lines	Restoring miRNA expression enhanced sensitivity	166
Wei	2016	miR-21	Up-regulation	5-FU	Cell lines	Inhibition of miR-21 enhanced sensitivity	163
Wang	2016	miR-320a	Up-regulation	5-FU	Cell lines	Regulation of resistance	165
Xiao	2014	miR-137	Down-regulation	5-FU	Cell lines	Restoring miRNA expression enhanced sensitivity	164
Zhao	2016	miR-221-3p	Up-regulation	5-FU	Cell lines	Regulation of resistance	167

[a]5-FU: 5-fluorouracil; miRNA: microRNA; NR: not reported.

Although modulation of miRNA expression to enhance sensitivity to oxaliplatin has been reported for other tumor types, the role of miRNA expression in resistance to oxaliplatin in PDAC remains to be explored.[168–170] Altogether, the reported studies reveal the critical role of miRNA expression in sensitivity to chemotherapy and enlighten several possibilities to modulate miRNA expression in order to overcome drug resistance.

15.4.2 miRNA Expression Profiles to Predict Response to Therapy in PDAC

Besides miRNAs as a novel therapeutic strategy to enhance sensitivity to chemotherapy, miRNA expression levels in tissue and blood could also be explored for their potential as predictive biomarkers. While most studies have identified miRNAs as prognostic biomarkers, therapy-specific miRNA expression profiles largely remain to be explored. High tissue expression levels of miR-21 correlated with shorter survival in patients treated with gemcitabine in the adjuvant setting as well as shorter survival in locally advanced PDAC treated with gemcitabine and capecitabine.[61,171,172] In another study, a high level of plasma miR-744 was reported to contribute to poorer progression-free survival in patients treated with gemcitabine-based therapy.[154] In patients with PDAC treated with 5-FU, miR-21 expression in cancer associated fibroblasts correlated with decreased overall survival.[173] In a neoadjuvant setting, miR-10b was overexpressed in tissue samples of patients with PDAC and lower levels of miR-10b corresponded to improved response to multimodality treatment.[174]

Taken together, several miRNAs have been identified as possible therapy-specific predictive biomarkers to current conventional chemotherapy, mostly for treatment with gemcitabine. With the recent advances towards circulating biomarkers, minimally invasive prediction of therapy response and tailored therapy administration accordingly will help to enhance outcomes for patients with PDAC. However, future prospective, randomized studies are necessary to strengthen the predictive potential and large-scale validation is needed before clinical implementation can be feasible.

15.4.3 Modification of miRNA Expression as Novel Therapy for PDAC

One of the major obstacles in drug resistance is the heterogeneity of tumor cells.[5] The ability of miRNAs to target multiple genes simultaneously can be used as an approach to modulate several pathways. Functional analyses of miRNAs revealed that certain miRNAs can promote malignant transformation whilst other miRNAs can exert tumor-suppressive functions in PDAC.[137,175,176] These features can be used to modulate miRNA expression profiles and explore novel miRNA-based therapeutics. Specifically, Neault *et al.* detected significantly reduced expression levels of miR-137 in pancreatic tumors. They demonstrated that miR-137 is involved in the p53 suppressor pathway and restoration of miR-137 expression levels might offer a new therapeutic window.[177] Others demonstrated that restoration of miR-34a expression resulted in increased sensitivity to chemotherapy in pancreatic cancer cells and less tumor formation *in vivo*.[178–180] miR-34a is a prominent tumor-suppressive miRNA, known to down-regulate Bcl-2 and Notch1/2 and up-regulation of miR-34a results in cell cycle arrest and apoptosis.[178,179] Further studies showed also the potential to simultaneously target multiple genes by RNA interference-based therapeutic strategies.[103] The combination of both restoration of miR-34a expression and inhibition of PLK1 was investigated to attack molecular defects and inhibit MYC. With a dual nanocarrier-based approach, this miRNA–siRNA combination efficiently inhibited tumor growth in an orthotopic mouse model.[103]

The rapid evolving field of experimental techniques leads to a more clinical based approach to enhance and predict sensitivity to anticancer drugs. By using patient-derived organoids and xenograft models, Gilles *et al.* revealed that six miRNAs (miR-21-5p, miR-146a-5p, miR-196a-5p, miR-196b-5p, let-7i-5p, and miR-29b-5p) are consistently up-regulated in PDAC.[181] In addition, they and others demonstrated that inhibition of miR-21, for example by anti-miR-21 oligonucleotides which specifically target pancreatic cancer receptors, resulted in less tumor growth.[148,181,182]

Taken together, the evolving field of miRNA-based therapeutics is very novel and of high potential. The strategies described suggest that miRNA-based therapeutics could be an efficient approach and this could be explored for many more miRNAs known to be involved in cancer progression. Collectively, this might aid to overcome therapy resistance and further improve outcomes for patients with PDAC.

15.5 Deregulated miRNAs in Resistance to Therapy in Lung Cancer

15.5.1 Impact of miRNA Deregulation in Lung Cancer

Lung cancer represents the first-ranked cause of cancer-related death worldwide.[1] Early diagnosis could allow appropriate curative treatment, but only 20–25% of patients present with localized disease. For this reason, we urgently need to identify easily detectable and minimally invasive biomarkers for early diagnosis and subsequent curative therapy. In recent years, different studies have shown that a circulating miRNA signature could potentially be used for non-small cell lung cancer (NSCLC) screening. Different studies reported that circulating microRNAs are able to identify patients with NSCLC from healthy subjects with an overall accuracy of 80–95%, mostly using a panel of differently expressed miRNAs in blood.[183–187] These studies demonstrate that circulating miRNAs could potentially be used for early detection of lung cancer, but further prospective studies are needed in order to apply these signatures for clinical use.

The relationship between miRNAs modulation and different cancer prognosis has been widely investigated. The most commonly reported prognostic miRNAs include the *let-7* family miRNA and miR-21, which have been studied extensively in lung cancer.[18,19,188,189] The decrease of expression of *let-7* family miRNA, which works as a tumor suppressor, has been associated with poor prognosis in lung cancer patients undergoing curative resection.[18] This pivotal study represents the first report of reduced expression of *let-7* in lung cancer and the potential prognostic impact and biological effects of such a miRNA alteration. The oncomiR miR-21 has also been related to lung cancer prognosis.[188] Significantly higher levels of miR-21 were found in more advanced stage tumors compared with early-stage tumors, suggesting a role of miR-21 in lung carcinogenesis.[188] In addition, different studies have created a miRNA signature associated with prognosis in lung cancer.[19,189–192] Nevertheless, further prospective validation in larger cohorts with standardized methodology is needed in order to implement current prognostic classification using miRNA signatures.

15.5.2 The Role of miRNAs in Resistance to Chemotherapy in Lung Cancer

15.5.2.1 Resistance to Cisplatin

Although significant advances have been made in NSCLC management, chemotherapy still represents the backbone of treatment of the large majority of patients.[193] In particular, chemotherapy is recommended as (i) adjuvant treatment for patients with stage II–III disease treated after radical resection; (ii) in patients with locally advanced unresectable disease with concomitant radiotherapy; and (iii) in advanced disease in patients without driver mutations and with PD-L1 expression $<50\%$.[193] For this reason, understanding mechanisms of resistance to chemotherapy represents a substantial goal in order to improve treatment strategies in this disease. miRNAs can cause drug resistance in NSCLC through different mechanisms.[194] In particular, several studies have demonstrated that miRNAs are implicated in cisplatin resistance. The study by Galluzzi *et al.* demonstrated that miR-181a and miR-630 modulate cell death induced by cisplatin in lung cancer A549 cells.[195] miR-181a and miR-630 modulated the mitochondrial/post-mitochondrial steps of the intrinsic pathway of apoptosis, including Bax oligomerization, mitochondrial transmembrane potential dissipation and the proteolytic maturation of caspase-9 and caspase-3.[195] A subsequent study displayed that miR-98 and miR-453 increased resistance to cisplatin by downregulating p53 expression in the A549 cell line.[196] The predictive role of miRNAs in resistance to platinum was further explored both in cell lines and in tissue samples. Expression of miR-21 was significantly increased in platinum-based chemotherapy-resistant patients.[197] Moreover, transfection of A549/CDDP with anti-miR-21 increased the expression of PTEN and decreased Bcl-2, underlining involvement of miR-21 in apoptotic pathways. Recently, eight miRNAs known to predict efficacy of platinum *in vitro* were evaluated in 111 tumor samples of lung adenocarcinoma.[198] Of those, decreased expression of miR-21, miR-125b, and miR-224 was independently associated response to platinum-based chemotherapy in patients with lung adenocarcinoma.[198] Although these results are promising, confirmation of the predictive role of other miRNAs in large cohorts of patients is mandatory.

15.5.2.2 Resistance to Taxanes

Microtubule-targeting agents such as docetaxel and paclitaxel are commonly used in advanced lung cancer as monotherapy or in combination with platinum.[193] Rui *et al.* reported that the expression of three miRNAs (miR-200b, miR-194 and miR-212) was significantly down-regulated while the expression of three other miRNAs (miR-192, miR-424 and miR-98) was significantly up-regulated in SPC-A1/docetaxel resistant cells, suggesting a peculiar miRNA expression profile in docetaxel-resistant NSCLC.[199] Down-regulation of miR-200b was confirmed in docetaxel-resistant SPC-A1/DTX cells and down-regulation of miR-200b seems to lead to overexpression of E2F3, which is implicated in cell cycle of proliferating cells.[200] Up-regulation of miR-135a was associated with resistance to apoptosis induced by paclitaxel, either *in vitro* or *in vivo*. Paclitaxel resistance seems to be associated with up-regulation of miR-135a in part determined by down-regulation of its target, the adenomatous polyposis coli gene (APC).[201] Down-regulation of miR-34c-5p increased resistance to paclitaxel-induced apoptosis in A549 cells. In particular, miR-34c-5p targets Bmf (Bcl-2-modifying factor) and c-myc, and this contributes to resistance to apoptosis induced by paclitaxel through p53 down-regulation.[202] Cui and collaborators demonstrated also that *let-7c* is significantly down-regulated in docetaxel-resistant SPC-A1/DTX cell line. Ectopic *let-7c* expression increased the *in vitro* and *in vivo* chemosensitivity of SPC-A1/DTX cells through enhanced apoptosis, reversal of EMT phenotypes, and inhibition of *in vivo* metastatic potential *via* inactivation of Akt phosphorylation[203] Furthermore, these authors found that Bcl-xL was significantly up-regulated in docetaxel-resistant adenocarcinoma, and its expression was inversely correlated with *let-7c* expression.[203]

15.5.3 Involvement of miRNAs in Resistance to Targeted Agents in Lung Cancer

15.5.3.1 Resistance to Anti-EGFR Treatment

In recent years, different targetable genetic alterations have been identified in NSCLC.[204] In particular, 10–15% of Caucasian patients and >30% of Asian patients with lung adenocarcinoma harbor activating mutation of epidermal growth factor receptor (EGFR) and about 3–7% the rearrangement of anaplastic lymphoma kinase (ALK) gene.[205,206] Although anti-EGFR and anti-ALK agents have significantly improved treatment of oncogene-addicted lung cancer, most of these tumors show acquired resistance during treatment.[193]

The most frequent mechanisms of acquired resistance to EGFR tyrosine kinase inhibitors (TKI) are represented by secondary gatekeeper T790M mutation (~50% of cases) and by amplification of MET oncogene (5–20% of cases).[207,208] Nevertheless, it is necessary to study in depth other possible mechanisms of resistance in order to improve current treatment strategies. EMT seems to be one of the mechanisms involved in anti-EGFR resistance. Bryant *et al.* reported that a signature of 13 miRNA genes potentially involved in EMT can predict response to EGFR-TKIs in cancer cell lines and discriminate primary and metastatic lesions.[209] Furthermore, miR-200c ectopic expression seems to modulate EMT and modestly restore sensitivity to erlotinib.[209] Similarly, Garofalo *et al.* identified specific miRNAs with an important role in gefitinib-induced apoptosis and in EMT.[210] In particular, miR-30b, miR-30c, miR-221, and miR-222 are modulated by EGFR and MET receptor, while miR-103 and miR-203 are regulated only by MET. The authors demonstrated that down-regulation of miR-30b, miR-30c, miR-221 and miR-222 led to apoptosis in gefitinib-sensitive cells by overexpression of BIM and APAF-1, while in gefitinib-resistant cells these miRNAs were not modulated by EGFR-TKIs due to MET overexpression.[210] The PI3K/AKT/mTOR pathway is involved in cell proliferation and invasion, but is also one of the mechanisms of acquired resistance to EGFR inhibitors in EGFR-positive lung cancer.[211] The study by Wang *et al.* demonstrated that miR-214 may regulate the acquired resistance to EGFR-TKIs in gefitinib-resistant cell line-HCC827/GR *via* PTEN/AKT signaling pathway.[212] In particular, miR-214 is significantly up-regulated in HCC827/GR, where miR-214 and PTEN are inversely expressed. Moreover, knockdown of miR-214 seems to modify PTEN expression and p-AKT and re-sensitize HCC827/GR to gefitinib.[212] Different studies have shown that miR-21 could be involved in resistance to EGFR-TKIs. *In vitro*, miR-21 expression has demonstrated a positive correlation with PI3K/Akt pathway and a negative correlation with PTEN expression.[213,214] Serum miR-21 expression in NSCLC patients treated with EGFR-TKIs is significantly higher at the time of acquiring resistance compared to baseline. In addition, high miR-21 and low PTEN expression have been associated with poor radiological response and short overall survival in patients undergoing EGFR-TKI treatment.[213,214]

15.5.3.2 Resistance to Anti-ALK Treatment

Crizotinib and alectinib are oral TKIs approved in ALK-rearranged patients.[193] As for anti-EGFR inhibitors, patients treated with ALK inhibitors can develop acquired resistance to therapy. Two main mechanisms of acquired resistance have been investigated: the first is represented by secondary mutations of ALK,[215] the second is ALK-independent and involves the activation of other transduction pathways.[216] There are few studies investigating the role of miRNAs in modulating resistance to anti-ALK agents. In crizotinib-resistant cell line (NCI-2228/CRI), miR-200c was shown to reverse EMT *via* targeting zinc finger E-box- binding homeobox 1 and up-regulation of E-chaderin, suggesting a role of miR-200c in mediating the sensitivity of NCI-2228/CRI to crizotinib.[217] Recently, Yun *et al.* investigated the epigenetic mechanisms underlying acquired resistance to ALK inhibitors. In an integrated analysis of ALK-positive cancer models, they showed that Histone H3 lysine 27 acetylation (H3K27ac) is altered during acquisition of resistance, and that enhancer remodeling induced repression of miR-34a and activation of its target genes, including AXL.[218] In an *in vitro* model, anti-ALK-resistant cells exhibited an EMT phenotype induced by down-regulation of miR-34a or miR-449a. An analysis on patient samples collected before and after treatment with ALK inhibitors documented that down-regulation of miR-34 or miR-449a and activation of AXL were mutually exclusive of ALK secondary mutations.[218] These results suggest that enhancer remodeling and altered expression of miRNAs play key roles in ALK resistance in lung cancer, and targeting epigenetic pathways could represent a strategy for overcoming acquired resistance to anti-ALK treatment.

In brief, deregulated miRNA expression profiles have been identified as diagnostic and prognostic biomarkers for lung cancer. In addition, miRNAs are implicated in resistance to both conventional chemotherapies as well as targeted agents. The specific potential of miRNA modulation to enhance sensitivity to anticancer agents remains to be explored.

15.6 Conclusion and Future Perspectives

miRNAs regulate major biological processes, including development, differentiation, apoptosis and proliferation. Thus, they have far-reaching effects by regulation of critical genes. miRNAs are linked to tumor initiation and progression and their expression profiles are often deregulated in patients with cancer. Therefore, they are rapidly evolving as diagnostic and prognostic biomarkers in various cancers. More recently, their implication in drug resistance has become of interest. Understanding the biological mechanisms behind drug resistance might help to enhance sensitivity to anticancer drugs and miRNAs have been shown to play critical roles in resistance. Deregulated miRNA expression profiles are linked to resistance to several anticancer drugs. This could be used to optimally predict response to therapy and tailor administration of drugs accordingly. Moreover, the ability of miRNA to target multiple genes makes them promising as a novel class of therapeutics. miRNA-based drugs could initiate antitumor effects by suppressing or restoring deregulated miRNA expression or enhance sensitivity in combination with conventional chemotherapies.

In conclusion, miRNAs hold great potential both as biomarkers and as novel therapeutic tools to overcome resistance and improve outcomes for patients with cancer. The use of miRNAs as predictive biomarkers to optimize personalized treatment and novel therapeutic strategies warrants further exploration. Future studies with standardized methods will help to define their disease- and therapy-specific roles and translate the use of these promising molecules towards clinical implementation.

Acknowledgements

This work was supported by The Bennink Foundation (The Netherlands), the Cancer Center Amsterdam Foundation (The Netherlands) and the Dutch Cancer Society (KWF-grant no. 10410, The Netherlands).

References

1. R. L. Siegel, K. D. Miller and A. Jemal, *Ca-Cancer J. Clin.*, 2017, **67**, 7–30.
2. G. Housman, S. Byler, S. Heerboth, K. Lapinska, M. Longacre, N. Snyder and S. Sarkar, *Cancers*, 2014, **6**, 1769–1792.
3. D. B. Longley and P. G. Johnston, *J. Pathol.*, 2005, **205**, 275–292.
4. C. Holohan, S. Van Schaeybroeck, D. B. Longley and P. G. Johnston, *Nat. Rev. Cancer*, 2013, **13**, 714–726.
5. D. J. Konieczkowski, C. M. Johannessen and L. A. Garraway, *Cancer Cell*, 2018, **33**, 801–815.
6. L. A. Garraway and P. A. Janne, *Cancer Discovery*, 2012, **2**, 214–226.
7. D. P. Bartel, *Cell*, 2009, **136**, 215–233.
8. Y. Lee, M. Kim, J. Han, K. H. Yeom, S. Lee, S. H. Baek and V. N. Kim, *EMBO J.*, 2004, **23**, 4051–4060.
9. L. He and G. J. Hannon, *Nat. Rev. Genet.*, 2004, **5**, 522–531.
10. H. Schwarzenbach, N. Nishida, G. A. Calin and K. Pantel, *Nat. Rev. Clin. Oncol.*, 2014, **11**, 145–156.
11. S. Jonas and E. Izaurralde, *Nat. Rev. Genet.*, 2015, **16**, 421–433.
12. J. Brennecke, D. R. Hipfner, A. Stark, R. B. Russell and S. M. Cohen, *Cell*, 2003, **113**, 25–36.
13. D. P. Bartel, *Cell*, 2004, **116**, 281–297.
14. K. Nairz, C. Rottig, F. Rintelen, E. Zdobnov, M. Moser and E. Hafen, *Dev. Biol.*, 2006, **291**, 314–324.
15. J. Tie and D. Fan, *Histol. Histopathol.*, 2011, **26**, 1353–1361.
16. G. A. Calin, C. D. Dumitru, M. Shimizu, R. Bichi, S. Zupo, E. Noch, H. Aldler, S. Rattan, M. Keating, K. Rai, L. Rassenti, T. Kipps, M. Negrini, F. Bullrich and C. M. Croce, *Proc. Natl. Acad. Sci. U. S. A.*, 2002, **99**, 15524–15529.
17. J. Lu, G. Getz, E. A. Miska, E. Alvarez-Saavedra, J. Lamb, D. Peck, A. Sweet-Cordero, B. L. Ebert, R. H. Mak, A. A. Ferrando, J. R. Downing, T. Jacks, H. R. Horvitz and T. R. Golub, *Nature*, 2005, **435**, 834–838.
18. J. Takamizawa, H. Konishi, K. Yanagisawa, S. Tomida, H. Osada, H. Endoh, T. Harano, Y. Yatabe, M. Nagino, Y. Nimura, T. Mitsudomi and T. Takahashi, *Cancer Res.*, 2004, **64**, 3753–3756.

19. N. Yanaihara, N. Caplen, E. Bowman, M. Seike, K. Kumamoto, M. Yi, R. M. Stephens, A. Okamoto, J. Yokota, T. Tanaka, G. A. Calin, C. G. Liu, C. M. Croce and C. C. Harris, *Cancer Cell*, 2006, **9**, 189–198.
20. H. Persson, A. Kvist, N. Rego, J. Staaf, J. Vallon-Christersson, L. Luts, N. Loman, G. Jonsson, H. Naya, M. Hoglund, A. Borg and C. Rovira, *Cancer Res.*, 2011, **71**, 78–86.
21. M. V. Iorio, R. Visone, G. Di Leva, V. Donati, F. Petrocca, P. Casalini, C. Taccioli, S. Volinia, C. G. Liu, H. Alder, G. A. Calin, S. Menard and C. M. Croce, *Cancer Res.*, 2007, **67**, 8699–8707.
22. J. Xu, J. Zhao and R. Zhang, *Sci. Rep.*, 2016, **6**, 38306.
23. Y. Murakami, T. Yasuda, K. Saigo, T. Urashima, H. Toyoda, T. Okanoue and K. Shimotohno, *Oncogene*, 2006, **25**, 2537–2545.
24. J. R. Puik, L. L. Meijer, T. Y. Le Large, M. M. Prado, A. E. Frampton, G. Kazemier and E. Giovannetti, *Pharmacogenomics*, 2017, **18**, 1343–1358.
25. M. Bloomston, W. L. Frankel, F. Petrocca, S. Volinia, H. Alder, J. P. Hagan, C. G. Liu, D. Bhatt, C. Taccioli and C. M. Croce, *JAMA*, 2007, **297**, 1901–1908.
26. G. A. Calin and C. M. Croce, *Nat. Rev. Cancer*, 2006, **6**, 857–866.
27. A. Esquela-Kerscher and F. J. Slack, *Nat. Rev. Cancer*, 2006, **6**, 259–269.
28. L. He, J. M. Thomson, M. T. Hemann, E. Hernando-Monge, D. Mu, S. Goodson, S. Powers, C. Cordon-Cardo, S. W. Lowe, G. J. Hannon and S. M. Hammond, *Nature*, 2005, **435**, 828–833.
29. S. Costinean, N. Zanesi, Y. Pekarsky, E. Tili, S. Volinia, N. Heerema and C. M. Croce, *Proc. Natl. Acad. Sci. U. S. A.*, 2006, **103**, 7024–7029.
30. A. L. Kasinski and F. J. Slack, *Cancer Res.*, 2012, **72**, 5576–5587.
31. G. A. Calin, M. Ferracin, A. Cimmino, G. Di Leva, M. Shimizu, S. E. Wojcik, M. V. Iorio, R. Visone, N. I. Sever, M. Fabbri, R. Iuliano, T. Palumbo, F. Pichiorri, C. Roldo, R. Garzon, C. Sevignani, L. Rassenti, H. Alder, S. Volinia, C. G. Liu, T. J. Kipps, M. Negrini and C. M. Croce, *N. Engl. J. Med.*, 2005, **353**, 1793–1801.
32. P. Landgraf, M. Rusu, R. Sheridan, A. Sewer, N. Iovino, A. Aravin, S. Pfeffer, A. Rice, A. O. Kamphorst, M. Landthaler, C. Lin, N. D. Socci, L. Hermida, V. Fulci, S. Chiaretti, R. Foa, J. Schliwka, U. Fuchs, A. Novosel, R. U. Muller, B. Schermer, U. Bissels, J. Inman, Q. Phan, M. Chien, D. B. Weir, R. Choksi, G. De Vita, D. Frezzetti, H. I. Trompeter, V. Hornung, G. Teng, G. Hartmann, M. Palkovits, R. Di Lauro, P. Wernet, G. Macino, C. E. Rogler, J. W. Nagle, J. Ju, F. N. Papavasiliou, T. Benzing, P. Lichter, W. Tam, M. J. Brownstein, A. Bosio, A. Borkhardt, J. J. Russo, C. Sander, M. Zavolan and T. Tuschl, *Cell*, 2007, **129**, 1401–1414.
33. M. V. Iorio and C. M. Croce, *J. Clin. Oncol.*, 2009, **27**, 5848–5856.
34. S. Volinia, G. A. Calin, C. G. Liu, S. Ambs, A. Cimmino, F. Petrocca, R. Visone, M. Iorio, C. Roldo, M. Ferracin, R. L. Prueitt, N. Yanaihara, G. Lanza, A. Scarpa, A. Vecchione, M. Negrini, C. C. Harris and C. M. Croce, *Proc. Natl. Acad. Sci. U. S. A.*, 2006, **103**, 2257–2261.
35. M. Galasso, S. K. Sandhu and S. Volinia, *Cancer J.*, 2012, **18**, 238–243.
36. R. Kumarswamy, I. Volkmann and T. Thum, *RNA Biol.*, 2011, **8**, 706–713.

37. E. Andres-Leon, I. Cases, S. Alonso and A. M. Rojas, *Sci. Rep.*, 2017, **7**, 46101.
38. Y. Toiyama, Y. Okugawa, J. Fleshman, C. Richard Boland and A. Goel, *Biochim. Biophys. Acta*, 2018, **1870**(2), 274–282.
39. X. Chen, Y. Ba, L. Ma, X. Cai, Y. Yin, K. Wang, J. Guo, Y. Zhang, J. Chen, X. Guo, Q. Li, X. Li, W. Wang, Y. Zhang, J. Wang, X. Jiang, Y. Xiang, C. Xu, P. Zheng, J. Zhang, R. Li, H. Zhang, X. Shang, T. Gong, G. Ning, J. Wang, K. Zen, J. Zhang and C. Y. Zhang, *Cell Res.*, 2008, **18**, 997–1006.
40. M. V. Iorio and C. M. Croce, *EMBO Mol. Med.*, 2012, **4**, 143–159.
41. N. Matamala, M. T. Vargas, R. Gonzalez-Campora, R. Minambres, J. I. Arias, P. Menendez, E. Andres-Leon, G. Gomez-Lopez, K. Yanowsky, J. Calvete-Candenas, L. Inglada-Perez, B. Martinez-Delgado and J. Benitez, *Clin. Chem.*, 2015, **61**, 1098–1106.
42. H. Wang, R. Peng, J. Wang, Z. Qin and L. Xue, *Clin. Epigenet.*, 2018, **10**, 59.
43. Z. Xie, X. Yin, B. Gong, W. Nie, B. Wu, X. Zhang, J. Huang, P. Zhang, Z. Zhou and Z. Li, *Cancer Prev. Res.*, 2015, **8**, 165–173.
44. D. Koppers-Lalic, M. Hackenberg, R. de Menezes, B. Misovic, M. Wachalska, A. Geldof, N. Zini, T. de Reijke, T. Wurdinger, A. Vis, J. van Moorselaar, M. Pegtel and I. Bijnsdorp, *OncoTarget*, 2016, **7**, 22566–22578.
45. T. Voigtlander, S. K. Gupta, S. Thum, J. Fendrich, M. P. Manns, T. O. Lankisch and T. Thum, *PLoS One*, 2015, **10**, e0139305.
46. P. Mestdagh, N. Hartmann, L. Baeriswyl, D. Andreasen, N. Bernard, C. Chen, D. Cheo, P. D'Andrade, M. DeMayo, L. Dennis, S. Derveaux, Y. Feng, S. Fulmer-Smentek, B. Gerstmayer, J. Gouffon, C. Grimley, E. Lader, K. Y. Lee, S. Luo, P. Mouritzen, A. Narayanan, S. Patel, S. Peiffer, S. Ruberg, G. Schroth, D. Schuster, J. M. Shaffer, E. J. Shelton, S. Silveria, U. Ulmanella, V. Veeramachaneni, F. Staedtler, T. Peters, T. Guettouche, L. Wong and J. Vandesompele, *Nat. Methods*, 2014, **11**, 809–815.
47. I. Garajova, T. Y. Le Large, E. Giovannetti, G. Kazemier, G. Biasco and G. J. Peters, *Methods Mol. Biol.*, 2016, **1395**, 163–187.
48. I. Babion, B. C. Snoek, M. A. van de Wiel, S. M. Wilting and R. D. M. Steenbergen, *J. Mol. Diagn.*, 2017, **19**, 625–637.
49. P. E. Blower, J. S. Verducci, S. Lin, J. Zhou, J. H. Chung, Z. Dai, C. G. Liu, W. Reinhold, P. L. Lorenzi, E. P. Kaldjian, C. M. Croce, J. N. Weinstein and W. Sadee, *Mol. Cancer Ther.*, 2007, **6**, 1483–1491.
50. P. E. Blower, J. H. Chung, J. S. Verducci, S. Lin, J. K. Park, Z. Dai, C. G. Liu, T. D. Schmittgen, W. C. Reinhold, C. M. Croce, J. N. Weinstein and W. Sadee, *Mol. Cancer Ther.*, 2008, **7**, 1–9.
51. S. Haenisch and I. Cascorbi, *Epigenomics*, 2012, **4**, 369–381.
52. M. M. Gottesman, T. Fojo and S. E. Bates, *Nat. Rev. Cancer*, 2002, **2**(1), 48–58.
53. H. J. Broxterman, K. J. Gotink and H. M. W. Verheul, *Drug Resist. Updates*, 2009, **12**(4–5), 114–126.
54. T. Fojo, *Drug Resist. Updates*, 2007, **10**(1–2), 59–67.
55. T. Zheng, J. Wang, X. Chen and L. Liu, *Int. J. Cancer*, 2010, **126**, 2–10.

56. M. T. M. van Jaarsveld, J. Helleman, E. M. J. J. Berns and E. A. C. Wiemer, *Int. J. Biochem. Cell Biol.*, 2010, **42**(8), 1282–1290.
57. P. J. Mishra and J. R. Bertino, *Pharmacogenomics*, 2009, **10**(3), 399–416.
58. S. H. Kaufmann and W. C. Earnshaw, *Exp. Cell Res.*, 2000, **256**(1), 42–49.
59. F. Meng, R. Henson, M. Lang, H. Wehbe, S. Maheshwari, J. T. Mendell, J. Jiang, T. D. Schmittgen and T. Patel, *Gastroenterology*, 2006, **130**(7), 2113–2129.
60. J. Li, H. Huang, L. Sun, M. Yang, C. Pan, W. Chen, D. Wu, Z. Lin, C. Zeng, Y. Yao, P. Zhang and E. Song, *Clin. Cancer Res.*, 2009, **15**(12), 3998–4008.
61. E. Giovannetti, N. Funel, G. J. Peters, M. Del Chiaro, L. A. Erozenci, E. Vasile, L. G. Leon, L. E. Pollina, A. Groen, A. Falcone, R. Danesi, D. Campani, H. M. Verheul and U. Boggi, *Cancer Res.*, 2010, **70**, 4528–4538.
62. P. Wang, F. Zou, X. Zhang, H. Li, A. Dulak, R. J. Tomko, J. S. Lazo, Z. Wang, L. Zhang and J. Yu, *Cancer Res.*, 2009, **69**(20), 8157–8165.
63. M. Folini, P. Gandellini, N. Longoni, V. Profumo, M. Callari, M. Pennati, M. Colecchia, R. Supino, S. Veneroni, R. Salvioni, R. Valdagni, M. G. Daidone and N. Zaffaroni, *Mol. Cancer*, 2010, **9**(1), 12.
64. M. L. Si, S. Zhu, H. Wu, Z. Lu, F. Wu and Y. Y. Mo, *Oncogene*, 2007, **26**(19), 2799–2803.
65. F. Fornari, M. Milazzo, P. Chieco, M. Negrini, G. A. Calin, G. L. Grazi, D. Pollutri, C. M. Croce, L. Bolondi and L. Gramantieri, *Cancer Res.*, 2010, **70**(12), 5184–5193.
66. S. Kim, J. L. Ui, N. K. Mi, E. J. Lee, Y. K. Ji, Y. L. Mi, S. Choung, J. K. Young and Y. C. Choi, *J. Biol. Chem.*, 2008, **283**(26), 18158–18166.
67. J. C. Reed, *Curr. Opin. Oncol.*, 1995, **7**, 541–546.
68. L. Xia, D. Zhang, R. Du, Y. Pan, L. Zhao, S. Sun, L. Hong, J. Liu and D. Fan, *Int. J. Cancer*, 2008, **123**(2), 372–379.
69. L. Chen, H. X. Yan, W. Yang, L. Hu, L. X. Yu, Q. Liu, L. Li, D. D. Huang, J. Ding, F. Shen, W. P. Zhou, M. C. Wu and H. Y. Wang, *J. Hepatol.*, 2009, **50**(2), 358–369.
70. M. Zhou, Z. Liu, Y. Zhao, Y. Ding, H. Liu, Y. Xi, W. Xiong, G. Li, J. Lu, O. Fodstad, A. I. Riker and M. Tan, *J. Biol. Chem.*, 2010, **285**(28), 21496–21507.
71. S. W. Lowe, S. Bodis, a. McClatchey, L. Remington, H. E. Ruley, D. E. Fisher, D. E. Housman and T. Jacks, *Science*, 1994, **266**(5186), 807–810.
72. B. Song, Y. Wang, Y. Xi, K. Kudo, S. Bruheim, G. I. Botchkina, E. Gavin, Y. Wan, A. Formentini, M. Kornmann, O. Fodstad and J. Ju, *Oncogene*, 2009, **28**(46), 4065–4074.
73. H. Hermeking, *Cancer Cell*, 2007, **12**(5), 414–418.
74. Y. Fujita, K. Kojima, N. Hamada, R. Ohhashi, Y. Akao, Y. Nozawa, T. Deguchi and M. Ito, *Biochem. Biophys. Res. Commun.*, 2008, **377**(1), 114–119.
75. L. Gramantieri, M. Ferracin, F. Fornari, A. Veronese, S. Sabbioni, C. G. Liu, G. A. Calin, C. Giovannini, E. Ferrazzi, G. L. Grazi, C. M. Croce, L. Bolondi and M. Negrini, *Cancer Res.*, 2007, **67**(13), 6092–6099.
76. F. Fornari, L. Gramantieri, C. Giovannini, A. Veronese, M. Ferracin, S. Sabbioni, G. A. Calin, G. L. Grazi, C. M. Croce, S. Tavolari, P. Chieco, M. Negrini and L. Bolondi, *Cancer Res.*, 2009, **69**, 5761–5767.

77. M. Ferracin, B. Zagatti, L. Rizzotto, F. Cavazzini, A. Veronese, M. Ciccone, E. Saccenti, L. Lupini, A. Grilli, C. De Angeli, M. Negrini and A. Cuneo, *Mol. Cancer*, 2010, **9**(1), 123.
78. E. Moussay, V. Palissot, L. Vallar, H. A. Poirel, T. Wenner, V. El Khoury, N. Aouali, K. Van Moer, B. Leners, F. Bernardin, A. Muller, P. Cornillet-Lefebvre, A. Delmer, C. Duhem, F. Ries, E. van Dyck and G. Berchem, *Mol. Cancer*, 2010, **9**(1), 115.
79. D. R. Cochrane, N. S. Spoelstra, E. N. Howe, S. K. Nordeen and J. K. Richer, *Mol. Cancer Ther.*, 2009, **8**(5), 1055–1066.
80. P. S. Mongroo and A. K. Rustgi, *Cancer Biol. Ther.*, 2010, **10**, 219–222.
81. D. R. Cochrane, E. N. Howe, N. S. Spoelstra and J. K. Richer, *J. Oncol.*, 2010, **2010**, 821717.
82. M. Maftouh, A. Avan, N. Funel, A. E. Frampton, H. Fiuji, S. Pelliccioni, L. Castellano, V. Galla, G. J. Peters and E. Giovannetti, *Nucleosides, Nucleotides Nucleic Acids*, 2014, **33**, 384–393.
83. V. Boni, N. Bitarte, I. Cristobal, R. Zarate, J. Rodriguez, E. Maiello, J. Garcia-Foncillas and E. Bandres, *Mol. Cancer Ther.*, 2010, **9**(8), 2265–2275.
84. W. H. Gmeiner, W. C. Reinhold and Y. Pommier, *Mol. Cancer Ther.*, 2010, **9**, 3105–3114.
85. C. Hebert, K. Norris, M. A. Scheper, N. Nikitakis and J. J. Sauk, *Mol. Cancer*, 2007, **6**(1), 5.
86. L. M. Boo, H. H. Lin, V. Chung, B. Zhou, S. G. Louie, M. A. O'Reilly, Y. Yen and D. K. Ann, *Cancer Res.*, 2005, **65**(15), 6622–6630.
87. J. B. Weidhaas, I. Babar, S. M. Nallur, P. Trang, S. Roush, M. Boehm, E. Gillespie and F. J. Slack, *Cancer Res.*, 2007, **67**(23), 11111–11116.
88. S. Josson, S. Y. Sung, K. Lao, L. W. K. Chung and P. A. S. Johnstone, *Prostate*, 2008, **68**(15), 1599–1606.
89. H. Zhu, H. Wu, X. Liu, B. R. Evans, D. J. Medina, C. G. Liu and J. M. Yang, *Biochem. Pharmacol.*, 2008, **76**(5), 582–588.
90. O. Kovalchuk, J. Filkowski, J. Meservy, Y. Ilnytskyy, V. P. Tryndyak, V. F. Chekhun and I. P. Pogribny, *Mol. Cancer Ther.*, 2008, **7**(7), 2152–2159.
91. Z. Li, S. Hu, J. Wang, J. Cai, L. Xiao, L. Yu and Z. Wang, *Gynecol. Oncol.*, 2010, **119**, 125–130.
92. G. D. Leonard, *Oncologist*, 2003, **8**(5), 411–424.
93. Y. Z. Pan, M. E. Morris and A. M. Yu, *Mol. Pharmacol.*, 2009, **75**, 1374–1379.
94. K. K. To, Z. Zhan, T. Litman and S. E. Bates, *Mol. Cell. Biol.*, 2008, **28**, 5147–5161.
95. Y. Tsuchiya, M. Nakajima, S. Takagi, T. Taniya and T. Yokoi, *Cancer Res.*, 2006, **66**(18), 9090–9098.
96. G. I. Murray, M. C. Taylor, M. C. E. McFadyen, J. A. McKay, W. F. Greenlee, M. D. Burke and W. T. Melvin, *Cancer Res.*, 1997, **57**(14), 3026–3031.
97. P. Honkakoski and M. Negishi, *Biochem. J.*, 2000, **347**(2), 321.
98. S. Takagi, M. Nakajima, T. Mohri and T. Yokoi, *J. Biol. Chem.*, 2008, **283**(15), 9674–9680.

99. C. Chakraborty, A. R. Sharma, G. Sharma, C. G. P. Doss and S. S. Lee, *Mol. Ther.--Nucleic Acids*, 2017, **8**, 132–143.
100. R. Rupaimoole and F. J. Slack, *Nat. Rev. Drug Discovery*, 2017, **16**, 203–222.
101. D. Pramanik, N. R. Campbell, C. Karikari, R. Chivukula, O. A. Kent, J. T. Mendell and A. Maitra, *Mol. Cancer Ther.*, 2011, **10**, 1470–1480.
102. M. Y. Shah, A. Ferrajoli, A. K. Sood, G. Lopez-Berestein and G. A. Calin, *EBioMedicine*, 2016, **12**, 34–42.
103. H. Gibori, S. Eliyahu, A. Krivitsky, D. Ben-Shushan, Y. Epshtein, G. Tiram, R. Blau, P. Ofek, J. S. Lee, E. Ruppin, L. Landsman, I. Barshack, T. Golan, E. Merquiol, G. Blum and R. Satchi-Fainaro, *Nat. Commun.*, 2018, **9**, 16.
104. J. Kota, R. R. Chivukula, K. A. O'Donnell, E. A. Wentzel, C. L. Montgomery, H. W. Hwang, T. C. Chang, P. Vivekanandan, M. Torbenson, K. R. Clark, J. R. Mendell and J. T. Mendell, *Cell*, 2009, **137**, 1005–1017.
105. J. C. Kaczmarek, P. S. Kowalski and D. G. Anderson, *Genome Med.*, 2017, **9**, 60.
106. T. Robb, G. Reid and C. Blenkiron, *Target. Oncol.*, 2017, **12**, 163–178.
107. J. K. Lam, M. Y. Chow, Y. Zhang and S. W. Leung, *Mol. Ther. – Nucleic Acids*, 2015, **4**, e252.
108. M. Nishimura, E. J. Jung, M. Y. Shah, C. Lu, R. Spizzo, M. Shimizu, H. D. Han, C. Ivan, S. Rossi, X. Zhang, M. S. Nicoloso, S. Y. Wu, M. I. Almeida, J. Bottsford-Miller, C. V. Pecot, B. Zand, K. Matsuo, M. M. Shahzad, N. B. Jennings, C. Rodriguez-Aguayo, G. Lopez-Berestein, A. K. Sood and G. A. Calin, *Cancer Discovery*, 2013, **3**, 1302–1315.
109. T. Ahmadzada, G. Reid and D. R. McKenzie, *Biophys. Rev.*, 2018, **10**, 69–86.
110. A. Judge and I. MacLachlan, *Hum. Gene Ther.*, 2008, **19**, 111–124.
111. S. F. Dowdy, *Nat. Biotechnol.*, 2017, **35**, 222–229.
112. O. A. Andreev, D. M. Engelman and Y. K. Reshetnyak, *Front. Physiol.*, 2014, **5**, 97.
113. J. S. Suk, Q. Xu, N. Kim, J. Hanes and L. M. Ensign, *Adv. Drug Delivery Rev.*, 2016, **99**, 28–51.
114. T. Yamamoto, Y. Teramura, T. Itagaki, Y. Arima and H. Iwata, *Sci. Technol. Adv. Mater.*, 2016, **17**, 677–684.
115. T. Catela Ivkovic, G. Voss, H. Cornella and Y. Ceder, *Cancer Lett.*, 2017, **407**, 113–122.
116. R. V. Benjaminsen, M. A. Mattebjerg, J. R. Henriksen, S. M. Moghimi and T. L. Andresen, *Mol. Ther.*, 2013, **21**, 149–157.
117. Y. S. Rhee, M. Sohn, B. H. Woo, B. C. Thanoo, P. P. DeLuca and H. M. Mansour, *AAPS PharmSciTech*, 2011, **12**, 1293–1301.
118. F. Danhier, E. Ansorena, J. M. Silva, R. Coco, A. Le Breton and V. Preat, *J. Controlled Release*, 2012, **161**, 505–522.
119. F. Y. Han, K. J. Thurecht, A. K. Whittaker and M. T. Smith, *Front. Pharmacol.*, 2016, **7**, 185.
120. Y. Xu, C. S. Kim, D. M. Saylor and D. Koo, *J. Biomed. Mater. Res., Part B*, 2017, **105**, 1692–1716.

121. C. Rolfo and E. Giovannetti, *Nat. Nanotechnol.*, 2018, **3**, 6–7.
122. R. L. Siegel, K. D. Miller and A. Jemal, *Ca-Cancer J. Clin.*, 2016, **66**, 7–30.
123. T. Kamisawa, L. D. Wood, T. Itoi and K. Takaori, *Lancet*, 2016, **388**, 73–85.
124. B. Ding, X. Gao, H. Li, L. Liu and X. Hao, *OncoTarget*, 2017, **8**, 28144–28153.
125. Q. Fu, F. Yang, T. Xiang, G. Huai, X. Yang, L. Wei, H. Yang and S. Deng, *Sci. Rep.*, 2018, **8**, 7933.
126. V. De Guire, R. Robitaille, N. Tetreault, R. Guerin, C. Menard, N. Bambace and P. Sapieha, *Clin. Biochem.*, 2013, **46**, 846–860.
127. K. Nakamura, K. Sawada, A. Yoshimura, Y. Kinose, E. Nakatsuka and T. Kimura, *Mol. Cancer*, 2016, **15**, 48.
128. N. Liu, N. Y. Chen, R. X. Cui, W. F. Li, Y. Li, R. R. Wei, M. Y. Zhang, Y. Sun, B. J. Huang, M. Chen, Q. M. He, N. Jiang, L. Chen, W. C. Cho, J. P. Yun, J. Zeng, L. Z. Liu, L. Li, Y. Guo, H. Y. Wang and J. Ma, *Lancet Oncol.*, 2012, **13**, 633–641.
129. X. Lai, M. Wang, S. D. McElyea, S. Sherman, M. House and M. Korc, *Cancer Lett.*, 2017, **393**, 86–93.
130. Y. F. Xu, B. N. Hannafon, Y. D. Zhao, R. G. Postier and W. Q. Ding, *OncoTarget*, 2017, **8**, 77028–77040.
131. Z. Pei, S. M. Liu, J. T. Huang, X. Zhang, D. Yan, Q. Xia, C. Ji, W. Chen, X. Zhang, J. Xu and J. Wang, *OncoTarget*, 2017, **8**, 22616–22624.
132. X. Sun, X. Zhou, Y. Zhang, X. Zhu and H. Liu, *Dis. Markers*, 2018, **2018**, 6292396.
133. A. E. Frampton, J. Krell, N. B. Jamieson, T. M. Gall, E. Giovannetti, N. Funel, M. Mato Prado, D. Krell, N. A. Habib, L. Castellano, L. R. Jiao and J. Stebbing, *Eur. J. Cancer*, 2015, **51**, 1389–1404.
134. P. Wald, X. S. Liu, C. Pettit, M. Dillhoff, A. Manilchuk, C. Schmidt, E. Wuthrick, W. Chen and T. M. Williams, *OncoTarget*, 2017, **8**, 73345–73361.
135. E. Giovannetti, A. van der Velde, N. Funel, E. Vasile, V. Perrone, L. G. Leon, N. De Lio, A. Avan, S. Caponi, L. E. Pollina, V. Galla, H. Sudo, A. Falcone, D. Campani, U. Boggi and G. J. Peters, *PLoS One*, 2012, **7**, e49145.
136. I. Negoi, S. Hostiuc, M. Sartelli, R. I. Negoi and M. Beuran, *Am. J. Surg.*, 2017, **214**, 515–524.
137. T. Imamura, S. Komatsu, D. Ichikawa, M. Miyamae, W. Okajima, T. Ohashi, J. Kiuchi, K. Nishibeppu, H. Konishi, A. Shiozaki, R. Morimura, H. Ikoma, T. Ochiai, K. Okamoto, H. Taniguchi and E. Otsuji, *Sci. Rep.*, 2017, **7**, 5708.
138. T. Y. Le Large, L. L. Meijer, M. M. Prado, G. Kazemier, A. E. Frampton and E. Giovannetti, *Expert Rev. Mol. Diagn.*, 2015, **15**, 1525–1529.
139. C. Caparello, L. L. Meijer, I. Garajova, A. Falcone, T. Y. Le Large, N. Funel, G. Kazemier, G. J. Peters, E. Vasile and E. Giovannetti, *World J. Gastroenterol.*, 2016, **22**, 6987–7005.
140. H. Oettle, P. Neuhaus, A. Hochhaus, J. T. Hartmann, K. Gellert, K. Ridwelski, M. Niedergethmann, C. Zulke, J. Fahlke, M. B. Arning, M. Sinn, A. Hinke and H. Riess, *JAMA*, 2013, **310**, 1473–1481.

141. D. D. Von Hoff, T. Ervin, F. P. Arena, E. G. Chiorean, J. Infante, M. Moore, T. Seay, S. A. Tjulandin, W. W. Ma, M. N. Saleh, M. Harris, M. Reni, S. Dowden, D. Laheru, N. Bahary, R. K. Ramanathan, J. Tabernero, M. Hidalgo, D. Goldstein, E. Van Cutsem, X. Wei, J. Iglesias and M. F. Renschler, *N. Engl. J. Med.*, 2013, **369**, 1691–1703.
142. J. P. Neoptolemos, D. H. Palmer, P. Ghaneh, E. E. Psarelli, J. W. Valle, C. M. Halloran, O. Faluyi, D. A. O'Reilly, D. Cunningham, J. Wadsley, S. Darby, T. Meyer, R. Gillmore, A. Anthoney, P. Lind, B. Glimelius, S. Falk, J. R. Izbicki, G. W. Middleton, S. Cummins, P. J. Ross, H. Wasan, A. McDonald, T. Crosby, Y. T. Ma, K. Patel, D. Sherriff, R. Soomal, D. Borg, S. Sothi, P. Hammel, T. Hackert, R. Jackson, M. W. Buchler and European Study Group for Pancreatic Cancer, *Lancet*, 2017, **389**, 1011–1024.
143. G. Yu, B. Jia, Y. Cheng, L. Zhou, B. Qian, Z. Liu and Y. Wang, *Am. J. Transl. Res.*, 2017, **9**, 5048–5055.
144. F. Liu, B. Liu, J. Qian, G. Wu, J. Li and Z. Ma, *Acta Biochim. Biophys. Sin.*, 2017, **49**, 520–529.
145. J. Xiong, D. Wang, A. Wei, N. Ke, Y. Wang, J. Tang, S. He, W. Hu and X. Liu, *OncoTarget*, 2017, **8**, 107500–107512.
146. P. Fan, L. Liu, Y. Yin, Z. Zhao, Y. Zhang, P. S. Amponsah, X. Xiao, N. Bauer, A. Abukiwan, C. C. Nwaeburu, J. Gladkich, C. Gao, P. Schemmer, W. Gross and I. Herr, *Cancer Lett.*, 2016, **373**, 130–137.
147. P. S. Amponsah, P. Fan, N. Bauer, Z. Zhao, J. Gladkich, J. Fellenberg and I. Herr, *Cancer Lett.*, 2017, **388**, 107–117.
148. Y. Li, Y. Chen, J. Li, Z. Zhang, C. Huang, G. Lian, K. Yang, S. Chen, Y. Lin, L. Wang, K. Huang and L. Zeng, *Cancer Sci.*, 2017, **108**, 1493–1503.
149. A. K. Chaudhary, G. Mondal, V. Kumar, K. Kattel and R. I. Mahato, *Cancer Lett.*, 2017, **402**, 1–8.
150. H. J. Yan, W. S. Liu, W. H. Sun, J. Wu, M. Ji, Q. Wang, X. Zheng, J. T. Jiang and C. P. Wu, *Dig. Dis. Sci.*, 2012, **57**, 3160–3167.
151. B. Cai, Y. An, N. Lv, J. Chen, M. Tu, J. Sun, P. Wu, J. Wei, K. Jiang and Y. Miao, *Oncol. Rep.*, 2013, **29**, 1769–1776.
152. H. Nagano, Y. Tomimaru, H. Eguchi, N. Hama, H. Wada, K. Kawamoto, S. Kobayashi, M. Mori and Y. Doki, *Int. J. Oncol.*, 2013, **43**, 1066–1072.
153. D. Takiuchi, H. Eguchi, H. Nagano, Y. Iwagami, Y. Tomimaru, H. Wada, K. Kawamoto, S. Kobayashi, S. Marubashi, M. Tanemura, M. Mori and Y. Doki, *Pancreatology*, 2013, **13**, 517–523.
154. M. Miyamae, S. Komatsu, D. Ichikawa, T. Kawaguchi, S. Hirajima, W. Okajima, T. Ohashi, T. Imamura, H. Konishi, A. Shiozaki, R. Morimura, H. Ikoma, T. Ochiai, K. Okamoto, H. Taniguchi and E. Otsuji, *Br. J. Cancer*, 2015, **113**, 1467–1476.
155. J. K. Park, E. J. Lee, C. Esau and T. D. Schmittgen, *Pancreas*, 2009, **38**, e190–199.
156. X. Xia, K. Zhang, G. Luo, G. Cen, J. Cao, K. Huang and Z. Qiu, *Am. J. Transl. Res.*, 2017, **9**, 1886–1895.
157. L. Huang, C. Hu, H. Cao, X. Wu, R. Wang, H. Lu, H. Li and H. Chen, *Cell. Physiol. Biochem.*, 2018, **47**, 747–758.

158. C. Liang, Z. Wang, Y. Y. Li, B. H. Yu, F. Zhang and H. Y. Li, *Tumor Biol.*, 2015, **36**, 9395–9403.
159. C. Liang, X. J. Yu, X. Z. Guo, M. H. Sun, Z. Wang, Y. Song, Q. X. Ni, H. Y. Li, N. Mukaida and Y. Y. Li, *OncoTarget*, 2015, **6**, 14440–14455.
160. Z. Liu, R. Du, J. Long, K. Guo, C. Ge, S. Bi and Y. Xu, *Chin. J. Cancer Res.*, 2015, **27**, 267–278.
161. S. A. Dhayat, W. A. Mardin, J. Seggewiss, A. J. Strose, C. Matuszcak, R. Hummel, N. Senninger, S. T. Mees and J. Haier, *PLoS One*, 2015, **10**, e0143755.
162. Y. Shen, Y. Pan, L. Xu, L. Chen, L. Liu, H. Chen, Z. Chen and Z. Meng, *Tumor Biol.*, 2015, **36**, 4525–4534.
163. X. Wei, W. Wang, L. Wang, Y. Zhang, X. Zhang, M. Chen, F. Wang, J. Yu, Y. Ma and G. Sun, *Cancer Med.*, 2016, **5**, 693–702.
164. J. Xiao, F. Peng, C. Yu, M. Wang, X. Li, Z. Li, J. Jiang and C. Sun, *Int. J. Clin. Exp. Pathol.*, 2014, **7**, 7442–7450.
165. W. Wang, L. Zhao, X. Wei, L. Wang, S. Liu, Y. Yang, F. Wang, G. Sun, J. Zhang, Y. Ma, Y. Zhao and J. Yu, *Sci. Rep.*, 2016, **6**, 27641.
166. L. M. Long, J. K. Zhan, H. Q. Wang, S. Li, Y. Y. Chen and Y. S. Liu, *Pathobiology*, 2017, **84**, 38–48.
167. L. Zhao, D. Zou, X. Wei, L. Wang, Y. Zhang, S. Liu, Y. Si, H. Zhao, F. Wang, J. Yu, Y. Ma and G. Sun, *Tumor Biol.*, 2016, **37**(12), 16053–16063.
168. C. Sun, F. J. Wang, H. G. Zhang, X. Z. Xu, R. C. Jia, L. Yao and P. F. Qiao, *World J. Gastroenterol.*, 2017, **23**, 1816–1827.
169. W. W. Ren, D. D. Li, X. Chen, X. L. Li, Y. P. He, L. H. Guo, L. N. Liu, H. X. Xu, L. P. Sun and X. P. Zhang, *Cell Death Dis.*, 2018, **9**, 547.
170. Y. Y. Jin, Q. J. Chen, K. Xu, H. T. Ren, X. Bao, Y. N. Ma, Y. Wei and H. B. Ma, *Mol. Cell. Biochem.*, 2016, **422**, 161–170.
171. S. Morinaga, Y. Nakamura, Y. Atsumi, M. Murakawa, K. Yamaoku, T. Aoyama, S. Kobayashi, M. Ueno, M. Morimoto, T. Yokose and Y. Miyagi, *Anticancer Res.*, 2016, **36**, 1083–1088.
172. K. Khan, D. Cunningham, C. Peckitt, S. Barton, D. Tait, M. Hawkins, D. Watkins, N. Starling, S. Rao, R. Begum, J. Thomas, J. Oates, V. Guzzardo, M. Fassan, C. Braconi and I. Chau, *OncoTarget*, 2016, **7**, 12672–12681.
173. T. R. Donahue, A. H. Nguyen, J. Moughan, L. Li, S. Tatishchev, P. Toste and J. J. Farrell, *J. Surg. Oncol.*, 2014, **110**, 952–959.
174. M. Preis, T. B. Gardner, S. R. Gordon, J. M. Pipas, T. A. Mackenzie, E. E. Klein, D. S. Longnecker, E. J. Gutmann, L. F. Sempere and M. Korc, *Clin. Cancer Res.*, 2011, **17**, 5812–5821.
175. Y. Zou, J. Li, Z. Chen, X. Li, S. Zheng, D. Yi, A. Zhong and J. Chen, *Carcinogenesis*, 2015, **36**, 676–684.
176. A. E. Frampton, L. Castellano, T. Colombo, E. Giovannetti, J. Krell, J. Jacob, L. Pellegrino, L. Roca-Alonso, N. Funel, T. M. Gall, A. De Giorgio, F. G. Pinho, V. Fulci, D. J. Britton, R. Ahmad, N. A. Habib, R. C. Coombes, V. Harding, T. Knosel, J. Stebbing and L. R. Jiao, *Gastroenterology*, 2014, **146**, 268–277 e218.

177. M. Neault, F. A. Mallette and S. Richard, *Cell Rep.*, 2016, **14**, 1966–1978.
178. Q. Ji, X. Hao, M. Zhang, W. Tang, M. Yang, L. Li, D. Xiang, J. T. Desano, G. T. Bommer, D. Fan, E. R. Fearon, T. S. Lawrence and L. Xu, *PLoS One*, 2009, **4**, e6816.
179. Q. L. Hu, Q. Y. Jiang, X. Jin, J. Shen, K. Wang, Y. B. Li, F. J. Xu, G. P. Tang and Z. H. Li, *Biomaterials*, 2013, **34**, 2265–2276.
180. S. D. Feng, Z. Mao, C. Liu, Y. S. Nie, B. Sun, M. Guo and C. Su, *OncoTargets Ther.*, 2017, **10**, 5591–5604.
181. M. E. Gilles, L. Hao, L. Huang, R. Rupaimoole, P. P. Lopez-Casas, E. Pulver, J. C. Jeong, S. K. Muthuswamy, M. Hidalgo, S. N. Bhatia and F. J. Slack, *Clin. Cancer Res.*, 2018, **24**, 1734–1747.
182. F. Sicard, M. Gayral, H. Lulka, L. Buscail and P. Cordelier, *Mol. Ther.*, 2013, **21**, 986–994.
183. X. Chen, Z. Hu, W. Wang, Y. Ba, L. Ma, C. Zhang, C. Wang, Z. Ren, Y. Zhao, S. Wu, R. Zhuang, Y. Zhang, H. Hu, C. Liu, L. Xu, J. Wang, H. Shen, J. Zhang, K. Zen and C. Y. Zhang, *Int. J. Cancer*, 2012, **130**, 1620–1628.
184. F. Bianchi, F. Nicassio, M. Marzi, E. Belloni, V. Dall'olio, L. Bernard, G. Pelosi, P. Maisonneuve, G. Veronesi and P. P. Di Fiore, *EMBO Mol. Med.*, 2011, **3**, 495–503.
185. F. Montani, M. J. Marzi, F. Dezi, E. Dama, R. M. Carletti, G. Bonizzi, R. Bertolotti, M. Bellomi, C. Rampinelli, P. Maisonneuve, L. Spaggiari, G. Veronesi, F. Nicassio, P. P. Di Fiore and F. Bianchi, *J. Natl. Cancer Inst.*, 2015, **107**, djv063.
186. M. B. Wozniak, G. Scelo, D. C. Muller, A. Mukeria, D. Zaridze and P. Brennan, *PLoS One*, 2015, **10**, e0125026.
187. E. Nadal, A. Truini, A. Nakata, J. Lin, R. M. Reddy, A. C. Chang, N. Ramnath, N. Gotoh, D. G. Beer and G. Chen, *Sci. Rep.*, 2015, **5**, 12464.
188. M. Saito, A. J. Schetter, S. Mollerup, T. Kohno, V. Skaug, E. D. Bowman, E. A. Mathe, S. Takenoshita, J. Yokota, A. Haugen and C. C. Harris, *Clin. Cancer Res.*, 2011, **17**, 1875–1882.
189. M. T. Landi, Y. Zhao, M. Rotunno, J. Koshiol, H. Liu, A. W. Bergen, M. Rubagotti, A. M. Goldstein, I. Linnoila, F. M. Marincola, M. A. Tucker, P. A. Bertazzi, A. C. Pesatori, N. E. Caporaso, L. M. McShane and E. Wang, *Clin. Cancer Res.*, 2010, **16**, 430–441.
190. X. Li, Z. An, P. Li and H. Liu, *Oncol. Lett.*, 2017, **14**, 2991–2995.
191. Z. Peng, L. Pan, Z. Niu, W. Li, X. Dang, L. Wan, R. Zhang and S. Yang, *OncoTarget*, 2017, **8**, 64143–64156.
192. Y. Lin, Y. Lv, R. Liang, C. Yuan, J. Zhang, D. He, X. Zheng and J. Zhang, *OncoTargets Ther.*, 2018, **11**, 29–36.
193. S. Novello, F. Barlesi, R. Califano, T. Cufer, S. Ekman, M. G. Levra, K. Kerr, S. Popat, M. Reck, S. Senan, G. V. Simo, J. Vansteenkiste, S. Peters and E. G. Committee, *Ann. Oncol.*, 2016, **27**, v1–v27.
194. C. Rolfo, E. Giovannetti, D. S. Hong, T. Bivona, L. E. Raez, G. Bronte, L. Buffoni, N. Reguart, E. S. Santos, P. Germonpre, M. Taron, F. Passiglia, J. P. Van Meerbeeck, A. Russo, M. Peeters, I. Gil-Bazo, P. Pauwels and R. Rosell, *Cancer Treat. Rev.*, 2014, **40**, 990–1004.

195. L. Galluzzi, E. Morselli, I. Vitale, O. Kepp, L. Senovilla, A. Criollo, N. Servant, C. Paccard, P. Hupe, T. Robert, H. Ripoche, V. Lazar, A. Harel-Bellan, P. Dessen, E. Barillot and G. Kroemer, *Cancer Res.*, 2010, **70**, 1793–1803.
196. S. Zhang, C. Zhang, Y. Li, P. Wang, Z. Yue and S. Xie, *Biomed. Pharmacother.*, 2011, **65**, 436–442.
197. W. Gao, X. Lu, L. Liu, J. Xu, D. Feng and Y. Shu, *Cancer Biol. Ther.*, 2012, **13**, 330–340.
198. X. Xu, S. Yu, W. Sun, X. Qin, Y. Chen, L. Zhou, R. Lou, S. Dong, B. Shen, J. Wu, J. Zang, H. Cao, M. Shi, Q. Zhang and J. Feng, *J. Cancer Res. Clin. Oncol.*, 2018, **144**, 431–438.
199. W. Rui, F. Bing, S. Hai-Zhu, D. Wei and C. Long-Bang, *J. Cell. Mol. Med.*, 2010, **14**, 206–214.
200. B. Feng, R. Wang, H. Z. Song and L. B. Chen, *Cancer*, 2012, **118**, 3365–3376.
201. A. Holleman, I. Chung, R. R. Olsen, B. Kwak, A. Mizokami, N. Saijo, A. Parissenti, Z. Duan, E. E. Voest and B. R. Zetter, *Oncogene*, 2011, **30**, 4386–4398.
202. S. Catuogno, L. Cerchia, G. Romano, P. Pognonec, G. Condorelli and V. de Franciscis, *Oncogene*, 2013, **32**, 341–351.
203. S. Y. Cui, J. Y. Huang, Y. T. Chen, H. Z. Song, B. Feng, G. C. Huang, R. Wang, L. B. Chen and W. De, *Mol. Cancer Res.*, 2013, **11**, 699–713.
204. The Cancer Genome Atlas Research Network, *Nature*, 2014, **511**, 543.
205. J. G. Paez, P. A. Janne, J. C. Lee, S. Tracy, H. Greulich, S. Gabriel, P. Herman, F. J. Kaye, N. Lindeman, T. J. Boggon, K. Naoki, H. Sasaki, Y. Fujii, M. J. Eck, W. R. Sellers, B. E. Johnson and M. Meyerson, *Science*, 2004, **304**, 1497–1500.
206. E. L. Kwak, Y. J. Bang, D. R. Camidge, A. T. Shaw, B. Solomon, R. G. Maki, S. H. Ou, B. J. Dezube, P. A. Janne, D. B. Costa, M. Varella-Garcia, W. H. Kim, T. J. Lynch, P. Fidias, H. Stubbs, J. A. Engelman, L. V. Sequist, W. Tan, L. Gandhi, M. Mino-Kenudson, G. C. Wei, S. M. Shreeve, M. J. Ratain, J. Settleman, J. G. Christensen, D. A. Haber, K. Wilner, R. Salgia, G. I. Shapiro, J. W. Clark and A. J. Iafrate, *N. Engl. J. Med.*, 2010, **363**, 1693–1703.
207. H. A. Yu, M. E. Arcila, N. Rekhtman, C. S. Sima, M. F. Zakowski, W. Pao, M. G. Kris, V. A. Miller, M. Ladanyi and G. J. Riely, *Clin. Cancer Res.*, 2013, **19**, 2240–2247.
208. J. A. Engelman, K. Zejnullahu, T. Mitsudomi, Y. Song, C. Hyland, J. O. Park, N. Lindeman, C. M. Gale, X. Zhao, J. Christensen, T. Kosaka, A. J. Holmes, A. M. Rogers, F. Cappuzzo, T. Mok, C. Lee, B. E. Johnson, L. C. Cantley and P. A. Janne, *Science*, 2007, **316**, 1039–1043.
209. J. L. Bryant, J. Britson, J. M. Balko, M. Willian, R. Timmons, A. Frolov and E. P. Black, *Br. J. Cancer*, 2012, **106**, 148–156.
210. M. Garofalo, G. Romano, G. Di Leva, G. Nuovo, Y. J. Jeon, A. Ngankeu, J. Sun, F. Lovat, H. Alder, G. Condorelli, J. A. Engelman, M. Ono, J. K. Rho, L. Cascione, S. Volinia, K. P. Nephew and C. M. Croce, *Nat. Med.*, 2011, **18**, 74–82.

211. C. Fumarola, M. A. Bonelli, P. G. Petronini and R. R. Alfieri, *Biochem. Pharmacol.*, 2014, **90**, 197–207.
212. Y. S. Wang, Y. H. Wang, H. P. Xia, S. W. Zhou, G. Schmid-Bindert and C. C. Zhou, *Asian Pac. J. Cancer Prev.*, 2012, **13**, 255–260.
213. B. Li, S. Ren, X. Li, Y. Wang, D. Garfield, S. Zhou, X. Chen, C. Su, M. Chen, P. Kuang, G. Gao, Y. He, L. Fan, K. Fei, C. Zhou and G. Schmit-Bindert, *Lung Cancer*, 2014, **83**, 146–153.
214. H. Shen, F. Zhu, J. Liu, T. Xu, D. Pei, R. Wang, Y. Qian, Q. Li, L. Wang, Z. Shi, J. Zheng, Q. Chen, B. Jiang and Y. Shu, *PLoS One*, 2014, **9**, e103305.
215. R. C. Doebele, A. B. Pilling, D. L. Aisner, T. G. Kutateladze, A. T. Le, A. J. Weickhardt, K. L. Kondo, D. J. Linderman, L. E. Heasley, W. A. Franklin, M. Varella-Garcia and D. R. Camidge, *Clin. Cancer Res.*, 2012, **18**, 1472–1482.
216. R. Katayama, A. T. Shaw, T. M. Khan, M. Mino-Kenudson, B. J. Solomon, B. Halmos, N. A. Jessop, J. C. Wain, A. T. Yeo, C. Benes, L. Drew, J. C. Saeh, K. Crosby, L. V. Sequist, A. J. Iafrate and J. A. Engelman, *Sci. Transl. Med.*, 2012, **4**, 120ra117.
217. H. X. Gao, L. Yan, C. Li, L. M. Zhao and W. Liu, *Mol. Med. Rep.*, 2016, **14**, 4135–4143.
218. M. R. Yun, S. M. Lim, S. K. Kim, H. M. Choi, K. H. Pyo, S. K. Kim, J. M. Lee, Y. W. Lee, J. W. Choi, H. R. Kim, M. H. Hong, K. Haam, N. Huh, J. H. Kim, Y. S. Kim, H. S. Shim, R. A. Soo, J. Y. Shih, J. C. Yang, M. Kim and B. C. Cho, *Cancer Res.*, 2018, **78**, 3350–3362.

CHAPTER 16

Nanoparticle Delivery of miRNA in Cancer

HAMADA MOKHLIS[a,b] AND BULENT OZPOLAT*[a,c]

[a]Department of Experimental Therapeutics, The University of Texas M.D. Anderson Cancer Center, Houston, Texas, 77030, USA; [b]Department of Pharmacology and Toxicology, Faculty of Pharmacy, The University of Al-Azhar, Cairo, Egypt; [c]Non-Coding RNA Center, The University of Texas M.D. Anderson Cancer Center, Houston, Texas, 77030, USA
*E-mail: bozpolat@mdanderson.org

16.1 Introduction

The discovery of therapeutic non-coding RNAs is considered one of the most exciting and significant breakthroughs in cell biology and the pathogenesis of human diseases, including cancer, as well as pharmaceutical development. MicroRNAs (miRNAs) are endogenous, single-stranded, non-coding small RNAs of about 22 nucleotides in length. The first miRNA was discovered in 1993 in *Caenorhabditis elegans* and participates in embryonic development.[1] To date, approximately 20 clinical trials have been initiated using miRNA-based therapeutics. miRNAs work as regulators of gene expression post-transcriptionally by binding to the seed or consensus sequences located in the 3′-untranslated region of the target messenger RNA (mRNA), thus inhibiting protein translation and/or reducing mRNA stability. A single gene can be simultaneously targeted by multiple miRNAs, while each miRNA can target multiple genes *via* similar seed sequence. To date, more than 2500

Drug Discovery Series No. 69
MicroRNAs in Diseases and Disorders: Emerging Therapeutic Targets
Edited by Philip V. Peplow, Bridget Martinez, George A. Calin and Aurora Esquela-Kerscher

Published by the Royal Society of Chemistry, www.rsc.org

miRNAs have been identified in the human genome. It is estimated that miRNAs regulate more than 50% of the genes in human genome and are involved in almost all fundamental cell processes including development, metabolism, cell proliferation, survival and apoptosis.[2] In addition, miRNAs can modulate disease pathways related to tumor growth, metastasis, angiogenesis, tumor microenvironment, cancer-immune system interactions, stromal cell interactions, and sensitivity to therapy.[2]

The expression patterns of miRNAs are usually altered in different development stages and various pathologic conditions such as senescence, cardiovascular diseases and cancer.[3,4] To date, more than 2200 differentially expressed miRNAs have been identified in 36 different human cancer types. Aberrant expression of miRNAs is often found in various types of human cancers due to dysfunction of the miRNA biogenesis process, hampered transcription of miRNA-encoding genes, or misregulation of mature miRNAs by factors such as circular RNAs.[3] Some of these altered miRNAs were noted to be markedly overexpressed in human tumors compared to normal tissues and shown to act as oncogenes, accelerating tumor growth and progression by affecting hallmarks of cancer. Meanwhile, a subgroup of miRNAs were notably decreased in cancer patients and shown to play tumor suppressive roles often inhibit the expression of proto-oncogenes.[3,4] The differential expression of miRNAs can be detected by polymerase chain reaction, Northern blotting, microarray, and deep sequencing, and has potential for clinical applications.[5,6]

16.2 miRNA-based Therapeutics

Studies demonstrated that overexpression of miRNAs acting as tumor suppressors by delivery of miRNA mimics and molecules inactivating pro-oncogenic miRNAs *via* antimiRs have shown promise in preclinical animal models for a variety of cancer types. Recently, several miRNA-based therapeutics entered clinical trials, including a mimic of the tumor suppressor miRNA miR-34a (MRX34), which reached phase I clinical trials for treating advanced cancers, and antimiRs targeting miR-122 (miravirsen), which reached phase II trials for treating hepatitis C. To date, approximately 20 clinical trials have been initiated using miRNA and small interfering RNA (siRNA)-based therapeutics. In this chapter, we will describe recent advances in the administration of miRNAs for cancer treatment and provide an overview of nanocarrier systems for the delivery of miRNA therapeutics to primary tumors and metastatic lesions. We also discuss the challenges of achieving safe and targeted delivery of miRNA therapeutics.

16.3 Challenges for Delivery of miRNAs

Although increasing amount of experimental data demonstrated significant potential of miRNAs in cancer therapy, their translation into clinical application has multiple challenges. One limitation for the usage of miRNAs in the clinic is the instability of miRNAs in biological fluids (*e.g.*, blood). These non-coding RNAs are susceptible to degradation by endonucleases and exonucleases, resulting in a short half-life and fast kidney clearance.[7] In general, oligonucleotides in the bloodstream have a half-life of only 2 min. However, suitable chemical substitutions can improve miRNA half-life to many hours.[8] Another major challenge for this technology is the potential off-target effects, some of which arise from activation of innate immune response caused by Toll-like receptor (TLR) activation. This immune reaction can be enhanced by a specific sequence, by chemical structure, or by certain modifications of the oligonucleotides.[9] Although the efficacy of miRNA therapy might be enhanced by this phenomenon in some pathologies,[10] off-target effects should be avoided to have safer miRNA therapeutics. Furthermore, the poor ability of miRNAs to cross the cell membrane is attributed to their polyanionic nature (negative charge), high molecular weight and size (13–14 kDa), and hydrophilic nature. In order to exhibit their silencing properties, miRNA must reach the cytoplasm of the target cells.[11] Another major barrier is the reticuloendothelial system (RES), in which the Kupffer cells of liver and spleen macrophages eliminate these oligonucleotides from the circulation system. Phagocytosis of the oligonucleotide results in a phagosome, which is then integrated into the lysosome and degraded by nucleases.[8] Nuclease activity in plasma and tissue degrade the oligonucleotide very rapidly. This phenomenon can be avoided by targeted delivery of nanoparticles (NPs) to cancer cells.[12]

To overcome these limitations, different approaches have been investigated. Chemical modifications have been shown to improve the stability of oligonucleotides, such as siRNA and miRNAs, and for some pathologies naked chemically-stabilized RNA interference (RNAi) sequences have been used.[13,14] There are ongoing clinical trials using naked siRNAs, but for systemic administration, a delivery system is required.[15] Even when the RNAi sequences are stabilized chemically and manage to survive the circulation, they are still unable to cross cellular membranes. The choice of an adequate delivery system can overcome poor cellular uptake, immune response stimulation upon *in vivo* administration,[16] and prevent sequestration of the oligonucleotides in cellular vesicles.[17] Therefore, several carriers have been developed for effective use of siRNA and miRNAs, including viral and non-viral vectors.

Viral vectors (*e.g.*, adenovirus) are efficient delivery systems based on the use of genetically modified viruses, exploiting their innate ability to introduce genetic material in eukaryotic cells.[18] However, induction of immune response limits their use in multiple times. Thus, recently, much effort has been dedicated to developing synthetic vehicles for non-coding RNAs delivery in humans. The non-viral vectors are mainly grouped in two categories according to composition: (1) lipid- and (2) polymer-based delivery systems.[19]

16.4 Types of Non-viral Nanoparticles

16.4.1 Lipid-based Nanovectors for miRNA Delivery

Liposomal formulations are historically the most successful vehicles for delivery of therapeutic cargo, including chemo-drugs, antigens, adjuvants, peptides, proteins, oligonucleotides, DNA, siRNAs, and miRNAs. Their high degree of biocompatibility by systemic administration have made them the most popular method for delivery of therapeutic agents. Liposomes are considered self-assembling and uni- or multi-layered nanoparticles are formed by dispersion of phospholipids with hydrophilic heads and hydrophobic anionic/cationic long chain tails, creating closed membrane structures.[20] Currently, there are more than eight United States Food and Drug Administration-approved liposomal formulations used in oncology (*e.g.*, liposomal doxorubicin, Doxil, approved in 1994).[21]

Liposomes offer several advantages, including (1) preferential accumulation in tumor tissues (passive targeting) and delivery of high concentrations of the payload into tumor tissues due to large junction gaps in the angiogenic endothelium lining; (2) protection of the therapeutic cargo from degradation by encapsulation; (3) high-affinity ligands (active targeting) enabling specific delivery of the cargo to tumor cells and microenvironment (Figure 16.1); and (4) safe and effective systemic delivery for *in vivo* applications in animals and humans.[26] Liposomes also provide a favorable pharmacokinetic profile

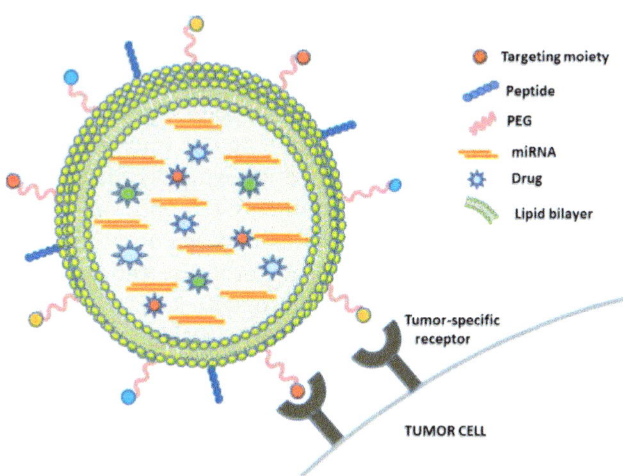

Figure 16.1 Tumor-targeting liposomes for the delivery of miRNAs and drugs. Liposomes can be coated with high-affinity ligands, including peptides, proteins, aptamers and antibody fragments for specific delivery of the therapeutic cargo to tumor cells and the microenvironment. Reproduced from ref. 107 with permission from Elsevier, Copyright 2014.

for the therapeutic payload and help maintain high plasma concentrations of drugs with hydrophobic features and poor bioavailability. For successful and robust *in vivo* delivery of oligonucleotides, such as antisense reagents (antagomiRs, antimiR inhibitors) and miRNA and siRNA mimics, using liposomal nanocarriers require optimization of lipid composition, particle size, charge, drug-to-lipid ratio, targeting moieties, encapsulation efficiency, and the manufacturing process. For longer-circulation half-life, liposomes can be PEGylated (coated with polyethylene glycol) to avoid elimination by cells of the RES. The PEGylated liposomes ('stealth liposomes') also provide a linker for attachment of tumor-targeting moieties (high-affinity ligands, *e.g.*, folate or peptides, aptamers, antibodies) for specific targeting of liposomes to cell surface receptors on targeted cells.

16.4.1.1 Cationic Lipid-based Liposomes

The cationic lipids are widely used non-viral delivery systems for *in vitro* transfections and delivering oligonucleotides, including plasmid DNA, antisense oligos, and siRNAs/shRNAs (short-hairpin RNAs) and miRNAs into normal and cancer cells. Therefore, there are a variety of cationic lipid-based transfection systems commercially available in the market for biomedical research (*i.e.*, Lipofectin, Lipofectamine, and HiPerFect). However, toxicity of cationic lipids is the major issue following systemic administration *in vivo* in mice and they have been shown to cause dose-dependent toxicity and pulmonary inflammation, hepatotoxicity, and a systemic interferon type I response due to activation of TLR and complement system.[21] It has also been demonstrated that cationic lipids are highly toxic to macrophages and other immune cells (ED_{50} < 50 nm L^{-1}), depending on the lengths of hydrocarbon chains, due to induction of reactive oxygen species and intracellular calcium levels.[22] However, this effect was not observed in mice treated with neutral or negatively charged particles. Conversely, negatively charged liposomes are taken up by phagocytic cells and may not result in optimal loading efficiency. Overall, the effectiveness of cationic liposomes as potential therapeutic molecules has been limited by their toxicity. Careful selection of lipids and formulations may help reduce the potential toxicities.

Many studies demonstrated that various miRNAs (as well as siRNA molecules) could be effectively delivered into tumor tissues *via* cationic liposomal formulations. Cationic liposomes such as DOTAP (1,2-dioleoyl-3-trimethylammonium-propane) and DOTMA (*N*-[1-(2,3-Dioleoyloxy)propyl]-*N*,*N*,*N*-trimethyl-ammonium methyl sulfate) can form complexes or lipoplexes with negatively charged DNA or siRNA to form nanoparticles by electrostatic interaction, providing *in vitro* a robust transfection efficiency.[61,62] Wu *et al.*[22] showed that a tumor suppressive miR-29b was delivered efficiently into lung tumors *via* cationic liposome, leading to significant reduction in their growth and clonogenicity. In a recent study, the delivery of miR-7 by cationic liposomes was shown to be as high as 30-fold compared to control microRNA and led to more effective inhibition of tumor growth in a lung adenocarcinoma tumor xenograft model in mice.[23] In addition, Shi *et al.*[24] developed a dual-drug delivery system based on cationic solid lipid nanoparticles incorporating miR-34a and paclitaxel for synergistic cancer therapy and demonstrated a significant decrease in tumor relapse in lung cancer models. Moreover, this formulation provided a stable nanoparticle system in blood and potentiated the efficacy of chemotherapy in a lung metastatic melanoma mouse model (B16F10-CD44+).[24]

A liposomal formulation of the naturally occurring tumor suppressor miR-34a was shown to lead to approximately 1000-fold increased miR-34a levels in orthotopic mouse models of liver cancer after systemic injection and a significant decrease in the mRNA levels of several miR-34a targets.[25] The administration of the formulated miR-34a mimic caused significant tumor growth inhibition in two different models of liver cancer and led to tumor regression in more than one-third of the animals, with no sign of immunostimulatory effects or dose-limiting toxicities. However, accumulation of the formulated miR-34a mimic was also noted in the spleen, lung, and kidney, suggesting the potential for therapeutic use in other cancers.[25] Furthermore, miR-34a reached phase I clinical trials in patients with advance cancers, including melanoma, renal, and hepatocellular cancer. Unfortunately, the clinical trials were terminated due to immunological adverse effects.

16.4.1.2 Neutral Lipid-based Liposomes

We recently developed novel liposomal nanoparticles to eliminate toxicity and overcome potential issues observed with cationic liposomes and negatively charged carriers. This delivery system is composed of 1,2-dimyristoyl-*sn*-glycero-3-phosphocholine (DMPC) and DSPE-PEG2000 (10 : 1 ratio) and highly effective in delivering siRNAs,[26] miRNA mimics, such as miR-34a, miR-603, and miR-873,[27–29] as well as antimiRs (*e.g.*, anti-miR-484) into tumor tissues in various tumor models in mice,[30] including breast and pancreatic cancers. We found loss of or reduced expression of miR-603 in triple negative breast cancer (TNBC) cell lines and patient tumors. Further studies revealed that miR-603 functions as a tumor suppressor, and loss of miR-603 expression leads to increased eukaryotic elongation factor 2 kinase (eEF2K) expression, contributing to the growth, invasion, and progression of TNBC. Thus, we therapeutically delivered miR-603 into *in vivo* TNBC xenografts in orthotopic mouse models by systemic administration of miR-603 encapsulated in liposomal nanoparticles (0.3 mg kg^{-1} once a week, *i.v.* injection). We found a significant inhibition of the target gene expression (eEF2K) and reduced tumor growth (Figure 16.2).[27] No toxic effects were observed during 4 weeks of the liposomal miR-603 treatment in mice. Mice appeared to be healthy and did not lose weight during the treatment, suggesting that liposomal miR-603-based gene therapy is safe and effective, and could be used for *in vivo* applications. Recently, we also used the same DMPC-based nanoliposomal approach for delivery of miR-873, which targets the *KRAS* oncogene using two different pancreatic

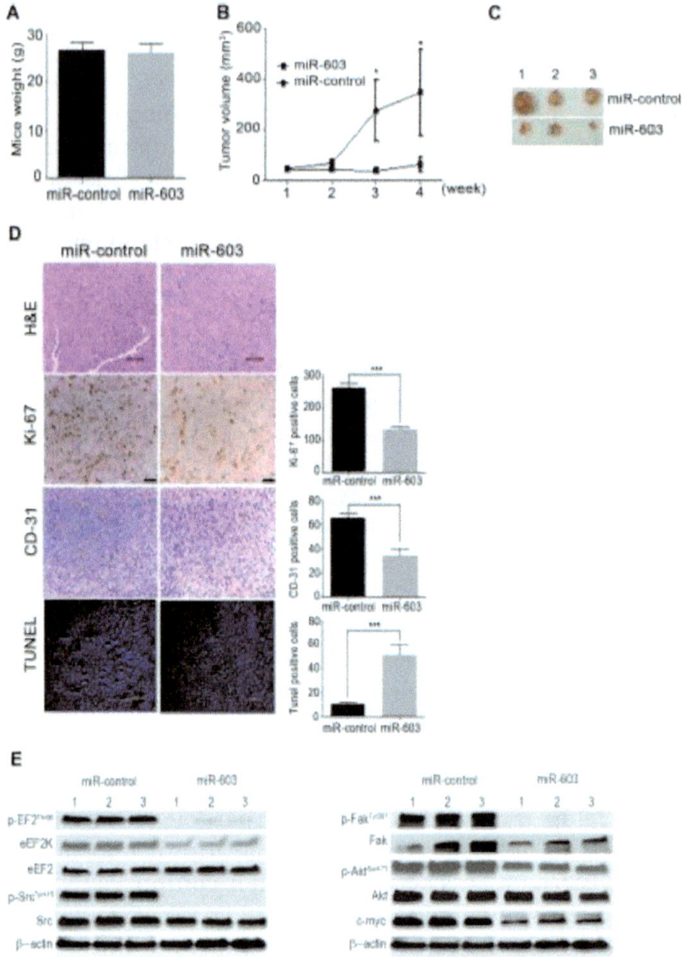

Figure 16.2 *In vivo* systemic administration of liposomal miR-603 nanoparticles inhibits growth triple negative breast cancer (TNBC) xenografts in mice.[27] (A, B, C) MDA-MB-436 cells were orthotopically injected into the mammary fat pad of female athymic nude mice. The mice were then treated with either nanoliposomal control miRNA (miR-control) or miR-603 nanoparticles (0.3 mg kg^{-1}; 8 μg per mouse intravenously twice per week for 4 weeks; five mice per group). Tumor volumes were measured weekly and shown as mean ± SD. miR-603 treatment did not cause any change in the mean mouse weight after 4 weeks. (D) Samples of MDA-MB-436 tumors from control and miR-603-treated mice were stained with hematoxylin and eosin. Scale bar = 100 μm. Tumor cell proliferation and microvessel density were analyzed by evaluating the expression of Ki-67 and CD-31, respectively, in tumor tissues by immunohistochemistry. *In vivo* induction of apoptosis was analyzed by TUNEL assay in MDA-MB-436 tumor xenografts. Magnification 20×. (E) Western blot analysis demonstrated that silencing of eukaryotic elongation factor 2 kinase (eEF2K) in MDA-MB-436 tumor samples reduced eEF2K, p-eEF2, p-Src, p-Akt, c-myc and p-Fak protein expression levels. β-Actin was used as a loading control. Reproduced from ref. 27 [https://doi.org/10.18632/oncotarget.14264] under the terms of a CC BY 3.0 license [https://creativecommons.org/licenses/by/3.0/].

cancer tumor xenograft mouse models (PANC1 and Mia-Paca-2). This approach demonstrated robust target (KRAS) inhibition and tumor growth inhibition.[29] More importantly, DMPC-based nanoliposomes did not cause any detectable distress or toxicity and treatment was safe for use in mice.[27–29]

Neutral-lipid(1,2-dioleoyl-*sn*-glycero-3-phosphatidylcholine-DOPC)-based liposomes (65 nm) have been shown to effectively deliver siRNAs[31] and miRNAs[32] at a dosing schedule of 0.15 mg kg^{-1} twice a week. Treatment with these type of nanoliposomal particles resulted in substantial reduction in the expression of the miRNA target genes. For example, DOPC-NPs incorporating miRNA-520d (at a dosing schedule of 0.15 mg kg^{-1} *i.v.* twice a week) reduced expression of EphA2 (Ephrin type-A receptor 2) in orthotopic models for ovarian cancer.[32] DOPC is currently being tested in a phase I clinical trial (clinicaltrials.gov identifier NCT01591356) at MD Anderson Cancer Center (Houston, TX, USA). Furthermore, DOPC-based neutral nanoliposomes did not cause any toxicity and were noted to be safe in mice and non-human primates (Rhesus macaques) by clinical chemistry and pathology. These reagents also did not exert any toxicity to normal cells, including fibroblast, bone marrow, and hematopoietic cells, making them highly attractive nanocarriers for systemic delivery.

16.4.1.3 *Exosomes*

Exosomes are natural cell-derived extracellular vesicles with a diameter of approximately 50–90 nm that originate from internal endocytic compartments and multi-vesicular bodies; they participate in intercellular communication.[33] Lipids and proteins are the main components of exosome membranes, which are enriched with lipid rafts.[34,35] In addition to the proteins, various nucleic acids have been identified in the exosomal lumen, including mRNAs, microRNAs, and other non-coding RNAs.[36] Among these molecules, miRNAs have attracted most attention, due to their regulatory roles in gene expression. The miRNAs in cell-released exosomes can circulate with the associated vehicles to reach neighboring cells and distant cells. After delivery into acceptor cells, exosomal miRNAs play functional roles. For example, exosomal miR-214, derived from the human microvascular endothelial cell line HMEC-1, stimulated migration and angiogenesis in neighboring HMEC-1 cells.[37]

16.4.1.4 *Solid Lipid Nanoparticles*

Lipid nanoparticles have many advantages, including biocompatibility, rapid cellular uptake, and potential for large-scale production.[38] A cationic lipid. DOTAP, has been used to formulate solid lipid nanoparticles for miRNA delivery and shown to serve as a promising carrier for miR-200c to breast cancer stem cells.[39] Solid lipid nanoparticles (SLNs) have been also used to deliver anti-miRNA oligonucleotides (AMOs) to inhibit action of oncogenic microRNAs. Other studies suggest that oncogenic miR-21 could be effectively silenced using AMO-SLNs, leading to significant inhibition of cell migration and invasion.[40]

16.4.2 Polymeric Micelles

Polymeric micelles have been largely identified as suitable carriers for, and distributors of, anticancer drugs. These micelles have an outer and inner core, which determine the different physicochemical properties of these nanocarriers.[41] For instance, their surface structure, hydrophobicity, and crystallinity[41] determine miRNA release. Nanoparticles containing polymers, lipids, hybrids, and metal/metal oxides provide marked opportunities for targeted delivery.[42] These nanoparticles easily accumulate at tumor sites because of an 'enhanced permeability and retention' (EPR) effect.[43]

16.4.3 Polymer-based Nanoparticles for miRNA Delivery

Polymer nanoparticles are defined as being colloidal systems, which range in size from ~5–10 nm to an upper limit of 1000 nm, although the range generally obtained is 100–500 nm.[44] Regarding their physicochemical properties, positively charged nanoparticles are optimal for cellular uptake. Thus, cationic polymers are the most abundantly used substances for non-viral delivery of nucleic acids.[45] The advantages of polymer nanoparticles as active substance delivery systems include high drug encapsulation efficiency, higher stability of encapsulating active substances, and biocompatibility with tissue and cells when prepared from biopolymers that are biocompatible or biodegradable.[45] Polymers are classified into natural and synthetic compounds that are available for nucleic acid complexation. Natural polymers have biocompatibility and biodegradability properties, whereas synthetic polymers have good mechanical properties and thermal stability. In addition, synthetic polymers can be provided in a wide range of shapes, whereas natural polymers are not easily prepared.[46]

16.4.3.1 Natural Polymers

The main natural polymers used for biomedical applications are collagen, chitin, keratin, chitosan, silk, and elastin, derived from animals; and starch, cellulose, and pectin, derived from plants.

16.4.3.1.1 Atelocollagen. Atelocollagen is a natural positively charged polymer and is used to form a complex with negatively charged miRNAs in order to enhance effectiveness and reduce off-target effects. miR-34a and atelocollagen complexes were shown to enhance miRNA delivery to human colon cancer cells and caused growth suppression.[47] Moreover, injections of this complex into mice bearing human colon cancer xenografts successfully inhibited tumor growth *in vivo*.[47]

16.4.3.1.2 Chitosan Nanoparticles. A novel approach for miRNA delivery is chitosan nanoformulations. The cationic nature of chitosan enables easy and fast complex formation with negatively charged miRNA, which increases the incorporation efficiency of microRNAs.[48] In addition, the hydroxyl and amino groups of chitosan allow for chemical conjugation of specific ligands tailored for targeted therapy.[49] Chitosan nanoparticles (CS NPs) have been designed to target hyaluronic acid (HA) receptors, which are overexpressed in cancer cells. HA-CS NPs incorporating miR-34a and doxorubicin (Dox) were delivered to breast cancer cell-derived xenograft tumors in athymic nude mice. Treatment resulted in increased inhibition of tumor growth and tumor volume when compared with delivery of Dox-NPs or free Dox administration,[50] suggesting that a chitosan-based delivery approach represents a potential *in vivo* delivery system for non-coding RNA molecules.

16.4.3.1.3 Protamine. Protamine sulfate is a clinically used, natural, positively charged polymer that has been used for the delivery of miRNAs. miR-203 complexed with protamine sulphate-nanodiamond nanoparticles suppressed proliferation of esophageal cancer cells.[51] Also, hyaluronic acid/protamine sulfate interpolyelectrolyte complexes (HP-IPECs) were shown to enhance delivery of miR-34a into triple-negative breast cancer cells.[52]

16.4.3.2 Synthetic Polymers

16.4.3.2.1 Polyethylenimine. Polyethylenimine (PEI) is a cationic polymer that has been widely utilized for non-viral transfections of siRNAs and miRNAs *in vitro* and *in vivo* because of its capacity to combine strong DNA compaction capacity with an intrinsic endosomolytic activity.[53] The complexes (polyplexes) produced from this interaction of PEI with oligonucleotides are more condensed, with a slightly net positive surface charge that enhances cellular uptake.[54] There are two types of PEI isomers, dictated by the linkage of the repeating ethylenimine units: (1) branched (bPEI), or (2) linear (lPEI) morphological isomers.[53] bPEI-derived vectors have been used to deliver oligonucleotides,[55] plasmid DNA (pDNA), and Epstein–Barr virus-based plasmid vectors,[56] as well as RNA and intact ribozymes.[56] A few studies demonstrated the utilization of PEI for miRNA delivery.[57] The PEG-PEI nanoparticles were used as a nanocarrier for miR-150 transfection in chronic myeloid leukemia cells. Polyurethane-short branchpolyethylenimine (PU-PEI) was used to deliver mir-145 into lung tumor tissues, resulting in reduction of tumor growth in a xenograft mouse model.[58]

16.4.3.2.2 PLGA Nanoparticles. Poly(D,L-lactide-*co*-glycolide) (PLGA) is one of the most successfully developed biodegradable and polymeric nanoparticles. PLGA has attracted considerable attention because of its features: (1) biodegradability and biocompatibility; (2) protection of drugs from degradation; (3) sustained release; and (4) the possibility to modify surface properties and provide stealthness.[59] Furthermore, PLGA nanoparticles coated with cell-penetrating peptide (penetratin) has been shown to deliver anti-miRs effectively. For example, PLGA nanoparticles were designed to silence miR-155, which is overexpressed in several lymphomas, and resulted in significant reduction of cell proliferation and tumor growth in mouse models for lymphoma.[60]

16.4.3.2.3 PEG Nanoparticles. When testing HA-coated PEG-PLGA nanoparticles conjugated with Dox and miR-542-3p, both improved targeting and increased the uptake of nanoparticles in TNBC cells.[61] Furthermore, delivery of PLGA-PEG nanoparticles encapsulating antisense-miR-21 and orlistat in combination with Dox significantly enhanced apoptotic effects in MDA-MB-231 and SKBR-3 TNBC cells.[62] Anti-sense miR-21 and miR-10b PEG complex was delivered successfully in TNBC by PLGA-*b*-PEG nanoparticles.[63]

16.4.3.3 Dendrimer Polymers

An additional polymer type used in the delivery of RNA oligonucleotides is dendrimers. Dendrimers are a new class of branched macromolecules with a tree-like topological structure and have been used for cancer detection and therapy.[64,65] They are gaining considerable attention in modern biomedicine due to their high water solubility, biocompatibility, polyvalence, and precise molecular weight.[66] Poly(amidoamine) PAMAM dendrimers were applied to deliver mir-7 into glioma tissues and they demonstrated high transfection capacity and successful therapeutic effect.[67] In a recent study, Wang *et al.*[68] developed PAMAM dendrimers and PEG-functionalized nanographene oxide (NGO)-conjugated dendrimers (NGO-PEG-dendrimer) for efficient delivery of anti-miR-21 into non-small cell lung cancer (NSCLC) and showed stronger inhibition of cell migration and invasion compared to bare (unmodified) dendrimers or Lipo2000 transfection reagents. When miR-126 was delivered to HUVEC cells through polyarginine and RGD (arginine-glycine-aspartic acid)-modified dendrimers, it significantly reduced cell proliferation and capillary tube formation.[69]

16.4.4 pH Low Insertion Peptide Monomers

pH low insertion peptide (pHLIP) is a small peptide that forms a transmembrane α-helix under acidic conditions such as those found in tumor microenvironments. A pH-sensitive antimiR conjugate called pHLIP–antimiR-155 has been shown to inhibit miR-155 in lymphoma mouse model[70] pHLIP and antimiR-155 facilitates the specific delivery of antimiR to cancer cells. Mice treated with pHLIP–antimiR-155 exhibited a significant reduction in tumor burden, resulting in prolonged survival. No significant toxicity was observed, suggesting that the clinical translation of this approach may be feasible.[70]

16.4.5 Inorganic Nanoparticles

Inorganic nanoparticles have great advantages for imaging and therapy due to their unique size and shape-dependent physical and chemical properties. These nanoparticles have ability to interact with multiple ligands such as drugs, oligonucleotides, and radioisotopes, which can be encapsulated inside the nanoparticles for molecular targeting, therapeutic delivery and multimodal imaging.[71]

16.4.5.1 Gold Nanoparticles

Among those particles, gold nanoparticles (AuNPs) have gained significant attention due to their relative non-toxic nature.[72] AuNPs have a strong binding affinity to thiol and amine functionalities in RNA and/or chemotherapeutics through Au–S and Au–N bonding. Many studies demonstrated successful utilization of AuNPs for miRNA delivery. In a recent study, Rui *et al.*[73] demonstrated that co-delivery of anti-miR-21 and DOX *via* near-infrared radiation-responsive hollow AuNPs produced successive release of this combination. Furthermore, this sequential delivery of anti-miR-21/DOX produced a synergistic apoptotic response in tumors, thereby enhancing anticancer efficacy eight-fold and increasing anti-cancer stem cell activity 50-fold. By using functionalized AuNPs in the study by Kim *et al.*,[74] *in vitro* anti-miR-29b delivery was shown to block miRNA activity, resulting in upregulation of its target myeloid cell leukemia-1 (MCL-1) in various cancer types including cervical cancer cells (HeLa), ovarian granulosa cells (KGN) and human embryonic kidney cells (293T).[74]

16.4.5.2 Magnetic Nanoparticles

Magnetic nanoparticles (MNPs), which include Fe_3O_4 nanospheres, have several advantages over other inorganic nanoparticles, such as tumor tracing,[75,76] because of their magnetism and utility in high-sensitivity real-time magnetic resonance imaging.[77] For example, antimiR-10b was loaded into dextran-coated MNPs functionalized with the tumor-targeting peptide RGD, and delivered to breast cancer cells to suppress breast cancer metastasis. After intravenous injection of anti-miR-10b MNPs into mice bearing orthotopic MDA-MB-231 tumors, the MNPs were shown to accumulate in the tumors and lymph nodes and prevented metastasis.[78]

16.4.5.3 Mesoporous Silica Nanoparticles

Mesoporous silica nanoparticles (MSNPs) exhibit several superior features compared to other inorganic nanoparticles; they have a unique structure with a tunable pore and particle size, resulting in a high specific surface area which can be easily functionalized, and most importantly are highly biocompatible.[79] Due to the negative charge of microRNA, cationic lipids are usually used to coat MSNPs on the surface to condense with microRNA.[80] Functionalization of MSNP was achieved by Bertucci *et al.*,[81] by introduction of an amine group to the surface of MSNP and conversion of the amine group into a carboxylic one, which enhanced the loading of the drug temozolomide (TMZ) in the pore system. Thus, once the MSNPs were filled with the drug, polyarginine-peptide nucleic acid of anti-miR221 (R8-PNA221) could finally be bound to the nanocarriers. The combined effect of TMZ and R8-PNA221 in the MSNPs shows the most effective induction of apoptosis (70.9% of apoptotic cells) thus far achieved in the TMZ-resistant T98G cell line.

16.5 Application of Nanoparticle Formulations to Deliver miRNAs to Specific Tumor Types

16.5.1 Breast Cancer

Breast cancer is accompanied with high morbidity and mortality and is the second leading cause of death in Western countries. A variety of nanoparticles have been developed, specifically targeting metastasized tumors of the breast.[82] Exosome-based studies revealed that miR-134 may act as a potential biomarker and therapeutic for breast cancer according to the amount of this microRNA delivered by this route. Delivery through mir-134-enriched exosomes inhibited TNBC cell migration and invasion by targeting STAT5B and Hsp90.[83] Wang *et al.*[52] revealed that efficient intracellular delivery of miR-34a to TNBCs was reached *via* encapsulation in a HP-IPEC system. In addition, the study of Hayward *et al.*[84] showed that a lipid nanoparticle platform surface coated with hyaluronic acid was used successfully for the delivery of mature tumor suppressor microRNA125a-5p to treat HER2-positive metastatic breast cancer. This system revealed marked therapeutic efficiency compared to a commercial transfection reagent (Lipofectamine).

16.5.2 Prostate Cancer

Various miRNAs, including miR-34a, miR-21, and *let-7c* have been implicated in prostate tumorigenesis.[85] Intrafemoral injection of chitosan-encapsulated miR-34a caused a two-fold decrease in bone tumor volume.[86] Intravenous delivery of miR-124 complexed with PEI-derived nanoparticles included was sufficient to inhibit prostate tumor growth and to increase tumor cell apoptosis in combination with enzalutamide.[87] Interestingly, effective delivery of anti-miR-21 oligonucleotides to prostate cancer cells *via* exosomes resulted in a significant downregulation of miR-21 levels and decreased motility of prostate cancer cells.[88]

16.5.3 Pancreatic Cancer

Nanoparticle-encapsulated delivery of miRNAs for pancreatic cancer treatment remains an unexplored field that has potential therapeutic value. Encapsulation of miR-145 in a magnetic nanoparticle formulation was delivered efficiently to the tumor site and downregulated the expression of oncogenic signaling, such as MUC13, HER2, and p-AKT, to inhibit pancreatic cancer growth and invasion.[89] Previously, tumor-suppressor miR-34a reconstitution was achieved using an antibody-modified liposome/polycation delivery system in a Panc-1 xenograft mouse model.[90] AuNPs with fluorophore-labeled hairpin DNA—so-called 'gold nanobeacons'—were used to target and silence miR-21, an endogenous miRNA involved in cancer development and chemoresistance.[91] The delivery of a PLGA-based nanoformulation of miR-150 increased its expression in Colo-357 and HPAF cells by 28- and 26-fold, respectively, compared with transfection of miRNA *via* Lipofectamine.[92]

16.5.4 Ovarian Cancer

Several aberrantly expressed miRNAs have been reported in ovarian cancer tumors and cell lines. miR-155 is downregulated in ovarian tumor-associated dendritic cells and is essential for optimal antigen presentation and activation of T-cells by dendritic cells.[93] PEI-based nanocomplexes were used to deliver miR-155 to tumor-associated dendritic cells, which increased the expression of miR-155 *in vitro*. Treatment ultimately resulted in increased antitumor immunity and thus increased survival of the mice (by 65%).[93] miR-124 is downregulated in ovarian cancer and acts as a tumor suppressor by targeting proteins such as MYC and increases the expression of p27, subsequently leading to cell cycle arrest at G1 phase because of the loss of p-Rb and decreased expression of the MYC protein.[94] Transfection of miR-124 in an ovarian cancer cell line reduced the invasive and migratory capability of ovarian cancer cells and increased two-fold their sensitivity to etoposide. While miR-124 is restored in ovarian cancer xenograft tumors using neutral-lipid DOPC-NPs, it resulted in a significant decrease in tumor weight alone and in combination with etoposide.[94]

16.5.5 Lung Cancer

Several studies demonstrated the feasibility of systemically delivering miRNA mimics and siRNAs simultaneously to lung adenocarcinoma cells using polymer-based nanoparticles in a mouse model of lung cancer and eliciting a potent antitumor response.[95] Using a liposomal formulation, miR-34a in combination with *let-7* had superior anti-tumorigenic effects in the lung cancer mouse model compared to treatment with miR-34a or *let-7* alone.[96] MiR-200c is a known negative regulator of ZEB1, which induces the epithelial–mesenchymal transition in cancer cells.[97] Liposomal nanoparticle-encapsulated miR-200c delivered to lung cancer cells induced the activation of oxidative stress response genes and enhanced the radiosensitivity of lung cancer cells up to 1.5 times in an *in vivo* mouse model.[98] MiR-29b is downregulated in NSCLC cells and directly targets oncogene cyclin-dependent kinase 6 (CDK6) to regulate cell cycle progression in these cells,[22,99] Cationic lipoplex-based delivery of miR-29b to lung cancer cells effectively reduced CDK6 expression by almost 54% and reduced tumor volume by almost 50% *in vivo*.[22]

16.5.6 Brain Cancer

The successful delivery of tumor-suppressor miRNA-conjugated nanoparticles to brain cells is feasible. Mesoporous silica nanoparticles containing polyarginine-peptide nucleic acid (R8-PNA) conjugates targeting miR-221 were used to treat TMZ-resistant glioma cells. These nanoparticles, combined with TMZ treatment, led to a significant increase in apoptosis.[81]

16.5.7 Hepatocellular Carcinoma

miR-34a targets the p53 signaling pathway to regulate cell cycle progression and apoptosis induction in hepatocellular carcinoma cells.[100] Systemic delivery of the liposomal miR34a mimic reduced the expression levels of several miR-34a target mRNAs and caused significant growth inhibition and even regression of the liver tumors.[101] Negatively charged liposomal delivery system coated with transferrin protein was utilized to deliver miR-221 antisense oligonucleotide (anti-miR-221) to HepG2 cells overexpressing the transferrin receptor.[102] Recently, miR-375/Dox-NPs were successfully synthesized and provided an excellent platform to deliver miR-375 and Dox effectively into hepatocellular carcinoma cells.[103]

16.5.8 Colorectal Cancer

In a recent study, anti-miR-155-loaded MSNPs modified with polymerized dopamine (PDA) and AS1411 aptamer (MSNPs-anti-miR-155-PDA-Apt) were used to inhibit miR-155 and enhance sensitivity of SW480 cell to 5-fluorouracil chemotherapy.[104] Also, mPEG-poly(amino acid)s was used as a novel nanocarrier for the delivery of miRNA-139-5p. The PEG-poly(amino acid)s/miRNA-139-5p nanoparticle complex was more effective at suppressing tumor growth and migration in mice with colorectal cancer than when animals were treated with miRNA-139-5p alone.[105] Interestingly, the study by Liang *et al.*[106] showed that use of a PLGA/PEI vector crosslinked with hyaluronic acid facilitated cellular uptake of miR-145. After local administration of PLGA/PEI/HA complexes, the miR-145 vector was successfully delivered into HCT-116 colon cancer cells as well as a xenograft tumor mouse model, resulting in marked antitumor effects.

16.6 miRNA-based Therapeutics in Human Clinical Trials

MRX34 was the first-in-class miRNA mimic therapy for cancer moved to clinical development (clinicaltrials.gov identifier NCT01829971). MRX is a liposomal formulation of the naturally occurring tumor suppressor miR-34a. Furthermore, miR-34a represents the first successful miRNA delivery and reached phase I clinical trials in treating melanoma patients as well as renal and hepatocellular cancer patients.[25] However, due to immune-related serious toxicity, this trial was terminated. This was an open-label, multicenter, dose-escalation study to investigate the safety, pharmacokinetics, and pharmacodynamics of miR34a in patients. MRX34 was administered daily for 5 days with 2 weeks off (total of 21 days) for three cycles followed by a no-treatment observation period.

16.7 Conclusions and Future Perspectives

Nanocarriers hold great potential for non-coding RNA-based cancer therapy, diagnosis and imaging. Non-coding RNAs, including miRNAs, play crucial roles in cancer pathogenesis and represent a diverse and novel class of targets and therapeutics for drug discovery and development. In addition to robust target silencing, toxicity should be carefully taken into consideration when using and designing nanocarriers. In many potential nanoparticles nanoliposomes seem to be the best candidates for the delivery of miRNA therapeutics due to successful results in preclinical models with no or limited toxicity and are currently being tested in clinical trials. Tumor-targeted nanoparticles are expected to significantly reduce miRNA-induced side effects and immune reactions, which was indeed observed in the recent MRX34 clinical trial. Enhancing the design of anti-miRNA and miRNA mimics, as well as evolving new delivery systems to target specific organs, tissues and cell types, will be critical when achieving this goal. The data presented here support the use of nanoparticles as promising miRNA carriers due to their ability to protect the loaded miRNAs from serum-nuclease induced degradation and to deliver miRNAs to their target successfully. Nanotechnology-based delivery systems have displayed fundamental impact in the evolution of miRNA therapeutics and its expected that continuing advancements will lead to effective application in clinical practice.

References

1. R. C. Lee, R. L. Feinbaum and V. Ambros, *Cell*, 1993, **75**, 843–854.
2. Y. Chi and D. Zhou, *J. Exp. Clin. Cancer Res.*, 2016, **35**, 43.
3. C. P. Bracken, H. S. Scott and G. J. Goodall, *Nat. Rev. Genet.*, 2016, **17**, 719–732.
4. S. Lin and R. I. Gregory, *Nat. Rev. Cancer*, 2015, **15**, 321–333.
5. G. A. Calin and C. M. Croce, *Nat. Rev. Cancer*, 2006, **6**, 857–866.
6. N. Rosenfeld, R. Aharonov, E. Meiri, S. Rosenwald, Y. Spector, M. Zepeniuk, H. Benjamin, N. Shabes, S. Tabak, A. Levy, D. Lebanony, Y. Goren, E. Silberschein, N. Targan, A. Ben-Ari, S. Gilad, N. Sion-Vardy, A. Tobar, M. Feinmesser, O. Kharenko, O. Nativ, D. Nass, M. Perelman, A. Yosepovich, B. Shalmon, S. Polak-Charcon, E. Fridman, A. Avniel, I. Bentwich, Z. Bentwich, D. Cohen, A. Chajut and I. Barshack, *Nat. Biotechnol.*, 2008, **26**, 462–469.

7. S. Akhtar and I. F. Benter, *J. Clin. Invest.*, 2007, **117**, 3623–3632.
8. R. Juliano, J. Bauman, H. Kang and X. Ming, *Mol. Pharm.*, 2009, **6**, 686–695.
9. A. Judge and I. MacLachlan, *Hum. Gene Ther.*, 2008, **19**, 111–124.
10. K. A. Whitehead, J. E. Dahlman, R. S. Langer and D. G. Anderson, *Annu. Rev. Chem. Biomol. Eng.*, 2011, **2**, 77–96.
11. J. Wang, Z. Lu, M. G. Wientjes and J. L. Au, *AAPS J.*, 2010, **12**, 492–503.
12. C. L. Esposito, L. Cerchia, S. Catuogno, G. De Vita, J. P. Dassie, G. Santamaria, P. Swiderski, G. Condorelli, P. H. Giangrande and V. de Franciscis, *Mol. Ther.*, 2014, **22**, 1151–1163.
13. M. A. Behlke, *Oligonucleotides*, 2008, **18**, 305–319.
14. K. A. Lennox and M. A. Behlke, *Gene Ther.*, 2011, **18**, 1111–1120.
15. N. S. Gandhi, R. K. Tekade and M. B. Chougule, *J. Controlled Release*, 2014, **194**, 238–256.
16. B. Yu, Y. Mao, L. Y. Bai, S. E. Herman, X. Wang, A. Ramanunni, Y. Jin, X. Mo, C. Cheney, K. K. Chan, D. Jarjoura, G. Marcucci, R. J. Lee, J. C. Byrd, L. J. Lee and N. Muthusamy, *Blood*, 2013, **121**, 136–147.
17. A. K. Varkouhi, M. Scholte, G. Storm and H. J. Haisma, *J. Controlled Release*, 2011, **151**, 220–228.
18. A. Baldo, E. van den Akker, H. E. Bergmans, F. Lim and K. Pauwels, *Curr. Gene Ther.*, 2013, **13**, 385–394.
19. A. Eldar-Boock, D. Polyak, A. Scomparin and R. Satchi-Fainaro, *Curr. Opin. Biotechnol.*, 2013, **24**, 682–689.
20. B. Aslan, B. Ozpolat, A. K. Sood and G. Lopez-Berestein, *J. Drug Targeting*, 2013, **21**, 904–913.
21. B. Ozpolat, A. K. Sood and G. Lopez-Berestein, *J. Intern. Med.*, 2010, **267**, 44–53.
22. Y. Wu, M. Crawford, Y. Mao, R. J. Lee, I. C. Davis, T. S. Elton, L. J. Lee and S. P. Nana-Sinkam, *Mol. Ther.--Nucleic Acids*, 2013, **2**, e84.
23. K. Rai, N. Takigawa, S. Ito, H. Kashihara, E. Ichihara, T. Yasuda, K. Shimizu, M. Tanimoto and K. Kiura, *Mol. Cancer Ther.*, 2011, **10**, 1720–1727.
24. S. Shi, L. Han, L. Deng, Y. Zhang, H. Shen, T. Gong, Z. Zhang and X. Sun, *J. Controlled Release*, 2014, **194**, 228–237.
25. A. Ganju, S. Khan, B. B. Hafeez, S. W. Behrman, M. M. Yallapu, S. C. Chauhan and M. Jaggi, *Drug Discovery Today*, 2017, **22**, 424–432.
26. Z. Hamurcu, A. Ashour, N. Kahraman and B. Ozpolat, *Oncotarget*, 2016, **7**, 16619–16635.
27. R. Bayraktar, M. Pichler, P. Kanlikilicer, C. Ivan, E. Bayraktar, N. Kahraman, B. Aslan, S. Oguztuzun, M. Ulasli, A. Arslan, G. Calin, G. Lopez-Berestein and B. Ozpolat, *Oncotarget*, 2017, **8**, 11641–11658.
28. R. Bayraktar, C. Ivan, E. Bayraktar, P. Kanlikilicer, N. Kabil, N. Kahraman, H. A. Mokhlis, D. Karakas, C. Rodriguez-Aguayo, A. Arslan, J. Sheng, S. T. Wong, G. Lopez-Berestein, G. A. Calin and B. Ozpolat, *Clin. Cancer Res.*, 2018, **24**, 4225–4241.
29. H. Mokhlis, R. Bayraktar, N. Kabil, A. Caner, N. Kahraman, C. Rodriguez-Aguayo, E. Zambalde, J. Sheng, K. Karagoz, P. Kanlikilicer,

A. Abdel Aziz, T. Abdelghany, A. Ashour, S. Wong, M. Gatza, G. Calin, G. Lopez-Berestein and B. Ozpolat, *Mol. Ther.--Nucleic Acids*, 2018, **14**, 301–317.
30. N. Kabil, H. Mokhlis, R. Bayraktar, C. Ivan, A. Caner, K. Karagoz, S. Wong, M. Gatza, G. Calin, G. Lopez-Berestein and B. Ozpolat, *Cancer Res.*, 2018, DOI: 10.1158/1538-7445.AM2018-504.
31. I. Tekedereli, S. N. Alpay, C. D. Tavares, Z. E. Cobanoglu, T. S. Kaoud, I. Sahin, A. K. Sood, G. Lopez-Berestein, K. N. Dalby and B. Ozpolat, *PLoS One*, 2012, **7**, e41171.
32. M. Nishimura, E. J. Jung, M. Y. Shah, C. Lu, R. Spizzo, M. Shimizu, H. D. Han, C. Ivan, S. Rossi, X. Zhang, M. S. Nicoloso, S. Y. Wu, M. I. Almeida, J. Bottsford-Miller, C. V. Pecot, B. Zand, K. Matsuo, M. M. Shahzad, N. B. Jennings, C. Rodriguez-Aguayo, G. Lopez-Berestein, A. K. Sood and G. A. Calin, *Cancer Discovery*, 2013, **3**, 1302–1315.
33. C. Thery, L. Zitvogel and S. Amigorena, *Nat. Rev. Immunol.*, 2002, **2**, 569–579.
34. M. Simons and G. Raposo, *Curr. Opin. Cell Biol.*, 2009, **21**, 575–581.
35. S. Mathivanan, H. Ji and R. J. Simpson, *J. Proteomics*, 2010, **73**, 1907–1920.
36. Y. Sato-Kuwabara, S. A. Melo, F. A. Soares and G. A. Calin, *Int. J. Oncol.*, 2015, **46**, 17–27.
37. B. W. van Balkom, O. G. de Jong, M. Smits, J. Brummelman, K. den Ouden, P. M. de Bree, M. A. van Eijndhoven, D. M. Pegtel, W. Stoorvogel, T. Wurdinger and M. C. Verhaar, *Blood*, 2013, **121**, 3997–4006.
38. S. N. He, Y. L. Li, J. J. Yan, W. Zhang, Y. Z. Du, H. Y. Yu, F. Q. Hu and H. Yuan, *Int. J. Nanomed.*, 2013, **8**, 2859–2869.
39. J. Liu, T. Meng, M. Yuan, L. Wen, B. Cheng, N. Liu, X. Huang, Y. Hong, H. Yuan and F. Hu, *Int. J. Nanomed.*, 2016, **11**, 6713–6725.
40. S. J. Shi, Z. R. Zhong, J. Liu, Z. R. Zhang, X. Sun and T. Gong, *Pharm. Res.*, 2012, **29**, 97–109.
41. N. Goonoo, A. Bhaw-Luximon, R. Ujoodha, A. Jhugroo, G. K. Hulse and D. Jhurry, *J. Controlled Release*, 2014, **183**, 154–166.
42. K. Hola, Z. Markova, G. Zoppellaro, J. Tucek and R. Zboril, *Biotechnol. Adv.*, 2015, **33**, 1162–1176.
43. S. Guo and L. Huang, *Biotechnol. Adv.*, 2014, **32**, 778–788.
44. D. Quintanar-Guerrero, E. Allemann, H. Fessi and E. Doelker, *Drug Dev. Ind. Pharm.*, 1998, **24**, 1113–1128.
45. U. Weirauch, D. Gutsch, S. Hobel and A. Aigner, *Methods Mol. Biol.*, 2013, **1049**, 443–465.
46. A. Namvar, A. Bolhassani, N. Khairkhah and F. Motevalli, *Biopolymers*, 2015, **103**, 363–375.
47. H. Tazawa, N. Tsuchiya, M. Izumiya and H. Nakagama, *Proc. Natl. Acad. Sci. U. S. A.*, 2007, **104**, 15472–15477.
48. M. Denizli, B. Aslan, L. S. Mangala, D. Jiang, C. Rodriguez-Aguayo, G. Lopez-Berestein and A. K. Sood, *Methods Mol. Biol.*, 2017, **1632**, 219–230.
49. H. Ragelle, G. Vandermeulen and V. Preat, *J. Controlled Release*, 2013, **172**, 207–218.

50. X. Deng, M. Cao, J. Zhang, K. Hu, Z. Yin, Z. Zhou, X. Xiao, Y. Yang, W. Sheng, Y. Wu and Y. Zeng, *Biomaterials*, 2014, **35**, 4333–4344.
51. M. Cao, X. Deng, S. Su, F. Zhang, X. Xiao, Q. Hu, Y. Fu, B. B. Yang, Y. Wu, W. Sheng and Y. Zeng, *Nanoscale*, 2013, **5**, 12120–12125.
52. S. Wang, M. Cao, X. Deng, X. Xiao, Z. Yin, Q. Hu, Z. Zhou, F. Zhang, R. Zhang, Y. Wu, W. Sheng and Y. Zeng, *Adv. Healthcare Mater.*, 2015, **4**, 281–290.
53. J. J. Zhao, J. Lin, T. Lwin, H. Yang, J. Guo, W. Kong, S. Dessureault, L. C. Moscinski, D. Rezania, W. S. Dalton, E. Sotomayor, J. Tao and J. Q. Cheng, *Blood*, 2010, **115**, 2630–2639.
54. O. Boussif, F. Lezoualc'h, M. A. Zanta, M. D. Mergny, D. Scherman, B. Demeneix and J. P. Behr, *Proc. Natl. Acad. Sci. U. S. A.*, 1995, **92**, 7297–7301.
55. M. Vogt, J. Munding, M. Gruner, S. T. Liffers, B. Verdoodt, J. Hauk, L. Steinstraesser, A. Tannapfel and H. Hermeking, *Virchows Arch.*, 2011, **458**, 313–322.
56. B. Weber, C. Stresemann, B. Brueckner and F. Lyko, *Cell Cycle*, 2007, **6**, 1001–1005.
57. C. Biray Avci, I. Ozcan, T. Balci, O. Ozer and C. Gunduz, *Cell Biol. Int.*, 2013, **37**, 1205–1214.
58. G. Y. Chiou, J. Y. Cherng, H. S. Hsu, M. L. Wang, C. M. Tsai, K. H. Lu, Y. Chien, S. C. Hung, Y. W. Chen, C. I. Wong, L. M. Tseng, P. I. Huang, C. C. Yu, W. H. Hsu and S. H. Chiou, *J. Controlled Release*, 2012, **159**, 240–250.
59. H. Sah, L. A. Thoma, H. R. Desu, E. Sah and G. C. Wood, *Int. J. Nanomed.*, 2013, **8**, 747–765.
60. I. A. Babar, C. J. Cheng, C. J. Booth, X. Liang, J. B. Weidhaas, W. M. Saltzman and F. J. Slack, *Proc. Natl. Acad. Sci. U. S. A.*, 2012, **109**, E1695–E1704.
61. S. Wang, J. Zhang, Y. Wang and M. Chen, *Nanomedicine*, 2016, **12**, 411–420.
62. A. Bhargava-Shah, K. Foygel, R. Devulapally and R. Paulmurugan, *Nanomedicine*, 2016, **11**, 235–247.
63. R. Devulapally, N. M. Sekar, T. V. Sekar, K. Foygel, T. F. Massoud, J. K. Willmann and R. Paulmurugan, *ACS Nano*, 2015, **9**, 2290–2302.
64. T. Li, M. Smet, W. Dehaen and H. Xu, *ACS Appl. Mater. Interfaces*, 2016, **8**, 3609–3614.
65. J. H. Myung, M. Roengvoraphoj, K. A. Tam, T. Ma, V. A. Memoli, E. Dmitrovsky, S. J. Freemantle and S. Hong, *Anal. Chem.*, 2015, **87**, 10096–10102.
66. A. Jurj, C. Braicu, L. A. Pop, C. Tomuleasa, C. D. Gherman and I. Berindan-Neagoe, *Drug Des., Dev. Ther.*, 2017, **11**, 2871–2890.
67. X. Liu, G. Li, Z. Su, Z. Jiang, L. Chen, J. Wang, S. Yu and Z. Liu, *Oncol. Rep.*, 2013, **29**, 1387–1394.
68. Y. Z. Wang, N. Hao, Q. M. Feng, H. W. Shi, J. J. Xu and H. Y. Chen, *Biosens. Bioelectron.*, 2016, **77**, 76–82.
69. W. D. Gray, R. J. Wu, X. Yin, J. Zhou, M. E. Davis and Y. Luo, *Biomacromolecules*, 2013, **14**, 101–109.

70. C. J. Cheng, R. Bahal, I. A. Babar, Z. Pincus, F. Barrera, C. Liu, A. Svoronos, D. T. Braddock, P. M. Glazer, D. M. Engelman, W. M. Saltzman and F. J. Slack, *Nature*, 2015, **518**, 107–110.
71. M. Swierczewska, S. Lee and X. Chen, *Mol. Imaging*, 2011, **10**, 3–16.
72. M. C. Daniel and D. Astruc, *Chem. Rev.*, 2004, **104**, 293–346.
73. M. Rui, Y. Qu, T. Gao, Y. Ge, C. Feng and X. Xu, *Int. J. Nanomed.*, 2017, **12**, 217–237.
74. J. H. Kim, J. H. Yeom, J. J. Ko, M. S. Han, K. Lee, S. Y. Na and J. Bae, *J. Biotechnol.*, 2011, **155**, 287–292.
75. G. Lin, W. Zhu, L. Yang, J. Wu, B. Lin, Y. Xu, Z. Cheng, C. Xia, Q. Gong, B. Song and H. Ai, *Biomaterials*, 2014, **35**, 9495–9507.
76. K. W. Huang, J. J. Chieh, H. E. Horng, C. Y. Hong and H. C. Yang, *Int. J. Nanomed.*, 2012, **7**, 2987–2996.
77. J. Huang, L. Bu, J. Xie, K. Chen, Z. Cheng, X. Li and X. Chen, *ACS Nano*, 2010, **4**, 7151–7160.
78. M. V. Yigit, S. K. Ghosh, M. Kumar, V. Petkova, A. Kavishwar, A. Moore and Z. Medarova, *Oncogene*, 2017, **36**, 2628.
79. A. Watermann and J. Brieger, *Nanomaterials*, 2017, **7**, 189.
80. H. Xue, Z. Yu, Y. Liu, W. Yuan, T. Yang, J. You, X. He, R. J. Lee, L. Li and C. Xu, *Int. J. Nanomed.*, 2017, **12**, 5271–5287.
81. A. Bertucci, E. A. Prasetyanto, D. Septiadi, A. Manicardi, E. Brognara, R. Gambari, R. Corradini and L. De Cola, *Small*, 2015, **11**, 5687–5695.
82. Z. Hussain, J. A. Khan and S. Murtaza, *Crit. Rev. Eukaryotic Gene Expression*, 2018, **28**, 163–175.
83. K. O'Brien, M. C. Lowry, C. Corcoran, V. G. Martinez, M. Daly, S. Rani, W. M. Gallagher, M. W. Radomski, R. A. MacLeod and L. O'Driscoll, *Oncotarget*, 2015, **6**, 32774–32789.
84. S. L. Hayward, D. M. Francis, P. Kholmatov and S. Kidambi, *J. Biomed. Nanotechnol.*, 2016, **12**, 554–568.
85. W. Y. Chen, S. Y. Liu, Y. S. Chang, J. J. Yin, H. L. Yeh, T. H. Mouhieddine, O. Hadadeh, W. Abou-Kheir and Y. N. Liu, *Oncotarget*, 2015, **6**, 441–457.
86. S. Gaur, Y. Wen, J. H. Song, N. U. Parikh, L. S. Mangala, A. M. Blessing, C. Ivan, S. Y. Wu, A. Varkaris, Y. Shi, G. Lopez-Berestein, D. E. Frigo, A. K. Sood and G. E. Gallick, *Oncotarget*, 2015, **6**, 29161–29177.
87. X. B. Shi, A. H. Ma, L. Xue, M. Li, H. G. Nguyen, J. C. Yang, C. G. Tepper, R. Gandour-Edwards, C. P. Evans, H. J. Kung and R. W. deVere White, *Cancer Res.*, 2015, **75**, 5309–5317.
88. A. H. Alhasan, P. C. Patel, C. H. Choi and C. A. Mirkin, *Small*, 2014, **10**, 186–192.
89. S. Setua, S. Khan, M. M. Yallapu, S. W. Behrman, M. Sikander, S. S. Khan, M. Jaggi and S. C. Chauhan, *J. Gastrointest. Surg.*, 2017, **21**, 94–105.
90. Q. L. Hu, Q. Y. Jiang, X. Jin, J. Shen, K. Wang, Y. B. Li, F. J. Xu, G. P. Tang and Z. H. Li, *Biomaterials*, 2013, **34**, 2265–2276.
91. J. Conde, J. Rosa, J. M. de la Fuente and P. V. Baptista, *Biomaterials*, 2013, **34**, 2516–2523.

92. S. Arora, S. K. Swaminathan, A. Kirtane, S. K. Srivastava, A. Bhardwaj, S. Singh, J. Panyam and A. P. Singh, *Int. J. Nanomed.*, 2014, **9**, 2933–2942.
93. J. R. Cubillos-Ruiz, J. R. Baird, A. J. Tesone, M. R. Rutkowski, U. K. Scarlett, A. L. Camposeco-Jacobs, J. Anadon-Arnillas, N. M. Harwood, M. Korc, S. N. Fiering, L. F. Sempere and J. R. Conejo-Garcia, *Cancer Res.*, 2012, **72**, 1683–1693.
94. E. G. Seviour, V. Sehgal, Y. Lu, Z. Luo, T. Moss, F. Zhang, S. M. Hill, W. Liu, S. N. Maiti, L. Cooper, R. Azencot, G. Lopez-Berestein, C. Rodriguez-Aguayo, R. Roopaimoole, C. V. Pecot, A. K. Sood, S. Mukherjee, J. W. Gray, G. B. Mills and P. T. Ram, *Oncogene*, 2016, **35**, 691–701.
95. W. Xue, J. E. Dahlman, T. Tammela, O. F. Khan, S. Sood, A. Dave, W. Cai, L. M. Chirino, G. R. Yang, R. Bronson, D. G. Crowley, G. Sahay, A. Schroeder, R. Langer, D. G. Anderson and T. Jacks, *Proc. Natl. Acad. Sci. U. S. A.*, 2014, **111**, E3553–E3561.
96. A. L. Kasinski, K. Kelnar, C. Stahlhut, E. Orellana, J. Zhao, E. Shimer, S. Dysart, X. Chen, A. G. Bader and F. J. Slack, *Oncogene*, 2015, **34**, 3547–3555.
97. R. Shi, H. Xiao, T. Yang, L. Chang, Y. Tian, B. Wu and H. Xu, *Front. Med.*, 2014, **8**, 456–463.
98. M. A. Cortez, D. Valdecanas, X. Zhang, Y. Zhan, V. Bhardwaj, G. A. Calin, R. Komaki, D. K. Giri, C. C. Quini, T. Wolfe, H. J. Peltier, A. G. Bader, J. V. Heymach, R. E. Meyn and J. W. Welsh, *Mol. Ther.*, 2014, **22**, 1494–1503.
99. D. W. Wu, N. Y. Hsu, Y. C. Wang, M. C. Lee, Y. W. Cheng, C. Y. Chen and H. Lee, *Oncogene*, 2015, **34**, 2072–2082.
100. Z. Xiao, C. H. Li, S. L. Chan, F. Xu, L. Feng, Y. Wang, J. D. Jiang, J. J. Sung, C. H. Cheng and Y. Chen, *Cancer Res.*, 2014, **74**, 6236–6247.
101. C. L. Daige, J. F. Wiggins, L. Priddy, T. Nelligan-Davis, J. Zhao and D. Brown, *Mol. Cancer Ther.*, 2014, **13**, 2352–2360.
102. W. Zhang, F. Peng, T. Zhou, Y. Huang, L. Zhang, P. Ye, M. Lu, G. Yang, Y. Gai, T. Yang, X. Ma and G. Xiang, *Int. J. Nanomed.*, 2015, **10**, 4825–4836.
103. Y. P. Fan, J. Z. Liao, Y. Q. Lu, D. A. Tian, F. Ye, P. X. Zhao, G. Y. Xiang, W. X. Tang and X. X. He, *Mol. Ther.--Nucleic Acids*, 2017, **7**, 181–189.
104. Y. Li, Y. Duo, J. Bi, X. Zeng, L. Mei, S. Bao, L. He, A. Shan, Y. Zhang and X. Yu, *Int. J. Nanomed.*, 2018, **13**, 1241–1256.
105. Y. Wang, F. Costanza, C. Li, A. Nimmagadda, D. Song and J. Cai, *J. Biomed. Nanotechnol.*, 2016, **12**, 1510–1519.
106. G. Liang, Y. Zhu, A. Jing, J. Wang, F. Hu, W. Feng, Z. Xiao and B. Chen, *Gene Ther.*, 2016, **23**, 829–838.
107. B. Ozpolat, A. K. Sood and G. Lopez-Berestein, *Adv. Drug Delivery Rev.*, 2014, **66**, 110–116.

Subject Index